Wetlands

Wetlands

Fourth Edition

William J. Mitsch

Professor
The Ohio State University
Columbus, Ohio

James G. Gosselink

Professor Emeritus
Louisiana State University
Baton Rouge, Louisiana

BICENTENNIAL
BICENTENNIAL
1807
⊗WILEY
2007
BICENTENNIAL
BICENTENNIAL

John Wiley & Sons, Inc.

Published by John Wiley & Sons, Inc., Hoboken, New Jersey
Published simultaneously in Canada

Wiley Bicentennial Logo: Richard J. Pacifico

For general information about our other products and services, please contact our Customer Care Department within the United States at (800) 762-2974, outside the United States at (317) 572-3993 or fax (317) 572-4002.

Wiley also publishes its books in a variety of electronic formats. Some content that appears in print may not be available in electronic books. For more information about Wiley products, visit our web site at www.wiley.com.

Library of Congress Cataloging-in-Publication Data:

Mitsch, William J.
    Wetlands / William J. Mitsch, James G. Gosselink. -- 4th ed.
        p. cm.
    Includes bibliographical references and index.
    ISBN 978-0-471-69967-5 (cloth)
    1. Wetland ecology--United States. 2. Wetlands--United States. 3.
Wetland management--United States. I. Gosselink, James G. II. Title.
    QH104.M57 2007
    577.68--dc22
        2007001732

Printed in the United States of America.

10 9 8 7 6 5 4 3 2

We dedicate this book to two important ecologists who heavily influenced wetland ecology and management—Howard T. Odum and Eugene P. Odum. Both brothers died in late summer 2002. We still see their influence on dozens of principles and concepts described in this book.

Contents

Preface

This is the fourth edition of *Wetlands*—we have done a new edition every 7 years since the first edition came out in 1986. The first important change in *Wetlands* 4th edition (referred to here as *Wetlands 4*) is that it is shorter than *Wetlands* 3rd edition—with 35 percent fewer pages and 14 chapters rather than 21. It is quite rare that a new edition of a book is smaller than its predecessor; but we had our reasons. The book was becoming encyclopedic and less of a textbook with every edition and yet we were still not covering every type of wetland in the ecosystem chapters. So we shortened the book by removing the seven wetland ecosystem chapters that were in the middle of the previous 3 editions of this book. We did so with great care and respect for the reputation that this so-called "wetland bible" has developed with its previous editions. Now, with those chapters eliminated, there will be less concern expressed by some that we left out their favorite wetland and much more opportunity to focus on the three remaining sections in *Wetlands 4*—an introduction to the extent, definitions and general features of wetlands of the world (called *Introduction*), wetland science (called *The Wetland Environment*), and the applied section called *Wetland Management*. In some cases, we moved important principles from the removed ecosystem chapters to one of the three sections in this new edition.

We added a new chapter to *Wetlands 4* on "Climate Change and Wetlands" (Chapter 10). This chapter includes new information, even up to our publishing date, from the International Panel on Climate Change's (IPCC's) 2007 reports. Since wetlands may be the linchpins of climate change, this may be the most important addition to the book in some time. Wetlands are affected by climate change probably more than any other ecosystem and they are also sources of important greenhouse gases, mainly methane and nitrous oxide. They also represent enormous storages of carbon—equivalent to 100 years or more of present-day fossil fuel emissions. Any climate drift could have major effects on those storages of carbon in the world.

Much greater international coverage is included in *Wetlands 4*. We merged the North American and "rest of the World" chapters from the previous edition to one chapter on "Wetlands of the World" (Chapter 3). We initiated or expanded coverage of the Great Plains Playas in the United States, the wetlands of Mexico and Central America, the Congolian Swamp and Sine Saloum Delta of Africa, the Western Siberian Lowlands, and a new wetland phenomenon in Asia—wetland parks such as XiXi National Wetland Park in eastern China and Gandau Nature Park in Taipai, Taiwan. Over 50 photographs of the world's wetlands are now found in Chapters 1 and 3. In addition, we have provided three new estimates of the extent of wetlands in the world and have updated our world wetland map. We have also documented the importance of coastal marshes in the Louisiana Delta after Hurricane Katrina, coastal mangroves as protective systems after the Indian Ocean tsunami of December 2004, and the Mesopotamian Marshland restoration in Iraq after its drainage in the 1990s.

Thirty-seven "boxes" or sidebars are another feature in *Wetlands 4*. These boxes include important footnote-type details in many of the chapters and case studies of wetland restoration (Chapter 12) and treatment wetlands (Chapters 13). These boxes and the shorter book should be welcome changes for college students who use this textbook for wetland ecology classes. Our students thought *Wetlands 3* was a bit much for one quarter or semester. Now it is more manageable.

The book is updated in every chapter. More than 200 new wetland publications are cited in this edition with over one hundred and seventy citations from 2000 or later to augment the classics from the last half of the 20th century. Many older citations, particularly those that would be hard to find, were eliminated. New or expanded subjects, in addition to climate change (Chapter 10) include seiches in wetlands (Chapter 4), anammox and dissimilatory nitrate reduction to ammonia (DNRA) in the wetland nitrogen cycle (Chapter 5), wetland plant hypertrophy (Chapter 6), the hydrogeomorphic wetland classification system (Chapter 8), waterfowl and wildlife management in wetlands (Chapter 9), the importance of wetlands in storm and tsunami abatement (Chapter 11), emergy analysis to quantify wetland values (Chapter 11), current status of mitigation wetlands in the United States (Chapter 12), and two important U.S. Supreme Court decisions in the 21st century related to wetland protection (Chapter 14). The status of the international Ramsar Convention on Wetlands is also brought up to date in Chapter 14.

On a personal note, we are pleased to share new wetland research results generated at the Wilma H. Schiermeier Olentangy River Wetland Research Park (ORWRP) on the campus of The Ohio State University. New findings from the experimental wetlands at the ORWRP are presented in boxes on the importance of hydrologic pulses on wetland function (Chapter 4), the importance of plant introduction in freshwater marsh succession (Chapter 7), the development of hydric soils in wetland creation (Chapter 12), and long-term water quality changes in flow-through riverine wetlands (Chapter 13).

We could not have completed this edition without help from many friends and colleagues. Anne Mischo provided dozens of new illustrations for *Wetlands 4* to supplement her beautiful work from *Wetlands 3*. We are honored to have a wetland photo from Jimmie Campbell, Columbus Ohio, on the cover of our book; we are even more thrilled that the picture was taken at the created wetlands at the Olentangy River Wetland Research Park (ORWRP) in Ohio. Ruthmarie Mitsch provided hours of editing and referencing as this edition was being completed. Cassie Tuttle assisted with literature reviews to find some of the new material in this book. Li Zhang kept the ORWRP running and provided a great sounding board for ideas to make this book better. We also appreciate the input, illustrations, or insight provided by the following (listed in alphabetical order): Jim Aber, Azzam Alwash, Chris Anderson, Mark Brown, Jean Cowan, Jenny Davis, Frank Day, John Day, Siobhan Fennessy, Max Finlayson, Brij Gopal, Wenshan He, Maria Hernandez, Carter Johnson, Wolfgang Junk, Jean-Claude LeFeuvre, Robin Lewis, Jianjian Lu, Pierrick Marion, Ken Mavuti, Andre Mauxion, Irv Mendelssohn, Thomas Nebbia, Nancy Rabalais, Bill Resch, Clayton Rubec, Kenneth Strait, Ralph Tiner, Louis Toth, Barry Warner, and Paul Whalen.

We also appreciate the professional effort on the part of editor Jim Harper and production manager Kerstin Nasdeo of John Wiley & Sons, Inc. It has been a pleasure to work with the Wiley operation since they purchased our original publisher Van Nostrand Reinhold in the mid-1990s and switched us to the Wiley brand.

William J. Mitsch   James G. Gosselink
Columbus, Ohio   Rock Island, Tennessee

June 2007

Wetlands

Part 1

Introduction

Wetlands: Human History, Use, and Science

Wetlands, landscape features found in almost all parts of the world, are known as "the kidneys of the landscape" and "ecological supermarkets" to bring attention to the important values they provide. Although many cultures have lived among and even depended on wetlands for centuries, the modern history of wetlands is fraught with misunderstanding and fear, as described in much of our Western literature. Wetlands have been destroyed at alarming rates throughout the developed and developing worlds. Now, as their many values are being recognized, wetland conservation and protection have become the norm in many parts of the world. Wetlands have properties that are not adequately covered by present terrestrial and aquatic ecology, making a case for wetland science as a unique discipline encompassing many fields, including terrestrial and aquatic ecology, chemistry, hydrology, and engineering. Wetland management, as the applied side of wetland science, requires an understanding of the scientific aspects of wetlands balanced with legal, institutional, and economic realities. As interest in wetlands has grown, so too have professional organizations and agencies that are concerned with wetlands, as well as the amount of journals and literature on wetland science.

Wetlands are among the most important ecosystems on Earth. In the great scheme of things, the swampy environment of the Carboniferous period produced and preserved many of the fossil fuels on which our society now depends. In more recent biological and human time periods, wetlands have been valuable as sources, sinks, and transformers of a multitude of chemical, biological, and genetic materials. Although the value of wetlands for fish and wildlife protection has been known for a century, some of the other benefits have been identified more recently.

Wetlands are sometimes described as "the kidneys of the landscape" because they function as the downstream receivers of water and waste from both natural and human sources. They stabilize water supplies, thus ameliorating both floods and drought. They have been found to cleanse polluted waters, protect shorelines, and recharge groundwater aquifers.

Wetlands also have been called "ecological supermarkets" because of the extensive food chain and rich biodiversity that they support. They play major roles in the landscape by providing unique habitats for a wide variety of flora and fauna. Now that we have become concerned about the health of our entire planet, wetlands are being described by some as important carbon sinks and climate stabilizers on a global scale.

These values of wetlands are now recognized worldwide and have led to wetland conservation, protection laws, regulations, and management plans. But our history with wetlands had been to drain, ditch, and fill them, never as quickly or as effectively as was undertaken in countries such as the United States beginning in the mid-1800s.

Wetlands have become the *cause célèbre* for conservation-minded people and organizations throughout the world, in part because they have become symptoms of our systematic dismantling of our water resources and in part because their disappearance represents an easily recognizable loss of natural areas to economic "progress." Scientists, engineers, lawyers, and regulators are now finding it both useful and necessary to become specialists in wetland ecology and wetland management in order to understand, preserve, and even reconstruct these fragile ecosystems. This book is for these aspiring wetland specialists, as well as for those who would like to know more about the structure and function of these unique ecosystems. It is a book about wetlands—how they work and how we manage them.

Human History and Wetlands

There is no way to estimate the impact humans have had on the global extent of wetlands except to observe that, in developed and heavily populated regions of the world, the impact has ranged from significant to total. The importance of wetland environments to the development and sustenance of cultures throughout human history, however, is unmistakable. Since early civilization, many cultures have learned to live in harmony with wetlands and have benefited economically from surrounding wetlands, whereas other cultures quickly drained the landscape. The ancient Babylonians, Egyptians, and the Aztec in what is now Mexico developed specialized systems of water delivery involving wetlands. Major cities of the world, such as Chicago and Washington, D.C., in the United States, Christchurch, New Zealand, and Paris, France, stand on sites that were once part wetlands. Many of the large airports (in Boston, New Orleans, and J. F. Kennedy in New York, to name a few) are situated on former wetlands.

While global generalizations are sometimes misleading, there was and is a propensity in Eastern cultures not to drain valuable wetlands entirely, as has been done in the West, but to work within the aquatic landscape, albeit in a heavily managed way. Dugan (1993) makes the interesting comparison between *hydraulic civilizations*

(European in origin) that controlled water flow through the use of dikes, dams, pumps, and drainage tile, partially because water was only seasonally plentiful, and *aquatic civilizations* (Asian in origin) that better adapted to their surroundings of water-abundant floodplains and deltas and took advantage of nature's pulses such as flooding. It is because the former approach of controlling nature rather than working with it is so dominant today that we find such high losses of wetlands worldwide.

Wetlands have been and continue to be part of many human cultures in the world. Coles and Coles (1989) referred to the people who live in proximity to wetlands and whose culture is linked to them as *wetlanders*. Some of these cultures and users of wetlands are illustrated in eighteen photographs in this chapter (Figures 1.1 through 1.18). Figures 1.1 through 1.7 show human cultures or settings around the world that have depended on wetlands, sometimes for centuries. Figures 1.8 through 1.11 show some of the many food products that are harvested from wetlands while Figures 1.12 through 1.16 illustrate the use of wetlands as sources of fuel, building materials, and even household goods. Most recently, wetlands have become the foci for ecotourism in many developing and developed parts of the world (Figure 1.17 through 1.18).

Sustainable Cultures in Wetlands

The Camarguais of southern France (Fig. 1.1), the Cajuns of Louisiana (Fig. 1.2), the Marsh Arabs of southern Iraq (Fig. 1.3), many Far Eastern cultures (Fig. 1.4), and the Native Americans in North America (Figs. 1.5 and 1.6) have lived in harmony with wetlands for hundreds if not thousands of years. These are the true wetlanders. For example, the Sokaogon Chippewa in Wisconsin have, for centuries, harvested and reseeded wild rice (*Zizania aquatica*) along the littoral zone of lakes and streams. They have a saying that "wild rice is like money in the bank." Wetlands were often used as places of cultural solitude and reverence, as with the Mont St. Michel, a Benedictine monastery, built between the 11th and 16th centuries in northern France (Fig. 1.7).

Food from Wetlands

Domestic wetlands such as rice paddies feed an estimated half of the world's population (Fig. 1.8). Countless other plant and animal products are harvested from wetlands throughout the world. Many aquatic plants besides rice such as Manchurian wild rice (*Zizania latifolia*) are harvested as vegetables in China (Fig. 1.9). Cranberries are harvested from bogs, and the industry continues to thrive today in North America (Fig. 1.10). Coastal marshes in northern Europe, the British Isles, and New England were used for centuries and are still used today for grazing of animals and hay production.

Wetlands can be an important source of protein. The production of fish in shallow ponds or rice paddies developed several thousands of years ago in China and Southeast Asia, and crayfish harvesting is still practiced in the wetlands of Louisiana and the Philippines. Shallow lakes and wetlands are an important provider of protein in many parts of sub-Saharan Africa (Fig. 1.11).

Figure 1.1   The Camargue region of southern France in the Rhone River delta is an historically important wetland region in Europe where Camarguais have lived since the Middle Ages. *(Photograph by Tom Nebbia, Horseshoe, North Carolina, reprinted by permission.)*

Figure 1.2   A Cajun lumberjack camp in the Atchafalaya Swamp of coastal Louisiana. American Cajuns are descendants of the French colonists of Acadia (present-day Nova Scotia, Canada), who were forced out of Nova Scotia by the English and moved to the Louisiana delta in the last half of the 18th century. Their society and culture flourished within the bayou wetlands. *(Photograph courtesy of the Louisiana Collection, Tulane University Library, New Orleans, reprinted by permission.)*

Peat and Building Materials

The Russians, Finns, Estonians, and Irish, among other cultures, have mined their peatlands for centuries, using peat as a source of energy on small-scale production (Fig. 1.12) and in large-scale extraction processes (Fig. 1.13). *Sphagnum* peat is now harvested for horticultural purposes throughout the world. In southwestern New Zealand, for example, surface *Sphagnum* has been harvested since the 1970s for export as a potting medium (Fig. 1.14). Reeds and even the mud from coastal and

Figure 1.3   The Marsh Arabs of southern Iraq lived for centuries on artificial islands in marshes at the confluence of the Tigris and Euphrates rivers. The marshes were mostly drained by Saddam Hussein in the 1990s and are now being restored (see Chapter 12).

Figure 1.4   Interior wetlands in Weishan County, Shandong Province, China, where approximately 60,000 people live amid wetland-canal systems and harvest aquatic plants for food and fiber. (*Photograph by W. J. Mitsch.*)

inland marshes have been used for thatching for roofs in Europe, Iraq, Japan, and China, as well as wall construction, fence material, lamps, and other household goods (Figs. 1.15 and 1.16). Coastal mangroves are harvested for timber, food, and tannin in many countries throughout Indo-Malaysia, East Africa, and Central and South America.

Figure 1.5   Native American "ricers" from the Sokaogon Chippewa Reservation poling and "knocking" wild rice (*Zizania aquatica*) as they have for hundreds of years on Rice Lake in Forest County, Wisconsin. *(Photograph by R. P. Gough, reprinted by permission.)*

Figure 1.6   Several Native American tribes have lived in and around the wetlands of southern Florida, including the Florida Everglades. These include the Calusa Indians, who disappeared as a result of imported European disease, and later the Seminole (Miccosukee) tribe that moved south to the Everglades in the 19th century while being pursued by the U.S. Army during the Seminole Indian wars. They never surrendered. The Miccosukee adapted to living in hammock-style camps spread throughout the Everglades and relied on fishing, hunting, and harvesting of native fruits from the hammocks. *(Photograph by W. J. Mitsch, panorama at Miccosukee Indian Village, Florida Everglades.)*

Figure 1.7   Mont St. Michel, a Benedictine monastery, built between the 11th and 16th centuries, sits amid the coastal mudflats and salt marshes between Normandy and Brittany in northwestern France. Entry to the island, now a UNESCO World Heritage site, is through a land bridge that crosses the wetlands. *(Photograph by A. Mauxion, reprinted by permission.)*

Figure 1.8   Rice production occurs in "managed" wetlands throughout Asia and other parts of the world. Half of the world's population is fed by rice paddy systems. *(Photograph by W. J. Mitsch.)*

Figure 1.9   Wetland plants such as *Zizania latifolia* are harvested and sold in markets such as this one in Suzhou, Jiangsu Province, China. This and several other aquatic plants are cooked and served as vegetables in China. *(Photograph by W. J. Mitsch.)*

Figure 1.10   Cranberry wet harvesting is done by flooding bogs in several regions of North America. The cranberry plant (*Vaccinium macrocarpon*) is native to the bogs and marshes of North America and was first cultivated in Massachusetts. It is now also an important fruit crop in Wisconsin, New Jersey, Washington, Oregon, and parts of Canada. *(Photograph courtesy of Ocean Spray Cranberries, Inc., Lakeville-Middleboro, Massachusetts.)*

Figure 1.11   Humans use the wetlands of sub-Saharan Africa for sustenance, as with this man fishing for lung fish (*Proptopterus aethiopicus*) in Lake Kanyaboli, western Kenya. *(Photograph by K. M. Mavuti, reprinted by permission.)*

Figure 1.12   Harvesting of peat or "turf" as a fuel has been a tradition in several parts of the world, as shown by this scene of "turf carts" in Ireland.

Wetlands and Ecotourism

A modern version of wetland use is through ecotourism. Wetlands have been the focus of several countries' attempts to increase tourist flow into their countries (Figs. 1.17 and 1.18). The Okavango Delta in Botswana is one of the natural resource jewels of Africa, and protection of this wetland for tourists and hunters has been a

Figure 1.13 Large-scale peat mining in Estonia. *(Photograph by W. J. Mitsch.)*

Figure 1.14 *Sphagnum* moss harvesting in Westland, South Island, New Zealand for gardens and potting of plants. *(Photograph by C. Pugsley, New Zealand Department of Conservation, Wellington, reprinted by permission.)*

priority in that country since the 1960s. Local tribes provide manpower for boat tours (in dugout canoes called mokoros) through the basin and assist with wildlife tours on the uplands as well. In Senegal, west Africa, there is keen interest in attracting European birder tourists to the mangrove swamps along the Atlantic coastline. The advantage of ecotourism as a management strategy is obvious—it provides income to

Figure 1.15 A "Wetland House" in the Ebro River Delta Region on the Mediterranean Sea, Spain. The walls are made from wetland mud, and the roof is thatched with reed grass and other wetland vegetation. *(Photograph by W. J. Mitsch.)*

Figure 1.16 Floor lamps developed from Yosi (reedgrass; *Phragmites australis*), Lake Biwa, Japan. *(Lamps designed by Mr. Morino; photograph by B. Cleveland, reprinted by permission.)*

Figure 1.17 Several rural communities exist in the vast, seasonally flooded Okavango Delta of northern Botswana in southern Africa. The wetlands attract tourists, as shown in this illustration, and also wildlife hunting, in addition to providing basic sustenance to these communities. *(Photograph by W. J. Mitsch.)*

Figure 1.18 Interest in the wetlands that surround Lake Biwa in Shiga Prefecture, Japan, is intense, as shown by this photograph of participants at a winter 2006 international wetlands forum. *(Photograph by W. J. Mitsch.)*

the country where the wetland is found without requiring or even allowing resource harvest from the wetlands. The potential disadvantage is that if the site becomes too popular, human pressures will begin to deteriorate the landscape and the very ecosystem that initially drew the tourism.

Literary References to Wetlands

With all of these valuable uses, not to mention the aesthetics of a landscape in which water and land often provide a striking panorama, one would expect wetlands to be revered by humanity; this has certainly not always been the case. Wetlands have been depicted as sinister and forbidding, and as having little economic value throughout most of history. For example, in the *Divine Comedy*, Dante describes a marsh of the Styx in Upper Hell as the final resting place for the wrathful:

Thus we pursued our path round a wide arc of that ghast pool,
Between the soggy marsh and arid shore,
Still eyeing those who gulp the marish [marsh] foul.

—Dante Alighieri

Centuries later, Carl Linnaeus, crossing the Lapland peatlands, compared that region to that same Styx of Hell:

Shortly afterwards began the muskegs, which mostly stood under water; these we had to cross for miles; think with what misery, every step up to our knees. The whole of this land of the Lapps was mostly muskeg, hinc vocavi Styx. Never can the priest so describe hell, because it is no worse. Never have poets been able to picture Styx so foul, since that is no fouler.

—Carl Linnaeus, 1732

In the 18th century, an Englishman who surveyed the Great Dismal Swamp on the Virginia–North Carolina border and is credited with naming it described the wetland as:

[a] horrible desert, the foul damps ascend without ceasing, corrupt the air and render it unfit for respiration.... Never was Rum, that cordial of Life, found more necessary than in this Dirty Place.
    —Colonel William Byrd III (1674–1744), "Historie of the Dividing Line Betwixt Virginia and North Carolina" in *The Westover Manuscripts*, written 1728–1736, Petersburg, VA; E. and J C. Ruffin, printers, 1841, 143 pp.

Even those who study and have been associated with wetlands have been belittled in literature:

Hardy went down to botanise in the swamp, while Meredith climbed towards the sun. Meredith became, at his best, a sort of daintily dressed Walt Whitman: Hardy became a sort of village atheist brooding and blaspheming over the village idiot.
    —G. K. Chesterton (1874–1936), Chapter 12 in *The Victorian Age in Literature*, Henry Holt and Company, New York, 1913

The English language is filled with words that suggest negative images of wetlands. We get *bogged down* in detail; we are *swamped* with work. Even the mythical *bogeyman*, the character featured in stories that frighten children in many countries, may be associated with European bogs. Grendel, the mythical monster in one of the oldest surviving pieces of Old English literature and Germanic epic, *Beowulf*, comes from the peatlands of present-day northern Europe:

> Grendel, the famous stalker through waste places, who held the rolling marshes in his sway, his fen and his stronghold. A man cut off from joy, he had ruled the domain of his huge misshapen kind a long time, since God had condemned him in condemning the race of Cain.
> —Beowulf, translated by William Alfred, *Medieval Epics*,
> The Modern Library, New York, 1993

Hollywood has continued the depiction of the sinister and foreboding nature of wetlands and their inhabitants, in the tradition of Grendel, with movies such as the classic *Creature from the Black Lagoon* (1954), a comic-book-turned-cult-movie *Swamp Thing* (1982), and its sequel *Return of the Swamp Thing* (1989). Even Swamp Thing, the man/monster depicted in Figure 1.19, evolved in the 1980s from a feared creature to a protector of wetlands, biodiversity, and the environment. But as long as wetlands remain more difficult to stroll through than a forest and more difficult to cross by boat than a lake, they will remain misunderstood ecosystems to the general public without a continued effort of education.

Wetland Destruction and Conservation

Prior to the mid-1970s, the drainage and destruction of wetlands were accepted practices around the world and were even encouraged by specific government policies. Wetlands were replaced by agricultural fields and by commercial and residential development. Had those trends continued, the resource would be in danger of extinction. Some countries and states such as New Zealand and California and Ohio in the United States have reported 90 percent loss of their wetlands. Only through the combined activities of hunters and anglers, scientists and engineers, and lawyers and conservationists has the case been made for wetlands as a valuable resource whose destruction has serious economic as well as ecological and aesthetic consequences for the nations of the world. This increased level of respect was reflected in activities such as the sale of federal "duck stamps" to waterfowl hunters that began in 1934 in the United States (Fig. 1.20); other countries such as New Zealand have followed suit. Approximately 2.1 million hectares (ha) of wetlands have been purchased or leased as waterfowl habitat by the U.S. duck stamp program alone since 1934. The U.S. government now supports a variety of other wetland protection programs through at least a dozen federal agencies; individual states have also enacted wetland protection laws or have used existing statutes to preserve these valuable resources.

That interest in wetland conservation, which first blossomed in the 1970s in the United States, has now spread around the world. The international Convention on

Figure 1.19 **The sinister image of wetlands, especially swamps, is often promoted in popular media such as Hollywood movies and comic books, although the man-turned-plant "Swamp Thing" is a hero as he fights injustice and even toxic pollution.** *(Swamp Thing #9 © DC Comics. All Rights Reserved. Used with Permission.)*

Wetlands, signed in Ramsar, Iran, in 1971, and referred to as the Ramsar Convention, is an intergovernmental treaty that provides the framework for national action and international cooperation for the conservation and wise use of wetlands around the world. More than 150 countries are participating in the agreement, with over 150 million ha of wetlands designated for inclusion in the Ramsar List of Wetlands

Figure 1.20 Federal Migratory Bird Hunting and Conservation Stamps are more commonly known as "Duck Stamps." They are produced by the U.S. Postal Service for the U.S. Fish & Wildlife Service and are not valid for postage. Originally created in 1934 as the federal licenses required for hunting migratory waterfowl, today income derived from their sale is used to purchase or lease wetlands. *Top:* First Duck Stamp from 1934 (Mallards); *Bottom:* 2005-06 duck stamp (Hooded Merganser).

of International Importance. The Convention's mission is "the conservation and wise use of all wetlands through local, regional and national actions and international cooperation, as a contribution towards achieving sustainable development throughout the world" (www.ramsar.org, 2006). Many other countries and nongovernmental organizations (NGOs) are now dedicated to preserving wetlands.

Wetland Science and Wetland Scientists

A specialization in the study of wetlands is often termed *wetland science* or *wetland ecology*, and those who carry out such investigations are called *wetland scientists* or *wetland ecologists*. The term *mire ecologist* has also been used. Some have suggested that the study of all wetlands be termed *telmatology* (*telma* being Greek for "bog"), a term originally coined to mean "bog science" (Zobel and Masing, 1987). No matter what the field is called, it is apparent that there are several good reasons for treating wetland ecology as a distinct field of study:

1. Wetlands have unique properties that are not adequately covered by present ecological paradigms and by fields such as limnology, estuarine ecology, and terrestrial ecology.
2. Wetland studies have begun to identify some common properties of seemingly disparate wetland types.
3. Wetland investigations require a multidisciplinary approach or training in several fields not routinely studied or combined in university academic programs.
4. There is a great deal of interest in formulating sound policy for the regulation and management of wetlands. These regulations and management approaches need a strong scientific underpinning integrated as wetland ecology.

A growing body of evidence suggests that the unique characteristics of wetlands—standing water or waterlogged soils, anoxic conditions, and plant and animal adaptations—may provide some common ground for study that is neither terrestrial ecology nor aquatic ecology. Wetlands provide opportunities for testing "universal" ecological theories and principles involving succession and energy flow, which were developed for aquatic or terrestrial ecosystems. For example, wetlands provided the setting for the successional theories of Clements (1916) and the energy flow approaches of Lindeman (1942). They also provide an excellent laboratory for the study of principles related to transition zones, ecological interfaces, and ecotones.

Our knowledge of different wetland types such as those discussed in this book is, for the most part, isolated in distinctive literatures and scientific circles. One set of literature deals with coastal wetlands, another with forested wetlands and freshwater marshes, and still another with peatlands. Very few investigators have analyzed the properties and functions common to all wetlands. This is probably one of the most exciting areas for wetland research because there is so much to be learned. Comparisons of wetland types have shown, for example, the importance of hydrologic flow-through for the maintenance and productivity of these ecosystems. The anoxic biochemical processes that are common to all wetlands provide another area for comparative research and pose many questions: What are the roles of different wetland types in local and global biochemical cycles? How do the activities of humans influence these cycles in various wetlands? What are the synergistic effects of hydrology, chemical inputs, and climatic conditions on wetland biological productivity? How can plant and animal adaptations to anoxic stress be compared in various wetland types?

The true wetland ecologist must be an ecological generalist because of the number of sciences that bear on those ecosystems. Knowledge of wetland flora and fauna, which are often uniquely adapted to a substrate that may vary from submerged to dry, is necessary. Emergent wetland plant species support both aquatic animals and terrestrial insects. Because hydrologic conditions are so important in determining the structure and function of the wetland ecosystems, a wetland scientist should be well versed in surface and groundwater hydrology. The shallow-water environment means that chemistry—particularly for water, sediments, soils, and water–sediment interactions—is an important science. Similarly, questions about wetlands as sources, sinks, or transformers of chemicals require investigators to be versed in many biological and chemical techniques. While the identification of wetland vegetation and animals requires botanical and zoological skills, backgrounds in microbial biochemistry and soil science contribute significantly to the understanding of the anoxic environment. Understanding adaptations of wetland biota to the flooded environment requires both biochemistry and physiology. If wetland scientists are to become more involved in the management of wetlands, some engineering techniques, particularly for wetland hydrologic control or wetland creation, need to be learned.

Wetlands are seldom, if ever, isolated systems. Rather, they interact strongly with adjacent terrestrial and aquatic ecosystems. Hence, a holistic view of these complex landscapes can be achieved only through an understanding of the principles of ecology, especially those that are part of ecosystem and landscape ecology and systems analysis. Finally, if wetland management involves the implementation of wetland policy, then training in the legal and policy-making aspects of wetlands is warranted.

Thousands of scientists and engineers are now studying and managing wetlands. Only a relatively few pioneers, however, investigated these systems in any detail prior to the 1960s. Most of the early scientific studies dealt with classical botanical surveys or investigations of peat structure. Several early scientific studies of peatland hydrology were also produced, particularly in Europe and Russia. Later, investigators such as Chapman, Teal, Sjörs, Gorham, Eugene and H. T. Odum, Weller, Patrick, and their colleagues and students began to use modern ecosystem and biogeochemical approaches in wetland studies (Table 1.1). Several research centers devoted to the study of wetlands have now been established in the United States, including the Sapelo Island Marine Institute in Georgia; the School of Coast and Environment at Louisiana State University; the H. T. Odum Center for Wetlands at the University of Florida; the Duke Wetland Center at Duke University; and the Wilma H. Schiermeier Olentangy River Wetland Research Park (ORWRP) at The Ohio State University. International laboratories such as the Harry Oppenheimer Okavango Research Centre (HOORC) in Botswana, Africa, have been established for the study of specific wetlands or wetland areas. In addition, a professional society now exists, the *Society of Wetland Scientists*, which has among its goals to provide a forum for the exchange of ideas within wetland science and to develop wetland science as a distinct discipline. *The International Association of Ecology* (INTECOL) has sponsored a major international wetland conference every four years somewhere in the world since 1980.

Wetland Managers and Wetland Management

Just as there are wetland scientists who are uncovering the processes that determine wetland functions and values, so too there are those who are involved, by choice or by vocation, in some of the many aspects of wetland management. These individuals, whom we call *wetland managers*, are engaged in activities that range from waterfowl production to wastewater treatment. They must be able to balance the scientific aspects of wetlands with a myriad of legal, institutional, and economic constraints to provide optimum wetland management. The management of wetlands has become

Table 1.1 Some pioneer researchers in wetland ecology and representative citations for their work

Wetland Type and Researcher	Country	Representative Citations
COASTAL MARSHES/MANGROVES		
Valentine J. Chapman	New Zealand	Chapman (1938, 1940)
John Henry Davis	USA	Davis (1940, 1943)
John M. Teal	USA	Teal (1958, 1962); Teal and Teal (1969)
Eugene P. Odum, Howard T. Odum	USA	E. P. Odum (1961); H. T. Odum et al. (1974)
D. S. Ranwell	UK	D. S. Ranwell (1972)
PEATLANDS/FRESHWATER WETLANDS		
C. A. Weber	Germany	Weber (1907)
Herman Kurz	USA	Kurz (1928)
A. P. Dachnowski-Stokes	USA	Dachnowski-Stokes (1935)
R. L. Lindeman	USA	Lindeman (1941, 1942)
Eville Gorham	UK/USA	Gorham (1956, 1961)
Hugo Sjörs	Sweden	Sjörs (1948, 1950)
G. Einar Du Rietz	Sweden	Du Rietz (1949, 1954)
P. D. Moore/D. J. Bellamy	UK	Moore and Bellamy (1974)
S. Kulczynski	Poland	Kulczynski (1949)
Paul R. Errington	USA	Errington (1957)
R. S Clymo	UK	Clymo (1963, 1965)
Milton Weller	USA	Weller (1981)
William H. Patrick	USA	Patrick and Delaune (1972)

increasingly important in many countries because government policy and wetland regulation seek to reverse historic wetland losses in the face of continuing draining or encroachment by agricultural enterprises and urban expansion. The simple act of being able to identify the boundaries of wetlands has become an important skill for a new type of wetland technician in the United States called a *wetland delineator*.

Private organizations such as *Ducks Unlimited, Inc.* and *The Nature Conservancy* have protected wetlands by purchasing thousands of hectares of wetlands throughout North America. Through the *Ramsar Convention* and an agreement jointly signed by the United States and Canada in 1986 called the *North American Waterfowl Management Plan*, wetlands are now being protected primarily for their waterfowl value on an international scale. In 1988, a federally sponsored *National Wetlands Policy Forum* (1988) in the United States raised public and political awareness of wetland loss and recommended a policy of "no net loss" of wetlands. This recommendation has stimulated widespread interest in wetland restoration and creation to replace lost wetlands, and "no net loss" has remained the policy of wetland protection in the United States since the late 1980s.

Subsequently, a National Research Council report in the United States (NRC, 1992) called for the fulfillment of an ambitious goal of gaining 4 million ha of wetlands by the year 2010, largely through the reconversion of crop and pasture land. Wetland creation for specific functions is an exciting new area of wetland management that needs trained specialists and may eventually stem the tide of loss and lead to an increase in this important resource. Another National Research Council report (NRC, 1995) reviewed the scientific basis for wetland delineation and classification, particularly as it related to the regulation of wetlands in the United States at that time, and yet another NRC (2001) study investigated the effectiveness of the national policy of mitigation of wetland loss in the United States.

The Wetland Scientific Literature

The increasing interest and emphasis on wetland science and management has been demonstrated by a veritable flood of books, reports, scientific studies, and conference proceedings, most in the last two decades of the 20th century and early 21st century. The journal citations in this book are only the tip of the iceberg of the literature on wetlands, much of which has been published since the mid-1980s. Two journals, *Wetlands* and *Wetlands Ecology and Management*, are now published to disseminate scientific and management papers on wetlands, and several other scholarly journals frequently publish papers on wetlands. Dozens of wetland meeting proceedings and journal special issues have been published from conferences on wetlands held throughout the world. Beautifully illustrated popular books and articles with color photographs have been developed on wetlands by Niering (1985), Littlehales and Niering (1991), Mitchell et al. (1992), Kusler et al. (1994), Rezendes and Roy (1996), and Lockwood and Gary (2005) on wetlands in North America; by McComb and Lake (1990) on Australian wetlands; by Mendelsohn and el Obeid (2004) on the Okavango River Delta in Africa; and by Finlayson and Moser (1991) and Dugan (1993) on wetlands of the world.

Government agencies and NGOs around the world have contributed significantly to the wetland literature and to our understanding of wetland functions and values. In the United States, the *U.S. Fish and Wildlife Service* has been involved in the classification and inventory of wetlands and has published a series of community profiles on various regional wetlands. The *U.S. Environmental Protection Agency* (U.S. EPA) has been interested in the impact of human activity on wetlands, and in wetlands as possible systems for the control of water pollution. Along with the *U.S. Army Corps of Engineers*, the U.S. EPA, especially through its Office of Wetlands, Oceans, and Watersheds (OWOW), the U.S. Fish and Wildlife Service, and the *Natural Resources Conservation Service* now are the primary wetland management agencies in the United States.

Wetland management organizations such as the *Association of State Wetland Managers* and the *Society of Wetland Scientists* focus on disseminating information on wetlands, particularly in North America. The *International Union for the Conservation of Nature and Natural Resources* (IUCN) and the *Ramsar Convention*, both based in Switzerland, have developed a series of publications on wetlands of the world. *Wetlands International* is the world's leading nonprofit organization concerned with the conservation of wetlands and wetland species. It comprises a global network of governmental and nongovernmental experts working on wetlands. Activities are undertaken in more than 120 countries worldwide. The headquarters for its Africa, Europe, Middle East (AEME) branch is located in Wageningen, The Netherlands.

Recommended Readings

Errington, P.L. 1957. *Of Men and Marshes*. The Iowa State University Press, Ames, Iowa.

Finlayson, M., and M. Moser, eds. 1991. *Wetlands. Facts on File*, Oxford, UK 224 pp.

Kusler, J., W. J. Mitsch, and J. S. Larson. 1994. Wetlands. *Scientific American* 270(1): 64–70.

Millenium Ecosystem Assessment. 2005. "Ecosystems and Human Well-Being: Wetlands and Water Synthesis." World Resources Institute, Washington, DC.

Teal, J., and M. Teal. 1969. *Life and Death in the Salt Marsh*. Little, Brown, Boston.

Wetland Definitions

Wetlands have many distinguishing features, the most notable of which are the presence of standing water for some period during the growing season, unique soil conditions, and organisms, especially vegetation, adapted to or tolerant of saturated soils. Wetlands are unique because of their hydrologic conditions and their role as ecotones between terrestrial and aquatic systems. Terms such as swamp, marsh, fen, and bog have been used in common speech for centuries to define wetlands and are frequently used and misused today. Formal definitions have been developed by several federal agencies in the United States, by scientists in Canada and the United States, and through an international treaty known as the Ramsar Convention. These definitions include considerable detail and are used for both scientific and management purposes. Wetlands are not easily defined, however, especially for legal purposes, because they have a considerable range of hydrologic conditions, because they are found along a gradient at the margins of well-defined uplands and deepwater systems, and because of their great variation in size, location, and human influence. No absolute answer to "What is a wetland?" should be expected, but legal definitions involving wetland protection are becoming increasingly comprehensive.

The most common questions that the uninitiated ask about wetlands are "What exactly is a wetland?" or "Is that the same as a swamp?" These are surprisingly good questions, and it is not altogether clear that they have been answered completely by wetland scientists and managers. Wetland definitions and terms are many and are often confusing or even contradictory. Nevertheless, definitions are important both for the scientific understanding of these systems and for their proper management. In the 19th century, when the drainage of wetlands was the norm, a wetland definition was unimportant because it was considered desirable to produce uplands from wetlands by draining them. In fact, the word "wetland" did not come into common use until the mid-20th century. One of the first references to the word was in the publication *Wetlands of the United States* (Shaw and Fredine, 1956). Before that time, wetlands were referred to by the many common terms that developed in the 19th century and before, such as swamp, marsh, bog, fen, mire, and moor. Even as the value of wetlands was being recognized in the early 1970s, there was little interest in precise definitions until it was realized that a better accounting of the remaining wetland resources was needed, and definitions were necessary to achieve that inventory.

When national and international laws and regulations pertaining to wetland preservation began to be written in the late 1970s, the need for precision became even greater as individuals recognized that definitions were having an impact on what they could or could not do with their land. The definition of a wetland, and by implication its boundaries (referred to as "delineation" in the United States), became important when society began to recognize the value of these systems and began to translate that recognition into laws to protect itself from further wetland loss. However, just as an estimate of the boundary of a forest, desert, or grassland is based on scientifically defensible criteria, so too should the definition of wetlands be based on scientific measures to as great a degree as possible. What society chooses to do with wetlands, once the definition has been chosen, remains a political decision.

Wetlands in the Landscape

Even after the ecological and economic benefits of wetlands were determined and became widely appreciated, wetlands have remained an enigma to scientists. They are difficult to define precisely, not only because of their great geographical extent, but also because of the wide variety of hydrologic conditions in which they are found. Wetlands are usually found at the interface of terrestrial ecosystems, such as upland forests and grasslands, and aquatic systems such as deep lakes and oceans (Fig. 2.1a), making them different from each yet highly dependent on both. They are also found in seemingly isolated situations, where the nearby aquatic system is often a groundwater aquifer (Fig. 2.1b). Sometimes these wetlands are referred to as *isolated wetlands*, a somewhat misleading term because they are usually connected hydrologically to groundwater and biologically through the movement of many mobile organisms. And of course, all wetland ecosystems are open to solar radiation. Because wetlands combine attributes of both aquatic and terrestrial ecosystems but are neither, they have fallen between the cracks of the scientific disciplines of terrestrial and aquatic ecology. They serve as sources, sinks, and transformers of nutrients; deepwater aquatic systems (at least lakes and oceans) are almost always sinks, and terrestrial systems are usually sources. Wetlands are also among the most productive ecosystems on the planet when compared to adjacent terrestrial and deepwater aquatic systems, but it is not correct to say that all wetlands are highly productive. Peatlands and cypress swamps are examples of low-productivity wetlands.

Distinguishing Features of Wetlands

We can easily identify a coastal salt marsh, with its great uniformity of grasses and its maze of tidal creeks, as a wetland. A cypress swamp, with majestic trees festooned with Spanish moss and standing in knee-deep water, provides an unmistakable image of a wetland. A northern *Sphagnum* bog, surrounded by tamarack trees that quake as people trudge by, is another easily recognized wetland. All of those sites have several features in common: (1) all have shallow water or saturated soil; (2) all accumulate organic plant material that decomposes slowly; and (3) all support a variety of plants and animals adapted to the saturated conditions. Wetland definitions, then, often include three main components:

1. Wetlands are distinguished by the presence of water, either at the surface or within the root zone.

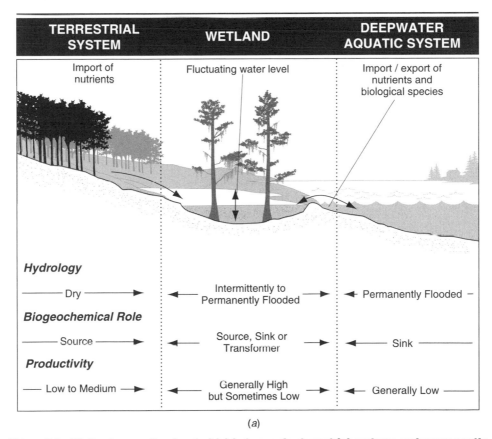

(a)

Figure 2.1 Wetlands are often located (a) between dry terrestrial systems and permanently flooded deepwater aquatic systems such as rivers, lakes, estuaries, or oceans or (b) as isolated basins with little outflow and no adjacent deepwater system.

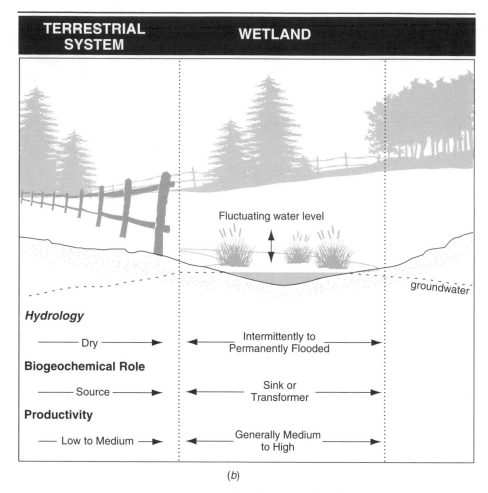

(b)

Figure 2.1 *(continued)*

2. Wetlands often have unique soil conditions that differ from adjacent uplands.
3. Wetlands support biota such as vegetation adapted to the wet conditions (*hydrophytes*) and, conversely, are characterized by an absence of flooding-intolerant biota.

This three-level approach to the definition of wetlands is illustrated in Figure 2.2. Climate and geomorphology define the degree to which wetlands can exist, but the starting point is the *hydrology*, which, in turn, affects the *physiochemical environment*, including the soils, which, in turn, determines with the hydrology what and how much *biota*, including vegetation, is found in the wetland. This model is reintroduced and discussed in more detail in Chapter 4.

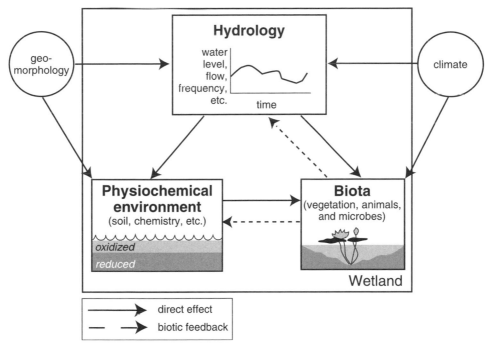

Figure 2.2 The three-component basis of a wetland definition: hydrology, physiochemical environment, and biota. From these components, the current approach to defining jurisdictional wetlands in the United States is based on three indicators—hydrology, soils, and vegetation. Note that these three components are not independent and that there is significant feedback from the biota.

The Difficulty of Defining Wetlands

Although the concepts of shallow water or saturated conditions, unique wetland soils, and vegetation adapted to wet conditions are fairly straightforward, combining these three factors to obtain a precise definition is difficult because of several characteristics that distinguish wetlands from other ecosystems yet make them less easy to define:

1. *Although water is present for at least part of the time, the depth and duration of flooding vary considerably from wetland to wetland and from year to year.* Some wetlands are continually flooded, whereas others are flooded only briefly at the surface or even just below the surface. Similarly, because fluctuating water levels can vary from season to season and year to year in the same wetland type, the boundaries of wetlands cannot always be determined by the presence of water at any one time.

2. *Wetlands are often located at the margins between deep water and terrestrial uplands and are influenced by both systems.* This ecotone position has been

suggested by some as evidence that wetlands are mere extensions of either the terrestrial or the aquatic ecosystem or both, and have no separate identity. Most wetland scientists, however, see emergent properties in wetlands not contained in either upland or deepwater systems.

3. *Wetland species (plants, animals, and microbes) range from those that have adapted to live in either wet or dry conditions (facultative), making difficult their use as wetland indicators, to those adapted to only a wet environment (obligate).*

4. *Wetlands vary widely in size, ranging from small prairie potholes of a few hectares in size to large expanses of wetlands several hundreds of square kilometers in area.* Although this range in scale is not unique to wetlands, the question of scale is important for their conservation. Wetlands can be lost in large parcels or, more commonly, one small piece at a time in a process called *cumulative loss.* Are wetlands better defined functionally on a large scale or in small parcels?

5. *Wetland location can vary greatly, from inland to coastal wetlands and from rural to urban regions.* Whereas most ecosystem types, for example, forests or lakes, have similar ecosystem structure and function, there are great differences among different wetland types such as coastal salt marshes, inland pothole marshes, and forested bottomland hardwoods.

6. *Wetland condition, or the degree to which the wetland has been modified by humans, varies greatly from region to region and from wetland to wetland.* In rural areas, wetlands are likely to be associated with farmlands, whereas wetlands in urban areas are often subjected to the impact of extreme pollution and altered hydrology associated with housing, feeding, and transporting a large population. Many wetlands can easily be drained and turned into dry lands by human intervention; similarly, altered hydrology or increased runoff can cause wetlands to develop where they were not found before. Some animals such as beavers, muskrats, and alligators can play a role in developing wetlands. Because wetlands are so easily disturbed, it is often difficult to identify them after such disturbances; this is the case, for example, with wetlands that have been farmed for many years.

Wetlands have been described as a halfway world between terrestrial and aquatic ecosystems, exhibiting some of the characteristics of each system. They form part of a continuous gradient between uplands and open water. As a result, the exact upper and the lower limits of wetlands are arbitrary boundaries in any definition. Consequently, few definitions adequately describe all wetlands. The problem of definition arises at the edges of wetlands, toward either wetter or drier conditions. How far upland and how infrequently should the land flood before we can declare that it is not a wetland? At the other edge, how far can we venture into a lake, pond, estuary, or ocean before we leave a wetland? Does a floating mat of vegetation define a wetland? What about a submerged bed of rooted vascular vegetation?

The frequency of flooding is another variable that has made the definition of wetlands particularly controversial. Some classifications include seasonally flooded bottomland hardwood forests, whereas others exclude them because they are dry for most of the year. Because wetland characteristics grade continuously from aquatic to terrestrial, there is no single, universally recognized definition of a wetland. This lack has caused confusion and inconsistencies in the management, classification, and inventorying of wetland systems, but, considering the diversity of types, sizes, locations, and conditions of wetlands in this country, inconsistencies should be no surprise.

Wetland Common Terms and Types

A number of common terms such as swamp, marsh, and mire have been used over the years to describe different types of wetlands (Table 2.1). The history of the use and misuse of these words has often revealed a decidedly regional or at least continental origin. Although the lack of standardization of terms is confusing, many of the old terms are rich in meaning to those familiar with them. They often bring to mind vivid images of specific kinds of ecosystems that have distinct vegetation, animals, and other characteristics. Each of the terms has a specific meaning to some people, and many are still widely used by both scientists and laypersons alike. A *marsh* is known by most as an herbaceous plant wetland. A *swamp*, however, has woody vegetation, either shrubs or trees. There are subtle differences among marshes. A marsh with significant (>30 cm) standing water throughout much of the year is often called a *deepwater marsh*. A shallow marsh with waterlogged soil or shallow standing water is sometimes referred to as a *sedge meadow* or a *wet meadow*. Intermediate between a marsh and a meadow is a *wet prairie*. Several terms are used to denote peat-accumulating systems. The most general term is *peatland*, which is generally synonymous with *moor* and *muskeg*. There are many types of peatlands, however, the most general being *fens* and *bogs*.

Within the international scientific community, these common terms do not always convey the same meaning relative to a specific type of wetland. In fact, some languages have no direct equivalents for certain kinds of wetlands. The word *swamp* has no direct equivalent in Russian because there are few forested wetlands there that are not simply a variety of peatlands. *Bog*, however, can easily be translated because bogs are a common feature of the Russian landscape. The word *swamp* in North America clearly refers to a wetland dominated by woody plants—shrubs or trees. In Europe, *reedswamps* are dominated by reed grass (*Phragmites*), a dense-growing but nonwoody plant. In Africa, what would be called a *marsh* in the United States is referred to as a *swamp*. A cutoff meander of a river is called a *billabong* in Australia (Shiel, 1994) and an *oxbow* in North America.

Even common and scientific names of plants and animals can become confusing on a global scale. *Typha* spp., a cosmopolitan wetland plant, is called *cattail* in America, *reedmace* in the United Kingdom, *bulrush* in Africa, *cumbungi* in Australia, and *raupo* or *bulrush* in New Zealand. True bulrush is still called *Scirpus* spp. by some

Table 2.1 Common terms used for various wetland types in the world

Billabong—Australian term for a riparian wetland that is periodically flooded by the adjacent stream or river.

Bog—A peat-accumulating wetland that has no significant inflows or outflows and supports acidophilic mosses, particularly *Sphagnum*.

Bottomland—Lowland along streams and rivers, usually on alluvial floodplains, that is periodically flooded. When forested, it is called a bottomland hardwood forest in the southeastern and eastern United States.

Carr—Term used in Europe for forested wetlands characterized by alders (*Alnus*) and willows (*Salix*). Cumbungi swamp—Cattail (*Typha*) marsh in Australia.

Dambo—A seasonally waterlogged and grass-covered linear depression in headwater zone of rivers with no marked stream channel or woodland vegetation. Term is ChiChewa (Central Africa) dialect meaning "meadow grazing."

Delta—A wetland-river-upland complex located where a river forms distributaries as it merges with the sea; there are also examples of inland deltas such as the Peace-Athabasca Delta in Canada and the Okavango Delta in Botswana (see Chapter 3).

Fen—A peat-accumulating wetland that receives some drainage from surrounding mineral soil and usually supports marshlike vegetation.

Lagoon—Term frequently used in Europe to denote a deepwater enclosed or partially opened aquatic system, especially in coastal delta regions.

Mangal—Same as mangrove.

Mangrove—Subtropical and tropical coastal ecosystem dominated by halophytic trees, shrubs, and other plants growing in brackish to saline tidal waters. The word "mangrove" also refers to the dozens of tree and shrub species that dominate mangrove wetlands.

Marsh—A frequently or continually inundated wetland characterized by emergent herbaceous vegetation adapted to saturated soil conditions. In European terminology, a marsh has a mineral soil substrate and does not accumulate peat. See also tidal freshwater marsh and salt marsh.

Mire—Synonymous with any peat-accumulating wetland (European definition); from the Norse word "myrr." The Danish and Swedish word for peatland is now "mose."

Moor—Synonymous with peatland (European definition). A highmoor is a raised bog; a lowmoor is a peatland in a basin or depression that is not elevated above its perimeter. The primitive sense of the Old Norse root is "dead" or barren land.

Muskeg—Large expanse of peatlands or bogs; particularly used in Canada and Alaska.

Oxbow—Abandoned river channel, often developing into a swamp or marsh.

Pakihi—Peatland in southwestern New Zealand dominated by sedges, rushes, ferns, and scattered shrubs. Most pakihi form on terraces or plains of glacial or fluvial outwash origin and are acid and exceedingly infertile.

Peatland—A generic term of any wetland that accumulates partially decayed plant matter (peat).

Playa—An arid- to semiarid-region wetland that has distinct wet and dry seasons. Term used in the southwest United States for shallow depressional recharge wetlands occurring in the Great Plains region of North America "that are formed through a combination of wind, wave, and dissolution processes" (Smith, 2003).

Pocosin—Peat-accumulating, nonriparian freshwater wetland, generally dominated by evergreen shrubs and trees and found on the southeastern coastal plain of the United States. The term comes from the Algonquin for "swamp on a hill."

Pothole—Shallow marshlike pond, particularly as found in the Dakotas and central Canadian provinces, the so-called prairie pothole region.

Raupo swamp—Cattail (*Typha*) marsh in New Zealand.

Reedmace swamp—Cattail (Typha) marsh in the UK.

Reedswamp—Marsh dominated by *Phragmites* (common reed); term used particularly in Europe.

Riparian ecosystem—Ecosystem with a high water table because of proximity to an aquatic ecosystem, usually a stream or river. Also called bottomland hardwood forest, floodplain forest, bosque, riparian buffer, and streamside vegetation strip.

Salt marsh—A halophytic grassland on alluvial sediments bordering saline water bodies where water level fluctuates either tidally or nontidally.

Table 2.1 (*continued*)

Sedge meadow—Very shallow wetland dominated by several species of sedges (e.g., *Carex, Scirpus, Cyperus*).

Slough—An elongated swamp or shallow lake system, often adjacent to a river or stream. A slowly flowing shallow swamp or marsh in the southeastern United States (e.g., cypress slough). From the Old English word "sloh" meaning a watercourse running in a hollow.

Swamp—Wetland dominated by trees or shrubs (U.S. definition). In Europe, forested fens and wetlands dominated by reed grass (*Phragmites*) are also called swamps (see reedswamp).

Tidal freshwater marsh—Marsh along rivers and estuaries close enough to the coastline to experience significant tides by nonsaline water. Vegetation is often similar to nontidal freshwater marshes.

Turlough—Areas seasonally flooded by karst groundwater with sufficient frequency and duration to produce wetland characteristics. They generally flood in winter and are dry in summer and fill and empty through underground passages. Term is specific for these types of wetlands found mostly in western Ireland.

Vernal pool—Shallow, intermittently flooded wet meadow, generally typical of Mediterranean climate with dry season for most of the summer and fall. Term is now used to indicate wetlands temporarily flooded in the spring throughout the United States.

Vleis—Seasonal wetland similar to a Dambo; term used in southern Africa.

Wad (pl. Wadden)—Unvegetated tidal flat originally referring to the northern Netherlands and northwestern German coastline. Now used throughout the world for coastal areas.

Wet meadow—Grassland with waterlogged soil near the surface but without standing water for most of the year.

Wet prairie—Similar to a marsh, but with water levels usually intermediate between a marsh and a wet meadow.

in North America and *Schoenoplectus* spp. in much of the rest of the world. *Scirpus fluviatilis* (river bulrush) is *Bolboschoenus fluviatilis* in much of the rest of the world. The Great Egret in North America is *Casmerodius albus*, whereas the Great Egret in Australia is *Ardea alba*. To further complicate matters, the Australian version of the Great Egret is called *Egretta alba* in New Zealand and is not called an egret at all but a White Heron.

Confusion in terminology occurs because of different regional or continental uses of terms for similar types of wetlands (Table 2.2). In North America, nonforested inland wetlands are often casually classified either as peat-forming, low-nutrient acid bogs or as marshes. European terminology, which is much older, is also much richer and distinguishes at least four different kinds of freshwater wetlands—from mineral-rich reed beds, called reedswamps, to wet grassland marshes, to fens, and, finally, to bogs or moors. To some, all of these wetland types are considered *mires*. According to others, mires are limited to peat-building wetlands. The European classification is based on the amount of surface water and nutrient inflow (rheotrophy), type of vegetation, pH, and peat-building characteristics.

Two points can be made about the use of common terms in classifying wetland types: First, the physical and biotic characteristics grade continuously from one of these wetland types to the next; hence, any classification based on common terms is, to an extent, arbitrary. Second, the same term may refer to different systems in different regions. The common terms continue to be used, even in the scientific literature; we simply suggest that they be used with caution and with an appreciation for an international audience.

Table 2.2 Comparison of terms used to describe similar inland nonforested freshwater wetlands

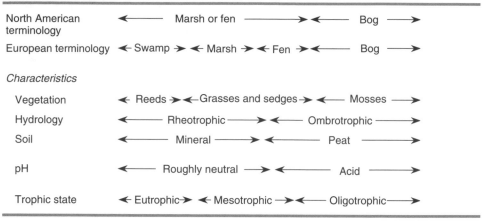

| North American terminology | ←——— Marsh or fen ———→ ←——— Bog ———→ |
| European terminology | ← Swamp → ← Marsh → ← Fen → ←——— Bog ———→ |

Characteristics

Vegetation	← Reeds → ←Grasses and sedges → ←——— Mosses ———→
Hydrology	←——— Rheotrophic ———→ ←——— Ombrotrophic ———→
Soil	←——— Mineral ———→ ←——— Peat ———→
pH	←——— Roughly neutral ———→ ←——— Acid ———→
Trophic state	← Eutrophic → ← Mesotrophic →←— Oligotrophic ———→

Formal Wetland Definitions

Precise wetland definitions are needed for two distinct interest groups: (1) wetland scientists and (2) wetland managers and regulators. The wetland scientist is interested in a flexible yet rigorous definition that facilitates classification, inventory, and research. The wetland manager is concerned with laws or regulations designed to prevent or control wetland modification and, thus, needs clear, legally binding definitions. Because of these differing needs, different definitions have evolved for the two groups. The discrepancy between the regulatory definition of *jurisdictional wetlands* and other definitions in the United States has meant, for example, that maps developed for wetland inventory purposes cannot be used for regulating wetland development. This is a source of considerable confusion to regulators and landowners.

Definitions that are more scientific in nature are presented in this section. Definitions that are more used in a legal sense are presented in the next section.

Early U.S. Definition: Circular 39 Definition

One of the earliest definitions of the term *wetlands* was presented by the U.S. Fish and Wildlife Service in 1956 in a publication that is frequently referred to as Circular 39 (Shaw and Fredine, 1956):

> The term "wetlands"... refers to lowlands covered with shallow and sometimes temporary or intermittent waters. They are referred to by such names as marshes, swamps, bogs, wet meadows, potholes, sloughs, and river-overflow lands. Shallow lakes and ponds, usually with emergent vegetation as a conspicuous feature, are included in the definition, but the permanent waters of streams, reservoirs, and deep lakes are not included. Neither are water areas that are so temporary as to have little or no effect on the development of moist-soil vegetation.

The Circular 39 definition (1) emphasized wetlands that were important as waterfowl habitats and (2) included 20 types of wetlands that served as the basis for the main wetland classification used in the United States until the 1970s (see Chapter 8). It thus served the limited needs of both wetland managers and wetland scientists.

U.S. Fish and Wildlife Service Definition

Perhaps the most comprehensive definition of wetlands was adopted by wetland scientists in the U.S. Fish and Wildlife Service in 1979, after several years of review. The definition was presented in a report entitled *Classification of Wetlands and Deepwater Habitats of the United States* (Cowardin et al., 1979):

> Wetlands are lands transitional between terrestrial and aquatic systems where the water table is usually at or near the surface or the land is covered by shallow water.... Wetlands must have one or more of the following three attributes: (1) at least periodically, the land supports predominantly hydrophytes; (2) the substrate is predominantly undrained hydric soil; and (3) the substrate is nonsoil and is saturated with water or covered by shallow water at some time during the growing season of each year.

This definition was significant for its introduction of several important concepts in wetland ecology. It was one of the first definitions to introduce the concepts of *hydric soils* and *hydrophytes*, and it served as the impetus for scientists and managers to define these terms more accurately (NRC, 1995). Designed for scientists as well as managers, it is broad, flexible, and comprehensive, and includes descriptions of vegetation, hydrology, and soil. It has its main utility in scientific studies and inventories and generally has been more difficult to apply to the management and regulation of wetlands. It is still frequently accepted and employed in the United States today and was, at one time, accepted as the official definition of wetlands by India. Like the Circular 39 definition, this definition serves as the basis for a detailed wetland classification and an updated and comprehensive inventory of wetlands in the United States. The classification and inventory are described in more detail in Chapter 8.

Canadian Wetland Definitions

Canadians, who deal with vast areas of inland northern peatlands, have developed a specific national definition of wetlands. Two definitions were formally published in the book *Wetlands of Canada* by the National Wetlands Working Group (1988). First, Zoltai (1988) defined a wetland as:

> Land that has the water table at, near, or above the land surface or which is saturated for a long enough period to promote wetland or aquatic processes as indicated by hydric soils, hydrophytic vegetation, and various kinds of biological activity which are adapted to the wet environment.

Zoltai (1988) also noted that "wetlands include waterlogged soils where in some cases the production of plant materials exceeds the rate of decomposition." He describes the wet and dry extremes of wetlands as:

- Shallow open waters, generally less than 2 m; and
- Periodically inundated areas only if waterlogged conditions dominate throughout the development of the ecosystem.

Tarnocai et al. (1988) offered a slightly reworded definition in that same publication as the basis of the Canadian wetland classification system. That definition, repeated by Zoltai and Vitt (1995) and Warner and Rubec (1997) in later years, remains the official definition of wetlands in Canada:

Land that is saturated with water long enough to promote wetland or aquatic processes as indicated by poorly drained soils, hydrophytic vegetation and various kinds of biological activity which are adapted to a wet environment.

These definitions emphasize wet soils, hydrophytic vegetation, and "various kinds" of other biological activity. The distinction between "hydric soils" in the Zoltai definition and "poorly drained soils" in the current, more accepted definition may be a reflection of the reluctance by some to use hydric soils exclusively to define wetlands. Hydric soils are discussed in more detail in Chapter 5.

U.S. National Academy of Sciences Definition

In the early 1990s, amid renewed regulatory controversy in the United States as to what constitutes a wetland, the U.S. Congress asked the private nonprofit National Academy of Sciences to appoint a committee through its principal operating agency, the National Research Council (NRC), to undertake a review of the scientific aspects of wetland characterization. The committee was charged with considering: (1) the adequacy of the existing definition of wetlands; (2) the adequacy of science for evaluating the hydrologic, biological, and other ways that wetlands function; and (3) regional variation in wetland definitions. The report produced by that committee two years later was entitled *Wetlands: Characteristics and Boundaries* (NRC, 1995) and included yet another scientific definition, referred to as a "reference definition" in that it was meant to stand "outside the context of any particular agency, policy or regulation" (NRC, 1995):

A wetland is an ecosystem that depends on constant or recurrent, shallow inundation or saturation at or near the surface of the substrate. The minimum essential characteristics of a wetland are recurrent, sustained inundation or saturation at or near the surface and the presence of physical, chemical, and biological features reflective of recurrent, sustained inundation or saturation. Common diagnostic features of wetlands are hydric soils and hydrophytic vegetation. These features will be present except where specific physiochemical, biotic, or anthropogenic factors have removed them or prevented their development.

Although little formal use has been made of this definition, it remains the most comprehensively developed scientific wetland definition. It uses the terms *hydric soils* and *hydrophytic vegetation*, as did the early U.S. Fish and Wildlife Service definition, but indicates that they are "common diagnostic features" rather than absolute necessities in designating a wetland.

An International Definition

The International Union for the Conservation of Nature and Natural Resources (IUCN) at the Convention on Wetlands of International Importance Especially as Waterfowl Habitat, better known as the *Ramsar Convention*, adopted the following definition of wetlands (Finlayson and Moser, 1991):

> Areas of marsh, fen, peatland or water, whether natural or artificial, permanent or temporary, with water that is static or flowing, fresh, brackish, or salt including areas of marine water, the depth of which at low tide does not exceed 6 meters.

This definition, which was adopted at the first meeting of the convention in Ramsar, Iran, in 1971, states that wetlands may incorporate riparian and coastal zones adjacent to the wetlands and islands or bodies of marine water deeper than 6 meters at low tide lying within the wetlands. This definition does not include vegetation or soil and extends wetlands to water depths of 6 meters or more, well beyond the depth usually considered wetlands in the United States and Canada. The rationale for such a broad definition of wetlands "stemmed from a desire to embrace all the wetland habitats of migratory water birds" (Scott and Jones, 1995).

Legal Definitions

When protection of wetlands began in earnest in the mid-1970s in the United States, there arose an almost immediate need for precise definitions that were based as much on closing legal loopholes as on science. Two such definitions have developed in U.S. agencies—one for the U.S. Army Corps of Engineers to enforce its legal responsibilities with a "dredge-and-fill" permit program in the Clean Water Act, and the other for the U.S. Natural Resources Conservation Service to administer wetland protection under the so-called swampbuster provision of the Food Security Act. Both agencies were parties to an agreement in 1993 to work together on administering a unified policy of wetland protection in the United States, yet the two separate definitions remain.

U.S. Army Corps of Engineers Definition

A U.S. government regulatory definition of wetlands is found in the regulations used by the U.S. Army Corps of Engineers for the implementation of a dredge-and-fill permit system required by Section 404 of the 1977 Clean Water Act amendments. That definition has now survived several decades in the legal world and is given as follows:

> The terms "wetlands" means those areas that are inundated or saturated by surface or ground water at a frequency and duration sufficient to support, and that under normal circumstances do support, a prevalence of vegetation typically adapted for life in saturated soil conditions. Wetlands generally include swamps, marshes, bogs, and similar areas. (33 CFR 328.3(b); 1984)

This definition replaced a 1975 definition that stated that "those areas that normally are characterized by the prevalence of vegetation that *requires* saturated soil conditions for growth and reproduction" (42 *Fed. Reg.* 3712X, July 19, 1977; italics added), because the Corps of Engineers found that the old definition excluded "many forms of truly aquatic vegetation that are prevalent in an inundated or saturated area, but that do not require saturated soil from a biological standpoint for their growth and reproduction." The words "normally" in the old definition and "that under normal circumstances do support" in the new definition were intended "to respond to situations in which an individual would attempt to eliminate the permit review requirements of Section 404 by destroying the aquatic vegetation..." (quotes from 42 *Fed. Reg.* 37128, July 19, 1977). The need to revise the 1975 definition illustrates how difficult it has been to develop a legally useful definition that also accurately reflects the ecological reality of a wetland site.

This legal definition of wetlands has been debated in the courts in several cases, some of which have become landmark cases. In one of the first court tests of wetland protection, the Fifth Circuit of the U.S. Court of Appeals ruled in 1972, in *Zabel v. Tabb*, that the U.S. Army Corps of Engineers has the right to refuse a permit for filling of a mangrove wetland in Florida. In 1975, in *Natural Resources Defense Council v. Callaway*, wetlands were included in the category "waters of the United States," as described by the Clean Water Act. Prior to that time, the Corps of Engineers regulated dredge-and-fill activities (Section 404 of the Clean Water Act) for navigable waterways only; since that decision, wetlands have been legally included in the definition of waters of the United States.

In 1985, the question of regulation of wetlands reached the U.S. Supreme Court for the first time. The Court upheld the broad definition of wetlands to include groundwater-fed wetlands in *United States v. Riverside Bayview Homes, Inc.* In that case, the Supreme Court affirmed that the U.S. Army Corps of Engineers had jurisdiction over wetlands that were adjacent to navigable waters, but it left open the question as to whether it had jurisdiction over nonadjacent wetlands (NRC, 1995). The legal definition of wetlands has since been involved in the U.S. Supreme Court twice more, in 2001 and in 2006; these cases are discussed in more detail in Chapter 14.

Food Security Act Definition

In December 1985, the U.S. Department of Agriculture, through its Soil Conservation Service [now known as the Natural Resources Conservation Service (NRCS)], was brought into the arena of wetland definitions and wetland protection by means of a provision known as "swampbuster" in the 1985 Food Security Act (see also Chapter 14). On agricultural land in the United States that, prior to December 1985, had been exempt from regulation, wetlands were now protected. As a result of this *swampbuster provision*, a definition, known as the NRCS or Food Security Act definition, was included in the Act (16 CFR 801(a)(16); 1985):

> The term "wetland" except when such term is part of the term "converted wetland" means land that—
>
> (A) has a predominance of hydric soils;
> (B) is inundated or saturated by surface or ground water at a frequency and duration sufficient to support a prevalence of hydrophytic vegetation typically adapted for life in saturated soil conditions; and
> (C) under normal circumstances does support a prevalence of such vegetation.
>
> For purposes of this Act and any other Act, this term shall not include lands in Alaska identified as having high potential for agricultural development which have a predominance of permafrost soils.

The emphasis on this agriculture-based definition is on hydric soils. The omission of wetlands that do not have hydric soils, while not invalidating this definition, makes it less comprehensive than some others, for example, the NRC (1995) definition. A curious feature of this definition is its wholesale exclusion of the largest state in the United States from the definition of wetlands. The exclusion of Alaskan wetlands that have a high potential for agriculture makes this definition even less of a scientific and more of a regulatory or even political definition. There is no scientific distinction between the characteristics of Alaskan wetlands and wetlands in the rest of the United States except for climatic differences and the presence of permafrost under many but certainly not all Alaskan wetlands (NRC, 1995).

Jurisdictional Wetlands

Since 1989, the term *jurisdictional wetland* has been used for legally defined wetlands in the United States to delineate those areas that are under the jurisdiction of Section 404 of the Clean Water Act or the swampbuster provision of the Food Security Act. The Army Corps of Engineers' definition cited previously emphasizes only one indicator, vegetative cover, to determine the presence or absence of a wetland. It is difficult to include soil information and water conditions in a wetland definition when its main purpose is to determine jurisdiction for regulatory purposes and there is little time to examine the site in detail. The Food Security Act definition, however, includes hydric soils as the principal determinant of wetlands.

It is likely that most of the wetlands that are considered "jurisdictional wetlands" by the preceding two legal definitions fit the scientific definition of wetlands. It is also just as likely that some types of wetlands, particularly those that have less chance of developing hydric soil characteristics or hydrophytic vegetation (e.g., riparian wetlands), would not be identified as jurisdictional wetlands with the legal definitions. And of course, excluding Alaskan wetlands "having high potential for agricultural development" from the Food Security Act definition has no scientific basis at all but is a political decision.

Those who delineate wetlands are interested in a definition that allows the rapid identification of a wetland and the degree to which it has been or could be altered. They are interested in the delineation of wetland boundaries, and establishing boundaries is facilitated by defining the wetland simply, according to the presence or absence of certain species of vegetation or aquatic life or the presence of simple indicators such as hydric soils. Several U.S. federal manuals spelling out specific methodologies for identifying jurisdictional wetlands were written or proposed in the 1980s and early 1990s. The manuals differed, however, in the prescribed ways these three criteria are proved in the field. The first of these manuals, written in 1987, is now the one most widely used to field-identify wetlands in the United States. All three manuals indicated that the three criteria for wetlands—namely, wetland hydrology, wetland soils, and hydrophytic vegetation—must be present. As illustrated in Figure 2.2, these three variables are not independent; strong evidence of long-term wetland hydrology, for example, should almost ensure that the other two variables are present. Furthermore, potentially other indicators of the physiochemistry and biota beyond hydric soils and hydrophytic vegetation may one day serve as useful indicators of wetlands.

Choice of a Definition

A wetland definition that will prove satisfactory to all users has not yet been developed because the definition of wetlands depends on the objectives and the field of interest of the user. Different definitions can be formulated by the geologist, soil scientist, hydrologist, biologist, ecologist, sociologist, economist, political scientist, public health scientist, and lawyer. This variance is a natural result of the differences in emphasis in the definer's training and of the different ways in which individual disciplines deal with wetlands. For ecological studies and inventories, the 1979 U.S. Fish and Wildlife Service definition has been and should continue to be applied to wetlands in the United States. Although somewhat generous in defining wetlands on the wet edge, the Ramsar definition is firmly entrenched in international circles. When wetland management, particularly regulation, is necessary, the U.S. Army Corps of Engineers' definition, as modified, is probably most appropriate.

Just as important as the precision of the definition of a wetland, however, is the consistency with which it is used. That is the difficulty we face when science and legal issues meet, as they often do, in resource management questions such as wetland conservation versus wetland drainage. Applying a comprehensive definition in a uniform and fair way requires a generation of well-trained wetland scientists and managers armed with a fundamental understanding of the processes that are important and unique to wetlands.

Recommended Readings

National Research Council (NRC). 1995. *Wetlands: Characteristics and Boundaries*. National Academy Press, Washington, DC. 306 pp.

Wetlands of the World

The extent of the world's wetlands is now thought to be from 7 to 10 million km², or about 5 to 8 percent of the land surface of the Earth. The loss of wetlands in the world is difficult to determine, but may be similar to the 50 percent loss rate for the lower 48 states of the United States, with higher rates of loss in Europe and parts of Australia, Canada, and Asia and lower rates in less developed regions like Africa, South America, and northern boreal regions. Estimated areas of wetlands in North America are 43.6 million hectares (ha) in the lower 48 states, 71 million ha in Alaska, and 127 million ha in Canada, representing in total about 30 percent of the world's wetlands. In this chapter we describe a number of important wetlands from around the world, including the Florida Everglades and the Louisiana Delta in the United States, the Pantanal in South America, the Okavango Delta and the Congolian Swamp in Africa, the Mesopotamian Marshlands in the Middle East, Australian billabongs, and wetlands in natural areas and parks in China. All of these wetlands are impacted by human activities to some extent, yet most remain functional ecosystems.

The Global Extent of Wetlands

Wetlands include the swamps, bogs, marshes, mires, fens, and other wet ecosystems found throughout the world. They are found on every continent except Antarctica and in every clime, from the tropics to the tundra (Fig. 3.1a). Any estimate of the extent of wetlands in the world is difficult and depends on the definition used as described in Chapter 2, as well as the pragmatic difficulty of quantifying wetlands in aerial and satellite images that are now the most common sources of data. It is now

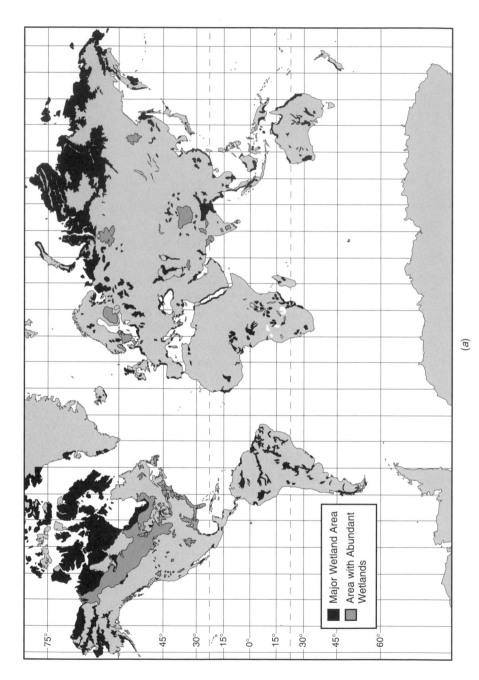

(a)

Figure 3.1 The wetlands of the world: a. general extent determined a composite from a number of separate sources, and b. distribution of wetlands with latitude based on data from Matthews and Fung (1987) and global lakes and wetlands database (GLWD) and gross wetland map from Lehner and Döll (2004).

Major Wetland Area

Area with Abundant Wetlands

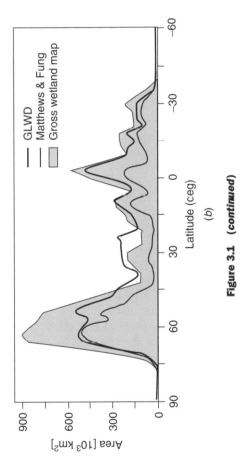

Figure 3.1 (*continued*)

Table 3.1   Comparison of estimates of extent of wetlands in the world by climatic zone

	Wetland Area ($\times 10^6$ km^2)						
Zone[a]	Maltby and Turner (1983)[b]	Matthews and Fung (1987)	Aselmann and Crutzen (1989)	Gorham (1991)	Finlayson and Davidson (1999)	Ramsar Convention Secretariat (2004)	Lehner and Döll (2004)
Polar/boreal	2.8	2.7	2.4	3.5	–	–	–
Temperate	1.0	0.7	1.1	–	–	–	–
Subtropical/tropical	4.8	1.9	2.1	–	–	–	–
Rice paddies	–	1.5	1.3	–	–	1.3	–
Total wetland area	8.6	6.8	6.9	–	12.8	7.2	8.2–10.1

[a]Definitions of polar, boreal, temperate, and tropical vary among studies.
[b]Based on Bazilevich et al. (1971).

fairly well established that most of the world's wetlands are found in both the boreal and tropical regions of the world and the least amount of wetlands found in temperate zones (Fig. 3.1b).

Based on several previous studies (Table 3.1), we now estimate the extent of the world's wetlands is 7 to 10 million km^2 , or about 5 to 8 percent of the land surface of the Earth. We estimate that number by deleting the high (Finlayson and Davidson, 1999) and low (Matthews and Fung, 1987) estimates in Table 3.1 and using the remaining numbers to provide a range. We believe that the estimate provided by Lehner and Döll (2004) of 8 to 10 million km^2 is the most detailed study on global wetland inventory that we have seen in the past 20 years and may be the most accurate. We do not include large lakes (e.g. Laurentian Great Lakes, Caspian Sea) or deepwater coastal systems (up to 6 m deep) in our preferred definition of wetlands. But because of the disparity of wetland definitions described in Chapter 2 and the more inclusive definition used by the Ramsar Convention, a more accurate estimate of the world's wetlands is probably not possible.

Earlier wetland estimates provided a narrower range. Maltby and Turner (1983), based on the work of Russian geographers, estimated that more than 6.4 percent of the land surface of the world, or 8.6 million km^2, is wetland. Almost 56 percent of this estimated total wetland area is found in tropical (2.6 million km^2) and subtropical (2.1 million km^2) regions. Wetlands occupy 1 million km^2 in sub-boreal regions (called temperate in Table 3.1), 2.6 million km^2 in boreal regions, and 0.2 million km^2 in polar regions.

Other estimates of the global extent of wetlands developed in the 1980s from studies of the role that wetlands play in global biogeochemical cycles, especially with regard to greenhouse gases. Using global digital databases (1 degree resolution), Matthews and Fung (1987) estimated that there were 5.3 million km^2 of wetlands in the world, with a higher percentage of wetlands being boreal and a far lower percentage of wetlands being subtropical and tropical than those estimated by Maltby and Turner (1983). Aselmann and Crutzen (1989, 1990) estimated that there were 5.6 million km^2 of natural wetlands in the world, with a higher amount and percentage of wetlands in the temperate region than given in either of the earlier estimates. They used regional wetland surveys and monographs rather than maps, which Matthews and Fung (1987) used to make their estimate. These two research groups estimated the coverage by rice paddies—1.3 to 1.5 million km^2—but did not include this in their total wetland area. By including rice fields, their estimates of the extent of the world's wetlands are 6.8 and 6.9 million km^2, respectively. Bogs and fens accounted for about 60 percent of the world's wetlands (3.35 million km^2) in the Matthews and Fung (1987) study, an estimate that is very close to Gorham's (1991) 3.46 million km^2 estimate for northern boreal and subarctic peatlands. Aselmann and Crutzen (1989) described bogs and fens as also occurring in both temperate (40–50° N) and tropical latitudes. Both Matthews and Fung (1987) and Aselmann and Crutzen (1989) showed a much lower extent of wetlands in tropical and subtropical regions than did Maltby and Turner (1983), although definitions of zones differ.

Finlayson and Davidson (1999) estimated that there were 12.8 million km^2 of wetlands using the international Ramsar definition described in Chapter 2. This estimate, 30% or more higher than the other estimates reported in the literature of the global extent of wetlands, was repeated in a Millenium Ecosystem Assessment (2005) report (co-authored by Finlayson and Davidson) on wetlands and water. This estimate includes all freshwater lakes, reservoirs, and rivers and near-shore marine ecosystems up to 6 m depth in the world, aquatic ecosystems that are not included in all wetland definitions. Ironically the Millenium Ecosystem Assessment (2005) report, in describing this high estimate, suggests that "it is well established that this estimate is an underestimate."

The Ramsar Convention (see Chapter 14) initiated a multiyear effort to encourage countries to perform national wetland inventories, but accurate summaries are not yet completed A meeting summary from 1999 (Ramsar Convention Secretariat, 2004) suggested that there are 5.9 million km^2 of wetlands in the world and 1.3 million km^2 of rice paddies, for a total of 7.2 million km^2. That report acknowledges that some wetland types, including salt marshes and coastal flats, are not included in this estimate. By comparison, over 1.5 million km^2 of wetlands has been registered with the Ramsar Convention on Wetlands of International Importance as of mid-2007. This represents only about one-quarter of the known area of the world's wetlands but also includes a large area of deepwater systems that would normally not be considered wetlands.

Lehner and Döll (2004) provide one of the most comprehensive and recent examinations of the global extent of wetlands. Their GIS-based Global Lakes and Wetlands Database (GLWD) system focused on three coordinated levels (1) large lakes and reservoirs, (2) smaller water bodies, and (3) wetlands. With the first two categories excluded, an estimate of 8.3–10.2 million km^2 of wetlands in the world was estimated. As with several of the other studies summarized above and in Table 3.1, they found the greatest proportion of wetlands in the northern boreal regions (peaking at $60°$ N latitude) and another peak of tropical wetlands exactly at the equator (Fig. 3.1b).

Worldwide Wetland Losses

The rate at which wetlands are being lost on a global scale is only now becoming clear, partially with the use of new technologies associated with satellite imagery. But there are still many vast areas of wetlands where accurate records have not been kept, and many wetlands in the world were drained centuries ago. Threats to wetlands take many forms, including drainage and hydrologic modification with dikes, dams, and levees. With more than 70 percent of the world's population living on or near coastlines, coastal wetlands have long been destroyed through a combination of excessive harvesting, hydrologic modification and seawall construction, coastal development, pollution, and other human activities. The loss of inland wetlands results from drainage for agriculture, forestry, and mosquito control; filling for residential, commercial, and industrial development; filling for solid-waste disposal; and mining of peat. Wetland losses are also caused indirectly by human activity such

Table 3.2 Loss of wetlands in various locations in the world

Location	Percent Loss
NORTH AMERICA	
United States[a]	53
Canada[b]	
Atlantic tidal and salt marshes	65
Lower Great Lakes–St. Lawrence River	71
Prairie potholes and sloughs	71
Pacific coastal estuarine wetlands	80
AUSTRALASIA	
Australia[c]	>50
Swan Coastal Plain	75
Coastal New South Wales	75
Victoria	33
River Murray basin	35
New Zealand[d]	>90
Philippine mangrove swamps[d]	67
CHINA[e]	60
EUROPE[f]	60

[a]Dahl (1990);
[b]National Wetlands Working Group (1988);
[c]Australian Nature Conservation Agency (1996);
[d]Dugan (1993);
[e]Lu (1995);
[f]loss due to agriculture (Revenga et al., 2000).

as construction of dams and roads and control of river dynamics. These impacts are discussed in more detail in Chapter 9.

It is probably safe to assume that (1) we are still losing wetlands at a fairly rapid rate globally, particularly in developing countries; and (2) we have lost about 50 percent of the original wetlands on the face of the Earth. There are some areas where the loss rate has been documented (Table 3.2). The estimate of about 53 percent loss of wetlands since European settlement in the lower 48 United States is fairly accurate. By 1985, 56 to 65 percent of wetlands in North America and Europe, 27 percent in Asia, 6 percent in South America, and 2 percent in Africa had been drained for intensive agriculture (Ramsar Convention Secretariat, 2004). Several regions of the world have lost considerable wetlands. A 90 percent loss of wetlands in New Zealand is documented. The loss rate of 60 percent from China is based on the present estimate of 250,000 km^2 of natural wetlands in the country out of a total of 620,000 km^2, including artificial wetlands such as rice paddies (Lu, 1995). Europe has lost an estimated 60 percent of its wetlands to agricultural conversion alone. Spain has lost more than 60 percent of its inland wetlands and Lithuania 70 percent of its total wetlands since 1970; Sweden drained 67 percent of its wetlands and ponds since the 1950s (Revenga et al., 2000).

North American Wetland Changes

The best and most recent estimate is that there are 43.6 million ha of wetlands in the lower 48 (conterminous) states of the United States (Table 3.3). In addition, there are an estimated 71 million ha of wetlands in Alaska The inclusion of Alaska in wetland surveys of the United States increases the wetland inventory in the United States by 160 percent. Combining these numbers with estimates of wetland areas from Canada and Mexico (described below), North America has about 2.5 million km^2 of wetlands or an estimated 30 percent of the world's wetlands.

Overall, 53 percent of the wetlands in the conterminous United States were estimated to have been lost from the 1780s to the 1980s (Table 3.4). Based on a review of several studies that have been made of wetland area in the United States, two general statements can be made: (1) estimates of the area of wetlands in the United States, while they vary widely, are becoming quite accurate (Table 3.3); and (2) most studies indicated a rapid rate of wetland loss in the United States prior to the mid-1970s and a steady but significant reduction in the loss rate since about the mid-1980s (Table 3.4).

The numbers initially vary widely for several reasons. First, the purposes of the inventories varied from study to study. Early wetland censuses, for example, Wright (1907) and Gray et al. (1924), were undertaken to identify lands suitable for drainage for agriculture. Later inventories of wetlands (Shaw and Fredine, 1956) were concerned with waterfowl protection. Only within the last three decades have wetland inventories considered all of the values of these ecosystems. Second, the definition and classification of wetlands varied with each study, ranging from simple terms to complex hierarchical classifications. Third, the methods available for estimating wetlands changed over the years or varied in accuracy. Remote sensing from aircraft

Table 3.3   Estimates of wetland area in the United States at different times

Period or Year of Estimate	Wetland Area ($\times 10^6$ HA)[a]	Reference
Presettlement	87	Roe and Ayres (1954)
Presettlement	86.2	USDA estimate, in Dahl (1990)
Presettlement	89.5	Dahl (1990)
1906	32[b]	Wright (1907)
1922	37 (total) 3 (tidal) 34 (inland)	Gray et al. (1924)
1940	39.4[c]	Whooten and Purcell (1949)
1954	30.1[d] (total) 3.8 (coastal) 26.3 (inland)	Shaw and Fredine (1956)
1954	43.8 (total) 2.3 (estuarine) 41.5 (inland)	Frayer et al. (1983)
1974	40.1 (total) 2.1 (estuarine) 38.0 (inland)	Frayer et al. (1983); Tiner (1984)
mid-1970s	42.8[e] (total) 2.2 (estuarine) 40.6 (inland)	Dahl and Johnson (1991)
mid-1980s	41.8[e] 2.2 (estuarine) 39.3 (inland)	Dahl and Johnson (1991)
1997	42.7 2.14 (estuarine) 40.56 (inland)	Dahl (2000)
2004	43.6 2.15 (estuarine) 41.45 (inland)	Dahl (2006)

[a]For 48 conterminous states unless otherwise noted.
[b]Does not include tidal wetlands or eight public land states in West.
[c]Outside of organized drainage enterprises.
[d]Only included wetlands important for waterfowl.
[e]Based on estimates of NWI classes for vegetated estuarine and palustrine wetlands.

and satellites is one example of a technique for wetland studies that was not generally available or used before the 1970s. Early estimates, in contrast, were often based on fragmentary records. Fourth, in a number of instances, the borders of geographical or political units changed between censuses, leading to gaps or overlaps in data. Finally, significant drainage of wetlands has occurred since the early estimates were made at the beginning of the 20th century.

Several states in the Midwestern United States (Illinois, Indiana, Iowa, Kentucky, Missouri, and Ohio) plus California all had wetland losses of more than 80 percent, principally for agricultural production (see Appendix A for individual state loss rates); these states collectively show a loss of 14.1 million ha of wetlands during the past

Table 3.4   Estimates of wetland changes in the conterminous United States. (All changes were losses until the most recent measurements, which indicated wetland gains.)

Period	Wetland Change			Reference
	million ha	ha/yr	Percent	
Presettlement–1980s	−47.3	−236,500	−53%	Dahl (1990)
1950s–1970s	−3.7	−185,000	−8.5%	Frayer et al. (1983)
1970s–1980s	−1.06	−105,700	−2.5%	Dahl and Johnson (1991)
1986–1997	−0.26	−23,700	−0.6%	Dahl (2000)
1998–2004	+0.19	+12,900	+0.44	Dahl (2006)

[a]Vegetated estuarine emergent wetland.

200 years, or 30 percent of the wetland loss of the entire conterminous United States. States with high densities of wetlands—Minnesota, Illinois, Louisiana, and Florida—had among the highest losses of total area of wetlands—2.6, 2.8, 3.0, and 3.8 million ha, respectively.

Estimates of wetland loss in the last 30 years suggest a substantial decrease in the wetland loss rate in the lower 48 states. Frayer et al. (1983) estimated a net loss from the 1950s to the 1970s of more than 3.7 million ha (8.5 percent loss), or an average annual loss of 185,000 ha. This loss represents a wetland area equivalent to the combined size of Massachusetts, Connecticut, and Rhode Island. Freshwater marshes and forested wetlands were hardest hit.

Wetland losses continued into the 1980s and 1990s, but the enactment of strong wetland protection laws in the mid-1970s and mid 1980s, combined with interest in wetland restoration and storm water pond creation, has had a dramatic effect. Wetland losses decreased from about 105,700 ha for the 1970s to 1980s (2.5 percent loss) to 23,700 ha from the mid-1980s to mid-1990s (0.6 percent loss). The loss changed to a gain of 12,900 ha of wetlands (0.44 percent gain) from 1998 to 2004, albeit mostly as gains in open water ponds (see Wetland Conversions box). While it has been difficult to document, wetland losses have been at least partially offset in area by a major effort in wetland restoration and creation in rural and suburban ponds during this period. The question remains as to whether these ponds and other additions to the wetland ledger are functioning wetlands.

Wetland Conversions: What wetlands are we really losing (and gaining)?

By themselves, estimates of net wetland losses or gains provide an incomplete picture of the dynamics of change. A more complete picture would show that human activities converted millions of hectares of wetlands from one class to another. Through these conversions, some wetland classes increased in area at the expense of other types. Considering the period from the mid-1970s to the mid-1980s, for example, swamps and forested riparian wetlands in the United States suffered the greatest loss, 1.4 million ha (Fig. 3.2). Although 800,000 ha were converted to agricultural and other land uses, large areas were converted to other wetland types: 292,000 ha to

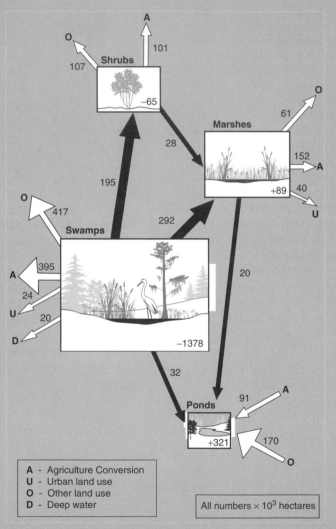

Figure 3.2 Wetland conversion in the conterminous United States, mid-1970s to mid-1980s. The figure shows how misleading the net change figures are. For example, although there was a net gain in freshwater marshes (89,000 ha), it occurred along with a loss of about 1,378,000 ha of swamps, some of which were converted to freshwater marshes. *(After Dahl and Johnson, 1991.)*

marshes, 195,000 ha to scrub and shrubs, and 32,000 ha to nonvegetated wetlands. Although shrub wetlands lost 208,000 ha to agriculture and other nonwetland uses, this was almost offset by the conversion of forested wetlands to shrubs, leaving a net loss of 65,000 ha. A net gain of 89,000 ha of marshes occurred despite a loss of 213,000 ha to agriculture and other land uses, because 320,000 ha of swamps and shrub wetlands changed to marshes. In this example, most of the scrub–shrub wetlands are probably areas recently cut over for their timber.

For the period of study 1998–2004 (Dahl, 2006), wetlands were estimated to actually increase in the United States by 12,900 hectares per year (ha/yr). The excitement of actually seeing an increase in wetlands for the first time in 200 years in the United States was dampened by the fact that this increase was a result of an increase of 46,900 ha/yr of freshwater ponds (13 percent increase). Furthermore, there was a net gain of 37,000 ha/yr of forested wetlands (1.1 percent increase), but these gains were balanced by losses of 60,800 ha/yr of scrub wetlands (4.9 percent decrease), 9,600 ha/yr of freshwater emergent marshes (0.5 percent decrease), and 2,240 ha/yr of estuarine emergent marshes (0.7 percent decrease). In essence, there were large gains in unvegetated ponds in human developments (farms, suburban developments, and even golf course ponds) and forested wetlands that were countered, respectively, by losses in marshes and shrub wetlands (many of which became forested wetlands). Describing wetland losses and gains is not a simple exercise.

Canada has about three times the area of wetlands than is found in the lower 48 states of the United States, or about 127 million ha of wetlands (about 14 percent of the country). Most of that area (111.3 million ha) is defined as peatland. The greatest concentration of Canadian wetlands can be found in the provinces of Manitoba and Ontario. The National Wetlands Working Group (1988), which provided a particularly comprehensive description of major regional wetlands in Canada, estimated that there were 22.5 million ha and 29.2 million ha, respectively, of wetlands in these two provinces, or about 41 percent of the total wetlands of Canada. Much of this total is boreal forested peatlands as bogs and fens, but there are also many shoreline marshes and floodplain swamps in the region.

Because of the vastness of Canada and its wetlands, and because the low-population regions have had less impact on wetland loss than the coastal and southern regions of Canada, there has been little attempt to summarize the loss of wetlands in Canada to one number, as has been the case for the conterminous United States. Locally, there are many regions of southern and coastal Canada where high rates of wetland loss have been experienced, and some detailed estimates do exist for the more populated regions of Canada. There has been a 65 to 80 percent loss of coastal marshes in the Atlantic and Pacific regions, respectively, a 71 percent loss of all wetlands in the lower

Table 3.5   Wetland losses near major urban centers in Quebec and Ontario, Canada

Urban Center Region (UCR)	UCR Area (km²)	Estimated Presettlement Wetland Area (km²)	Wetland Area in 1981 (km²)	Percentage Loss
Chicoutimi–Jonquiere (QC)	398	114	22	81
Montreal (QC)	3,148	1,945	228	88
Quebec City (QC)	1,000	405	168	58
Hamilton (ON)	863	102	18	82
Kitchener (ON)	1,482	101	63	37
London (ON)	972	27	7.7	70
Oshawa (ON)	191	18	4.9	73
Ottawa–Hull (ON,QC)	2,864	868	297	66
St. Catherines–Niagara (ON)[a]	1,282	522	119	77
Sudbury (ON)	1,174	148	80	45
Thunder Bay (ON)	529	104	73	30
Toronto (ON)	2,732	172	23	87
Windsor (ON	403	422	11	97

[a]1976 data.
Source: National Wetlands Working Group (1988).

Great Lakes, and a 71 percent loss of wetlands in the prairie pothole region (Table 3.2). Even higher loss rates have occurred in the major urban areas of Canada (Table 3.5). The most extensive wetland loss has occurred in southern Ontario, Canada's most populated region, particularly from Windsor on the west toward and past Toronto on the east, where 80 to more than 90 percent wetland loss is common. Farther north to Quebec City, Quebec, and farther west to Thunder Bay, Ontario, loss rates are lower. Few data are available on the conversion of wetlands to other uses in rural areas, even in eastern Canada. However, studies have suggested that 32 percent of the tidal marshes along the St. Lawrence Estuary have been converted to agricultural use and that, on the St. Lawrence River between Cornwall and Quebec, there was a 7 percent loss in wetland area from 1950 to 1978 alone (National Wetlands Working Group, 1988).

Regional Wetlands of the World

The remainder of this chapter describes some of the regionally important wetlands found around the world (Fig. 3.3). We cannot possibly include every major wetland in the world in this section, but we chose to present a wide diversity of international wetlands and attempted to give a broad range of wetland ecosystems. Each of these regional wetland areas or specific wetlands has or had a significant influence on the culture and development of its region. Some areas, such as the Florida Everglades, have had the luxury of major investigations by wetland scientists or books written for both academicians and the public. These studies and books have taught us much about wetlands and have identified much of their intrinsic values.

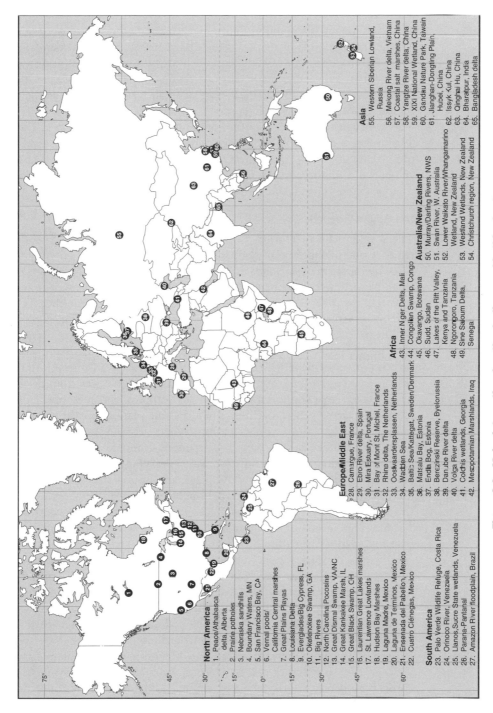

North America

1. Peace/Athabasca delta, Alberta
2. Prairie potholes
3. Nebraska sandhills
4. Boundary Waters, MN
5. San Francisco Bay, CA
6. Vernal pools/ California Central marshes
7. Great Plains Playas
8. Louisiana Delta
9. Everglades/Big Cypress, FL
10. Okefenokee Swamp, GA
11. Big Rivers
12. North Carolina Pocosins
13. Great Dismal Swamp, VA/NC
14. Great Kankakee Marsh, IL
15. Great Black Swamp, OH
16. Laurentian Great Lakes marshes
17. St. Lawrence Lowlands
18. Hudson Bay Marshes
19. Laguna Madre, Mexico
20. Laguna de Terminos, Mexico
21. Ensenada del Pabellon, Mexico
22. Cuatro Ciénegas, Mexico

South America

23. Palo Verde Wildlife Refuge, Costa Rica
24. Orinoco River, Venezuela
25. Llanos,Sucre State wetlands, Venezuela
26. Parana-Pantanal
27. Amazon River floodplain, Brazil

Europe/Middle East

28. Camargue, France
29. Ebro River delta, Spain
30. Mira Estuary, Portugal
31. Bay of Mont St. Michel, France
32. Rhine delta, The Netherlands
33. Oostwaardersplassen, Netherlands
34. Wadden Sea
35. Baltic Sea/Kattegat, Sweden/Denmark
36. Matsalu Bay, Estonia
37. Endla Bog, Estonia
38. Berezinski Reserve, Byelorussia
39. Danube River delta
40. Volga River delta
41. Colchis wetlands, Georgia
42. Mesopotamian Marshlands, Iraq

Africa

43. Inner Niger Delta, Mali
44. Congolian Swamp, Congo
45. Okavango, Botswana
46. Sudd, Sudan
47. Lakes of the Rift Valley, Kenya and Tanzania
48. Ngorongoro, Tanzania
49. Sine Saloum Delta, Senegal

Australia/New Zealand

50. Murray/Darling Rivers, NWS
51. Swan River, W. Australia
52. Lower Waikato River/Whangamarino Wetland, New Zealand
53. Westland Wetlands, New Zealand
54. Christchurch region, New Zealand

Asia

55. Western Siberian Lowland, Russia
56. Mekong River delta, Vietnam
57. Coastal salt marshes, China
58. Yangtze River delta, China
59. XiXi National Wetland, China
60. Gandau Nature Park, Taiwan
61. Jianghan-Dongting Plain, Hubei, China
62. Issyk Kul, China
63. Qinghai Hu, China
64. Bharatpur, India
65. Bangladesh delta

Figure 3.3 Major international wetlands discussed in this chapter.

North America

Many regions in the United States and Canada support, or once supported, large, contiguous wetlands or many smaller and more numerous wetlands. Some are often large, heterogeneous wetland areas, such as the Okefenokee Swamp in Georgia and Florida, that defy categorization as one type of wetland ecosystem. Others can also be large regions containing a single class of small wetlands, such as the prairie pothole region of Manitoba, Saskatchewan, and Alberta in Canada and the Dakotas and Minnesota in the United States. Some regional wetlands, such as the Great Dismal Swamp on the Virginia–North Carolina border, have been drastically altered since presettlement times, and others, such as the Great Kankakee Marsh of northern Indiana and Illinois and the Great Black Swamp of northwestern Ohio, have virtually disappeared as a result of extensive drainage programs.

The Florida Everglades

The southern tip of Florida, from Lake Okeechobee southward to the Florida Bay, harbors one of the unique regional wetlands in the world. The region encompasses three major types of wetlands in its 34,000-km^2 area: the Everglades, the Big Cypress Swamp, and the coastal mangroves and Florida Bay (Fig. 3.4). The water that passes through the Everglades on its journey from Lake Okeechobee is often referred to conceptually as a "river of grass" that is often only centimeters in depth and 80 km wide. The Everglades is dominated by sawgrass (*Cladium jamaicense*), which is actually a sedge, not a grass. The expanses of sawgrass, which can be flooded by up to a meter of water in the wet season (summer) and burned in a fire in the dry season (winter/spring), are interspersed with deeper water sloughs and tree islands, or *hammocks*, that support a vast diversity of tropical and subtropical plants, including hardwood trees, palms, orchids, and other air plants. To the west of the sawgrass Everglades is the Big Cypress Swamp, called big because of its great expanse, not because of the size of the trees. The swamp is dominated by cypress (*Taxodium* spp.) interspersed with pine flatwoods and wet prairie (Fig. 3.5). It receives about 125 cm of rainfall per year but does not receive major amounts of overland flow as the Everglades does. The third major wetland type, mangroves, forms impenetrable thickets where the sawgrass and cypress swamps meet the saline waters of the coastline.

Since about half of the original Everglades (Fig. 3.4a) has been lost to agriculture (the Everglades Agricultural Area) in the north and to urban development in the east (Fig. 3.4b), concern for the remaining Everglades has been extended to the quality and quantity of water delivered to the Everglades through a series of canals and water conservation areas. The Everglades is currently the site of one of the largest wetland restoration efforts in the United States. The project includes the expertise of all major federal and state environmental agencies and universities in the region, as well as a commitment of $8 billion by the federal government and the state of Florida. The comprehensive restoration blueprint includes plans for improving the water quality as it leaves the agricultural areas and for modifying the hydrology to conserve and restore habitat for declining populations of wading birds such as the wood stork and the white ibis. North of the Everglades, there is a renewed effort to

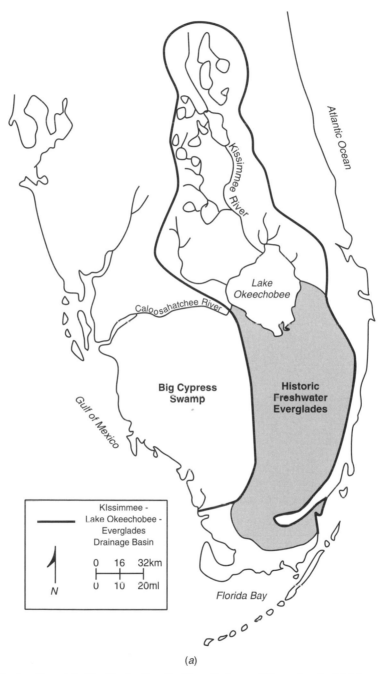

Figure 3.4 Southern Florida illustrating the Everglades and its watershed (a) in presettlement times and (b) in present conditions with cities, drainage works, water conservation areas (WCAs), and the Everglades Agricultural Area (EAA). *(After Light and Dineen, 1994.)*

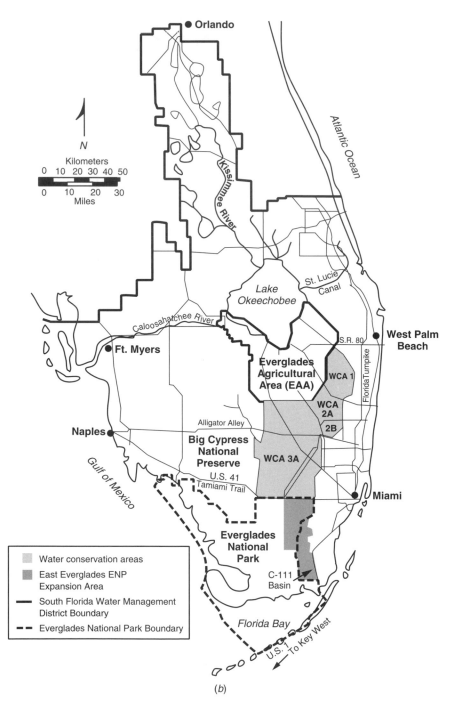

Orlando

Kissimmee River

Atlantic Ocean

N

Kilometers
0 10 20 30 40 50
0   10   20   30
Miles

Lake Okeechobee

St. Lucie Canal

Caloosahatchee River

West Palm Beach

S.R. 80

Ft. Myers

Everglades Agricultural Area (EAA)

WCA 1

Florida Turnpike

WCA 2A

2B

Alligator Alley

Naples

Big Cypress National Preserve

WCA 3A

U.S. 41
Tamiami Trail

Gulf of Mexico

Miami

Everglades National Park

C-111 Basin

Water conservation areas

East Everglades ENP Expansion Area

South Florida Water Management District Boundary

Everglades National Park Boundary

Florida Bay

U.S. 1 To Key West

(b)

Figure 3.4 (continued)

Figure 3.5   The Florida Everglades area also includes extensive forested wetland swamps, such as Corkscrew Swamp in the Big Cypress Swamp in southwestern Florida. *(Photo by W. J. Mitsch.)*

restore the ecological functions of the Kissimmee River, including many of its back-swamp areas. This river feeds Lake Okeechobee, which, in turn, spills over to the Everglades.

Numerous popular books and articles, including the classic *The Everglades: River of Grass* by Marjory Stoneman Douglas (1947), have been written about the Everglades and its natural and human history. Davis (1940, 1943) gives some of the earliest and best descriptions of the plant communities in southern Florida. A textbook specific to the Florida Everglades is now in its second edition (Lodge, 2006), and a wonderful historical account of the many attempts to manage, drain, and restore the Everglades, called *The Swamp*, was written by Grunwald (2005). The wetlands and south Florida have been through a history of several drainage attempts, a land-grab boom, a hurricane in 1926 that killed 400 people, a massive water management system developed by the U.S. Army Corps of Engineers, and today's attempt to restore the hydrology and part of the Everglades to something resembling what it was before.

Okefenokee Swamp

The Okefenokee Swamp on the Atlantic Coastal Plain of southeastern Georgia and northeastern Florida is a 1,750-km^2 mosaic of several different types of wetland communities. It is believed to have been formed during the Pleistocene or later when ocean water was impounded and isolated from the receding sea by a sand ridge (now referred to as the Trail Ridge) that kept water from flowing directly toward the Atlantic. The swamp forms the headwaters of two river systems: the Suwannee River, which flows southwest through Florida to the Gulf of Mexico, and the St. Mary's River, which flows southward and then eastward to the Atlantic Ocean.

Much of the swamp is now part of the Okefenokee National Wildlife Refuge, established in 1937 by Congress. The Okefenokee is named for an Indian word meaning "Land of Trembling Earth" because of the numerous vegetated floating islands that dot the wet prairies. Six major wetland communities comprise the Okefenokee Swamp: (1) pond cypress forest, (2) emergent and aquatic bed prairie, (3) broad-leaved evergreen forest, (4) broad-leaved shrub wetland, (5) mixed cypress forest, and (6) black gum forest. Pond cypress (*Taxodium distichum* var. nutans), black gum (*Nyssa sylvatica* var. *biflora*), and various evergreen bays (e.g., *Magnolia virginiana*) are found in slightly elevated areas where water and peat deposits are shallow. Open areas, called prairies, include lakes, emergent marshes of *Panicum* and *Carex*, floating-leaved marshes of water lilies (e.g., *Nuphar* and *Nymphaea*), and bladderwort (*Utricularia*). Fires that actually burn peat layers are an important part of this ecosystem and have recurred in a 20- to 30-year cycle when water levels became very low. Many people believe that the open prairies represent early successional stages, maintained by burning and logging, of what would otherwise be a swamp forest. A complete scientific description of the Okefenokee Swamp is contained in several papers in a book compiled by Cohen et al. (1984).

The Pocosins of the Carolinas

Pocosins are evergreen shrub bogs found on the Atlantic Coastal Plain from Virginia to northern Florida. These wetlands are particularly dominant in North Carolina, where an estimated 3,700 km² remained undisturbed or only slightly altered in 1980, whereas 8,300 km² were drained for other land uses between 1962 and 1979 alone (Richardson et al., 1981). The word *pocosin* comes from the Algonquin phrase for "swamp on a hill." In successional progression and in nutrient-poor acid conditions, pocosins resemble bogs typical of much colder climes and, in fact, were classified as bogs in a 1954 National Wetlands Survey. A typical pocosin ecosystem in North Carolina is dominated by evergreen shrubs and pine (*Pinus serotina*). Pocosins are found "growing on waterlogged, acid, nutrient poor, sandy or peaty soils located on broad, flat topographic plateaus, usually removed from large streams and subject to periodic burning" (Richardson et al., 1981). Draining and ditching for agriculture and forestry have affected pocosins in North Carolina. Proposed peat mining and phosphate mining could cause serious losses of these wetlands. Summaries of the ecological, economic, and legal aspects of pocosin management are included in *Pocosin Wetlands*, edited by Richardson (1981), and in a review paper by Richardson (1983).

Great Dismal Swamp

The Great Dismal Swamp is one of the northernmost "southern" swamps on the Atlantic Coastal Plain and one of the most studied and romanticized wetlands in the United States. The swamp covers approximately 850 km^2 in southeastern Virginia and northeastern North Carolina near the urban sprawl of the Norfolk–Newport News–Virginia Beach metropolitan area. It once extended over 2,000 km^2. The swamp has been severely affected by human activity during the past 200 years. Draining, ditching, logging, and fire played a role in diminishing its size and altering its ecological communities. The Great Dismal Swamp was once primarily a magnificent bald cypress–gum swamp that contained extensive stands of Atlantic white cedar (*Chamaecyparis thyoides*). Although remnants of those communities still exist today, much of the swamp is dominated by red maple (*Acer rubrum*), and mixed hardwoods are found in drier ridges. In the center of the swamp lies Lake Drummond, a shallow, tea-colored, acidic body of water. The source of water for the swamp is thought to be underground along its western edge as well as surface runoff and precipitation. Drainage occurred in the Great Dismal Swamp as early as 1763 when a corporation called the Dismal Swamp Land Company, which was owned in part by George Washington, built a canal from the western edge of the swamp to Lake Drummond to establish farms in the basin (Fig. 3.6). That effort, like several others in the ensuing years, failed, and Mr. Washington went on to help found a new country. Timber companies, however, found economic reward in the swamp by harvesting the cypress and cedar for shipbuilding and other uses. One of the last timber companies that owned the swamp, the Union Camp Corporation, gave almost 250 km^2 of the swamp to the federal government to be maintained as a national wildlife refuge.

At least one book, *The Great Dismal Swamp* (Kirk, 1979), describes the ecological and historical aspects of this important wetland. The extent and management of Atlantic white cedar, a dominant species in the Great Dismal Swamp, are presented by Sheffield et al. (1998).

The Swamp Rivers of the South Atlantic Coast

The Atlantic Coastal Plain, extending from North Carolina to the Savannah River in Georgia, is a land dominated by forested wetlands and marshes and cut by large rivers that drain the Piedmont and cross the Coastal Plain in a northwest–southeast direction to the ocean. These rivers include the Roanoke, Chowan, Little Pee Dee, Great Pee Dee, Lynches, Black, Santee, Congaree, Altamaha, Cooper, Edisto, Combahee, Coosawhatchie, and Savannah, as well as a host of smaller tributaries. Extensive bottomland hardwood forests and cypress swamps line these rivers and spread into the lowlands between them. Interspersed among these forests are hundreds of Carolina Bays, small elliptical-shaped lakes of uncertain origin surrounded by or overgrown with marshes and forested wetlands (Lide et al., 1995). The origin of these lake–wetland complexes, of which there are more than 500,000 along the eastern Coastal Plain, has been suggested to be meteor showers, wind, or groundwater flow (D. C. Johnson, 1942; H. T. Odum, 1951; Prouty, 1952; Savage, 1983). Along the coast, freshwater tides on the lower rivers formerly overflowed extensive forests, but many of these were

Figure 3.6   Washington's Ditch in the Great Dismal Swamp in eastern Virginia. This ditch was part of an unsuccessful effort began by George Washington to drain the swamp for commercial reasons in the mid-18th century. *(Photograph by Frank Day, reprinted with permission.)*

cleared in the early 1800s to establish rice plantations. Most of the rice plantations have since been abandoned, and the former fields are now extensive freshwater marshes that have become a paradise for ducks and geese. The estuaries at the mouths of the rivers support the most extensive salt marshes on the Southeast Coast.

In 1825, Robert Mills wrote of Richland County, South Carolina: "What clouds of miasma, invisible to sight, almost continually rise from these sinks of corruption, and who can calculate the extent of its pestilential influence?" (quoted by Dennis, 1988). At that time, only 10,000 ha of the 163,000-ha county were being cultivated. Almost all the rest was a vast, untouched swamp. Our appreciation of these swamps has changed dramatically since that time, and parts of this swamp are now the Congaree Swamp National Monument and the Francis Beidler Forest; the latter includes the world's largest virgin cypress–tupelo (*Taxodium–Nyssa*) swamp and is now an Audubon sanctuary. Both preserves contain extensive stands of cypress more than 500 years old that escaped the logger's ax in the late 1800s.

The Prairie Potholes

A significant number of small wetlands, primarily freshwater marshes, are found in a 780,000-km^2 region in the states of North Dakota, South Dakota, and Minnesota and in the Canadian provinces of Manitoba, Saskatchewan, and Alberta (Fig. 3.7).

Figure 3.7   Oblique aerial view of prairie pothole wetlands, showing many small ponds surrounded by wetland vegetation, in the middle of large agricultural fields. *(File photograph, U.S. Fish and Wildlife Service, Northern Prairie Wildlife Research Center, Jamestown, North Dakota.)*

It has been estimated that there are only about 10 percent of the original wetlands remaining from presettlement times. These wetlands, called *prairie potholes*, were formed by glacial action during the Pleistocene. This region is considered one of the most important wetland regions in the world because of its numerous shallow lakes and marshes, its rich soils, and its warm summers, which are optimum for waterfowl. Wet-and-dry cycles are a natural part of the ecology of these prairie wetlands. In fact, many of the prairie potholes might not exist if there were no periodic dry periods. In some cases, dry periods of 1 to 2 years every 5 to 10 years are required to maintain emergent marshes. Another feature of this wetland region is the occasional presence of saline wetlands and lakes caused by high evapotranspiration/precipitation ratios. Salinities as high as 370 parts per thousand (ppt) have been recorded for some hypersaline lakes in Saskatchewan. It is estimated that 50 to 75 percent of all the waterfowl originating in North America in any given year comes from this region. More than half of the original wetlands in the prairie pothole region have been drained or altered, primarily for agriculture. An estimated 500 km^2 of prairie pothole wetlands in North Dakota, South Dakota, and Minnesota were lost between 1964 and 1968 alone. However, major efforts to protect the remaining prairie potholes are progressing. Thousands of square kilometers of wetlands have been purchased under the U.S. Fish and Wildlife Service Waterfowl Production Area program in North Dakota alone since the early 1960s. The Nature Conservancy and other private foundations have also purchased many wetlands in the region.

The Nebraska Sandhills and Great Plains Playas

South of the prairie pothole region is an irregular-shaped region of 52,000 km² in northern Nebraska described as "the largest stabilized dune field in the Western Hemisphere" (Novacek, 1989). These Nebraska sandhills, which constitute one-fourth of the state, represent an interesting and sensitive coexistence of wetlands, agriculture, and a very important aquifer-recharge area. The area was originally mixed-grass prairie composed of thousands of small wetlands in the interdunal valleys. Much of the region is now used for farming and rangeland agriculture, and many of the wetlands in the region have been preserved, even though the vegetation is often harvested for hay or grazed by cattle. The Ogallala Aquifer is an important source of water for the region and is recharged to a significant degree through overlying dune sands and to some extent through the wetlands. It has been estimated that there are 558,000 ha of wetlands in the Nebraska sandhills, many of which are interconnected wet meadows or shallow lakes that contain water levels determined by both runoff and regional water table levels. The wetlands in the region have been threatened by agricultural development, especially pivot irrigation systems that cause a lowering of the local water tables despite increased wetland flooding in the vicinity of the irrigation systems. Like the prairie potholes to the north, the Nebraska sandhill wetlands are important breeding grounds for numerous waterfowl, including about 2 percent of the Mallard breeding population in the north-central flyway.

Smith (2003) has argued that many of the wetlands that occur in Nebraska, particularly in southwestern Nebraska, could be defined as *playas* (see definition in Table 2.1) because of their seasonal flooding patterns in a semi-arid environment in the Great Plains Region. Most of the playas in North America are found in the Southern Great Plains that includes western Texas, southern New Mexico, southeastern Colorado, and southwestern Kansas. These temporarily or seasonally flooded wetlands are characterized as being depressional (i.e., isolated) and recharge (i.e., they recharge groundwater; see Chapter 4) wetland basins. It is estimated that there are over 25,000 playas in the United States Great Plains (Sabin and Holliday, 1995) and that they cover 1,800 km² in what is otherwise a semi-arid to arid agricultural landscape (Smith, 2003).

The Great Kankakee Marsh

For all practical purposes, this wetland no longer exists, although, until about 100 years ago, it was one of the largest marsh-swamp basins in the interior United States. Located primarily in northwestern Indiana and northeastern Illinois, the Kankakee River basin is 13,700 km² in size, including 8,100 km² in Indiana, where most of the original Kankakee Marsh was located. From the river's source to the Illinois line, a direct distance of only 120 km, the river originally meandered through 2,000 bends along 390 km, with a nearly level fall of only 8 cm per km. Numerous wetlands, primarily wet prairies and marshes, remained virtually undisturbed until the 1830s, when settlers began to enter the region. The naturalist Charles Bartlett (1904) described the wetland as follows:

More than a million acres of swaying reeds, fluttering flags, clumps of wild rice, thick-crowding lily pads, soft beds of cool green mosses, shimmering ponds and black mire and trembling bogs—such is Kankakee Land. These wonderful fens, or marshes, together with their wide-reaching lateral extensions, spread themselves over an area far greater than that of the Dismal Swamp of Virginia and North Carolina.

The Kankakee region was considered a prime hunting area until the wholesale draining of the land for crops and pasture began in the 1850s. The Kankakee River and almost all of its tributaries in Indiana were channelized into a straight ditch in the late 19th century and early 20th century. In 1938, the Kankakee River in Indiana was reported to be one of the largest drainage ditches in the United States; the Great Kankakee Marsh was essentially gone by then. Early accounts of the region were given by Bartlett (1904) and Meyer (1935). More recently, there has been a major effort to restore parts of the Great Kankakee Marsh in northwestern Indiana.

The Great Black Swamp

Another vast wetland of the Midwest that has ceased to exist is the Great Black Swamp in what is now northwestern Ohio. The Great Black Swamp (Fig. 3.8) was once a combination of marshland and forested swamps that extended about 160 km long and 40 km wide in a southwesterly direction from the lake and covered an estimated 4,000 km^2. The bottom of an ancient extension of Lake Erie, the Black Swamp was named for the rich, black muck that developed in areas where drainage was poor as a result of several ridges that existed perpendicular to the direction of

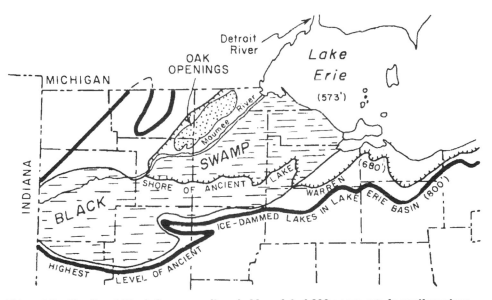

Figure 3.8 The Great Black Swamp as it probably existed 200 years ago in northwestern Ohio. Essentially none of this 4,000-km^2 wetland remains. (After Forsyth, 1960.)

the flow to the lake. There are numerous accounts of the difficulty that early settlers and armies (especially during the War of 1812) had in negotiating this region, and few towns of significant size have developed in the location of the original swamp. One account of travel through the region in the late 1700s suggested that "man and horse had to travel mid-leg deep in mud" for three days just to cover a distance of only 50 km (Kaatz, 1955). As with many other wetlands in the Midwest, state and federal drainage acts led to the rapid drainage of this wetland, until little of it was left by the beginning of the 20th century. Only one small example of an interior forested wetland and several coastal marshes (about 150 km^2) remain of the original western Lake Erie wetlands.

The Louisiana Delta

As the Mississippi River reaches the last phase of its journey to the Gulf of Mexico in southeastern Louisiana, it enters one of the most wetland-rich regions of the world. The total area of marshes, swamps, and shallow coastal lakes covers more than 36,000 km^2. As the Mississippi River distributaries reach the sea, forested wetlands give way to freshwater marshes and then to salt marshes. The salt marshes are some of the most extensive and productive in the United States (Fig. 3.9a), and depend on the influx of fresh water, nutrients, sediments, and organic matter from upstream swamps. Freshwater and saltwater wetlands has been decreasing at a rapid rate in coastal Louisiana, amounting to a total wetland loss of 4,800 km^2 since the 1930s and annual loss rates between 60 and 100 km^2 yr^{-1} (Day et al., 2005, 2007). These losses have been attributed to both natural and artificial causes although the main cause has been the isolation of the river from the delta (Day et al., 2007).

Characteristic of the riverine portion of the delta, the Atchafalaya River, a distributary of the Mississippi River, serves as both a flood-relief valve for the Mississippi River and a potential captor of its main flow. The Atchafalaya Basin by itself is the third-largest continuous wetland area in the United States and contains 30 percent of all the remaining bottomland forests in the entire lower Mississippi alluvial valley. The river passes through this narrow 4,700-km^2 basin for 190 km, supplying water for 1,700 km^2 of bottomland forests and cypress–tupelo swamps and another 260 km^2 of permanent bodies of water. The Atchafalaya Basin, contained within a system of artificial and natural levees, has had a controversial history of human intervention. Its flow is controlled by structures located where it diverges from the Mississippi River main channel, and it has been dredged for navigation and to prevent further infilling of the basin by Mississippi River silt. It has been channelized for oil and gas production. The old-growth forests were logged at the beginning of the 20th century, and the higher lands are now in agricultural production.

Another frequently studied wetland area in the delta is the Barataria Bay estuary in Louisiana, an interdistributary basin of the Mississippi River that is now isolated from the Mississippi River by a series of flood-control levees. This basin, 6,500 km^2 in size, contains 700 km^2 of wetlands, including cypress–tupelo swamps, bottomland hardwood forests, marshes, and shallow lakes.

(a)

(b)

Figure 3.9 **(a) Coastal marshlands of the Mississippi River Delta in southern Louisiana; breakup of marshes caused by land subsidence and lack of sediment inputs from the river is evident. (b) Diversion structure on the Mississippi River designed to divert river water into the Louisiana delta to assist in sediment deposition and marsh recovery.** *(Photographs by W. J. Mitsch.)*

The U.S. Army Corps of Engineers, in cooperation with other federal and state agencies, began designing a comprehensive strategy for conservation and restoration of the delta. This project, first known as Project 2050 and later as the Louisiana Coastal Area (LCA) plan, was to be funded to the tune of billions of dollars by federal and state funds, and would provide barrier island reconstruction and major river diversions of fresh water and sediments from the Mississippi River into the coastal wetlands and bays (Fig. 3.9b). Then in late August 2005, Hurricanes Katrina and Rita battered the Louisiana coastline and destroyed much of the city of New Orleans (see Costanza et al., 2006; Day et al., 2007), prompting a redirection of some funds from wetland restoration to levee construction. The delta restoration plan is described in more detail in Chapter 12 on wetland creation and restoration.

San Francisco Bay

One of the most altered and most urbanized wetland areas in the United States is San Francisco Bay in northern California. The marshes surrounding the bay covered more than 2,200 km^2 when the first European settlers arrived. Almost 95 percent of these marshes have since been destroyed. The ecological systems that make up San Francisco Bay range from deep, open water to salt and brackish marshes. The salt marshes are dominated by Pacific cordgrass (*Spartina foliosa*) and pickleweed (*Salicornia virginica*), and the brackish marshes support bulrushes (*Scirpus* spp.) and cattails (*Typha* spp.). Soon after the beginning of the Gold Rush in 1849, the demise of the bay's wetlands began. Industries such as agriculture and salt production first used the wetlands, clearing the native vegetation and diking and draining the marsh. At the same time, other marshes were developing in the bay as a result of rapid sedimentation. The sedimentation was caused primarily by upstream hydraulic mining. Sedimentation and erosion continue to be the greatest problems encountered in the remaining tidal wetlands.

Great Lakes Wetlands/St. Lawrence Lowlands

The Canadian marshes of this region, especially those along the Great Lakes in Ontario and in the St. Lawrence lowlands of Ontario and Quebec (Fig. 3.10), are important habitats for migratory waterfowl. Several of the notable wetlands in the region include Long Point and Point Pelee on northern Lake Erie, the St. Clair National Wildlife Area on Lake St. Clair along the Great Lakes in southern Ontario, and the Cap Tourmente National Wildlife Area and Lac St.-François along the upper St. Lawrence River in eastern Ontario and southwestern Quebec.

Cap Tourmente, a 2,400-ha tidal freshwater marsh complex located about 50 km northeast of Quebec City, was the first wetland in Canada designated as a Ramsar site of international importance (Fig. 3.11). It consists of both intertidal mud flats and freshwater marshes, as well as nontidal marshes, swamps, shrub swamps, and peatlands. The Cap Tourmente freshwater tidal marshes are subjected to heavy tidal flooding, with tidal amplitudes of 4.1 m at mean tides and 5.8 m during spring tides. The Cap Tourmente National Wildlife Area has a wide range of communities, including 400 ha of tidal marsh, 100 ha of coastal meadow, 700 ha of agricultural land, and 1,200 ha of forest. *Scirpus americanus* (American bulrush) marshes of the St. Lawrence, such as those found at Cap Tourmente, are restricted to the freshwater tidal portion of the river, with only 4,000 ha remaining in the entire region. Although increasing numbers of Greater Snow Geese have led to a depletion of *Scirpus* rhizomes, which may eventually cause a deterioration of the marshes at Cap Tourmente, the snow

Figure 3.10 Wetland scientist botanizing in the *Scirpus americanus* marsh adjacent to the St. Lawrence River near Quebec City, Canada. *(Photograph by W. J. Mitsch.)*

Figure 3.11 Snow geese at Cap Tourmente National Wildlife Area, Quebec, Canada. *(Photograph by Robbie Sproule; provided by Creative Commons licence.)*

geese remain one of the notable features of this wetland during the migratory season. Tens of thousands of Greater Snow Geese migrate in both the spring and fall and feed on the bulrushes.

Marshes of this region are considered the least stable class of wetlands and are significantly changed by variations in rainfall, flooding patterns, lake levels, and herbivory from one year to the next. The marshes along the Great Lakes are generally diked and heavily managed to buffer them from the year-to-year fluctuations in lake levels. This temperate region in Canada also has a considerable number of hardwood forested swamps dominated by red and silver maples (*Acer rubrum* and *A. saccharinum*) and ash (*Fraxinus* spp.). Without human intervention, these swamps are quite stable; however, logging has been frequent, including clear-cutting. A clear-cut swamp is often replaced by a marsh, and the successional pattern starts all over again.

Canada's Central and Eastern Province Peatlands

The peatlands of northern Ontario and Manitoba are extensive regions that are used less by waterfowl and more by a wide variety of mammals, including moose, wolf, beaver, and muskrat. Wild rice (*Zizania palustris*), a common plant in littoral zones of boreal lakes, is often harvested for human consumption. Some of the boreal wetlands are mined for peat that is used for horticultural purposes or fuel. Fens of the region can be quite stable and are fairly common; bogs are also stable in this region, but less common. Radiocarbon dating of the bottom peat layers in Quebec bogs suggests that they began as fens between 9,000 and 5,500 years ago. Once formed, open bogs are quite stable, and forested bogs are even more stable, although they can revert to open bogs if fire occurs.

Hudson–James Bay Lowlands

A large wetland complex is found in northern Ontario and Manitoba and the eastern Northwest Territories, wrapping around the southern shore of the Hudson Bay (Fig. 3.12) and its southern extension, James Bay. These Hudson–James Bay lowlands are part of the vast subarctic wetland region of Canada, which stretches from the Hudson Bay northwestward to the northwestern corner of Canada and into Alaska and covers 760,000 km^2 of Canada (Zoltai et al., 1988). This region has been described as the region with the highest density and percent cover of wetlands in North America (76 to 100 percent) (Abraham and Keddy, 2005). One of the largest and best-described wetland sites in this region is the 24,000-km^2 Polar Bear Provincial Park in northern Ontario. Two additional sanctuaries of note are located in the southern James Bay: the Hannah Bay Bird Sanctuary and the Moose River Bird Sanctuary, which total 250 km^2. The region is dominated by extensive areas of mud flats, intertidal marshes, and supertidal meadow marshes, which grade into peatlands, interspersed with small lakes, thicket swamps, forested bogs and fens, and open bogs, fens, and marshes away from the shorelines. A band of the high subarctic wetland region along the southern shore of the bay is dominated by sedges (*Carex* spp.), cotton grasses (*Eriophorum* spp.), and clumps of birches (*Betula* spp.). The more southerly low subarctic wetland region is made up of low, open bogs, sedge–shrub

Figure 3.12 Extensive peatlands and marshes of Hudson Bay lowlands. *(Photograph by C. D. A. Rubec, reprinted with permission.)*

fens, moist sedge-covered depressions, and open pools and small lakes separated by ridges of peat, lichen–peat-capped hummocks, raised bogs, and beach ridges. Even though the tidal range from the Hudson Bay is small, the gradual slope of land allows much tidal inundation of flats that vary from 1 to 5 km in width. Low-energy coasts with wide coastal marshes occur in the southern James Bay; high-energy coasts with sand flats and sand beaches are found along the Hudson Bay shoreline itself. Isostatic rebound following glacial retreat has resulted in the emergence of land from the bay at a rate of 1.2 m per century for the past 1,000 years, the greatest rate of glacial rebound in North America.

The coastal marshes, intertidal sand flats, and river mouths of the Hudson–James Bay lowlands serve as breeding and staging grounds for a large number of migratory waterfowl, including the Lesser Snow Goose, which was once in danger of disappearing but is now flourishing; Canada Goose; Black Duck; Pintail; Green-winged Teal; Mallard; American Wigeon; Shoveler; and Blue-winged Teal. The western and southwestern coasts of the Hudson and James bays form a major migration pathway for many shorebird species as well, including Red Knot, Short-billed Dowitcher, Dunlin, Greater Yellowlegs, Lesser Yellowlegs, Ruddy Turnstone, and Black-bellied Plover.

**Figure 3.13   Athabasca River in the Peace–Athabasca Delta in Jaspar National Park.
(Photographs by AudeVivere; courtesy of Wikimedia Commons.)**

The wetlands of Polar Bear Provincial Park provide nesting habitat for Red-throated, Arctic, and Common loons; American Bittern; Common and Red-breasted Merganser; Yellow Rail; Sora; Sandhill Crane; and several gulls and terns.

Peace–Athabasca Delta

The Peace–Athabasca Delta in Alberta, Canada (Fig. 3.13), is the largest freshwater inland boreal delta in the world and is relatively undisturbed by humans. It actually comprises three deltas: the Athabasca River delta (1,970 km²), the Peace River delta (1,684 km²), and the Birch River delta (168 km²). It is one of the most important waterfowl nesting and staging areas in North America and is the staging area for breeding ducks and geese on their way to the MacKenzie River lowlands, Arctic river deltas, and Arctic islands. The major lakes of the delta are very shallow (0.6–3.0 m) and have a thick growth of submerged and emergent vegetation during the growing season. The delta consists of very large flat areas of deposited sediments with some outcropping islands of the granitic Canadian Shield. The site has the following types of wetlands: emergent marshes, mud flats, fens, sedge meadows, grass meadows, shrub–scrub wetlands, deciduous forests of balsam (*Populus balsamifera*) and birch (*Betula* spp.), and coniferous forests dominated by white and black spruce (*Picea glauca* and *P. mariana*). Owing to the shallow water, high fertility, and relatively long growing season for that latitude, the area is an abundant food source of particular importance during drought years on the prairie potholes to the south. All four major North American flyways cross the delta, with the most important being the Mississippi and central flyways.

Studies in the early 1970s showed that water levels on the delta required regulating to mitigate the effects of the Bennett Dam, and weirs were subsequently constructed at Riviere des Rochers and Revillon Coupe. The dam, located upstream on the Peace River in British Columbia, was constructed in 1967 and caused a significant drop in water flow to the delta, resulting in insufficient water levels to fill the numerous perched basins in the area (Healey, 1994). The effects of the reduced water flow as a result of this dam construction have been mitigated by the construction of weirs on the Peace River tributaries, which have nearly restored natural summer peak water levels in the delta. The amplitude of seasonal and annual fluctuations, however, is still less than under the natural water flow regime prior to the construction of the dam (Healey, 1994). Between 1976 and 1986, about 30 to 40 percent of the delta experienced severe drying, and woody vegetation was rapidly colonizing.

At least 215 species of birds, 44 species of mammals, 18 species of fish, and thousands of species of insects and invertebrates are found in the delta. Up to 400,000 birds use this wetland in the spring, and more than 1 million birds in autumn. Waterfowl species recorded in the delta area include Lesser Snow Goose, White-fronted Goose, Canada Goose, Tundra Swan, all four species of the loon, all seven species of North American grebe, and 25 species of duck. The world's entire population of the endangered Whooping Crane nests in the northern part of the delta area. The site also contains the largest undisturbed grass and sedge meadows in North America, which support an estimated 10,000 wood and plains buffalo.

Wetlands of Mexico

Because of extensive arid regions in its interior, Mexico was initially underrepresented in the number and area of Ramsar wetlands of international important, with only seven Ramsar sites designated in 2001 (Pérez-Arteaga et al., 2002). That situation has changed since then, with 65 Ramsar sites covering 5.3 million ha now located in Mexico. Many of the priority wetland sites in Mexico are associated with coastal sites on the Gulf of Mexico and the Pacific Ocean. Mexico has an estimated 1.6 million ha of wetlands on its coasts, with 74,800 ha on the Pacific coast and 674,500 ha on the Gulf of Mexico. Coastal wetlands in Mexico include about 118 major wetlands complexes and at least another 538 smaller systems representing a wide variety of types (Contreras-Espinosa and Warner, 2004).

One of the largest wetland sites includes the 0.7-million-ha Laguna de Términos in Campeche. This Laguna de Términos protection area includes mangrove swamps on its coastline as well as coastal dune vegetation, freshwater swamps, flooded vegetation, lowland forest, palms, spiny scrubs, forests, secondary forests, and sea grass beds. The Ensenada del Pabellon on the Gulf of California on the Pacific Coast was estimated to account for almost 10 percent of the birds wintering in Mexico (Pérez-Arteaga et al., 2002). Laguna Madre on the Gulf of Mexico just south of the Texas coastline is another important coastal wetland in Mexico, with 200,000 ha of shallow water and mudflats and 42,000 ha of sea grass beds (dominated by *Halodule wrightii*).

Other important Mexican wetlands are those localized in the arid north region of the country in the Sonoran and Chihuhuan deserts (Davis et al., 2002; Souza et al.,

2004). The Cuatro Ciénegas basin is a small valley (about 1,500 km^2) in Coahuila State that is formed by mountain ranges of the Sierra Madre Oriental. Although the basin is in one of the driest areas of the Chihuhuan Desert, it is estimated to contain more than 200 springs and other associated habitats with many endemic organisms. Six types of aquatic habitats occur in the basin: *pozas* (small springs), *lagunas* (larger spring-fed lakes), *playa* lakes (large lakes fed by surface runoff but without outlets), *ciénegas* (shallow swamps), and manmade channels (Dinger et al., 2005). Because of the basin's biological distinctiveness, the Mexican government declared Cuatro Ciénegas as a natural protected area in 1984.

Central and South America

There are extensive and relatively understudied tropical and subtropical wetlands throughout Central and South America. Some of the more significant ones are located on the South American map in Figure 3.14 and are discussed here.

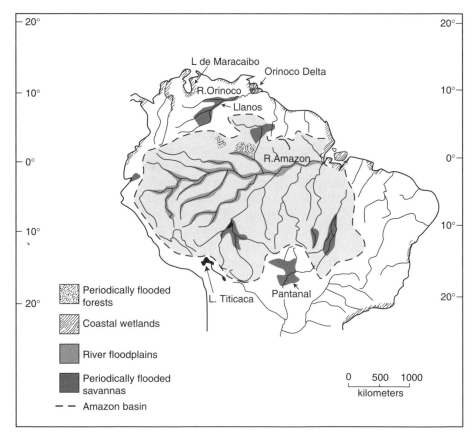

Figure 3.14 Major wetland areas of tropical South America. *(After Junk, 1993.)*

Central American Wetlands

Although poorly mapped, there are an estimated 40,000 km² of wetlands in Central America (Ellison, 2004). Mangrove swamps occur on both coastlines and cover 6,500 to 12,000 km² in Central America. Forested freshwater wetlands, the most common type of wetland in Central America (Fig. 3.15a), cover an estimated

(a)

(b)

Figure 3.15   Freshwater wetlands of Costa Rica, Central America, include (a) riverine forested freshwater wetlands in continuously wet eastern region, and (b) seasonal flooded freshwater marsh at the Palo Verde National Park during the dry season in western region of Costa Rica. *(Photographs by W. J. Mitsch.)*

15,000 km^2 of land. One type of forested wetlands—Palm swamps dominated by *Raphia taedigera*—account for 1.2 percent of the land cover of Costa Rica alone, particularly in the Atlantic lowlands. There are also some freshwater marshes (1,000–2,000 km^2) in Central America, often dominated by floating aquatic plants (*Azolla, Salvinia, Pistia, Eichhornia crassipes*) rather than emergent plants. These same floating aquatic plants often dominate wastewater treatment wetlands in Central America (Nahlik and Mitsch, 2006).

Rivers on the Pacific Coast side of Central America are shorter and more seasonal than their counterparts on the Caribbean side of the isthmus. As a result of this and the prevailing climate, which causes more even monthly distribution of precipitation on the Caribbean Sea (Atlantic) side than on the Pacific side, wetlands near the Pacific Coast tend to be very seasonal with wet summers and dry winters. One of the most important wetlands in this Central American setting is a seasonal, freshwater marsh at the Palo Verde National Park in Costa Rica. The 500-ha tidal freshwater marsh receives rainwater, agricultural runoff, and overflow water from the Tempisque River during the wet season, which, in turn, discharges to the Gulf of Nicoya about 20 km downstream of the wetland on Costa Rica's Pacific Coast. The marsh dries out almost completely by March during the dry season (Fig. 3.15b). It was habitat for about 60 resident and migratory birds (McCoy and Rodriguez, 1994), and thousands of migrating Black-bellied Whistling Ducks and Blue-winged Teal and hundreds of Northern Shoveler, American Wigeon, and Ring-necked Ducks visited the wetland during the dry season. More recently, after cattle were removed because of the marsh's designation as a wildlife refuge, the marsh was completely taken over by cattail (*Typha domingensis*), which covered 95 percent of the marsh by the late 1980s. This is a common problem in wetlands throughout the world, where clonal dominants such as *Typha* tend to choke off any other vegetation and make a poor habitat for many waterfowl and other birds. Curiously, the diversity of birds was partially maintained because of cattle grazing, which was permitted until 1980. Site managers have recently tried to reintroduce cattle grazing, burning, disking, below-water mowing, and mechanical crushing to control the *Typha*. The only method that was consistently successful was crushing the cattails (McCoy and Rodriguez, 1994).

The Orinoco River Delta

The Orinoco River delta of Venezuela was explored by Columbus during one of his early voyages. It covers 36,000 km^2 and is dominated along its brackish shoreline by magnificent mangrove forests (Fig. 3.16). The Orinoco Delta economy is based on cattle ranching, with the cattle being shipped out during the high-water season, as well as on cacao production and palm heart canning. The delta's indigenous population practices subsistence farming and fishing, and exports salted fish to the population centers bordering the region (Dugan, 1993). Although some regions are protected and conservation efforts have been made by government and industry, grazing and illegal hunting have been detrimental to the area's flora and fauna.

Figure 3.16   Mangroves of the Orinoco River delta in Venezuela. *(Reprinted from Mitsch et al., 1994a, with permission from Elsevier.)*

The Llanos

The western part of the Orinoco River basin in western Venezuela and northern Colombia (Fig. 3.14) is a very large (450,000 km^2) sedimentary basin called the Llanos. This region represents one of the largest inland wetland areas of South America. The Llanos has a winter wet season coupled with a summer dry season, which causes it to be a wetland dominated by savanna grasslands and scattered palms rather than floodplain forests typical of the Orinoco Delta (Junk, 1993). The region is an important wading-bird habitat and is rich with such animals as the caiman (*Caiman crocodilus*), the giant green anaconda (*Eunectes murinus*), and the red piranha (*Serrasalmus nattereri*). It supports about 470 bird species, although only one species is considered endemic. Dominant mammals include the giant anteater (*Myrmecophaga tridactyla*) and the abundant capybara (*Hydrochaeris hydrochaeris*).

The Pantanal

One of the largest regional wetlands in the world is the Gran Pantanal of the Paraguay–Paraná River basin and Mato Grosso and Mato Grosso do Sul, Brazil (Por, 1995; da Silva and Girard, 2004; Junk and de Cunha, 2005; Harris et al., 2005), located almost exactly in the geographic center of South America (Fig. 3.14). The wetland complex is 160,000 km², four times the size of the Florida Everglades, with about 130,000 km² of that area flooded annually. The annual period of flooding (called the *cheia*) from March through May supports luxurious aquatic plant and animal life and is followed by a dry season (called the *seca*) from September through November, when the Pantanal reverts to vegetation typical of dry savannas. There are also specific terms for the period of rising waters (*enchente*) from December through February and the period of declining waters (*vazante*) from June through August. There is also an asynchronous pattern to flooding in the Pantanal: while maximum rainfall and upstream flows occur in January, water stage does not peak until May in downstream reaches.

Just as in the Everglades cycle of wet and dry seasons, the biota spread across the landscape during the wet season and concentrate in fewer wet areas in a food chain frenzy during the dry season. Even though the Pantanal is one of the least-known regions of the globe, it is legendary for its bird life (Fig. 3.17). The Pantanal has been described as the "bird richest wetland in the world" with 463 species of birds recorded there (Harris et al., 2005). There are 13 species of herons and egrets, 3 stork species, 6 ibis and spoonbill species, 6 duck species, 11 rail species, and 5 kingfisher species. Wetland birds also include the Anhinga and the magnificent symbol of the Pantanal, the Jabiru, the largest flying bird of the Western Hemisphere. In addition, the wetland supports abundant populations of the jacaré, or caiman, a relative of the North American crocodile, and the large rodent capybara (*Hydrochoerus hydrochaeris*).

Figure 3.17 The seasonally flooded Pantanal of South America is a haven to abundant wildlife including over 450 species of birds including egrets, herons, and the Jabiru (*Jabiru mycteria*), intermixed with jacaré or caiman (*Caiman yacare*) and, during the dry period, cattle. (*Photograph by W. J. Mitsch.*)

The threats to the Pantanal are many, but, until recently, there was a semibalance between human use of the Pantanal region, particularly for cattle ranching during the dry season, and the ecological functions of the region. The ecological health of the Pantanal, however, is in a state of developmental uneasiness. Some of the rivers are polluted with metals, particularly mercury, from gold-mining activity and by agrochemicals from farms. Although the Pantanal provides tourist revenues, it is also the site of illegal wildlife trafficking and cocaine smuggling. In such a vast and remote wetland, law enforcement is physically difficult and prohibitively expensive. The Hidrovia Project, an international project to develop the Paraguay and Paraná rivers into a 3,340-km waterway, was agreed to by five South American countries in 1988 at the First International Meeting for the Development of the Paraguay–Paraná Waterway in Campo Grande, Brazil. Since then, there has been a flurry of activity by both developers and environmentalists on this project. The political and economic will to see this project completed is as great as the uncertainties of the effects that this massive project will have on the Pantanal (Gottgens et al., 2001; Junk and Nunes de Cunha, 2005).

The Amazon

Vast wetlands are found along many of the world's rivers, well before they reach the sea, especially in tropical regions. The Amazon River in Brazil (Fig. 3.14) is one of the best examples; wetlands cover about 20 to 25 percent of the 7 million km^2 Amazon basin (Junk and Piedade, 2004, 2005). It is considered one of the world's major rivers, with a flow that results in about one-sixth to one-fifth of all the fresh water in the world. Deforestation from development threatens many Amazon aquatic ecosystems and has great social ramifications for people displaced in the process. Some of the floodplain forested wetlands of the Amazon, which are estimated to cover about 300,000 km^2, undergo flooding with flood levels reaching 5 to 15 m or more (see Chapter 4). During the flood season, it is possible to boat around the canopy of trees (Fig. 3.18a). Many Amazonian streams and rivers are characterized as being either "black water" or "white water" (Fig. 3.18b), with the former dominated by dissolved humic materials and low dissolved materials and the latter dominated by suspended sediments derived from the eroding Andes mountains. Floodplains on the white water or high-sediment rivers (called várzea) are nutrient rich while floodplains on the black water streams (called igapó) are nutrient poor (Junk and Piedade, 2005).

Europe

Mediterranean Sea Deltas

The saline deltaic marshes of the mostly tideless Mediterranean Sea are among the most biologically rich in Europe. The Rhone River delta created France's most important wetland, the Camargue (Fig. 3.19; see also Chapter 1), an expanse of wetlands centered around the 9,000-ha Étang du Vaccarès. This land is home to the free-roaming horses celebrated in literature and film; here, too, is a species of

(a)

(b)

Figure 3.18 Two views of the Amazon River (a) When the Amazon River is flooded annually, it is possible to boat around the treetops. *(Photograph by W. Junk, reprinted by permission.)* (b) A contrast of white and black waters at the confluence of the Amazon (white or turbid waters on left) and Jutai (black clear waters on right) rivers. *(Photograph by W.J. Mitsch.)*

bull that inhabited Gaul several thousand years ago before being driven south by encroaching human settlements. The Camargue is also home to one of the world's 25 major flamingo nesting sites, and France's only such site. The sense of mystery and the feeling for space and freedom pervading the Camargue are linked with the gypsies, who have gathered at Les Saintes-Maries-de-la-Mer since the 15th century, as well as with the Camarguais cowboys, the *gardians*, who ride their herds over the lands (see Fig. 1.1).

Figure 3.19   The Camargue of the Rhone River delta in southern France is highly affected by a Mediterranean climate of hot, dry summers and cool, wet winters. *(Photograph by W. J. Mitsch.)*

Aquatic plants and plant communities differ distinctly from those of northern Europe or tropical Africa, as the landscape transitions from dune to lagoon, to marshland, to grassland, and then to forest. Current set-aside agricultural policies in Europe call for restoration of some of the rice fields in the Camargue, and some restoration of former wetlands along rivers in the region has already taken place (Mesléard et al., 1995; Mauchamp et al., 2002).

A principal delta on the Spanish Mediterranean coast is the Ebro Delta, located halfway between Barcelona and Valencia and fed by the Ebro River, which flows hundreds of kilometers through arid landscape to the sea. The delta itself, covered with extensive and ancient rice paddies, also has salt marshes dominated by several species of *Salicornia* and other halophytes. Lagoons are populated with a wide variety of avian species. Some restoration of rice paddies to *Phragmites* marshes has been attempted in the delta (Comin et al., 1997).

Rhine River Delta

The Rhine River is a highly managed river and a major transportation artery in Europe. The Netherlands, which comes from *Nederland*, meaning "low country," is, essentially, the Rhine River delta, and although the Dutch language did not even have a word for "wetlands," the English word was adopted in the 1970s. The Netherlands is one of the most hydraulically controlled locations on Earth (Fig. 3.20). It is estimated that 16 percent of the Netherlands is wetland; the Dutch have warmed to the idea of the importance of wetlands and have registered 7 percent

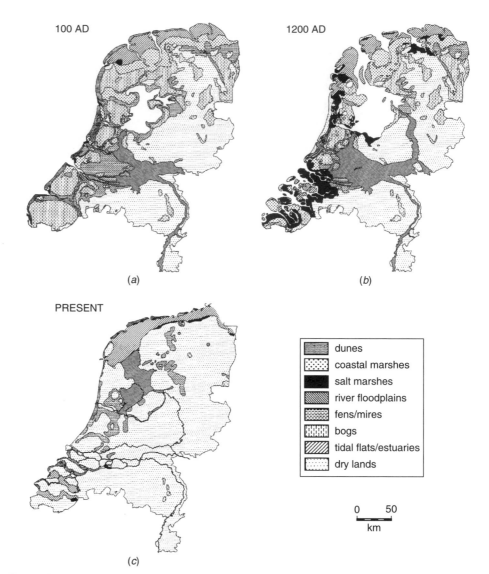

100 AD

1200 AD

(a)

(b)

PRESENT

	dunes
	coastal marshes
	salt marshes
	river floodplains
	fens/mires
	bogs
	tidal flats/estuaries
	dry lands

0   50
km

(c)

Figure 3.20   Estimated extent of wetlands in the present-day Netherlands and Rhine River delta in: (a) 100 A.D., (b) 1200 A.D., and (c) present day. *(From Wolff, 1993, pp. 3–5, reprinted with kind permission from Springer Science and Business Media.)*

of the country as internationally important wetlands with the Ramsar Convention on Wetlands of International Importance. Today, several governmental initiatives are designed to encourage some water to enter, or at least remain, on the lands, in great contrast to earlier Dutch traditions of controlling water in this close-to-sea-level environment.

Figure 3.21 Konik horses (descended from the Tarpan wild horses of Western Europe) are among the unusual features of the Oostvaardersplassen, one of the largest and best known created wetlands in the Netherlands. It was originally designed for industrial development and is now one of the Netherlands' best birding locations. *(Photograph by W. J. Mitsch.)*

Earlier in the 20th century, thousands of hectares were reclaimed from the Zuiderzee; today, some of these areas are reverting back to wetlands. For example, beginning in 1968, the Oostvaardersplassen in the Flevoland Polder, originally created as a site for industrial development, was artificially flooded in order to create a wildlife sanctuary. The 5,600-ha site is now a habitat for birds such as herons, cormorants, and spoonbills (250 bird species have been recorded there, 90 of which have bred there), as well as for Konik horses, descended from the original Tarpan wild horses of Western Europe (Fig. 3.21). Cattle have been crossbred from Scottish, Hungarian, and Camarguais breeds in an effort to re-create the original oxen of Europe. The Oostvaardersplassen is now one of the most popular places in the Netherlands for bird watching, and this wetland, less than 40 years old, has become a national treasure.

Coastal Marshes, Mud Flats, and Bays of Northern Europe

Extensive salt marshes and mud flats are found along the Atlantic Ocean and the North Sea coastlines of Europe from the Mira Estuary in Portugal to the Wadden Sea of the Netherlands, Germany, and Denmark. These marshes contrast with the extensive salt marshes of North America, which stretch from the Bay of Fundy in Canada to southern Florida and the Gulf of Mexico, in dominant vegetation, tidal inundation, and sediment transport. One of the better-known coastal wetland areas in France is at the Normandy–Brittany border near the world-famous abbey of Mont St. Michel, perched atop a promontory in a bay of the English Channel and accessible to pilgrims and tourists by day only, until the tides turn it into an island. Some of the most extensive salt marshes of Europe are found surrounding the abbey (Fig. 1.7). There has been a 60 percent drainage of coastal marshes since the beginning of the

Figure 3.22 Ancient drainage ditch several kilometers from the current shoreline of the Wadden Sea in Schleswig-Holstein, Germany. By developing drainage ditches, the Dutch and Germans encroached gradually on the Wadden Sea over the centuries, producing more arable land. *(Photograph by W. J. Mitsch.)*

20th century in this region, but now coastal wetlands are better protected, even though sheep grazing is still commonly practiced on these marshes. At nearby Le Vivier-sur-Mer, mussels are grown on *bouchots* (mussel beds created by sinking poles into the mud flats) in the shelter of a 30-km dike built in the 11th century.

The Wadden Sea, making up over 8,000 km^2 of shallow water, extensive tidal mud flats, marsh, and sand, is considered by some to be Western Europe's most important coastal wetlands. Over the past five centuries or more, drainage of the coastal land created hundreds of square kilometers of arable land as the local residents created more and more land out of the sea (Fig. 3.22). The wetlands extend for more than 500 km along the coasts of Denmark (10 percent), Germany (60 percent), and the Netherlands (30 percent), supporting North Sea fisheries, including almost 50 percent of the North Sea's sole, 60 percent of brown shrimp, 80 percent of plaice, and nearly all herring for some part of their life cycle (Dugan, 1993).

Numerous bays surround the Baltic Sea and adjacent seas in northern Europe, many with extensive wetlands, although most of the rivers that feed these brackish seas are relatively small. Matsalu Bay, a water meadow and reed marsh in northwestern

Estonia, has been known for years as a very important bird habitat. The wetland covers about 500 km², with much of that designated the Matsalu State Nature Preserve. Great numbers of birds, as many as 300,000 to 350,000, including swans, mallards, pintails, coots, geese, and cranes, stay in the Matsalu wetland during migration in the spring.

In many bays adjacent to southern Sweden, both in the Baltic Sea and in the Kattegat, there are significant eutrophication problems caused by excessive nitrogen from agricultural practices and from the loss of wetlands, drainage of streams, and destruction of riparian zones. For example, Lanholm Bay, a small (300 km²) bay in southwestern Sweden on the Kattegat, experienced significant blooms of *Cladophora* in the 1970s, which has affected fisheries and tourism in the region. This was followed by the development of a much more extensive low-oxygen condition (hypoxia) caused by excessive plankton production in large parts of the Kattegat (22,000 km²) in the 1980s (Fleischer et al., 1994). A goal of 50 percent reduction of nitrogen loading has been proposed for streams entering the coastal waters of Sweden, and restored wetlands are being considered as a cost-effective measure for reducing the nitrogen loads.

Southeastern Europe Inland Deltas

There are many important wetlands around the world that form not as coastal deltas, but as inland deltas or coastal marshes along large bodies of brackish and freshwater systems. There are several significant inland deltas in southeastern Europe. The 6,000-km² Danube River delta, one of the largest and most natural European wetlands, has been degraded by drainage and by activities related to agricultural development, gravel extraction, and dumping. The delta occurs where the Danube River spills into the Black Sea, spreading its sediments over 4,000 km². Plans to dike the delta and grow rice and corn ended when the communist regime of Nicolae Ceausescu ended in 1990 (Schmidt, 2001). Now there is significant international research in the delta, and plans continue for its restoration. Much of the restoration has been simple—restore the natural hydrology by breeching dams and reconnecting waterways. The Danube Delta supports 320 species of birds and is the home of white water lilies, oak-ash forests, and floating marshes of *Phragmites australis*.

On the edge of the Caspian Sea, the Volga River forms one of the world's largest inland deltas (19,000 km²), a highly "braided" delta over 120 km in length and spreading over 200 km at the sea's edge. The most extensive wetland area occurs in the delta of the Caspian Sea as the sea declined in water level, creating extensive *Phragmites* marshes and water lotus (*Nelumbo nucifera*) beds (Fig. 3.23). A large percentage of the world's sturgeon comes from the Caspian Sea, and the delta is a wintering site in mild winters for water birds and a major staging area for a broad variety of water bird, raptor, and passerine species. A series of dams destroyed the river's natural hydrology, and heavy industrial and agricultural pollution, as well as sea-level decline in the Caspian Sea, are making an impact.

Yet another lowland inland delta is the Colchis wetlands of eastern Georgia, a 13,000-km² region of subtropical alder (*Alnus glutinosa, A. barbata*) swamps and

Figure 3.23 Lotus bed in the Volga Delta, Russia. *(Photograph by C. M. Finlayson, reprinted by permission.)*

sedge–rush–reed marshes created by tectonic settling plus backwaters from the rivers discharging into the eastern Black Sea. This wetland is found in an area of great mythological interest because it is supposedly where Jason and the Argonauts (the Greek story of Argonautica as told by Apollonius) "hid their ship in a bed of reeds" (Grant, 1962) as they attempted to claim the Golden Fleece from the King of Colchis. These wetlands must indeed be long-lived!

European Peatlands

A good portion of the world's peatlands are found in the Old World, where peatlands spread across a significant portion of Ireland, Scandinavia, Finland, northern Russia, and many of the former Soviet republics. There are about 960,000 km^2 of peatlands in Europe, or about 20 percent of Europe. About 60 percent of those peatlands have been altered for agriculture, forestry, and peat extraction (Vasander et al., 2003), and about 25 percent of the peatlands are in the Baltic Sea Basin. The Endla Bog in Estonia (Fig. 3.24) and the Berezinski Bog in Byelorussia are but two examples of many peatlands that have been protected as nature preserves and are in semi-natural states in this region of Europe. The 76,000-ha Berezinski reservation in northeastern Byelorussia is over half peatland and predominantly forested peatland dominated by pine (*Pinus*), birch (*Betula*), and black alder (*Alnus*).

Africa

An abundance of wetlands are found in sub-Saharan Africa (Fig. 3.25). The vast size of some of these major wetlands is far beyond what the Western world experiences,

Figure 3.24   The Endla Bog in central Estonia. *(Photograph by W. J. Mitsch.)*

with examples such as the Inner Niger Delta of Mali (320,000 km² when flooded), the Congolian Swamp Forests (190,000 km²), the Sudd of the Upper Nile (more than 30,000 km² when flooded), and the Okavango Delta in Botswana (28,000 km²).

Okavango Delta

One of the great seasonally pulsed inland deltas of the world, the Okavango Delta (28,000 km²) forms at the convergence of the Okavango River and the sands of the Kalahari in Botswana, Africa (Fig. 3.26). The wetland has a dramatic seasonal pulse with the water surface expanding from 2500 to 4000 km² in February to March to a peak of 6000 to 12,000 km² in August to September (McCarthy et al., 2004; Ramberg et al., 2006a,b; Ringrose et al., 2007). The system is thus divided into three major hydrologic zones: permanent swamp, seasonally flooded floodplains, and occasional floodplains (Fig. 3.26a). There is little to no surface outflow from this inland delta, and infiltration to the groundwater from the seasonally floodplain is very rapid during the 90 to 175 days of flooding (Ramberg et al., 2006a). The wetland has a web of channels, islands, and lagoons supporting crocodiles, elephants, lions, hippos, and water buffalo, and more than 400 bird species (Fig. 3.26b). Several species of tilapia and bream spawn in the Okavango Delta, contributing to the 71 species of fish found in the streams and floodplains of the delta (Ramberg et al., 2006b). Many of northern Botswana's diverse tribes find refuge there. Most of the inhabitants depend on the delta's resources. Like so many other wetlands, however, the Okavango is threatened by the increased burning (fires are natural in the Okavango), clearing associated with crop production and livestock grazing, and possible plans by upstream countries to use some of the Okavango River water. Tourism is an issue here, too, as in many other wetland sites; ecotourism is the largest single employer in Maun, located on the edge of the delta. Maun also benefits economically from the wetland's

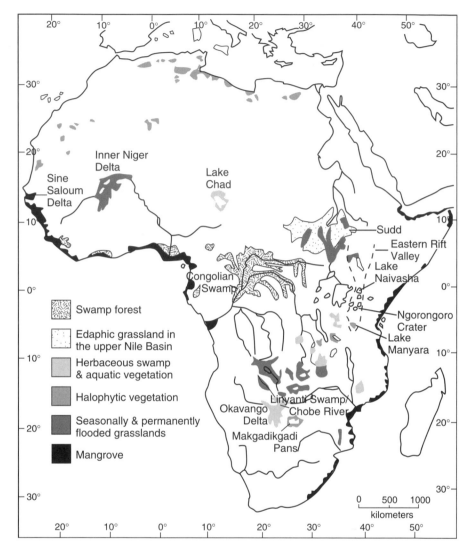

Figure 3.25   Map of major wetland areas of Africa.

water lily tubers, bulrush roots, palm hearts, and palm wine, made from the sap of the *Hyphaene* palm. Fencing, roofing, and wall materials are also derived from the wetlands.

Congolian Swamp Forests

This 190,000-km^2 region in Congo and the Democratic Republic of Congo (formerly Zaire) is one of the largest yet least studied swamp forests in the world (Campbell, 2005). This freshwater tropical African wetland includes swamp forests, flooded

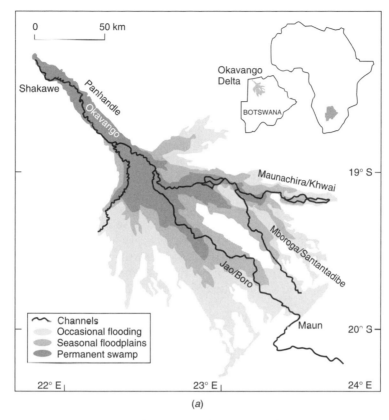

(a)

(b)

Figure 3.26 The Okavango Delta of Botswana, southern Africa: (a) Map of wetland showing permanent wetlands and seasonal and occasional floodplains; and (b) Photo of floating-leaved aquatics, mostly the day water lily *Nymphaea nouchali* var. *caerulea* and the African Jacana (*Actophilornis africanus*) in a permanently flooded stream. *(Map from Ramberg et al., 2006a; photograph by W. J. Mitsch.)*

savannas, and floating prairies on its rivers and streams. The Congolian Swamp Forest is found on the banks of the middle reaches of the Congo River in a large depression in equatorial Africa called the *cuvette centrale congolaise*. The Congo River has the second highest flow of any river in the world and, along with its tributaries, provides the water that supports these forested alluvial swamps. In the wet season, the forests are flooded to a depth of 0.5 to 1.0 m; during the dry season, they often lack standing water. Human population is low in the region, and the people that live in the region are involved in hunting and fishing in the forest and its rivers. The eastern portion of the Congolian Swamp Forest in the Democratic Republic of Congo is generally thought to be more diverse than the western region in the Congo. Large tracts of the forest remain free of logging because of their relative isolation (Minnemeyer, 2002).

The swamp forest has significant populations of western lowland gorilla (*Gorilla gorilla gorilla*) and forest elephants (*Loxodonta africana cyclotis*) along its waterways. The area also supports forest buffalo (*Sycerus caffer nanus*) and chimpanzee, and many bird species, although there are only are two near-endemic bird species: the African river-martin and the Congo martin. Lac Télé-Likouala-aux-Herbes Reserve, a 4,390-km^2 Ramsar site in the Republic of Congo, is the home of a mythical 9-m-long dinosaur-like animal called Mokele Mbembe (WWF, 2001).

East Africa Tropical Marshes

Several wetlands that form around tropical lakes in Africa are typical of what are called "swamps" in Old World usage of the word but would be "marshes" in New World terminology (see Chapter 2). These highly productive wetland margins tend to be dominated by tropical species of cattail (*Typha domingensis*) and papyrus (*Cyperus papyrus*), often with mats of floating plants (*Eichhornia crassipes* and *Salvinia molesta*). Many lakes and wetlands are found along the 6,500-km Rift Valley of eastern Africa. Not far from Nairobi on the floor of the Rift Valley, Lake Naivasha is one of the most studied tropical lakes in East Africa. The area provides a home for nearly the entire range of ducks and herons found in eastern Africa. Vegetation changes in Lake Naivasha have been caused by a combination of water-level fluctuations, the introduction of crayfish (*Procambarus clarkii*), and the physical effects of floating rafts of *Eichhornia crassipes* (Harper et al., 1995).

Other vast wetlands are found in the Rift Valley in northern Tanzania, including the shorelines of Lake Manyara (Fig. 3.27) and the wetlands of Ngorongoro Crater (Fig. 3.28). Two swamps, Mandusi Swamp and Gorigor Swamp, and one lake, Lake Makat, are found in the caldera. The abundant wildlife of Ngorongoro was summarized by the East African/German conservationist Bernhard Grzimek, who stated, "It is impossible to give a fair description of the size and beauty of the Crater, for there is nothing with which one can compare it. It is one of the wonders of the World" (Hanby and Bygott, 1998).

West Africa Mangrove Swamps

Extensive mangrove swamps are found on Africa's tropical and subtropical coastlines. One example on the Atlantic Ocean west coast of Africa, about 150 km south of

Figure 3.27 Wildlife is abundant in the Rift Valley lakes and wetlands. This photo, showing wildebeests, monkeys, and yellow-billed storks (*Ibis Ibis*), is along Lake Manyara, Tanzania, one of the southernmost lakes along the Rift Valley. *(Photograph by W. J. Mitsch.)*

Figure 3.28 Waterfowl in the wetlands of the Ngorongoro Crater, northern Tanzania, including Yellow-billed Duck (*Anas undulata*), Red-billed Duck (*A. erythrorhynchos*), and Egyptian Goose (*Alopochen aegyptiaca*). *(Photograph by W. J. Mitsch.)*

Figure 3.29 **African Reef Heron (*Egretta gularis*) in mangrove prop roots in Sine Saloum Delta, Senegal. (*Photograph by W.J. Mitsch.*)**

Dakar, is the Sine Saloum Delta in Senegal (Vidy, 2000), a vast (180,000 ha) almost untouched expanse of mangrove swamps (Fig. 3.29). These mangrove swamps, as well as similar mangroves at the Senegal River delta around St. Louis to the north, and in the coastal reaches of the Gambia River in Gambia to the south, support a wide variety of bird life, mammals, and four species of breeding turtles in what is otherwise an extremely arid climate. Birds included several species of herons and egrets as well as the Great While Pelican (*Pelecanus onocrotalus*). To the east, where the the delta meets the arid uplands, salt pans or "tannes" develop where little vegetation is supported because of excessive salinities. The mangrove system is distinguished by the lack of permanent river flow, related to the Sahelian drought dating to the 1970s, and for that reason, the Sine Saloum is termed a "reverse estuary," meaning that salinity increases going upstream. But desertification is also due to mismanagement of natural resources. The region is only lightly populated, and local people support themselves with fishing, salt production, and peanut farming. The mangrove forests have suffered from overexploitation for the wood they provide for housing and charcoal as well as from conversion to rice fields. UNESCO and other international agencies are encouraging both ecotourism in the region and adaptation of oyster farming techniques to better fit the mangrove system, and the creation of village "green belts." Many shell islands built well above the intertidal zone are found throughout the delta and attest to Neolithic civilizations in this delta indicating a long human history here.

Figure 3.30   The Mesopotamian marshland in southern Iraq. *(Photograph by Azzam Alwash, reprinted with permission.)*

Middle East

Mesopotamian Marshlands

The crown jewel wetlands of the Middle East are the Mesopotamian marshlands of southern Iraq and Iran. These wetlands are in an arid region of the world and exist at the confluence of the Tigris and Euphrates rivers. The watersheds of both the Euphrates and the Tigris are predominantly in the countries of Turkey, Syria, and Iraq. The Tigris–Euphrates Basin has had water control projects for over six millennia. The Mesopotamian wetlands (Fig. 3.30) are the largest wetland ecosystem in the Middle East, have been the home to the Marsh Arabs for 5,000 years (see Fig. 1.3), and support a rich biodiversity (UNEP, 2001). Since 1970, the wetlands have been damaged dramatically. The Mesopotamian wetlands once were 15,000 to 20,000 km^2 in area but were drained in the 1980s and 1990s to less than 10 percent of that extent. There was a 30 percent decline just in the period 2000 to 2002. The draining of the wetlands was the result of human-induced changes. Upstream dams and drainage systems constructed in the 1980 and 1990s drastically altered the river flows and have eliminated the flood pulses that sustained the wetlands (UNEP, 2001). Turkey alone constructed more than a dozen dams on the upper rivers. But the main cause of the disappearance of the wetlands was water control structures built by Iraq from 1991 to 2002 (Altinbilek, 2004).

These marshes, dominated by *Phragmites australis*, are located on the intercontinental flyway of migratory birds and provide wintering and staging areas for waterfowl. Two-thirds of West Asia's wintering waterfowl have been reported to live in the marshes. Globally threatened wildlife, which have been recorded in the marshes, include 11 species of birds, 5 species of mammals, 2 species of amphibians and reptiles, 1 species of fish, and 1 species of insect. The drying of the marshes has had a devastating effect on wildlife (UNEP, 2001). Restoration efforts in these marshlands are discussed in Chapter 12.

Australia/New Zealand

Eastern Australian Billabongs

Australia's wetlands are distinctive for their seasons of general dryness caused by high evaporation rates and low rainfall. Wetlands do occur on the Australian mainland, but only where the accumulation of water is possible, generally on the eastern and western portions of the continent. Thus, there are not many permanent wetlands—most are intermittent and seasonal. Furthermore, because of the high evaporation rates, saline wetlands and lakes are not uncommon. A particular feature in eastern Australia is the *billabong* (Shiel, 1994), a semipermanent pool that develops from an overflowing river channel (Fig. 3.31). Although found throughout Australia, billabongs are best concentrated along the Murray and Darling rivers in southeastern Australia. There are about 1,400 wetlands representing 32,000 ha in four watersheds alone in New South Wales. Billabongs support a variety of aquatic plants, are a major habitat for birds and fish, and are often surrounded by one of many species of eucalyptus, especially the river red gum *Eucalyptus camaldulensis*. The billabongs serve as refuges for aquatic animals during the dry season, when the rivers come close to drying.

Figure 3.31   A billabong of New South Wales, Australia, showing bulrushes, river red gum *Eucalyptus camaldulensis* in the background, and invasive *Salvinia molesta* on the water's surface.

Figure 3.32   Freshwater wetland in the Swan Coastal Plain, Western Australia. *(Photograph by J. Davis, reprinted by permission.)*

Western Australia Wetlands

The Mediterranean-type climate of southwestern Australia docs, in fact, favor a wide variety of wetlands, which are especially important to the waterfowl that are separated from the rest of the continent by vast expanses of desert. Swamps arc numerous, and many can be found inland or just above the saline wetlands of the tidal rivers and bays of the Swan Coastal Plain, near Perth (Fig. 3.32). Nevertheless, it is estimated that 75 percent of the wetlands in the Swan Coastal Plain in southwestern Australia have been lost (Chambers and McComb, 1994).

New Zealand Wetlands

For a small country, New Zealand has a wide variety of wetland types (Johnson and Gerbeaux, 2004). However, New Zealand has lost 90 percent of its wetlands, amounting to over 300,000 ha. The western region of South Island, called Westland, is sometimes humorously called "Wetland" because of the enormous amount of rain it receives (2–10 m annually) due to its location between the Tasman Sea to the west and the Southern Alps to its east. It is thus, not surprisingly, the location of a great variety of coastal wetlands. Grand Kahikatea (*Dacrycarpus dacrydiodes*) or "white pine" forested wetlands (Fig. 3.33), reminiscent of the bald cypress swamps of the southeastern United States, are found throughout Westland and also on North Island. *Pakihi* (peatlands) are found on both North Island and South Island.

One of the largest wetlands in North Island is Whangamarino Wetland (Fig. 3.34), a 7,300-ha peatland and seasonally flooded swamp adjacent to the Waikato River, New Zealand's largest river (Clarkson, 1997; Shearer and Clarkson, 1998).

Figure 3.33   Kahikatea Swamp in the background with Okarito Lagoon in the foreground in western New Zealand. The Kahikatea tree (*Dacrycarpus dacrydiodes*) is locally called "white pine." These white pine forests once dominated both coastlines of New Zealand. *(Photograph by W. J. Mitsch.)*

Figure 3.34   Peatland in the lower Waikato River basin, about 60 km south of Aukland, New Zealand. Circular ponds with earthen paths are hunting ponds. *(Photograph by W. J. Mitsch.)*

Management issues facing this and other peatlands in the area are reduced inundation by the river, silt deposition from agricultural development, increased fire frequency over presettlement times, and invading willows and other exotics. Flax (*Phormium tenax*) swamps and raupo (*Typha orientalis*) marshes are also common in New Zealand. Willows (*Salix* spp.) are generally considered undesirable woody invaders to many of these wetlands.

Asia

Western Siberian Lowlands

One of the largest contiguous wetland areas in the world is the region of central Russia bordered by the Kara Sea of the Arctic Ocean to the north, the Ural Mountains to the west, Kazakhstan to the south. The area is referred to the Western Siberian Lowland and encompasses about 2.7 million km^2, about 787,000 km^2 of which is peatland (Solomeshch, 2004). The region also has more than 800,000 lakes. Precipitation is relatively low (<600 mm/yr) but evapotranspiration is even lower (<400 mm/yr), leading to excess moisture that creates the peatlands. Part of this region includes the Bi-Ob' region of central Russia, a large floodplain on the Ob River between Kazakhstan to the south and the Ob River's estuary to the north on the Kara Sea. This valley of channels, floodplain lakes, and river distributaries is actually an inland delta caused more by decreased sea levels than by deposited sediments. The region has been described as "the largest single breeding area for waterfowl in Eurasia" (Dugan, 1993). One of the greatest values of these peatlands could be carbon sequestration. It has been estimated that these wetlands alone have an average carbon accumulation of 22.8 Tg/yr or about 24 to 35% of the global accumulation rate of all northern peatlands (Solomeshch, 2004).

Indian Freshwater Marshes

The world's second most populous nation is India. It is slightly more than one-third the size of the United States, but has more than three times as many people. Droughts, soil erosion, overgrazing, and desertification are common. Agriculture employs two-thirds of the labor force, based in and around the alluvial plains and coastal zones on 55 percent of the land. The wetlands are under intense pressure for farm expansion, water control, and urbanization. Flooding cycles on alluvial valleys have been aggravated by these developments, resulting in "natural" disasters to humans and habitat alike. A few conservation wetlands remain under moderate protection, sometimes as remnants of the formerly upper-class lands. Keoladeo National Park in Bharatpur (Fig. 3.35) is an example, where the hunting reserve is now a protected area of international significance. About 850 ha of the park are wetlands. The local economy benefits from tourism and also collects or illegally harvests products from the area. The protected wildlife heritage includes migratory species from northern Asia. In all, more than 350 species of birds, 27 mammals, 13 amphibians, 40 fish, and 90 wetland flowering plants are found in the park (Prasad et al., 1996).

Figure 3.35   Keoladeo National Park, Bharatpur, India, during flooding season. *(Photograph by B. Gopal, reprinted with permission.)*

Southeast Asian River Deltas

More than 80 percent of Asian wetlands are located in seven countries: Indonesia, China, India, Papua New Guinea, Bangladesh, Myanmar, and Vietnam. The diversity of Asia's wetlands is reflected in its intertidal mud flats, swamp forests, natural lakes, open marshes, arctic tundra, and mangrove forests (recognized as one of the most productive ecosystems in the world—yielding over 70 direct and indirect uses of the forest or its products—but now threatened by logging). The snowfields and glaciers of the Himalayas are the birthplace of many of the world's well-known rivers, including the Ganges, the Indus, the Mekong, and the Yangtze. The Mekong, Southeast Asia's longest river, begins in the Tibetan Plateau, enters its lower basin at the boundary of Myanmar, Laos, and Thailand, and then flows to the ocean through one of the world's great deltas. The basin catchment area is more than 600,000 km^2 and includes Laos, Cambodia, Thailand, and Vietnam. There has been little coordination among these countries concerning the basin's management, especially with regard to the extensive wetlands in the Mekong Delta region. Problems stemming from devegetation and drainage during the war years have been exacerbated by more recent efforts at agricultural intensification, urbanization, industrialization, and dam and reservoir construction. Even drained soil became acidic (pH < 3) when sulfur-rich soils oxidized, making them unsuitable for agriculture. Current research is focused on the mangrove wetlands of the Mekong Delta, especially because they are an important source of fuel and medicine. Restoration of a freshwater portion of the

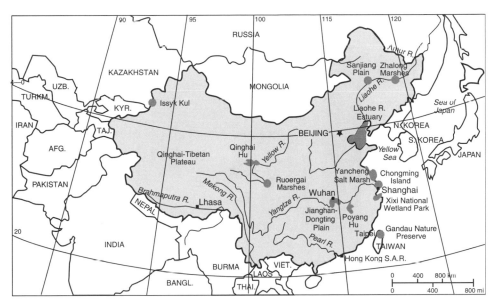

Figure 3.36   Wetlands of China discussed in this chapter.

Mekong Delta, known as the Dong Thap Muoi ("Plain of Reeds"), continues with international assistance.

Wetlands of China

The total area of wetlands in China ranks as Asia's highest (Lu, 1990; Chen, 1995), with an estimated 625,000 km² of wetlands, 250,000 km² of which are natural, with the rest artificial wetlands such as rice paddies and fish ponds. Natural wetlands thus comprise about 2.5 percent of the country (Lu, 1995). There are several important wetland sites in China, and some of them are discussed here (Fig. 3.36). Few of the wetlands are preserved in semi-pristine conditions as is done in the West as habitat alone; most wetlands in China provide fish, cattle, grain, duck, and other food, as well as habitat and recreation benefits, in a symbiotic relationship between humans and nature. However, an estimated 40 percent of the wetlands in China that have been designated as wetlands of international importance are "under moderate to severe threat from conversion to agricultural land, increased siltation due to catchment degradation, pollution, over-fishing, and hunting" (Parish and Elliott, 1990).

Many of the important wetlands of China are found in the lower and delta regions of the Changjiang (Yangtze) River (Fig. 3.37), the Zhujiang (Pearl) River, and the Liaohe River. Because these regions are among the most populated in the world, very few natural wetlands remain, having been converted to rice paddies or fish ponds. But many new wetlands are being created by accretion of sediments, such as those on the downstream (east) and upstream (west) coasts of 1400-km² Chongming Island in the Yangtze Delta in Shanghai (Fig. 3.38). Chongming Island is the third-largest island

Figure 3.37 Aquaculture site in the reeds of the Taihu (Tai Lake) shoreline in the lower Yangtze River delta of Eastern China. *(Photograph by W. J. Mitsch.)*

Figure 3.38 Marshes on the eastern extent of Chongming Island in the Yangtze River near Shanghai, China. *(Photograph by W. J. Mitsch.)*

Figure 3.39   Xixi National Wetland Park, Hangzhou China. *(Photograph by W. He, reprinted with permission.)*

in China and supports a human population of 600,000. In many cases, wetlands and reed (*Phragmites*) fields are connected hydrologically with fish ponds and rice paddies to enhance food and fiber production (Ma et al., 1993).

China has become interested in establishing urban wetland parks that provide scenery and relaxation for the public in an aquatic setting. Xixi National Wetland Park (Fig. 3.39), located in the suburbs of Hangzhou China in eastern China, covers an area of about 350 ha. The park includes streams for boating and several marshes, swamps, and ponds. It is the first formal national wetland park in China and provides a semi-natural water-land park opened to the public. Much of the park is restored on former fish ponds and rice paddies.

In contrast to the marshes found on the European and eastern North American coastlines, many of China's coastal marshes are of more recent origin. As described by Chung (1994), both *Spartina anglica* and *S. alterniflora* were introduced in China in an effort to restore the eastern coastline, particularly near Yancheng in Jiangsu province. After 30 years of experiments and full-scale projects, many of these salt marsh plantations have proved to be successful, both ecologically and economically. Almost 40,000 ha have been restored by Professor Chung-Hsin Chung and his colleagues at Nanjing University for accretion for agricultural use, for mitigation of saline soils, as a source of green manure and animal fodder, and for seashore stabilization (Qin et al., 1998). Now there is some second guessing because *Spartina alterniflora* is replacing the native *Phragmites australis* in many brackish marshes; it is now viewed as an invasive species and is being controlled or monitored in some coastal wetlands (He et al., 2007).

There are also extensive inland wetlands associated with the Yangtze River, particularly in the Jianghan–Dongting Plain in the middle of the river valley near Wuhan, Hubei Province, and Poyang Hu in northern Jiangxi province. The Jianghan–Dongting Plain is approximately 10,000 km^2 of former marshes and lakes that has been extensively drained and diked, yet consistently suffers crop damage due to excessive water. Integrating these backwater areas with the Yangtze River as they once were is probably impossible, but converting wet areas from rice and other crops to wetland crops such as lotus (*Nelumbo nucifera*) and wild rice stem (*Zizania latifolia*) has been suggested as a viable "ecological" approach (Bruins et al., 1998). Poyang Hu is the largest lake in China but varies considerably with season. The lake shrinks to less than 1,000 km^2 in the dry season and grows to 4,000 km^2 in the late summer rainy season. The lake is connected to the Yangtze River with a 1-km-long channel that allows natural overflow. The lake's basin is one of China's most important rice-producing regions, but because of regular flooding, Jiangxi Province is among the poorest in China.

China has several major inland wetland areas in the western mountain and desert regions. The sources of several major rivers can also be found in the Qinghai–Tibetan Plateau—the Yellow, the Yangtze, the Indus, and the Ganges—along with high-altitude lakes and bogs. Most of the plateau's larger lakes are saline, and, at 458,000 ha, Qinghai Hu (Lake) is the largest (Fig. 3.40). As this whole area is experiencing desiccation, the lakes are shrinking, most recently at a rate of 12 cm per year in depth. Nevertheless, these wetlands are habitat for millions of migratory and resident birds comprising over 160 species (Lu, 1990). One of the world's great mountain lakes, Issyk Kul, is a brackish wetland lying in a basin of the Tianshan mountain chain along the China–Kazakhstan border (Fig. 3.41). Some of these wetlands, while important havens for migratory waterfowl, are also important for regional fisheries and for supplying reeds used in the manufacture of paper (Lu, 1990).

Urban Wetland Park, Taiwan

Similar to the emerging urban wetland parks in China, a site worth mentioning where wetlands have been brought before a large urban population is the 57-ha Gandau Nature Park in Taipei, Taiwan (Fig. 3.42). The wetland site, which is very popular with local environmental and bird-watching groups, includes a major bird-watching gallery and several paved and unpaved pathways. It forms along a major bend of the Keelung River in Taipei. The wetland supports mostly created freshwater wetland ponds at the Gandau Nature Park and several hectares of saline mangrove forest in the adjacent Gandau nature reserve along the river.

Figure 3.40   A bird Island in Qinghai Hu, western China. This lake is in an arid region of China and is the largest saltwater lake in the country. *(Photograph by J. Lu, reprinted by permission.)*

Figure 3.41   River and wetlands in the desert areas of Xinjiang, China. The river is fed from the Tianshan Mountains. *(Photograph by J. Lu, reprinted by permission.)*

Figure 3.42 View from nature center building at Gandau Park, Taipei, Taiwan. *(Photograph by W. J. Mitsch.)*

Recommended Readings

Dugan, P. 1993. *Wetlands in Danger*. Michael Beasley, Reed International Books, London. 192 pp.

Fraser, L.H. and P.A. Keddy, eds. 2005. *The World's Largest Wetlands: Ecology and Conservation*. Cambridge University Press, Cambridge, UK. 488 pp.

Grunwald, M. 2006. *The Swamp: The Everglades, Florida, and the Politics of Paradise*. Simon & Schuster, New York. 450 pp.

Lodge, T. E. 2005. *The Everglades Handbook: Understanding the Ecosystem*, 2nd ed. CRC Press, Boca Raton, FL. 302 pp.

McComb, A. J., and P. S. Lake. 1990. *Australian Wetlands*. Angus and Robertson, London. 258 pp.

Mitsch, W. J., ed. 1994. *Global Wetlands: Old World and New*. Elsevier, Amsterdam. 967 pp.

Part 2

The Wetland Environment

Wetland Hydrology

Hydrologic conditions are extremely important for the maintenance of a wetland's structure and function. They affect many abiotic factors, including soil anaerobiosis, nutrient availability, and, in coastal wetlands, salinity. These, in turn, determine the biota that develop in a wetland. Finally, completing the cycle, biotic components are active in altering the wetland hydrology and other physicochemical features. The hydroperiod, or hydrologic signature of a wetland, is the result of the balance between inflows and outflows of water (called the water budget), the wetland basin geomorphology, and the subsurface conditions. The hydroperiod can have dramatic seasonal and year-to-year variations, yet it remains the major determinant of wetland processes. The major components of a wetland's water budget include precipitation, evapotranspiration, overbank flooding in riparian wetlands, other surface flows, groundwater fluxes, and tides or seiches in coastal wetlands. Simple determinations of the hydroperiod, water budget, and turnover time in wetland studies can contribute to a better understanding of wetland function. Hydrology affects species composition and richness, primary productivity, organic accumulation, and nutrient cycling in wetlands. In general, productivity is high in wetlands that have high flow-through of water and nutrients or in wetlands with pulsing hydroperiods. Decomposition in wetlands is slower in anaerobic standing water than it is under wet-dry conditions. Although many wetlands export organic carbon, this cannot be generalized even within one wetland type.

The hydrology of a wetland creates the unique physiochemical conditions that make such an ecosystem different from both well-drained terrestrial systems and deepwater aquatic systems. Hydrologic pathways such as precipitation, surface runoff, groundwater, tides, and flooding rivers transport energy and nutrients to and from wetlands. Water depth, flow patterns, and duration and frequency of flooding, which are the result of all of the hydrologic inputs and outputs, influence the biochemistry of the soils and are major factors in the ultimate selection of the biota of wetlands. Biota ranging from microbial communities to vegetation to waterfowl are all constrained or enhanced by hydrologic conditions. An important point about wetlands—one that is often missed by ecologists who begin to study these systems—is this: *Hydrology is probably the single most important determinant of the establishment and maintenance of specific types of wetlands and wetland processes.* An understanding of rudimentary hydrology should be in the repertoire of any wetland scientist.

The Importance of Hydrology in Wetlands

Wetlands are transitional between terrestrial and open-water aquatic ecosystems. They are transitional in terms of spatial arrangement, for they are usually found between uplands and aquatic systems (see Fig. 2.1a). They are also transitional in the amount of water they store and process, and in other ecological processes that result from the water regime. Wetlands form the aquatic boundary of the habitats of many terrestrial plants and animals; they also form the terrestrial edge for many aquatic plants and animals. Hence, small changes in hydrology can result in significant biotic changes.

The starting point for the *hydrology* of a wetland is the climate and basin geomorphology (Fig. 4.1). All things being equal, wetlands are more prevalent in cool or wet climates than in hot or dry climates. Cool climates have less water loss from the land via evapotranspiration, whereas wet climates have excess precipitation. The second important factor is the geomorphology of the landscape and basin. Steep terrain tends to have fewer wetlands than flat or gently sloping landscapes. Isolated basins have different potential for wetlands than do tidal-fed or river-fed environments. When climate, basin geomorphology, and hydrology are considered as one unit, it is referred to as a wetland's *hydrogeomorphology*. Fig. 4.1 illustrates that the hydrology of a wetland directly modifies and changes its *physiochemical environment* (chemical and physical properties), particularly oxygen availability and related chemistry such as nutrient availability, pH, and toxicity (e.g., the production of hydrogen sulfide). Hydrology also transports sediments, nutrients, and even toxic materials into wetlands, thereby further influencing the physiochemical environment. Except in nutrient-poor wetlands such as bogs, water inputs are the major source of nutrients to wetlands. Hydrology also causes water outflows from wetlands that often remove biotic and abiotic material, such as dissolved organic carbon, excessive salinity, toxins, and excess sediments and detritus. Some modifications in the physicochemical environment, such as the buildup of sediments, can modify the hydrology by changing the basin geometry or affecting the hydrologic inflows or outflows (pathway A in Fig. 4.1).

Modifications of the physiochemical environment, in turn, have a direct impact on the *biota* in the wetland. When hydrologic conditions in wetlands change even slightly, the biota may respond with massive changes in species composition and richness and in ecosystem productivity. Biota such as emergent aquatic plants adapt

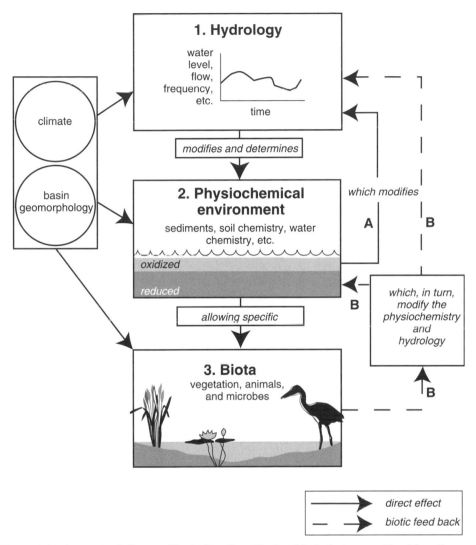

Figure 4.1 Conceptual diagram illustrating the effects of hydrology on wetland function and the biotic feedbacks that affect wetland hydrology. Pathways A and B are feedbacks to the hydrology and physiochemistry of the wetland.

to the anoxia in the sediments, although most vascular plant species are excluded by the anoxia. The level of nutrients in the sediments determines productivity and which species will dominate that productivity. Animals adapted to shallow water and this vegetation cover will, in turn, flourish. Microbes able to metabolize in anoxic conditions dominate the reduced sediments, while aerobic microorganisms survive in a thin layer of oxidized sediments and in the water column if oxygen is present there. When hydrologic patterns remain similar from year to year, a wetland's biotic structural and functional integrity may persist for many years.

Biotic Control of Wetland Hydrology

Just as many other ecosystems exert feedback (cybernetic) control of their physical environments, wetland biota are not passive to their hydrologic conditions. Feedback loop B in Figure 4.1 shows that the biotic components of wetlands, in turn, can control the hydrology and chemistry of their environment through a variety of mechanisms. Microbes, in particular, catalyze virtually all chemical changes in wetland soils and, thus, control nutrient availability to plants and even the production of phytotoxins such as sulfides. Plants, animals, and microbes that use these essential biological feedback mechanisms have been formally recognized in the ecological literature as *ecosystem engineers* (Jones et al., 1994, 1997; Alper, 1998). Plants cause changes in their physical environment through processes such as peat building, sediment trapping, nutrient retention, water shading, and transpiration. Wetland vegetation influences the hydrologic conditions of the physicochemical environment by binding sediments to reduce erosion, by trapping sediments, by interrupting water flows, and by building peat. Accumulated sediments and organic matter, in turn, interrupt water flows and can eventually decrease the duration and frequency by which the wetlands are flooded. Bogs build peat to the point at which they are no longer influenced at the surface by the inflow of mineral waters. Some trees in some southern swamps save water by their deciduous nature, their seasonal shading, and their relatively slow rates of transpiration. In more temperate climates, trees that invade shallow marshes and vernal pools can decrease water levels during the growing season by increasing transpiration, thus allowing even more woody plants to take over. Removal of these trees in what appears to be a dry forest sometimes surprisingly causes standing water and marsh vegetation to reappear.

Several animals are particularly noted for their contributions to hydrologic modifications and subsequent changes in wetlands. The exploits of beavers (*Castor canadensis*) in much of North America in both creating and destroying wetland habitats are well known. They build dams on streams, backing up water across great expanses and creating wetlands where none existed before. In colonial times, beaver populations covered the entire American continent north of Mexico before they were drastically reduced by fur trappers. Beavers have been an important causal force in the creation of the Great Dismal Swamp of Virginia and North Carolina and are even suggested as a factor contributing to altered global carbon biogeochemistry (Johnston, 1994). Hey and Philippi (1995) estimated that a population of 40 million beavers could have accounted for 207,000 km² of beaver ponds (wetlands) in the upper Mississippi and Missouri river basins before European trappers entered the region and that, with the demise of the beaver, only 1 percent of those beaver ponds exist today and probably a similar percentage of wetlands remain as well.

Muskrats (*Ondatra zibethicus*) burrow through wetlands, changing flow patterns and sometimes water levels directly. They harvest large amounts of emergent vegetation for their food and to build winter lodges, thereby opening up large areas of marshes. Geese, especially Canada geese (*Branta canadensis*) and several varieties of Snow geese (*Chen* spp.), cause *eat-outs*, or major wetland vegetation removal by herbivory, in many parts of the world. Newly planted wetlands are particularly susceptible to Canada geese eat-outs in North America. By removing vegetation cover, these herbivores reset the successional status of the wetlands and, thus, have a major impact on wetland hydrology.

The American alligator (*Alligator mississippiensis*) is known for its role in the Florida Everglades in constructing "gator holes" that serve as oases for fish, turtles, snails, and other aquatic animals during the dry season. In all of these cases, the biota of the ecosystem have contributed to their own survival, to the survival of other species, and to the elimination of others by influencing the ecosystem's hydrology and other physical characteristics.

Studies of Wetland Hydrology

Many early wetland investigations that dealt with hydrology explored the relationships between hydrologic variables (usually water depth) and wetland productivity or species composition. Many case studies of the hydrology of wetlands have been published; fewer studies have described in detail the hydrologic characteristics within specific wetland types. An exception to this has been the study of northern peatlands, for which a wealth of literature exists, including work from the former Soviet Union, the British Isles, and North America, particularly the Lake Agassiz region of northern Minnesota. Some of the more notable hydrology studies for other types of wetlands in the United States have included salt marshes in New England (Hemond and Fifield, 1982), cypress swamps in Florida (Heimburg, 1984), large-scale wetland complexes at the Okefenokee Swamp in Georgia (Rykiel, 1984; Hyatt and Brook, 1984), and the Okavango Delta of Botswana (Ramberg et al., 2006a). A major contribution to the understanding of wetland hydrology resulted from papers published from a 1989 symposium on wetland hydrogeology held at the 28th International Geological Congress in Washington, D.C. (Winter and Llamas, 1993).

Wetland Hydroperiod

The *hydroperiod* is the seasonal pattern of the water level of a wetland and is the wetland's hydrologic signature. It characterizes each type of wetland, and the constancy of its pattern from year to year ensures a reasonable stability for that wetland. It defines the rise and fall of a wetland's surface and subsurface water by integrating all of the inflows and outflows. The hydroperiod is also influenced by physical features of the terrain and by proximity to other bodies of water.

Many terms are used to describe qualitatively a wetland's hydroperiod (Table 4.1). These terms such as *seasonally flooded* or *intermittently flooded* are specific in their meaning and should be used with care and with sufficient data in describing a wetland's hydroperiod. For wetlands that are not subtidal or permanently

Table 4.1 Definitions of wetland hydroperiods

TIDAL WETLANDS

Subtidal—permanently flooded with tidal water
Irregularly exposed—surface exposed by tides less often than daily
Regularly flooded—alternately flooded and exposed at least once daily
Irregularly flooded—flooded less often than daily

NONTIDAL WETLANDS

Permanently flooded—flooded throughout the year in all years
Intermittently exposed—flooded throughout the year except in years of extreme drought
Semipermanently flooded—flooded during the growing season in most years
Seasonally flooded—flooded for extended periods during the growing season, but usually no surface
  water by end of growing season
Saturated—substrate is saturated for extended periods during the growing season, but standing
  water is rarely present
Temporarily flooded—flooded for brief periods during the growing season, but water table is
  otherwise well below surface
Intermittently flooded—surface is usually exposed with surface water present for variable periods
  without detectable seasonal pattern

Source: After Cowardin et al. (1979).

flooded, the amount of time that a wetland is in standing water is called the *flood duration*, and the average number of times that a wetland is flooded in a given period is known as the *flood frequency*. Both terms are used to describe periodically flooded wetlands such as coastal salt marshes and riparian wetlands.

Typical hydroperiods for a diverse set of wetlands are shown in Figure 4.2. A coastal salt marsh has a hydroperiod of semidiurnal flooding and dewatering superimposed on a twice-monthly pattern of spring and ebb tides (Fig. 4.2a). Wetlands along coastlines often show some of this same spring-and-ebb pulsing (Fig. 4.2b), whereas others reflect seasonal water-level changes of freshwater inflows and the water levels of the ocean itself (Fig. 4.2c). Hydroperiods of coastal lacustrine wetlands along the Laurentian Great Lakes in the United States and Canada vary considerably, depending on whether pumps and water management are used or whether the marshes are open to the seasonal patterns of river flows and lake levels (Fig. 4.2d). In fact, the hydroperiods of these wetlands, when managed as hunting clubs for waterfowl production, are actually managed to be dry when the normal season calls for wet and wet when the seasonal pattern calls for dry conditions. Water levels for interior wetlands such as the prairie potholes of North America vary considerably from year to year (see the next section), with differences depending on climate variability (Fig. 4.2e). Wetlands affected by groundwater tend to have water levels that are less seasonally variable (Fig. 4.2f).

Some of the most seasonally variable wetlands are the vernal pools of central California, where surface water essentially disappears in this Mediterranean-type climate for all but four or five months (Fig. 4.2g). Cypress domes in central Florida have standing water during the wet summer season and dry periods in the late autumn and early spring (Fig. 4.2h). Low-order riverine wetlands such as the alluvial swamps in the southeastern United States respond sharply to local rainfall events rather than to general seasonal patterns (Fig. 4.2i). The hydroperiods of many bottomland

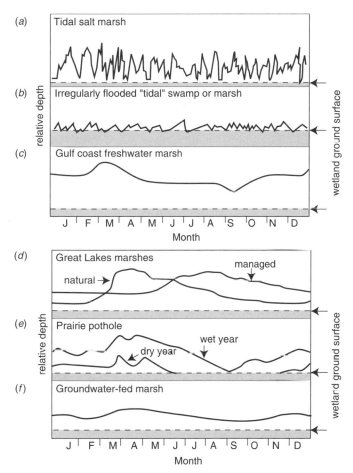

Figure 4.2 Hydroperiods for several different wetlands, presented in approximately the same relative scale: (a) tidal salt marsh, Rhode Island; (b) irregularly flooded "tidal" swamp or marsh; (c) Gulf Coast freshwater marsh, Louisiana; (d) Great Lakes marshes, northern Ohio (natural and managed); (e) prairie pothole marsh with little groundwater flow (dry and wet years); (f) groundwater-fed prairie pothole marsh; (g) vernal pool, California; (h) sub-tropical cypress dome, Florida; (i) alluvial swamp, North Carolina; (j) bottomland hardwood forest, northern Illinois; (k) mineral soil swamp, Ontario, Canada; (l) rich fen, North Wales; (m) pocosin or Carolina Bay, North Carolina; (n) tropical floodplain forest, Amazon River, Manaus, Brazil. (After Nixon and Oviatt, 1973; Mitsch et al., 1979b; Gilman, 1982; Junk, 1982; Zedler, 1987; Mitsch, 1989; van der Valk, 1989; Brinson, 1993b; Woo and Winter, 1993.)

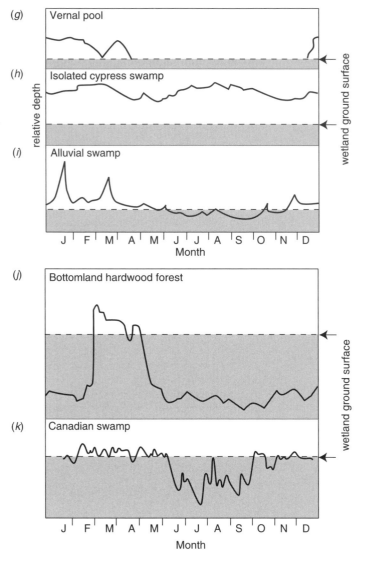

Figure 4.2 (*continued*)

hardwood forests and swamps in colder climates have distinct periods of surface flooding in the winter and early spring due to snow and ice conditions followed by spring floods but otherwise have a water table that can be a meter or more below the surface (Fig. 4.2j and k).

Peatlands in cooler climates can have hydroperiods with little pronounced seasonal fluctuation, as in the fen from North Wales in Figure 4.2l. If peatlands such as the pocosins of North Carolina are located in regions of warm summers, significant

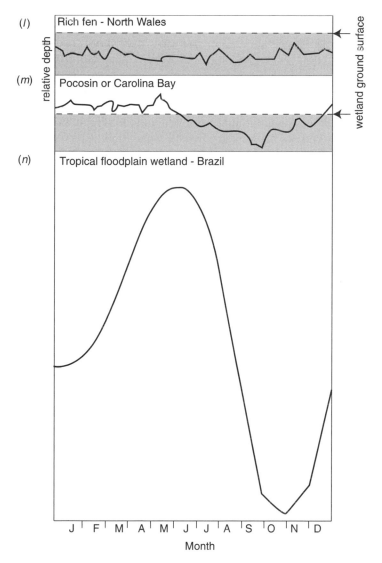

Figure 4.2 (*continued*)

patterns of seasonal water-level change will occur (Fig. 4.2m). The most dramatic hydroperiods result from high-order rivers that are more influenced by seasonal patterns of precipitation throughout a large watershed than by local precipitation, leading to a more predictable and seasonally distinct hydroperiod. For example, the annual fluctuation of water in the tropical floodplain forests along the Amazon River is a predictable seasonal pattern that can include a seasonal fluctuation in water level of 5 to 10 m caused by flooding of upstream rivers (Fig. 4.2n).

Year-to-Year Fluctuations

The hydroperiod is not the same each year but varies statistically according to climate and antecedent conditions. Great variability can be seen from year to year for some wetlands, as illustrated in Figure 4.3 for a prairie pothole regional wetland in Canada and for the Big Cypress Swamp region of south Florida. In the pothole region, a wet-dry cycle of 10 to 20 years is seen; spring is almost always wetter than fall, but depths vary significantly from year to year (Fig. 4.3a). Figure 4.3b illustrates cases of an even seasonal rainfall pattern for the Big Cypress Swamp between 1957 and 1958, which caused a fairly stable hydroperiod throughout the year, and a significant dry season during 1970 and 1971, which caused the hydroperiod to vary about 1.5 m between high and low water. A three-year study of groundwater levels in a red maple swamp shows dramatically different growing season water levels from year to year (Fig. 4.4). Water is near or at the surface during high precipitation periods (last half of first year and entire second year) while dry low-water conditions are mainly driven by seasonal evapotranspiration in the swamp accelerated by groundwater loss during tree transpiration.

Pulsing Water Levels

Water levels in most wetlands are generally not stable but fluctuate seasonally (riparian wetlands), daily or semi-daily (types of tidal wetlands), or unpredictably (wetlands in low-order streams and coastal wetlands with wind-driven tides). In fact, wetland hydroperiods that show the greatest differences between high and low water levels such as those seen in riverine wetlands are often caused by flooding "pulses" that occur seasonally or periodically (Junk et al., 1989). These pulses nourish the riverine wetland with additional nutrients and carry away detritus and waste products. Pulse-fed wetlands are often the most productive wetlands and are the most favorable for exporting materials, energy, and biota to adjacent ecosystems. Despite this obvious fact, many wetland managers, especially those who manage wetlands for waterfowl, often attempt to control water levels by isolating formerly open wetlands with dikes and pumps (Mitsch, 1992b). Fredrickson and Reid (1990) stated that "Because the goal of many [wetland] management scenarios is to counteract the effects of seasonal and long-term droughts, a general tendency is to restrict water level fluctuations in managed wetlands. This misconception is based on the fact that most wetland wildlife requires water for most stages in their life cycles." Kushlan (1989) argues that, because avian fauna that use wetlands often possess adaptations to fluctuating water levels, active manipulation of water levels may be appropriate in artificially managed wetlands. A seasonally fluctuating water level, then, is the rule, not the exception, in most wetlands. Unfortunately, using dikes and levees to control water levels in managed marshes also tends to restrict water inflows and outflows through the managed area, reducing both the range of fluctuations and the water renewal rates.

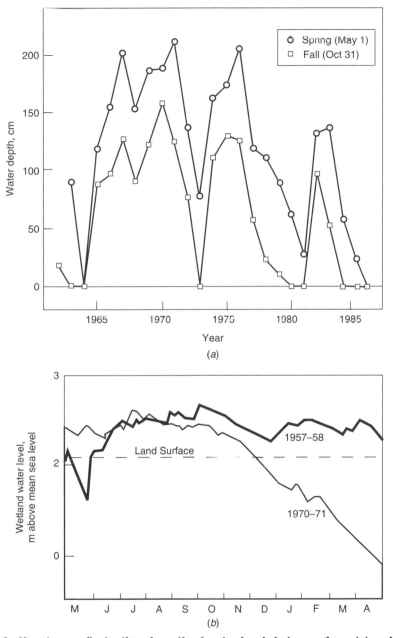

Figure 4.3 Year-to-year fluctuations in wetland water levels in two regions: (a) spring and fall water depths for 25 years in shallow open-water wetlands in the prairie pothole region of southwestern Saskatchewan, Canada, and (b) wet and dry year hydrographs for the Big Cypress Swamp region of the Everglades, southwestern Florida. *(a. after Kantrud et al., 1989; Millar, 1971; b. after Freiberger, 1972; Duever, 1988.)*

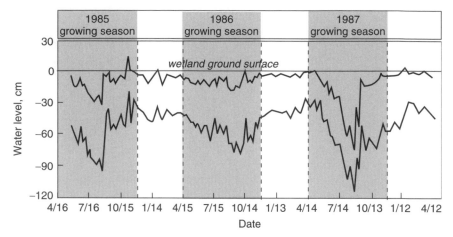

Figure 4.4 **Relative water levels in two seasonally saturated red maple swamps in Rhode Island, United States, for 1985–1987. Growing seasons precipitation amounts for 1985, 1986, and 1987 were 104, 76, and 59 cm, respectively.** *(After Golet et al., 1993.)*

The Wetland Water Budget

The hydroperiod, or hydrologic state of a given wetland, can be summarized as being a result of the following factors:

1. The balance between the inflows and outflows of water
2. The surface contours of the landscape
3. Subsurface soil, geology, and groundwater conditions

The first condition defines the *water budget* of the wetland, whereas the second and the third define the capacity of the wetland to store water. The general balance between water storage and inflows and outflows, illustrated in Figure 4.5, is expressed as

$$\frac{\Delta V}{\Delta t} = P_n + S_i + G_i - ET - S_o - G_o \pm T \tag{4.1}$$

where

$$
\begin{aligned}
V &= \text{volume of water storage in wetlands} \\
\Delta V/\Delta t &= \text{change in volume of water storage in wetland per unit time, } t \\
P_n &= \text{net precipitation} \\
S_i &= \text{surface inflows, including flooding streams} \\
G_i &= \text{groundwater inflows} \\
ET &= \text{evapotranspiration} \\
S_o &= \text{surface outflows} \\
G_o &= \text{groundwater outflows} \\
T &= \text{tidal inflow }(+)\text{ or outflow }(-)
\end{aligned}
$$

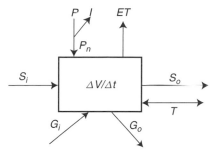

Figure 4.5 Generalized water budget for a wetland with corresponding terms as in Equation 4.1. *P*, precipitation; *ET*, evapotranspiration; *I*, interception; *P_n*, net precipitation; *S_i* surface inflow; *S_o* surface outflow; *G_i* groundwater inflow; *G_o* groundwater outflow; $\Delta V/\Delta t$, Change in storage per unit time; *T*, tide or seiche.

The average water depth, d, at any one time, can further be described as

$$d = \frac{V}{A} \tag{4.2}$$

where

A = wetland surface area

Each of the terms in Equation 4.1 can be expressed in terms of depth per unit time (e.g., cm/yr) or in terms of volume per unit time (e.g., m^3/yr).

Examples of Water Budgets

Equation 4.1 serves as a useful summary of the major hydrologic components of any wetland water budget. Examples of hydrologic budgets for several wetlands are illustrated in Figure 4.6. The terms in the equation vary in importance according to the type of wetland observed; furthermore, not all terms in the hydrologic budget apply to all wetlands (Table 4.2). There is a large variability in certain flows, particularly in surface inflows and outflows, depending on the openness of the wetlands. An alluvial cypress swamp in southern Illinois received a gross inflow of floodwater from one flood that was more than 50 times the gross precipitation for the entire year (Fig. 4.6a). Even the net surface inflow from that flood (the water left behind after the flooding river receded) was three times the precipitation input for the entire year. Surface and groundwater inflows to a coastal Lake Erie marsh in northern Ohio were estimated to be almost 20 times the precipitation for a major part of a drought year (Fig. 4.6b), and tides contributed 10 times the precipitation to a black mangrove swamp in Florida (Fig. 4.6c).

In contrast to these inflow-dominated wetlands, surface inflow is approximately equal to the precipitation inflow in the prairie pothole marshes of North Dakota (Fig. 4.6d), considerably less than the precipitation for the Okefenokee Swamp in

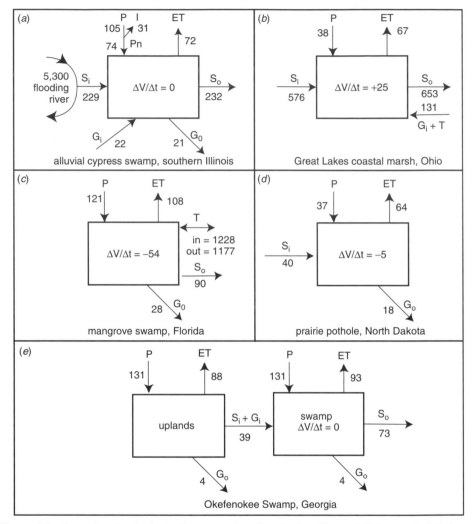

Figure 4.6 Annual water budgets for several wetlands. See Figure 4.5 for symbol definitions. All values are expressed in centimeters per year (cm/yr) except (b), which is March–September only. (After Mitsch, 1979; Mitsch and Reeder, 1992; Twilley, 1982; Shjeflo, 1968; Rykiel, 1984; Gilman, 1982; Pride et al., 1966; Hemond, 1980; Richardson, 1983; data for j. from P. Wolski.)

Georgia (Fig. 4.6e) and a rich fen in North Wales (Fig. 4.6f), and essentially nonexistent in the upland Green Swamp of central Florida (Fig. 4.6g), a bog in Massachusetts (Fig. 4.6h), and a pocosin wetland of North Carolina (Fig. 4.6i). In most of these examples, the change in storage is small or zero, indicating that the water level at the end of the study period (usually an annual cycle) is close to where it was at the beginning of the study period.

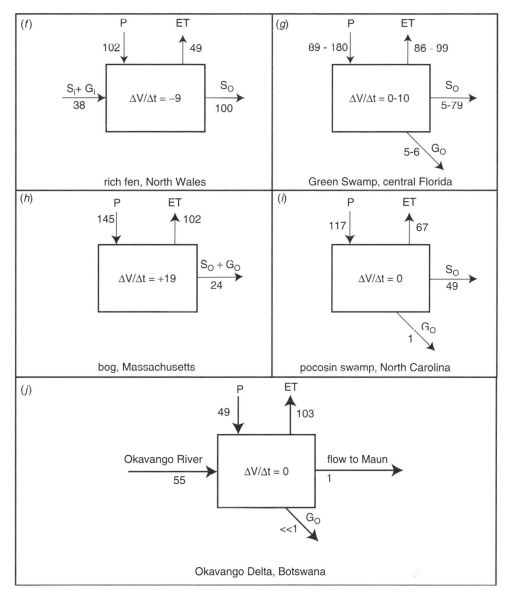

Figure 4.6 (*continued*)

The water budget for the tropical Okavango Delta in southern Africa (Botswana) has been investigated for many years. Figure 4.6j represents the average conditions for the past 36 years (P. Wolski, pers. comm.). The data show clearly that the Okavango River input, when averaged over the entire delta, is about equivalent to the rainfall over this vast area. Furthermore, the budget shows that essentially all of the inputs

Table 4.2   Major components of hydrologic budgets for wetlands

Component	Pattern	Wetlands Affected
Precipitation	Varies with climate, although many regions have distinct wet and dry seasons	All
Surface inflows and outflows	Seasonally, often matched with precipitation pattern or spring thaw; can be channelized as streamflow or nonchannelized as runoff; includes river flooding of alluvial wetlands	Potentially all wetlands except ombrotrophic bogs; riparian wetlands, including bottomland hardwood forests and other alluvial wetlands, are particularly affected by river flooding.
Groundwater	Less seasonal than surface inflows and not always present	Potentially all wetlands except ombrotrophic bogs and other perched wetlands
Evapotranspiration	Seasonal with peaks in summer and low rates in winter. Dependent on meterorological, physical, and biological conditions in wetlands	All
Tides	One to two tidal periods per day; flooding frequency varies with elevation	Tidal freshwater and salt marshes; mangrove swamps

are balanced by a loss of evapotranspiration in this semi-arid climate and only about 1 percent of the water now leaves the wetland region to the downstream village of Maun.

Residence Time—How Long Does Water Stay in a Wetland?

A generally useful concept of wetland hydrology is that of the *renewal rate* or *turnover rate* of water, defined as the ratio of throughput to average volume within the system:

$$t^{-1} = \frac{Q_t}{V} \tag{4.3}$$

where

t^{-1} = renewal rate (time^{-1})
Q_t = total inflow rate (volume/time)
V = average volume of water storage in wetland

Few measurements of renewal rates have been made in wetlands, although it is a frequently used parameter in limnological studies. Chemical and biotic properties are often determined by the openness of the system, and the renewal rate is an index of this because it indicates how rapidly the water in the system is replaced. The reciprocal of the renewal rate is the *turnover time* or *residence time* (*t*, sometimes called *detention time* by engineers for constructed wetlands), which is a measure of the average time that water remains in the wetland. The theoretical residence time, as calculated by Equation 4.3, is often much longer than the actual residence time of water flowing through a wetland, because of nonuniform mixing. Because there are often parts of wetland where waters are stagnant and not well mixed, the theoretical residence time (t) estimate should be used with caution when estimating the hydrodynamics of wetlands.

Precipitation

Wetlands occur most extensively in regions where *precipitation*, a term that includes rainfall and snowfall, is in excess of losses such as evapotranspiration and surface runoff (Fig. 4.7). The dividing line between excess precipitation in the eastern United States and precipitation deficit in the western part of the country is generally the Mississippi River. Wetland-rich regions such as the eastern provinces of Canada have 50 to 60 cm/yr of excess precipitation (precipitation less evaporative losses), whereas regions of the southwestern United States have precipitation deficits of 100 cm or more and generally few wetlands. Exceptions to this generality occur in coastal salt marshes fed by tides or in arid regions where riparian wetlands depend more on river flow than on local precipitation.

The fate of precipitation that falls on wetlands with forested, shrub, or emergent vegetation is shown in Figure 4.8. When some of the precipitation is retained by the vegetation cover, particularly in forested wetlands, the amount that actually passes through the vegetation to the water or substrate below is called *throughfall*. The amount of precipitation that is retained in the overlying vegetation canopy is called *interception*. Interception depends on several factors, such as the total amount of precipitation, the intensity of the precipitation, and the character of the vegetation, including the stage of vegetation development, the type of vegetation (e.g., deciduous or evergreen), and the strata of the vegetation (e.g., tree, shrub, or emergent macrophyte). The percentage of precipitation that is intercepted in forests varies between 8 and 35 percent. One review cites a median value of 13 percent for several studies of deciduous forests and 28 percent for coniferous forests (Dunne and Leopold, 1978). The water budget in Figure 4.6a, for example, illustrates that 29 percent of precipitation in a forested wetland was intercepted by a canopy dominated by *Taxodium distichum* a deciduous conifer.

Little is known about the interception of precipitation by emergent herbaceous macrophytes, but it probably is similar to that measured in grasslands or croplands. Essentially, in those systems, interception at maximum growth can be as high as that in a forest (10 to 35 percent of gross precipitation). An interesting hypothesis about interception and the subsequent evaporation of water from leaf surfaces is that, because the same amount of energy is required whether water evaporates from the surface of a leaf or is transpired by the plant, the evaporation of intercepted water is not "lost" because it may reduce the amount of transpiration loss that occurs (Dunne

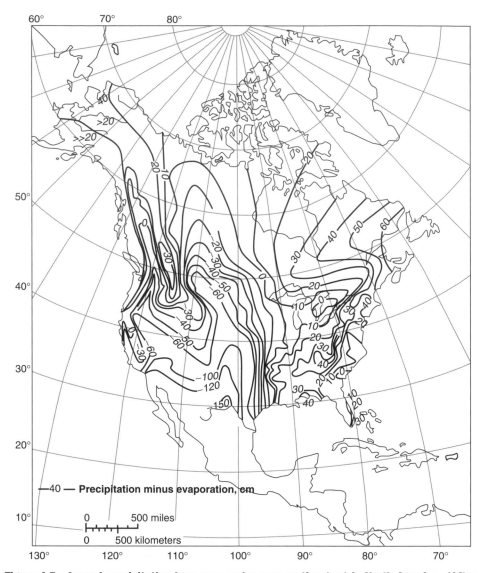

Figure 4.7 Annual precipitation less open-water evaporation (cm) in North America. (*After Winter and Woo, 1990; Woo and Winter, 1993.*)

and Leopold, 1978). This argues that wetlands with high and low interception may be similar in overall water loss to the atmosphere.

Another term related to precipitation, *stemflow*, refers to water that passes down the stems of the vegetation (Fig. 4.8). This flow is generally a minor component of the water budget of a wetland. For example, Heimburg (1984) found that stemflow

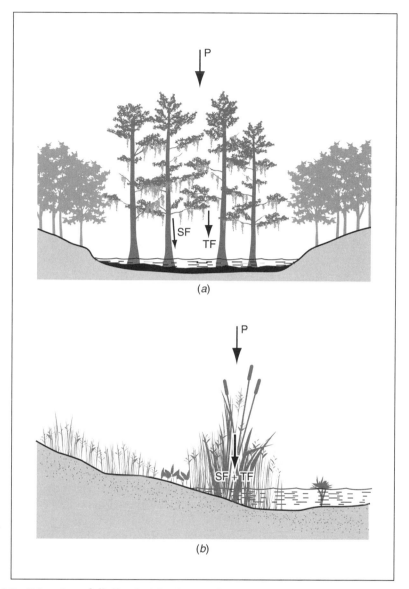

Figure 4.8   Fate of precipitation in (a) a forested wetland and (b) a marsh. *P*, precipitation; *TF*, throughfall; *SF*, stemflow.

was, at maximum, 3 percent of throughfall in cypress dome wetlands in northcentral Florida.

These terms are related in a simple water balance as follows:

$$P = I + TF + SF \qquad (4.4)$$

where

$$P = \text{total precipitation}$$
$$I = \text{interception}$$
$$TF = \text{throughfall}$$
$$SF = \text{stemflow}$$

The total amount of precipitation that actually reaches the water's surface or substrate of a wetland is called the net precipitation (P_n) and is defined as

$$P_n = P - I \qquad\qquad (4.5)$$

Combining Equations 4.4 and 4.5 yields the most commonly used form for estimating net precipitation in wetlands

$$P_n = TF + SF \qquad\qquad (4.6)$$

Surface Flow

Watersheds and Runoff

The percentage of precipitation that becomes surface flow depends on several variables, with climate being the most important (Fig. 4.9). Humid cool regions such as the Pacific Northwest, western British Columbia, and the northeastern Canadian provinces have 60 to 80 percent of precipitation converted to runoff. In the arid southwestern United States, less than 10 percent of the already low precipitation becomes runoff. This difference is related, in large part, to the higher temperatures in the arid Southwest, which translate into higher evapotranspiration rates, greater soil moisture deficits, and higher soil infiltration rates than in the Northeast. Even though runoff in arid regions is small relative to that in humid areas, it does contribute streamflow, which is an important part of a riparian wetland's water budget. Wetlands can be receiving systems for surface water flows (*inflows*), or surface water streams can originate in wetlands to feed downstream systems (*outflows*). Surface outflows are found in many wetlands that are located in the upstream reaches of a watershed. These wetlands are often important water flow regulators for downstream rivers. Some wetlands have surface outflows that develop only when their water stages exceed a critical level.

Wetlands are subjected to surface inflows of several types. *Overland flow* is nonchannelized sheet flow that usually occurs during and immediately following rainfall or a spring thaw, or as tides rise in coastal wetlands. A wetland influenced by a drainage basin may receive channelized *streamflow* during most or all of the year. Wetlands are often an integrated part of a stream or river, for example, as instream freshwater marshes or riparian bottomland forests. Wetlands that form in wide, shallow expanses of river channels or floodplains adjacent to them are greatly influenced by the seasonal streamflow patterns of the river. Wetlands can also receive

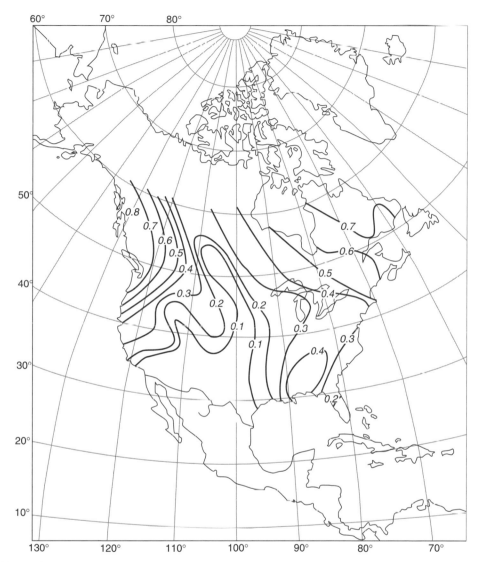

Figure 4.9 Ratio of mean annual runoff to mean annual precipitation for North America. A higher number indicates a higher percentage of precipitation that becomes runoff and streamflow. *(After Hare, 1980; Woo and Winter, 1993.)*

surface inflow from seasonal or episodic pulses of flood flow from adjacent streams and rivers that may otherwise not be connected hydrologically with the wetland. Coastal saline and brackish wetlands are also significantly influenced by freshwater runoff and streamflow (in addition to tides) that contribute nutrients and energy to the wetland and often ameliorate the effects of soil salinity and anoxia.

Surface inflow from a drainage basin into a wetland is usually difficult to estimate without a great deal of data. Nevertheless, it is often one of the most important sources of water in a wetland's hydrologic budget. The direct runoff component of streamflow refers to rainfall during a storm that causes an immediate increase in streamflow. An estimate of the amount of precipitation that results in direct runoff, or *quickflow*, from an individual storm can be determined from the following equation:

$$S_i = R_p P A_w \qquad (4.7)$$

where

S_i = direct surface runoff into wetland (m^3 per storm event)
R_p = hydrologic response coefficient
P = average precipitation in watershed (m)
A_w = area of watershed draining into wetland (m^2)

This equation states that the flow is proportional to the volume of precipitation ($P \times A_w$) on the watershed feeding the wetland in question. The values of R_p, which represent the fraction of precipitation in the watershed that becomes direct surface runoff, range from 4 to 18 percent for small watersheds in the eastern United States. As suggested by Figure 4.9, R_p increases with latitude. Otherwise, slope and type of vegetation appear to have little influence on R_p in a watershed with a mature forest cover. As the following paragraph suggests, land use and soil type can strongly influence runoff.

While Equation 4.7 predicts the volume of direct runoff caused by a storm event, in some cases wetland scientists and managers might be interested in calculating the peak runoff (*flood peak*) into a wetland caused by a specific rainfall event. Although this is generally a difficult calculation for large watersheds, a formula with the unlikely name of the *rational runoff method* is a widely accepted and useful way to predict peak runoff for watersheds less than 80 ha in size. The equation is given by

$$S_{i(pk)} = 0.278 C I A_w \qquad (4.8)$$

where

$S_{i(pk)}$ = peak runoff into wetland (m^3/s)
C = rational runoff coefficient (see Table 4.3)
I = rainfall intensity (mm/h)
A_w = area of watershed draining into wetland (km^2)

The coefficient C, which ranges from 0 to 1 (Table 4.3), depends on the upstream land use. Concentrated urban areas have a coefficient ranging from 0.5 to 0.95, and rural areas have lower coefficients that greatly depend on soil type, with sandy soils lowest ($C = 0.1 - 0.2$) and clay soils highest ($C = 0.4 - 0.5$).

Table 4.3 Values of the rational runoff coefficient C used to calculate peak runoff

		C
URBAN AREAS		
Business areas:	high-value districts	0.75–0.95
	neighborhood districts	0.50–0.70
Residential areas:	single-family dwellings	0.30–0.50
	multiple-family dwellings	0.40–0.75
	suburban	0.25–0.40
Industrial areas:	light	0.50–0.80
	heavy	0.60–0.90
Parks and cemeteries		0.10–0.25
Playgrounds		0.20–0.35
Unimproved land		0.10–0.30
RURAL AREAS		
Sandy and gravelly soils:	cultivated	0.20
	pasture	0.15
	woodland	0.10
Loams and similar soils:	cultivated	0.40
	pasture	0.35
	woodland	0.30
Heavy clay soils; shallow soils over bedrock:	cultivated	0.50
	pasture	0.45
	woodland	0.40

Source: Dunne and Leopold (1978).

Channelized Streamflow

Channelized streamflow into and out of wetlands is described simply as the product of the cross-sectional area of the stream (A_x) and the average velocity (v) and can be determined through stream velocity measurements in the field:

$$S_i \text{ or } S_o = A_x v \tag{4.9}$$

where

$$
\begin{aligned}
S_i, S_o &= \text{surface channelized flow into or out of wetland (m}^3/\text{s)} \\
A_x &= \text{cross-sectional area of stream (m}^2) \\
v &= \text{average velocity (m/s)}
\end{aligned}
$$

The velocity can be determined in several ways, ranging from handheld velocity meter readings taken at various locations in the stream cross-section to the floating-orange technique where the velocity of a floating orange or similar fruit (which is 90 percent or more water and therefore floats but just beneath the water surface) is timed as it goes downstream. If a continuous or daily record of streamflow is needed, then a *rating curve* (Fig. 4.10), a plot of instantaneous streamflow (as

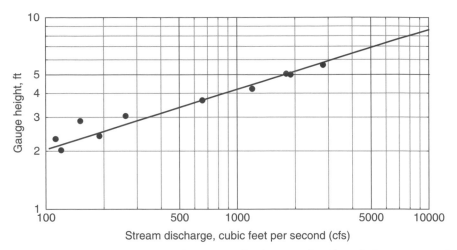

Figure 4.10   Rating curve for streamflow determination as a function of stream stage. 100 cfs = 2.832 m³/s. *(After Dunne and Leopold, 1978.)*

estimated using Eq. 4.9) versus stream elevation or stage, is useful. If this type of rating curve is developed for a stream (the basis of most hydrologic streamflow gauging stations operated by the U.S. Geological Survey), then a simple measurement of the stage in the stream can be used to determine the streamflow. Because hydrographs generally assume a constant water gradient, caution should be taken in using this approach for streams flowing into wetlands to ensure that no "backwater effect" of the wetland's water level will affect the stream stage at the point of measurement.

Weir Flow Measurement

When a weir or other control structure is used at the outflow of a wetland (Fig. 4.11), the outflow of a wetland can also be estimated to be a function of the water level in the wetland itself according to the equation:

$$S_o = xL^y \tag{4.10}$$

where

S_o = surface outflow
L = wetland water level above a control structure crest

(level at which flow just begins)

x, y = calibration coefficients

Figure 4.11 Control structures such as the V-notched weir shown here can be used for measuring surface water flow in small streams into or out of wetlands. (Photograph by W. J. Mitsch.)

If a control structure such as a rectangular or V-notched weir is used to measure the outflow from a wetland, standard equations of the form of Equation 4.10 can be obtained from water measurement manuals (e.g., U.S. Department of Interior, 1984). Care should be taken to calibrate standard weir equations with actual measurements of streamflow and water level.

When an estimate of surface flow into or out of a riverine wetland is needed and no stream velocity measurements are available, the *Manning equation* can often be used if the slope of the stream and a description of the surface roughness are known:

$$S_i \text{ or } S_o = \frac{A_x R^{2/3} s^{0.5}}{n} \tag{4.11}$$

Table 4.4 Roughness coefficients (*n*) for Manning equation used to determine streamflow in natural streams and channels

Stream Conditions	Manning Coefficient, *n*
Straightened earth canals	0.02
Winding natural streams with some plant growth	0.035
Mountain streams with rocky stream bed	0.040–0.050
Winding natural streams with high plant growth	0.042–0.052
Sluggish streams with high plant growth	0.065
Very sluggish streams with high plant growth	0.112

Source: Chow (1964) and R. Lee (1980).

where

n = roughness coefficient (Manning coefficient; see Table 4.4)
R = hydraulic radius (m) [cross-sectional area divided by the wetted perimeter; this is an estimate of the relative portion of the stream cross section (and, hence, flow volume) in contact with the stream bed]
s = channel slope (dimensionless)

The equation states that flow is proportional to stream cross-section, as modified by the roughness of the stream bed and the proportion of flow in contact with that bed. Although the potential exists for their use in wetland studies, the roughness coefficients given in Table 4.4 and the Manning equation (Eq. 4.11) have not been used very often. The relationship is particularly useful for estimating streamflow where velocities are too slow to measure directly, and to estimate flood peaks from high-water marks on ungauged streams. These circumstances are common in wetland studies.

Floods and Riparian Wetlands

A special case of surface flow occurs in wetlands that are in floodplains adjacent to rivers or streams and are occasionally flooded by those rivers or streams. These ecosystems are often called *riparian wetlands*. The flooding of these wetlands varies in intensity, duration, and number of floods from year to year, although the probability of flooding is fairly predictable. In the eastern and midwestern United States and in much of Canada, a pattern of winter or spring flooding caused by rains and sudden snowmelt is often observed (Fig. 4.12). When river flow begins to overflow onto the floodplain, the streamflow is referred to as *bankfull discharge*. A hydrograph of a stream that flooded its riparian wetlands above bankfull discharge for several months in the spring is shown in Figure 4.13. There is a remarkable consistency in the hydrographs of riparian streams in midwestern United States, in that they tend to overflow their banks at intervals between 1 and 2 years for bankfull discharge, with an average of approximately 1.5 years (Leopold et al., 1964; see BOX).

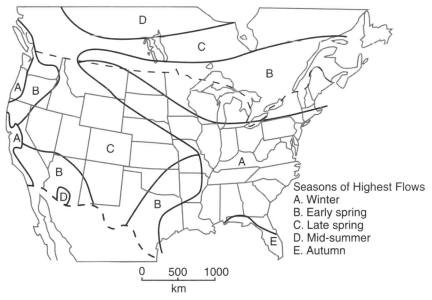

Figure 4.12 Periods of maximum streamflow in North American streams and rivers. (After Beaumont, 1975.)

Seasons of Highest Flows
A. Winter
B. Early spring
C. Late spring
D. Mid-summer
E. Autumn

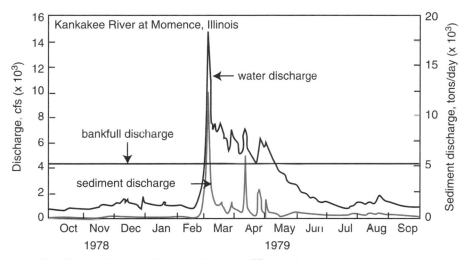

Figure 4.13 River hydrograph from northeastern Illinois, showing discharge and sediment load of the river and discharge at which a riparian wetland is flooded (bankfull discharge). 1,000 cfs = 28.32 m³/s. (After Bhowmik et al., 1980.)

133

Recurrence Intervals

The *recurrence interval* is the average interval between the occurrence of floods at a given or greater stage (depth). The inverse of the recurrence interval is the average probability of flooding in any one year. Figure 4.14 indicates that a stream in midwestern and southern United States will overflow its banks onto the adjacent riparian forest with an average recurrence interval of 1.5 years (or a probability of 1/1.5, or 67 percent, of overbank flooding in any one year). Stated another way, these rivers, on average, overflow their banks in 2 out of every 3 years. Figure 4.14 also illustrates that flow that is twice that of bankfull discharge occurs at recurrence intervals of approximately 5 years; this flow, however, results in only a 40 percent greater river depth over bankfull depth on the floodplain. This predictable relationship suggests that in natural stream systems, the size of a stream channel is related to the hydraulic energy that scours the stream bed.

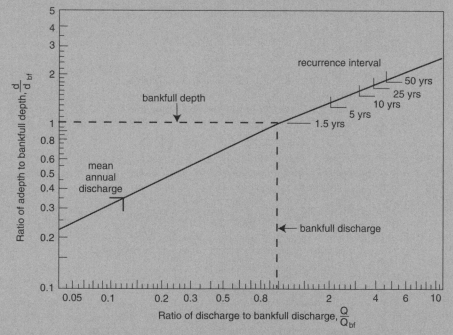

Figure 4.14 Relationships among streamflow (discharge), stream depth, and recurrence interval for streams and rivers in the midwestern and southern United States. *Q*, stream discharge; Q_{df}, bankfull discharge; *d*, stream discharge; d_{df}, bankfull depth (depth of river with floodplain is initially flooded). *(After Leopold et al., 1964.)*

Groundwater

Recharge and Discharge Wetlands

Groundwater can heavily influence some wetlands, whereas in others it may have hardly any effect at all. The influence of wetland recharge and discharge on groundwater resources has often been cited as one of the most important attributes of wetlands, but it does not hold for all wetland types; nor is there sufficient experience with site-specific studies to make many generalizations. Groundwater inflow results when the surface water (or groundwater) level of a wetland is lower hydrologically than the water table of the surrounding land (called a *discharge wetland* by geologists, who generally view their water budget from a groundwater, not a wetland, perspective). Wetlands can intercept the water table in such a way that they have only inflows and no outflows, as shown for a prairie marsh in Figure 4.15a. Another type of discharge wetland, called a *spring* or *seep* wetland, is often found at the base of steep slopes where the groundwater surface intersects the land surface (Fig. 4.15b). This type of wetland can be an isolated low point in the landscape; more often, it discharges excess water downstream as surface water or as groundwater, as shown in the riparian wetland in Figure 4.15c.

When the water level in a wetland is higher than the water table of its surroundings, groundwater will flow out of the wetland (called a *recharge wetland*; Fig. 4.15d). When a wetland is well above the groundwater of the area, the wetland is referred to as being *perched* (Fig. 4.15e). This type of wetland, also referred to as a *surface water depression wetland* by Novitzki (1979), loses water only through infiltration into the ground and through evapotranspiration. Tidally influenced wetlands often have significant groundwater inflows that can reduce soil salinity and keep the wetland soil wet even during low tide (Fig. 4.15f).

A final type of wetland, one that is fairly common, is little influenced by or has little influence on groundwater. Because wetlands often occur where soils have poor permeability, the major source of water can be restricted to surface water runoff, with losses occurring only through evapotranspiration and other surface outflows. This type of wetland often has fluctuating hydroperiods and intermittent flooding (e.g., some prairie potholes; Fig. 4.2e), and standing water is dependent on seasonal surface inflows. If, however, such a wetland were to be influenced by groundwater, its water level would be better buffered against dramatic seasonal changes, or at least it would be semipermanently flooded (Fig. 4.2f).

Groundwater patterns for the four types of hydrologic settings for freshwater wetlands as described by Novitski (1979, 1982) are illustrated in Figure 4.16 and summarized here:

1. *Surface water depression wetland.* This type of wetland is dominated by surface runoff and precipitation, with little groundwater outflow due to a layer of low-permeability soils. This is similar to the perched wetland type described in Fig. 4.15e, where the wetland is separated from the water table by an unsaturated zone.

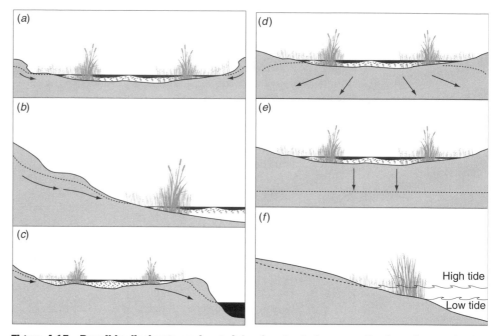

Figure 4.15 Possible discharge–recharge interchanges between wetlands and groundwater systems including (a) marsh as a depression receiving groundwater flow ("discharge wetland"); (b) groundwater spring or seep wetland or groundwater slope wetland at the base of a steep slope; (c) floodplain wetland fed by groundwater; (d) marsh as a "recharge wetland" adding water to groundwater; (e) perched wetland or surface water depression wetland; (f) groundwater flow through a tidal wetland. Dashed lines indicate groundwater level.

2. *Surface water slope wetland.* This type of wetland is generally found in alluvial soil adjacent to a lake or stream and is fed, to some degree, by precipitation and surface runoff but, more important, by overbank flooding from the adjacent stream, river, or lake. Hydroperiods of these wetlands match the seasonal patterns of the adjacent bodies of water, with relatively rapid wetting and drying. Some groundwater recharge is possible, but that groundwater shortly discharges back to the stream, river, or lake.

3. *Groundwater depression wetland.* These are the groundwater discharge wetlands described previously (Fig. 4.15a), where the swamp is in a depression low enough to intercept the local groundwater table. These kinds of swamps can occur in coarse-textured glaciofluvial deposits, where the interchange between groundwater and surface water is enhanced by relatively coarse soil material. Water-level fluctuations in these types of swamps are less dramatic than fluctuations in surface flow wetlands because of the relative stability of the groundwater levels.

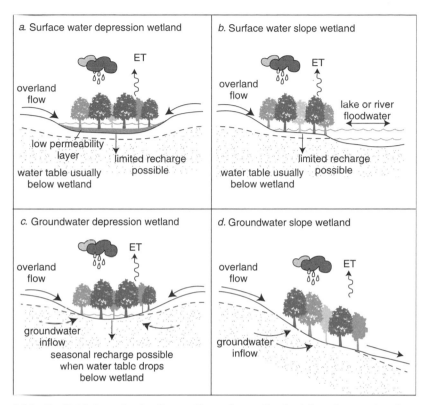

Figure 4.16 Novitski groundwater flow patterns for wetlands: (a) surface water depression, (b) surface water slope, (c) groundwater depression, and (d) groundwater slope. Dashed lines indicate groundwater level. (After Golet et al., 1993.)

4. *Groundwater slope wetland.* Wetlands often develop on slopes or hillsides where groundwater discharges to the surface as springs and seeps. Groundwater flow into these wetlands can be continuous or seasonal, depending on the local geohydrology and on the evapotraspiration rates of the wetland and adjacent uplands.

Darcy's Law

The flow of groundwater into, through, and out of a wetland is often described by *Darcy's law*, an equation familiar to groundwater hydrologists. This law states that the flow of groundwater is proportional to (1) the slope of the piezometric surface (the hydraulic gradient) and (2) the hydraulic conductivity, or *permeability*, the capacity of the soil to conduct water flow. In equation form, Darcy's law is given as

$$G = kA_x s \tag{4.12}$$

where

$$G = \text{flow rate of groundwater (volume per unit time)}$$
$$k = \text{hydraulic conductivity or permeability (length per unit time)}$$
$$A_x = \text{groundwater cross-sectional area perpendicular to the direction of flow}$$
$$s = \text{hydraulic gradient (slope of water table or piezometric surface)}$$

Despite the importance of groundwater flows in the budgets of many wetlands, there is poor understanding of groundwater hydraulics in wetlands, particularly in those that have organic soils. The hydraulic conductivity can be predicted for some peatland soils from their bulk density or fiber content, both of which can easily be measured (Fig. 4.17). In general, the conductivity of organic peat decreases as the fiber content decreases through the process of decomposition. Water can pass through fibric, or poorly decomposed, peats a thousand times faster than it can through more decomposed sapric peats. The type of plant material that makes up the peat is also important. Peat composed of the remains of grasses and sedges such as *Phragmites* and *Carex*, for example, is more permeable than the remains of most mosses, including sphagnum. The hydraulic conductivity of peat can vary over several orders of magnitude, showing a range almost as great as the range for mineral soil

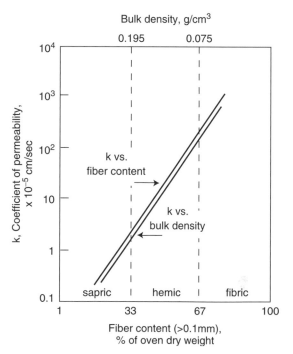

Figure 4.17 Permeability of peatland soil as a function of fiber content and bulk density. (After Verry and Boelter, 1979.)

Table 4.5   Typical hydraulic conductivity for wetland soils compared with other soil materials

Wetland or Soil Type	Hydraulic Conductivity, k(cm/s $\times 10^{-5}$)	Reference
NORTHERN PEATLANDS		
Highly humified blanket bog, UK	0.02–0.006	Ingram (1967)
Fen, Russia		
Slightly decomposed	500	Romanov (1968)
Moderately decomposed	80	
Highly decomposed	1	
Carex fen, Russia		
0–50 cm deep	310	Romanov (1968)
100–150 cm deep	6	
North American peatlands (general)		
Fibric	>150	Verry and Boelter (1979)
Hemic	1.2–150	
Sapric	<1.2	
COASTAL SALT MARSH		
Great Sippewissett Marsh, Massachusetts (vertical conductivity)		Hemond and Fifield (1982)
0–30 cm deep	1.8	
High permeability zone	2,600	
Sand–peat transition zone	9.4	
NONPEAT WETLAND SOILS		
Cypress dome, Florida		
Clay with minor sand	0.02–0.1	Smith (1975)
Sand	30	
Okefenokee Swamp watershed, Georgia	3.4–834	Hyatt and Brook (1984)
MINERAL SOILS (general)		
Clay	0.05	Linsley and Franzini (1979)
Limestone	5.0	
Sand	5000	

Source: Partially after Rycroft et al. (1975).

between clay ($k = 5 \times 10^{-7}$ cm/s) and sand ($k = 5 \times 10^{-2}$ cm/s) (Table 4.5). There has been some disagreement over the appropriate methods for measuring hydraulic conductivity in wetlands and about whether Darcy's law applies to flow through organic peat (Hemond and Goldman, 1985; Kadlec, 1989).

Evapotranspiration

The water that vaporizes from water or soil in a wetland (*evaporation*), together with moisture that passes through vascular plants to the atmosphere (*transpiration*), is called *evapotranspiration*. The meteorological factors that affect evaporation and transpiration are similar as long as there is adequate moisture, a condition that almost always exists in most wetlands. The rate of evapotranspiration is proportional to the difference between the vapor pressure at the water surface (or at the leaf surface) and the vapor pressure in the overlying air. This is described in a version of *Dalton's law:*

$$E = cf(u)(e_w - e_a) \tag{4.13}$$

where

$$E = \text{rate of evaporation}$$
$$c = \text{mass transfer coefficient}$$
$$f(u) = \text{function of wind speed, } u$$
$$e_w = \text{vapor pressure at surface, or saturation vapor pressure at wet surface}$$
$$e_a = \text{vapor pressure in surrounding air}$$

Evaporation and transpiration are enhanced by the same meteorological conditions, such as solar radiation or surface temperature, that increase the value of the vapor pressure at the evaporating surface and by factors such as decreased humidity or increased wind speed that decrease the vapor pressure of the surrounding air. This equation assumes an adequate supply of water for capillary movement in the soil or for access by rooted plants. When the water supply is limited (not a frequent occurrence in wetlands), evapotranspiration is limited as well. Transpiration can also be physiologically limited in plants through the closing of leaf stomata despite adequate moisture during periods of stress such as anoxia.

Direct Measurement of Wetland Evapotranspiration

Several direct measurement techniques can be used in wetlands to determine evapotranspiration. The classical reference method is the measurement of evaporation from a water-filled pan, usually by measuring the weight loss, by measuring the volume required to replace lost water over a period of time, or by measuring the drop in water level. This is generally considered a measurement of potential evaporation, since the evaporating surface is saturated. The method is tedious and the results often poorly correlated with actual evaporation from vegetated surfaces, because the transpiration, unsaturated soils, winds, and shading effects of the plant canopy all influence the rate, often in unknown ways (see discussion on pages 143–144). However, pan evaporation provides a reference evaporation rate for comparison with other techniques. Furthermore, because wetland soils tend to be saturated most of the time, the pan method may be more accurate for wetlands than for terrestrial environments.

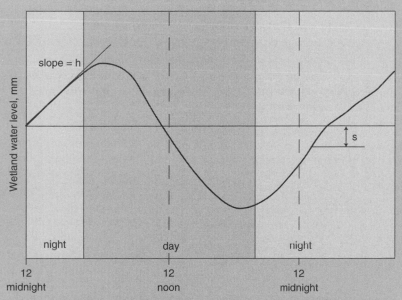

Figure 4.18 Diurnal water fluctuation in some wetlands can be used to estimate evapotranspiration as in Equation 4.14.

Wetland evapotranspiration can also be estimated by measuring the change in water level of the water in the wetland itself. This method, illustrated in Figure 4.18, can be calculated as follows:

$$ET = S_y(24h \pm s) \tag{4.14}$$

where

ET = evapotranspiration (mm/day)
S_y = specific yield of aquifer (unitless)
   = 1.0 for standing-water wetlands
   = <1.0 for groundwater wetlands
h = hourly rise in water level from midnight to 4:00 A.M. (mm/h)
s = net fall (+) or rise (−) of water table or water surface in one day

The pattern assumes active "pumping" of water by vegetation during the day and a constant rate of recharge equal to the midnight-to-4:00-A.M. rate. This method also assumes that evapotranspiration is negligible around midnight and that the water table around this time approximates the daily mean. The water level is usually at or near the root zone in many wetlands, a necessary condition for this method to measure evapotranspiration accurately (Todd, 1964).

Empirical Estimates of Wetland Evapotranspiration

Thornthwaite Equation

Evapotranspiration can be determined with any number of empirical equations that use easily measured meteorological variables. One of the most frequently used empirical equations for evapotranspiration from terrestrial ecosystems, which has been applied with some success to wetlands, is the *Thornthwaite equation* for potential evapotranspiration:

$$ET_i = 16(10T_i/I)^a \qquad (4.15)$$

where

$$ET_i = \text{potential evapotranspiration for month } i \text{ (mm/month)}$$
$$T_i = \text{mean monthly temperature } (^\circ C)$$
$$I = \text{local heat index } \sum_{i=1}^{12}(T_i/5)^{1.514}$$
$$a = (0.675 \times I^3 - 77.1 \times I^2 + 17{,}920 \times I + 492{,}390) \times 10^{-6}$$

Penman Equation

A second empirical relationship that has had many applications in hydrologic and agricultural studies but relatively few in wetlands is the *Penman equation* (Penman, 1948; Chow, 1964). This equation, based on both Dalton's law and the energy budget approach, is given as

$$ET = \frac{\Delta H + 0.27 E_a}{\Delta + 0.27} \qquad (4.16)$$

where

$$ET = \text{evapotranspiration (mm/day)}$$
$$\Delta = \text{slope of curve of saturation vapor pressure versus mean air temperature (mmHg/}^\circ C)$$
$$H = \text{net radiation (cal/cm}^2\text{-day)}$$
$$= R_t (1 - a) - R_b$$
$$R_t = \text{total shortwave radiation}$$
$$a = \text{albedo of wetland surface}$$
$$R_b = \text{effective outgoing longwave radiation} = f(T^4)$$
$$E_a = \text{term describing the contribution of mass transfer to evaporation}$$
$$= 0.35 (0.5 + 0.00625u)(e_w - e_a)$$
$$u = \text{wind speed 2 m above ground (km/day)}$$
$$e_w = \text{saturation vapor pressure of water surface at mean air temperature (mmHg)}$$
$$e_a = \text{vapor pressure in surrounding air (mmHg)}$$

The Penman equation was compared with the pan evaporation (multiplied by a factor of 0.8) and other methods at natural enriched fens in Michigan and constructed wetlands in Nevada by Kadlec et al. (1988). They found that the Penman equation, like the Thornthwaite equation, generally underpredicted evapotranspiration from the Michigan wetland but agreed within a few percent with other measurement techniques for the Nevada wetlands.

Hammer and Kadlec Equation

Another empirical relationship for describing summer evapotranspiration using solar energy was developed by Scheffe (1978) and was described by Hammer and Kadlec (1983). The equation, which was used individually for sedge, willow, leatherleaf, and cattail vegetation covers, is

$$ET = a + bR_t + cT_a + dH_r + eu \tag{4.17}$$

where

a, b, c, d, e = correlation coefficients
R_t = incident shortwave radiation (measured by pyranograph)
T_a = air temperature
H_r = relative humidity
u = wind speed

The equation gives estimates that are better than some more frequently used evapotranspiration relationships. When the results of using this model were compared to actual measurements, the radiation term was shown to dominate (Hammer and Kadlec, 1983).

Because of the many meteorological and biological factors that affect evapotranspiration, none of the many empirical relationships is entirely satisfactory for estimating wetland evapotranspiration. Several comparisons of approaches to measuring evapotranspiration have been attempted (Lott and Hunt, 2001; Rosenberry et al., 2004). One finding has been that empirical estimates of potential evapotranspiration (PET), such as those determined from the Penman equation, generally underestimate true wetland evapotranspiration during the growing season (Lott and Hunt, 2001), possibly due to limitation of the Penman equation for describing surface roughness. A comparison of an energy budget method for estimating evapotranspiration at a wetland in North Dakota with 12 empirical ET equations found that most of the empirical methods gave reasonable approximations of evapotranspiration (Rosenberry et al., 2004). The Thornthwaite equation, the simplest method investigated as it only requires air temperature, worked relatively well and may provide the most accurate measurement per instrument cost.

Effects of Vegetation on Wetland Evapotranspiration

A question about evapotranspiration from wetlands that does not elicit a uniform answer in the literature is, "Does the presence of wetland vegetation increase or decrease the loss of water compared to that which would occur from an open body of water?" Data from individual studies are conflicting. Obviously, the presence of vegetation retards evaporation from the water surface, but the question is whether the transpiration of water through the plants equals or exceeds the difference. Eggelsmann (1963) found evaporation from bogs in Germany to be generally less than that from open water except during wet summer months. In studies of evapotranspiration from small bogs in northern Minnesota, Bay (1967) found it to be 88 percent to 121 percent of open-water evaporation. Eisenlohr (1976) reported 10 percent lower evapotranspiration from vegetated prairie potholes than from nonvegetated potholes in North Dakota. Hall et al. (1972) estimated that a stand of vegetation in a small New Hampshire wetland lost 80 percent more water than did the open water in the wetland. In a forested pond cypress dome in north-central Florida, Heimburg (1984) found that swamp evapotranspiration was about 80 percent of pan evaporation during the dry season (spring and fall) and as low as 60 percent of pan evaporation during the wet season (summer). S. L. Brown (1981) found that transpiration losses from pond cypress wetlands were lower than evaporation from an open-water surface even with adequate standing water.

In the arid West, it has been a longstanding practice to conserve water for irrigation and other uses by clearing riparian vegetation from streams. In this environment where groundwater is often well below the surface but within the rooting zone of deep-rooted plants, trees "pump" water to the leaf surface and actively transpire even when little evaporation occurs at the soil surface.

The conflicting measurements and the difficulty of measuring evaporation and evapotranspiration led Linacre (1976) to conclude that neither the presence of wetland vegetation nor the type of vegetation had major influences on evaporation rates, at least during the active growing season. Bernatowicz et al. (1976) also found little difference in evapotranspiration among several species of vegetation. The general unimportance of plant species variation on overall wetland water loss is probably a reasonable conclusion for most wetlands, although it is clear that the type of wetland ecosystem and the season are important considerations. Ingram (1983), for example, found that fens have about 40 percent more evapotranspiration than do treeless bogs and that evaporation from the bogs is less than potential evapotranspiration in the summer and greater than potential evapotranspiration in the winter.

In some cases the type of vegetation in the wetland does matter. When trees are removed from some forested swamps where the soil is hydric but there is little surface flooding, standing water may return and, with it, herbaceous marsh vegetation. This resets a hydrologic succession; woody plants are able to reinvade the marsh during dry years and reestablish the site back to a forested wetland.

Tides

The periodic and predictable tidal inundation of coastal salt marshes, mangroves, and freshwater tidal marshes is a major hydrologic feature of these wetlands. The tide acts as a stress by causing submergence, saline soils, and soil anaerobiosis; it acts as a subsidy by removing excess salts, reestablishing aerobic conditions, and providing

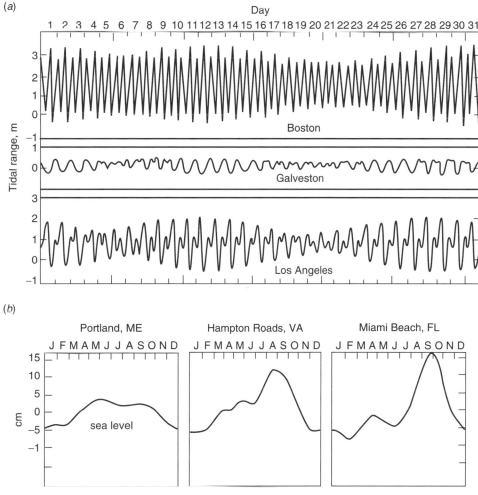

Figure 4.19 Patterns of tides: (a) daily tides for a month and (b) seasonal changes in mean monthly sea level for several locations in North America. *(After Emery and Uchupi, 1972.)*

nutrients. Tides also shift and alter the sediment patterns in coastal wetlands, causing a uniform surface to develop.

Typical tidal patterns for several coastal areas of the United States are shown in Figure 4.19a. Seasonal as well as diurnal patterns exist in the tidal rhythms. Annual variations of mean monthly sea level are as great as 25 cm (Fig. 4.19b). Tides also have significant bimonthly patterns, because they are generated by the gravitational pull of the moon and, to a lesser extent, the sun. When the sun and the moon are in line and pull together, which occurs almost every two weeks, *spring tides*, or tides of the greatest amplitude, develop. When the sun and the moon are at right angles, *neap tides*, or tides of least amplitude, occur. Spring tides occur roughly at full and new moons, whereas neap tides occur during the first and third quarters.

Tides vary more locally than regionally. The primary determinant is the coastline configuration. In North America, tidal amplitudes vary from less than 1 m along the Texas Gulf Coast to several meters in the Bay of Fundy in Canada. Tidal amplitude can actually increase as one progresses inland in some funnel-shaped estuaries. Typically, on a rising tide, water flows up tidal creek channels until the channels are bankfull. It overflows first at the upstream end, where tidal creeks break up into small creeklets that lack natural levees. The overflowing water spreads back downstream over the marsh surface. On falling tides, the flows are reversed. At low tides, water continues to drain through the natural levee sediments into adjacent creeks because these sediments tend to be relatively coarse; in the marsh interior, where sediments are finer, drainage is poor and water is often impounded in small depressions in the marsh.

Seiches

While inland wetlands are nontidal by definition, periodic water-level fluctuations in wetlands adjacent to large freshwater lakes do occur as a result of short-term water-level seiches or "wind tides." These are a common occurrence in wetlands adjacent to large lakes such as the Laurentian Great Lakes in the United States and Canada (Fig. 4.20). When wind has a persistent direction, particularly in a long fetch across a lake, water "piles up" on the downwind side of the lake, causing high-water events for wetlands in that location. When the wind shifts or dies down, the high water is released and flows to the opposite shoreline, causing a secondary wind-relaxation seiche there and lower-than-normal water in the original high-water location.

Effects of Hydrology on Wetland Function

The effects of hydrology on wetland structure and function can be described with a complicated series of cause-and-effect relationships. A conceptual model of the general effects of hydrology in wetland ecosystems was shown in Figure 4.1. The effects are shown to be primarily on the chemical and physical aspects of the wetlands, which, in turn, influence the biotic components of the ecosystem. The biotic components then have a feedback effect on hydrology. Several principles underscoring the importance of hydrology in wetlands can be elucidated from studies that have been conducted to date. These principles, discussed later, are as follows:

1. Hydrology leads to a unique vegetation composition but can limit or enhance species richness.
2. Primary productivity and other ecosystem functions in wetlands are often enhanced by flowing conditions and a pulsing hydroperiod and are often depressed by stagnant conditions.

3. Accumulation of organic material in wetlands is controlled by hydrology through its influence on primary productivity, decomposition, and export of particulate organic matter.

4. Nutrient cycling and nutrient availability are both significantly influenced by hydrologic conditions.

Species Composition and Richness

Hydrology is a two-edged sword for species composition and diversity in wetlands. It acts as a limit or a stimulus to species richness, depending on the hydroperiod and physical energies. At a minimum, the hydrology acts to select water-tolerant vegetation

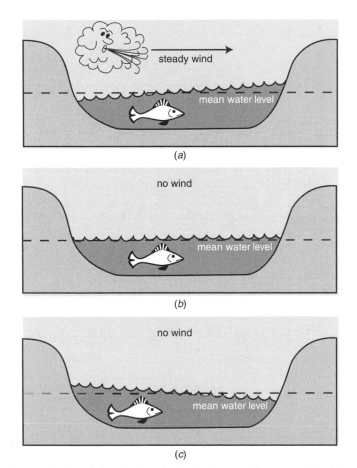

Figure 4.20   Concept of a seiche: (a-c) a wind-relaxation seiche caused by (a) a steady wind that (b) relaxes or shifts directions from initial wind set and (c) results in an oppositely directed tilt; (d) water levels in Ohio (Toledo and Cleveland) and New York (Buffalo) coastlines of Lake Erie during an April 1979 storm and subsequent wind-relaxation seiche. (After Korgen, 1995.)

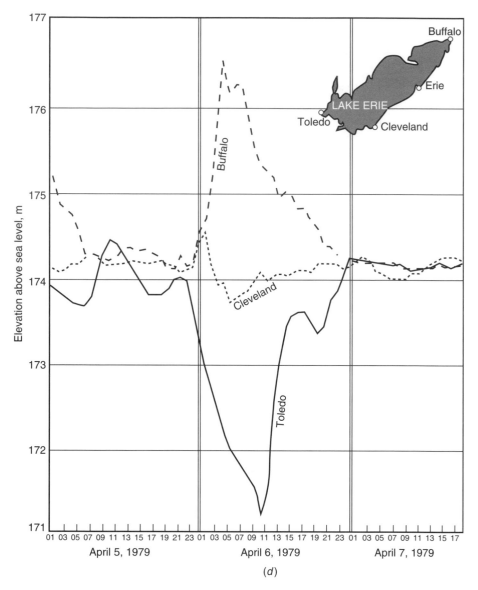

Figure 4.20 (continued)

in both freshwater and saltwater conditions and to exclude flood-intolerant species. Of the thousands of vascular plants that are on Earth, relatively few have adapted to waterlogged soils. (These adaptations are discussed in more detail in Chapter 6.) Although it is difficult to generalize, many wetlands that sustain long flooding durations have lower species richness in vegetation than do less frequently flooded or pulsing areas. Waterlogged soils and the subsequent changes in oxygen content

Table 4.6   Relationship between hydrologic regime and species richness in northern Minnesota peatlands

	Tree	Shrub	Field Herbs	Grasses and Ferns	Ground Layer	Total	Flow Conditions
				Species Present			
1. Rich swamp forest	6	16	28	11	10	71	Good surface flow; minerotrophic
2. Poor swamp forest	3	14	17	12	5	51	Downstream from 1; not adapted to strong water flow
3. Cedar string bog and fen	3	10	10	12	4	39	Better drainage than 2
4. Larch string bog and fen	3	9	9	12	4	37	Similar to 3; sheet flow
5. Black spruce feather moss forest	2	9	2	2	10	25	Gentle water flow on semiconvex template
6. Sphagnum bog	2	8	2	1	7	20	Isolated; little standing water
7. Sphagnum heath	2	6	2	2	5	17	Wet, soggy, and on convex template

Source: After Gosselink and Turner (1978) and Heinselman (1970).

and other chemical conditions significantly limit the number and the types of rooted plants that can survive in this environment.

In general, species richness, at least in the vegetation community, increases as flowthough or pulsing hydrology increases (Table 4.6). Flowing water can be thought of as a stimulus to diversity, probably caused by its ability to renew minerals and reduce anaerobic conditions. Hydrology also stimulates diversity when the action of water and transported sediments creates spatial heterogeneity, opening up additional ecological niches. When rivers flood riparian wetlands or when tides rise and fall on coastal marshes, erosion, scouring, and sediment deposition sometimes create niches that allow diverse habitats to develop. However, flowing water can also create a relatively uniform surface that might enable monospecific stands of *Typha* or *Phragmites* to dominate a freshwater marsh or *Spartina* to dominate a coastal marsh. Keddy (1992b) likened water-level fluctuations in wetlands to fires in forests. They eliminate one growth form of vegetation (e.g., woody plants) in favor of another (e.g., herbaceous species) and allow regeneration of species from buried seeds (see Chapter 7).

Primary Productivity

In general, the "openness" of a wetland to hydrological fluxes is probably one of the most important determinants of potential primary productivity. For example, peatlands that have flowthrough conditions (fens) have long been known to be more productive than stagnant raised bogs. Some studies have found that wetlands in stagnant (nonflowing) or continuously deep water have low productivities, whereas wetlands that are in slowly flowing strands or are open to flooding rivers have high productivities.

This relationship between hydrology and ecosystem primary productivity has been investigated most extensively for forested wetlands. Figure 4.21 shows a set of similar typical "Shelford-type" limitation curves that have been suggested in separate studies to explain the importance of hydrology on forested wetland productivity. In a study of cypress tree productivity in Florida, Mitsch and Ewel (1979) concluded that growth was low in pure stands of cypress (characterized by deep standing water) and in cypress–pine associations (characterized by dry conditions). The productivity of cypress was high in cypress–tupelo associations that were characterized by moderately wet conditions, in cypress–hardwood associations that were characterized by moderately wet conditions, and in cypress–hardwood associations that have fluctuating water levels characteristic of alluvial river swamps (Fig. 4.21a).

Conner and Day (1982) suggested a similar relationship between swamp productivity and hydrologic conditions and produced a similar curve, including some data points (Fig. 4.21b). Golet et al. (1993) developed a similar curve for radial growth of *Acer rubrum* as a function of hydrologic conditions for six swamps in Rhode

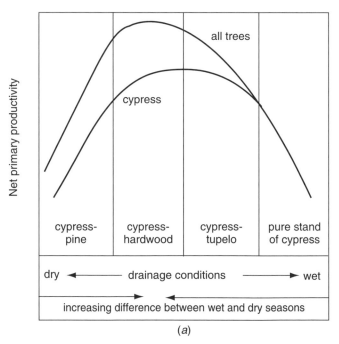

(a)

Figure 4.21   Relationships between swamp productivity and hydrologic conditions: (a) for cypress swamps in north-central Florida, (b) between flooding regime and net primary productivity of Louisiana swamps, and (c) between radial growth of red maple (*Acer rubrum*) and annual water level for six Rhode Island red maple swamps over six years. *(a after Mitsch and Ewel, 1979; b after Conner and Day, 1982; c after Golet et al., 1993.)*

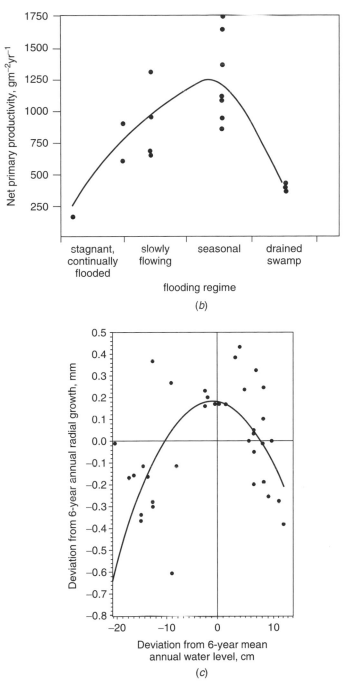

(b)

flooding regime

(c)

Figure 4.21 (*continued*)

Island (Fig. 4.21c). They showed that productivity was greatest when the water level was average and that productivity, as measured by radial growth, was lower when conditions were either wetter or drier than normal. All of the curves in Figure 4.21 suggest that the highest productivity occurs in systems that are neither very wet nor too dry but that have either average hydrologic conditions or seasonal hydrologic pulsing.

The subsidy-stress model of H. T. Odum (1971) and E. P. Odum (1979a), later refined as the *pulse stability* concept by all three Odums (W. E. Odum et al., 1995), includes concepts that potentially apply well to the effects of hydrology on wetland productivity. Seasonal pulsing of flood water can be both a subsidy and a stress, whether the wetland is a salt marsh or mangrove swamp subject to twice-per-day flooding or a riparian wetland subject to seasonal river pulses. Pulsing is frequent in nature, and ecosystems such as bottomland forests and salt marshes appear to be well adapted to taking advantage of this subsidy. Despite this clear theoretical basis for understanding the effects of hydrology on productivity, it has been difficult to confirm or deny these theories in practice.

The model developed by Mitsch and Rust (1984; Fig. 4.22) may explain the difficulty in ascribing a direct relationship between vascular plant productivity and

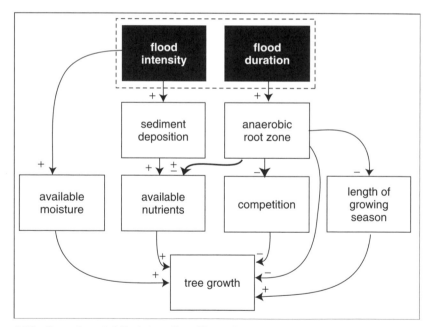

Figure 4.22 Causal model that describes the major causes for increases and decreases in individual tree growth in riparian floodplain forests. Plus (+) sign indicates a positive effect; minus sign (−) indicates a negative effect. *(After Mitsch and Rust, 1984.)*

Table 4.7   Selected macrophyte measurements at peak biomass from diked (hydrologically isolated) and undiked wetlands of Ohio's coastal Lake Erie

	Average ± Standard Error	
Measure of Vegetation Structure	Diked Wetlands ($N = 6$)	Undiked Wetlands ($N = 4$)
Biomass, g dry weight/m^2	897 ± 277	473 ± 149
# species/plot[a]	1.7 ± 0.3	1.4 ± 0.3
# stems/m^2	597 ± 211	241 ± 59

[a]Only species greater than 10 percent by weight per plot. Plots were 0.5 m^2 randomly placed in each wetland (3 to 6 per wetland).
Source: Mitsch (1992b) and Mitsch et al. (1994b).

hydrologic conditions. While flood intensity increases available moisture and nutrients, longer flood durations increase stresses caused by an anaerobic root zone and can actually decrease the length of the growing season. In effect, "subsidies and stresses may occur simultaneously and cancel one another" (Megonigal et al., 1997). In the Mitsch-Rust model, flood intensity and duration affect moisture, available nutrients, anaerobiosis, and even length of growing season in a complex and nonlinear "push-pull" arrangement.

The influence of hydrologic conditions on freshwater marsh productivity is less certain. If peak biomass or similar measures are used as indicators of marsh productivity, some studies have shown the classical stimulation of vegetation along the water's edge, whereas other studies have indicated a higher macrophyte productivity in sheltered, nonflowing marshes than in wetlands that are open to flowing conditions or coastal influences. For example, consistently higher macrophyte biomass was found in wetlands isolated from surface fluxes with artificial dikes than in wetlands that were open to coastal fluxes along Lake Erie (Table 4.7). Several explanations are possible: (1) The coastal fluxes may also be serving as a stress as well as a subsidy on the macrophytes; (2) the open marshes may be exporting a significant amount of their productivity; and (3) the diked wetlands have more predictable hydroperiods. Similar results were found in a hydrologic pulsing experiment in central Ohio, where simulated river floods caused a decrease in macrophyte primary productivity because of a flushing effect (see BOX; also Fig. 4.23a).

Conversely, an earlier study of the influence of flowthrough conditions on water column primary productivity of constructed marshes found that, after two years of experimentation, water column (phytoplankton and submerged aquatics) productivity was higher in steady-flow but high-flow wetlands compared to steady-flow but low-flow wetlands (Fig. 4.23b). While macrophyte productivity may take many years to respond to the difference in hydrology, water column productivity, which is often caused by attached and planktonic algae, responds relatively quickly to changing hydrologic conditions.

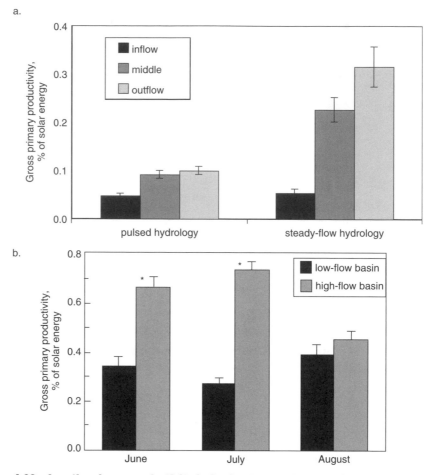

Figure 4.23 Aquatic primary productivity in freshwater marshes as a function of hydrologic conditions: (a) pulsed flooding vs. steady-flow hydrology at the Olentangy River Wetland Research Park, central Ohio; (b) high-flow and low-flow conditions at the Des Plaines Wetland Demonstration Project, northeastern Illinois. (a. after Tuttle and Mitsch, in review; b. after Cronk et al., 1994.)

Full-Scale Wetland Hydrology Experiment—Effects of Pulsing on Riverine Wetlands

The Wilma H. Schiermeier Olentangy River Wetland Research Park at The Ohio State University has been the site of multiyear experiments on the effect of river pulsing on wetland function. Two 1-ha experimental wetlands were subjected to pumped river pulses in one year (2004) and steady-flow conditions in

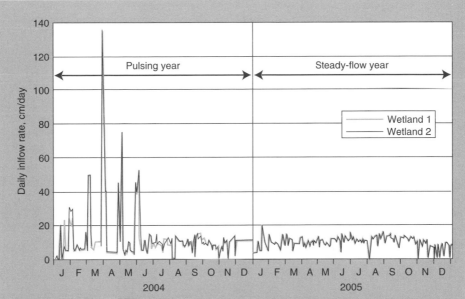

Figure 4.24   Pumped inflow rates to two 1-ha experimental wetlands at Olentangy River Wetland Research Park pulsing experiment of 2004–05. The first year had six scheduled 7-day floods in the first 6 months of the year; the second year had no scheduled floods but only steady-flow conditions. Flow into the two wetlands averaged 43 m/yr for the pulsing year and 39 m/yr for the steady-flow year.

the second year (2005) (Fig. 4.24). Floods were scheduled for the first week of the month for January through June 2004 to enable researchers to be ready for measurements. Mean inflow rates were increased to about 50 cm/day for the flood and then reduced to flows of about 7 cm/day between floods. During the last 6 months of 2004 and all of 2005, a mean flow of about 10 cm/day was maintained. Results of this pulsing experiment are summarized in Table 4.8. Aquatic primary productivity and macrophyte productivity were significantly lower during the river-pulse year than during the steady-flow year due to flushing effects and scouring of macrophytes caused by pulsing turbulence. Aquatic gross primary productivity was normalized for solar radiation and was 0.08 percent of solar energy during pulses and 0.20 percent during steady-flow conditions. There were conflicting effects of pulsing on production of two "greenhouse gases" produced by wetlands—methane and nitrous oxide. Methane fluxes in intermittently flooded zones of the wetlands were 30 percent of those in permanently flooded wetland areas (Altor and Mitsch, 2006), while nitrous oxide emissions in intermittently flooded zones were double those in permanently flooded areas (Hernandez and Mitsch, 2006).

Table 4.8   Effects of hydrologic pulsing on riparian wetland productivity and biogeochemistry in Olentangy River Wetland Research Park pulsing experiment of 2004–05 as shown in Figure 4.24

Wetland productivity

- Lower water column productivity during pulses
- Lower macrophyte net primary productivity/more organic export during pulsing year;

Wetland biogeochemistry

- Uneven seasonal effect of pulsing on sedimentation
- Higher denitrification rates in permanently flooded low marsh areas than in intermittently flooded edges
- Higher overall denitrification during pulsing year compared to steady-flow year
- Higher nitrous oxide emissions on pulsed intermittently flooded edges of wetlands than in permanently flooded areas
- Lower methane on pulsed intermittently flooded edges of wetlands than in permanently flooded areas
- Increased nitrate-nitrogen retention and organic nitrogen export between pulses

Sources: Mitsch et al. (2005c), Altor and Mitsch (2006), Hernandez and Mitsch (2006, 2007), Tuttle et al. (in review)

Overall denitrification rates in the wetlands decreased by 24 percent from the year when river pulses occurred to the steady-flow year.

Coastal wetlands subject to frequent tidal action are generally more productive than those that are only occasionally inundated. A comparison of several Atlantic Coast salt marshes, for example, showed a direct relationship between tidal range (as a measure of water flux) and end-of-season peak biomass of *Spartina alterniflora* (Fig. 4.25). Apparently, vigorous tides increase the nutrient subsidy and cause a flushing of toxic materials such as salt. Freshwater tidal wetlands are even more productive than saline tidal wetlands, because they receive the energy and nutrient subsidy of tidal flushing while avoiding the stress of saline soils.

Organic Accumulation and Export

Wetlands can accumulate excess organic matter as a result of either increased primary productivity (as described previously) or decreased decomposition and export. Notwithstanding the discrepancies from short-term litter decomposition studies, peat accumulates to some degree in all wetlands as a result of these processes. The effects of hydrology on decomposition pathways are even less clear than the effects on primary

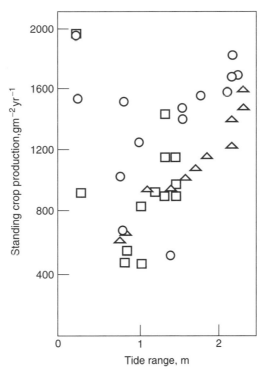

Figure 4.25   Production of *Spartina alterniflora* versus mean tidal range for several Atlantic Coast salt marshes. Different symbols indicate different data sources. (*After Steever et al., 1976.*)

productivity discussed previously. Probably the lack of agreement among the many studies published on the subject results from the complexity of the decomposition process. In general, decomposition of organic detritus requires electron donors (usually oxygen, but alternate chemicals such as sulfate or nitrate may be effective under anoxic conditions), moisture, inorganic nutrients, and microorganisms capable of metabolizing in the specific environment concerned. The observed rate of organic decomposition is also influenced by the ambient temperature and by the activity of macrodetritivores that shred the plant remains and/or repackage it as bacterially inoculated fecal pellets. Hydrology modifies many of these variables; for example, moisture depends on the flooding regime, flowing water carries oxygen and nutrients, while in stagnant water oxygen is rapidly depleted and nutrients are transformed to more or less available forms. Given this complexity, it is not surprising that the results of short-term *in situ* decomposition studies often disagree.

The importance of hydrology for organic carbon export is obvious. A generally higher rate of export is to be expected from wetlands that are open to the flowthrough of water. Riparian wetlands often contribute large amounts of organic detritus to streams, including macro-detritus such as whole trees. There is also considerable

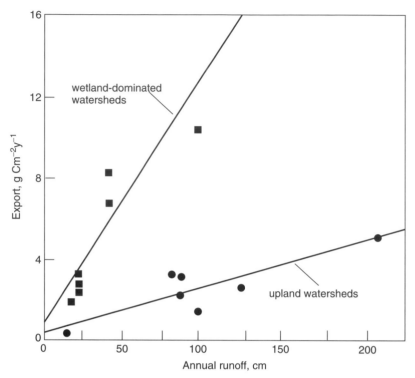

Figure 4.26 **Organic carbon export from wetland-dominated watersheds compared with non-wetland watersheds.** *(After Mulholland and Kuenzler, 1979.)*

evidence that watersheds that drain wetland regions export more organic material but retain more nutrients than do watersheds that do not have wetlands (Fig. 4.26). For example, the slope of the line in Figure 4.26 for a swamp-draining watershed is much steeper than that for upland watersheds, indicating a much greater organic carbon concentration in runoff as well as greater export for a give runoff from the wetland-dominated watersheds. Salt marshes and mangrove swamps are also considered major exporters of their productivity by most, but the generality of this concept is not fully accepted by coastal ecologists. Hydrologically isolated wetlands such as northern peatlands have much lower organic export.

Nutrient Cycling

Nutrients are carried into wetlands by the hydrologic inputs of precipitation, river flooding, tides, and surface and groundwater inflows. Outflows of nutrients are controlled primarily by the outflow of water. These hydrologic/nutrient flows are also important determinants of wetland productivity and decomposition (see previous sections). Intrasystem nutrient cycling is generally, in turn, tied to pathways such as primary productivity and decomposition. When productivity and decomposition rates are high, as in flowing water or pulsing hydroperiod wetlands, nutrient cycling is rapid. When productivity and decomposition processes are slow, as in isolated ombrotrophic bogs, nutrient cycling is also slow.

The hydroperiod of a wetland has a significant effect on nutrient transformations, on the availability of nutrients to vegetation, and on loss from wetland soils of nutrients that have gaseous forms (see Chapter 5). Thus, nitrogen availability and loss are affected in wetlands by the reduced conditions that result from waterlogged soil. Typically, a narrow oxidized surface layer develops over the anaerobic zone in wetland soils, causing a combination of reactions in the nitrogen cycle—nitrification and denitrification—that may result in substantial losses of dinitrogen gas to the atmosphere. Furthermore, ammonium nitrogen is usually the form of nitrogen most available to plants in wetland soils, because the anaerobic environment favors the reduced ionic form over the nitrate common in agricultural soils.

Flooding of wetland soil, by altering both the pH and the redox potential of the soil, influences the availability of other nutrients. The pH of both acid and alkaline soils tends to converge on a pH of 7 when they are flooded (see Chapter 5). The redox potential, a measure of the intensity oxidation or reduction of a chemical or biological system, indicates the state of oxidation (and, hence, availability) of several nutrients. Phosphorus is known to be more soluble under anaerobic conditions, partly because of the hydrolysis and reduction of ferric and aluminum phosphates to more soluble compounds. The availability of major ions such as potassium and magnesium and several trace nutrients such as iron, manganese, and sulfur is also affected by hydrologic conditions in the wetlands.

Techniques for Wetland Hydrology Studies

It is curious that so little attention has been paid to hydrologic measurements in wetland studies, despite the importance of hydrology in ecosystem function. A great deal of information can be obtained with only a modest investment in supplies and equipment. A diagram summarizing many of the hydrology measurements typical for developing a wetland's water budget is given in Figure 4.27. Water levels can be recorded continuously with a water-level recorders or data loggers or during site visits with a staff gauge. With records of water level, all of the following hydrologic parameters can be determined: hydroperiod, frequency of flooding, duration of flooding, and water depth. Water-level recorders can also be used to determine the change in storage in a water budget, as in Equation 4.1.

Evapotranspiration measurements are more difficult to obtain, but several empirical relationships, such as the Thornthwaite equation, use meteorological variables. Evaporation pans can also be used to estimate total evapotranspiration from wetlands, although pan coefficients are highly variable. Evapotranspiration of continuously flooded nontidal wetlands can also be determined by monitoring the diurnal water-level fluctuation.

Precipitation or throughfall or both can be measured by placing a statistically adequate number of rain gauges in random locations throughout the wetland or by

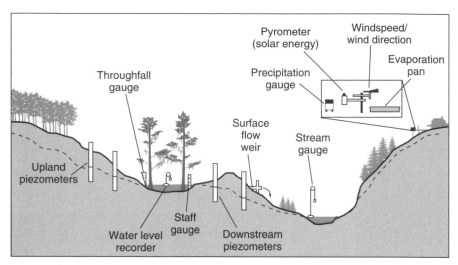

Figure 4.27 Placement of hydrology instruments in the landscape to estimate a water budget for a floodplain wetland.

utilizing weather station data. Surface runoff to wetlands can usually be determined as the increase in water level in the wetland during and immediately following a storm after net precipitation has been subtracted. Weirs can be constructed on more permanent streams to monitor surface water inputs and outputs.

Groundwater flows are usually the most difficult and most costly hydrologic flows to measure accurately. In some cases, clusters of shallow monitoring wells, placed around a wetland, will help indicate the direction of groundwater flow and the slope of the water or hydraulic gradient as required in Equation 4.12. The wells are called *piezometers* when they are only partially screened, and thus measure the piezometric head of an isolated part of the groundwater rather than being screened through the entire length of the well and thus measuring the surface water aquifer. Piezometers can be installed by professional well-drilling companies or, for low-budget installations, can generally be installed with augers or as well points. Estimates of permeability or hydraulic conductivity are then required to quantify the flows. Permeability can be estimated through *in situ* pump tests using the wells or through laboratory analysis of intact soil cores. The variability of results among different hydraulic conductivity measuring techniques suggests that caution should be used in taking these numbers.

If a wetland is a perched or a recharge wetland, seepage can be estimated either through a water budget approach (e.g., subtracting evapotranspiration losses from water-level decreases when there are no other inflows or outflows) or by using half-barrel seepage meters. Other methods available to measure groundwater flows in wetlands include the use of stable isotopes, generally $^{18}O/^{16}O$ or $^{2}H/^{1}H$, because of the propensity of the lighter isotope in each case to evaporate more readily, allowing water to be "tagged" according to its source (Hunt et al., 1996). Groundwater flow models have also been used to estimate the flow of groundwater into and out of wetlands with some success (Hunt et al., 1996; Koreny et al., 1999).

The uncertainty in the scientific literature concerning many wetland processes (e.g., the rates of organic matter decomposition discussed earlier) is often closely related to unquantified hydrologic parameters. Thus, careful attention to quantification of pertinent hydrologic parameters in wetland research studies is virtually certain to improve our understanding of the ecological processes that control wetlands.

Recommended Readings

Winter, T. C., and M. R. Llamas, eds. 1993. "Hydrogeology of Wetlands." Special Issue of *Journal of Hydrology* 141:1–269.

Wetland Biogeochemistry

Wetland biogeochemistry features a combination of chemical transformations and chemical transport processes not shared by many other ecosystems. Wetland soils, known as hydric soils, are formed when oxygen is cut off due to the presence of water, causing chemically reduced conditions. They can be organic soils or mineral soils. Hydric mineral soils can be identified through redox concentrations, redox depletions, and reduced matrices. Transformations of nitrogen, sulfur, iron, manganese, carbon, and phosphorus occur as a result of these anaerobic conditions. Some transformations cause toxic conditions, as with the production of hydrogen sulfide, whereas others, such as denitrification and methanogenesis, allow emissions of chemicals to the atmosphere. Many transformations in wetlands are mediated by microbial populations that are adapted to the anaerobic environment, while processes such as those in the phosphorus cycle are dominated by physical and chemical processes. Wetlands can be sources, sinks, or transformers of chemicals and are often chemically coupled to adjacent ecosystems by the export of organic materials, although the direct effects on adjacent ecosystems have been difficult to quantify.

The transport and transformation of chemicals in ecosystems, known as *biogeochemical cycling*, involve a great number of interrelated physical, chemical, and biological processes. The diverse hydrologic conditions in wetlands discussed in the previous chapter markedly influence biogeochemical processes. These processes result not only in changes in the chemical forms of materials but also in the spatial movement of materials within wetlands, as in water–sediment exchange and plant uptake, and with surrounding ecosystems, as in organic exports. These processes, in turn, determine overall wetland productivity. The interrelationships among hydrology, biogeochemistry, and response of wetland biota were summarized in Figure 4.1.

The biogeochemistry of wetlands can be divided into (1) *intrasystem cycling* through various transformation processes and (2) the exchange of chemicals between a wetland and its surroundings. Although no transformation processes are unique to wetlands, the permanent or intermittent flooding of these ecosystems causes certain processes to be more dominant in wetlands than in either upland or deep aquatic ecosystems. For example, while *anaerobic*, or oxygenless, conditions are sometimes found in other ecosystems, they prevail in wetlands. Wetland soils are characterized by waterlogged conditions during part or all of the year, which produce reduced conditions, which, in turn, have a marked influence on several biochemical transformations unique to anaerobic conditions.

This intrasystem cycling, along with hydrologic conditions, influences the degree to which chemicals are transported to or from wetlands. An ecosystem is considered biogeochemically *open* when there is an abundant exchange of materials with its surroundings. When there is little movement of materials across the ecosystem boundary, it is biogeochemically *closed*. Wetlands can fall into either category. For example, wetlands such as bottomland forests and tidal salt marshes have a significant exchange of minerals with their surroundings through river flooding and tidal exchange, respectively. Other wetlands such as ombrotrophic bogs and cypress domes have little material exchange except for gaseous matter that passes into or out of the ecosystem. These latter systems depend more on intrasystem cycling than on throughput for their chemical supplies.

Wetland Soils

Types and Definitions

Wetland soils are both the medium in which many of the wetland chemical transformations take place and the primary storage of available chemicals for most wetland plants. They are often described as *hydric soils*, defined by the U.S. Department of Agriculture's Natural Resources Conservation Service (NRCS, 1998) as "soils that formed under conditions of saturation, flooding or ponding long enough during the growing season to develop anaerobic conditions in the upper part." Wetland soils are of two types: (1) *mineral soils* or (2) *organic soils*. Nearly all soils have some organic material, but when a soil has less than 20 to 35 percent organic matter (on a dry-weight basis), it is considered a mineral soil.

Organic soils and organic soil materials are defined under either of two conditions of saturation:

1. Soils are saturated with water for long periods or are artificially drained and, excluding live roots, (a) have 18 percent or more organic carbon if the mineral fraction is 60 percent or more clay, (b) have 12 percent or more organic carbon if the mineral fraction has no clay, or (c) have a proportional content of organic carbon between 12 and 18 percent if the clay content of the mineral fraction is between 0 and 60 percent (Fig. 5.1); or

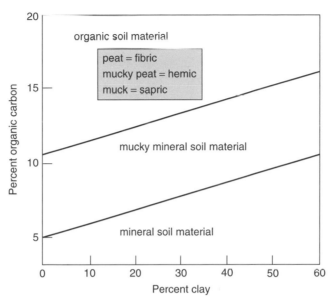

Figure 5.1 Percentage organic carbon required for a soil material to be called organic, mucky mineral soil, or mineral soil material. *(After NRCS, 1998.)*

2. Soils are never saturated with water for more than a few days and have 20 percent or more organic carbon.

For an estimate of organic carbon when organic matter content is known,

$$\%C_{org} = \%OM/2 \tag{5.1}$$

where

$\%C_{org}$ = percentage of organic carbon
$\%OM$ = percentage of organic matter

Any soil material that is not included in the preceding definition is considered mineral soil material. Where mineral soils occur in wetlands, such as in some freshwater marshes or riparian forests, they generally have a soil profile made up of horizons, or layers. The upper layer of wetland mineral soils is often organic peat composed of partially decayed plant materials.

Although the preceding definition of organic soil is applicable to many types of wetlands, particularly to northern peatlands, *peat*, a generic term for relatively undecomposed organic soil, is not usually that strictly defined. Most peats contain less than 20 percent unburnable inorganic matter (and therefore usually contain more than 80 percent burnable organic material, which is about 40 percent organic carbon). Some soil scientists, however, allow up to 35 percent unburnable inorganic matter (approximately 33 percent organic carbon), and commercial operations sometimes allow 55 percent unburnable material (22 percent organic carbon).

Table 5.1 Comparison of mineral and organic soils in wetlands

	Mineral Soil	Organic Soil
Organic content (percent)	Less than 20 to 35	Greater than 20 to 35
Organic carbon (percent)	Less than 12 to 20	Greater than 12 to 20
pH	Usually circumneutral	Acid
Bulk density	High	Low
Porosity	Low (45–55%)	High (80%)
Hydraulic conductivity	High (except for clays)	Low to high[a]
Water holding capacity	Low	High
Nutrient availability	Generally high	Often low
Cation exchange capacity	Low, dominated by major cations	High, dominated by hydrogen ion
Typical wetland	Riparian forest, some marshes	Northern peatland

[a]See Chapter 4.

Organic soils are different from mineral soils in several physicochemical features other than the percentage of organic carbon (Table 5.1):

1. *Bulk density and porosity.* Organic soils have lower bulk densities and higher water-holding capacities than do mineral soils. Bulk density, defined as the dry weight of soil material per unit volume, is generally 0.2 to 0.3 g/cm^3 when the organic soil is well decomposed, although peatland soils composed of *Sphagnum* moss can be extremely light, with bulk densities as low as 0.04 g/cm^3. By contrast, mineral soil bulk density generally ranges between 1.0 and 2.0 g/cm^3. Bulk density is low in organic soils because of their high porosity, or percentage of pore spaces. Peat soils generally have at least 80 percent pore spaces and are, thus, 80 percent water by volume when flooded. Mineral soils generally range from 45 to 55 percent total pore space, regardless of the amount of clay or texture.

2. *Hydraulic conductivity.* Both mineral and organic soils have wide ranges of possible hydraulic conductivities; the latter depends on their degree of decomposition (see Table 4.5 and Fig. 4.17). Organic soils may hold more water than mineral soils, but, given the same hydraulic conditions, they do not necessarily allow water to pass through more rapidly.

3. *Nutrient availability.* Organic soils generally have more minerals tied up in organic forms unavailable to plants than do mineral soils. This follows from the fact that a greater percentage of the soil material is organic. This does not mean, however, that there are more total nutrients in organic soils; very often, the opposite is true in wetland soils. For example, organic soils can be extremely low in bioavailable phosphorus or iron content—enough to limit plant productivity.

4. *Cation exchange capacity.* Organic soils have a greater cation exchange capacity, defined as the sum of exchangeable cations (positive ions) that a soil can hold. Figure 5.2 summarizes the general relationship between organic

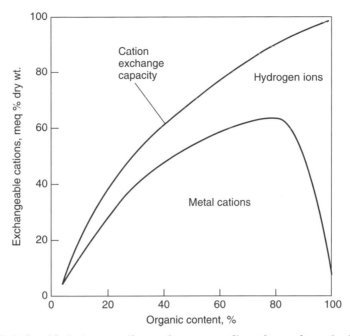

Figure 5.2   Relationship between cation exchange capacity and organic content for wetland soils. For low organic content (mineral soils), the cation exchange capacity is saturated by metal cations; when organic content is high, the exchange capacity is dominated by hydrogen ions. (After Gorham, 1967.)

content and cation exchange capacity of soils. Mineral soils have a cation exchange capacity that is dominated by the major metal cations (Ca^{2+}, Mg^{2+}, K^+, and Na^+). As organic content increases, both the percentage and the amount of exchangeable hydrogen ions increase. For *Sphagnum* moss peat, the high cation capacity may be caused by long chain polymers of uronic acid (Clymo, 1983).

Organic Wetland Soil

Organic soil is composed primarily of the remains of plants in various stages of decomposition and accumulates in wetlands as a result of the anaerobic conditions created by standing water or poorly drained conditions. Two of the more important characteristics of organic soil, including soils commonly termed peat and *muck*, are the botanical origin of the organic material and the degree to which it is decomposed. Several of the properties that have been discussed, including bulk density, cation exchange capacity, hydraulic conductivity, and porosity, are often dependent on these characteristics. Therefore, it is often possible to predict the range of the physical properties of an organic soil if the origin and state of decomposition can be observed in the field or laboratory.

Botanical Origin

The botanical origin of the organic material can be (1) mosses, (2) herbaceous material, and (3) wood and leaf litter. For most northern peatlands, the moss is usually *Sphagnum*, although several other moss species can dominate if the peatland is receiving inflows of mineral water. Organic soils can originate from herbaceous grasses such as reed grass (*Phragmites*), wild rice (*Zizania*), and salt marsh cord-grass (*Spartina*), or from sedges such as *Carex* and *Cladium*. Organic soils can also be produced in freshwater marshes by plant fragments from several nongrass and nonsedge plants, including cattails (*Typha*) and water lilies (*Nymphaea*). In forested wetlands, the peat can be a result of woody detritus or leaf material or both. In northern peatlands, the material can originate from birch (*Betula*), pine (*Pinus*), or tamarack (*Larix*), and in southern deepwater swamps, the organic horizon can be composed of material from cypress (*Taxodium*) or water tupelo (*Nyssa*) trees.

Decomposition

The state of decomposition, or *humification*, of wetland soils is the second key characteristic of organic peat. As decomposition proceeds, albeit at a very slow rate in flooded conditions, the original plant structure is changed physically and chemically until the resulting material little resembles the parent material. As peat decomposes, bulk density increases, hydraulic conductivity decreases, and the quantity of larger (>1.5 mm) fiber particles decreases as the material becomes increasingly fragmented. Chemically, the amount of peat "wax," or material soluble in nonpolar solvents, and lignin increase with decomposition, whereas cellulose compounds and plant pigments decrease. When some wetland plants such as salt marsh grasses die, the detritus rapidly loses a large percentage of its organic compounds through leaching. These readily soluble organic compounds are thought to be easily metabolized in adjacent aquatic systems.

Classification and Characteristics

Organic soils (*histosols*) are classified into four groups, three of which are considered hydric soils:

1. *Saprists* (muck). Two-thirds or more of the material is decomposed, and less than one-third of plant fibers are identifiable.
2. *Fibrists* (peat). Less than one-third of material is decomposed, and more than two-thirds of plant fibers are identifiable.
3. *Hemists* (mucky peat or peaty muck). Conditions fall between saprist and fibrist soil.
4. *Folists*. Organic soils caused by excessive moisture (precipitation > evapotranspiration) that accumulate in tropical and boreal mountains; these soils are not classified as hydric soils because saturated conditions are the exception rather than the rule.

Organic soil is generally dark in color, ranging from the dark black soils characteristic of mucks such as those found in the Everglades in Florida to the dark brown color of partially decomposed peat from northern bogs.

Mineral Wetland Soil

Mineral soils, when flooded for extended periods, develop certain characteristics that allow for their identification. These characteristics are collectively called *redoximorphic features*, defined as features formed by the reduction, translocation, and/or oxidation of iron and manganese oxides (Vepraskas, 1995).

The development of redoximorphic features in mineral soils is mediated by microbiological processes. The rate at which they are formed depends on three conditions, all of which must be present:

1. Sustained anaerobic conditions
2. Sufficient soil temperature (5°C is often considered "biological zero," below which much biological activity ceases or slows considerably; see description of biological zero and its importance to wetland science by Rabenhorst, 2005)
3. Organic matter, which serves as a substrate for microbial activity

Reduced Matrices and Redox Depletions

One characteristic of many hydric mineral soils that are semi-permanently or permanently flooded is the development of black, gray, or sometimes greenish or blue-gray color as the result of a process known as *gleying*. This process, also known as *gleization*, is the result of the chemical reduction of iron (see Iron and Manganese Transformations later in this chapter). When soils are not saturated with water, iron (ferric = Fe^{3+}) oxides are the principal chemicals that give the soil its typical red, brown, yellow, or orange color. Manganese (Mn^{3+} or Mn^{4+}) oxides give the soil a black color. When soils are flooded and become reduced, the iron is reduced to a soluble form of iron (ferrous = Fe^{2+}) and the maganese is reduced to its soluble manganous (Mn^{2+}) form. These soluble forms of iron and manganese can be leached out of the soil, leaving the natural (gray or black) color of the parent sand, silt, or clay, called the matrix. A similar term used to describe these reduced soils is *redox depletions*—iron is reduced and then depleted from the soil matrix. In a similar manner, *clay depletions* occur when clay is selectively removed along root channels after iron and manganese oxides have been depleted, only to redeposit as clay coatings on soil particles below the clay depletions (Vepraskas, 1995).

Oxidized Rhizosphere

Another characteristic of some mineral wetland soils is the presence of an *oxidized rhizosphere* (also called *oxidized pore linings*) that results from the capacity of many hydrophytes to transport oxygen through above-ground stems and leaves to below-ground roots (Fig. 5.3). Excess oxygen, beyond the root's metabolic needs, diffuses from the roots to the surrounding soil matrix, forming deposits of oxidized

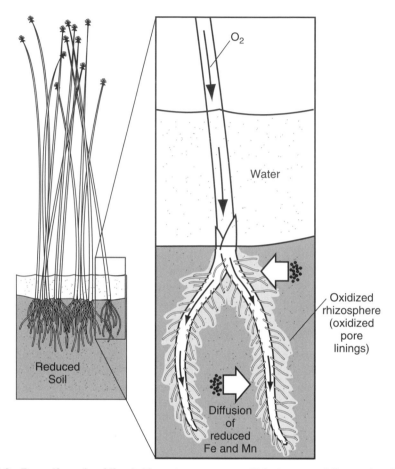

Figure 5.3 Formation of oxidized rhizospheres, or pore linings, around the roots of a wetland plant caused by the transport of excess oxygen by wetland plants to their roots. When the plant dies, pore linings of iron and manganese oxides remain. (After Vepraskas, 1995.)

iron along small roots. When a wetland soil is examined, these oxidized rhizosphere deposits can often be seen as thin traces through an otherwise dark matrix.

Redox Concentrations

Mineral soils that are seasonally flooded, particularly by alternate wetting and drying, develop spots of highly oxidized materials called *mottles* or *redox concentrations* (Fig. 5.4). Mottles and redox concentrations are orange/reddish-brown (because of iron) or dark reddish-brown/black (because of manganese) spots seen throughout an otherwise gray (gleyed) soil matrix and suggest intemittently exposed soils with spots of iron and manganese oxides in an otherwise reduced environment. Mottles are relatively insoluble, enabling them to remain in soil long after it has been drained.

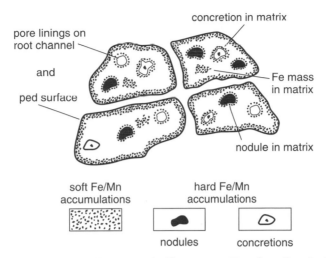

Figure 5.4 Different kinds of redox concentrations, or mottles, in soil peds (soil macroparticles), including nodules and concretions, iron masses in soil matrix (also called reddish mottles), and pore linings on root channel (also called oxidized rhizospheres). *(After Vepraskas, 1995.)*

Modern Nomenclature

A revised set of terms defining redoximorphic features has been devised by soil scientists to describe indicators of hydric soils, or more properly, to identify an *aquic condition*—the condition in which soils are saturated with water, are reduced, and display redoximorphic features. The term *aquic condition* was introduced in the early 1990s to better reconcile field techniques that used soil colors (e.g., iron reduction or oxidation) with the former term *aquic moisture regime*—any soil that was saturated with water and chemically reduced such that no dissolved oxygen was present. The redoximorphic features that can be used to identify aquic conditions are (Vepraskas, 1995):

1. *Redox concentrations.* Accumulation of iron and manganese oxides (formerly called mottles) in at least three different structures (Fig. 5.4):
   a. *Nodules and concretions.* Firm to extremely firm irregularly shaped bodies with diffuse boundaries
   b. *Masses.* Formerly called "reddish mottles"
   c. *Pore linings.* Formerly included "oxidized rhizosphores" (Fig. 5.3)
2. *Redox depletions.* Low-chroma (≤ 2) bodies with high values (≥ 4) including:
   a. *Iron depletions.* Sometimes called "gray mottles" or "gley mottles"; these are low-chroma bodies
   b. *Clay depletions.* Contain less iron, manganese, and clay than adjacent soils
3. *Reduced matrices.* Low-chroma soils (because of presence of Fe^{2+} *in situ* that change color if exposed to air and iron is oxidized to Fe^{3+}

Mineral Hydric Soil Determination

In practice, the determination of whether a mineral soil is a hydric soil is a complicated process, but it is often done by determining soil color relative to a standard color chart called the Munsell soil color chart (Fig. 5.5). Soils that contain *low chromas* (as indicated by the color chips on the left-hand side of the color chart in Fig. 5.5b) indicate hydric soils. Soils that contain bright reds, browns, yellows, or oranges are nonhydric. In general, a chroma of 2 or less on the Munsell color chart is necessary for a soil to be classified as a hydric soil. These color charts are commonly used in the United States to identify the presence of hydric soils for the delineation of wetlands.

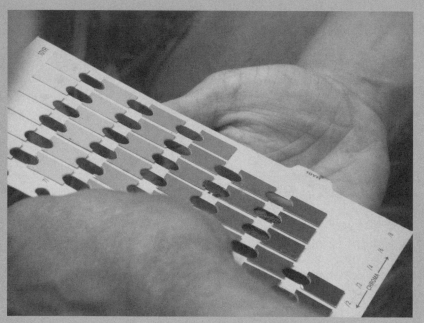

Figure 5.5 Hydric soils can be identified by comparing the soil color with standard soil color charts such as the Munsell® Soil Color Chart shown here. The hue, given in the upper right-hand corner of the chart (10YR in this case), indicates the relation to standard spectral colors such as red (R) or yellow (Y). The value notation (vertical scale) indicates the soil lightness (darker with lower value), and the chroma (horizontal scale) indicates the color strength or purity, with grayer soils to the left. Chromas of 2 or less generally indicate hydric soils. *(Photo by L. Zhang, reprinted with permission.)*

Reduction/Oxidation in Wetlands

When soils, whether mineral or organic, are inundated with water, anaerobic conditions usually result. When water fills the pore spaces, the rate at which oxygen can diffuse through the soil is drastically reduced. Diffusion of oxygen in an aqueous solution has been estimated at 10,000 times slower than oxygen diffusion through a porous medium such as drained soil. This low diffusion rate leads relatively quickly to anaerobic, or reduced, conditions, with the time required for oxygen depletion on the order of several hours to a few days after inundation begins (Fig. 5.6). The rate at which the oxygen is depleted depends on the ambient temperature, the availability of organic substrates for microbial respiration, and sometimes the chemical oxygen demand from reductants such as ferrous iron. The resulting lack of oxygen prevents plants from carrying out normal aerobic root respiration and strongly affects the availability of plant nutrients and toxic materials in the soil. As a result, plants that grow in anaerobic soils generally have some specific adaptations to this environment (see Chapter 6).

It is not always true that oxygen is totally depleted from the soil water of wetlands. There is usually a thin layer of oxidized soil, sometimes only a few millimeters thick,

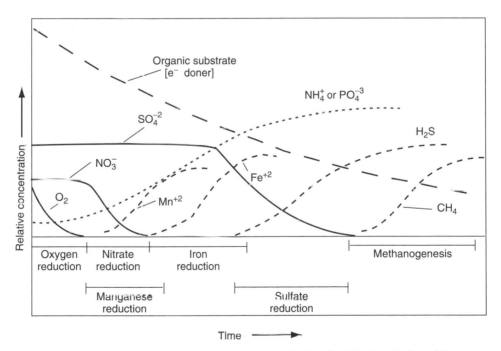

Figure 5.6 Sequence in time of transformations in soil after flooding, beginning with oxygen depletion and followed by nitrate and then sulfate reduction. Increases are seen in reduced manganese (manganous), reduced iron (ferrous), hydrogen sulfide, and methane. Note the gradual decrease in organic substrate (electron donor) and increases in available ammonium (NH_4^+) and phosphate (PO_4^{3-}) ions. The graph can also be interpreted as relative concentrations with depth in wetland soils. (After Reddy and D'Angelo, 1994.)

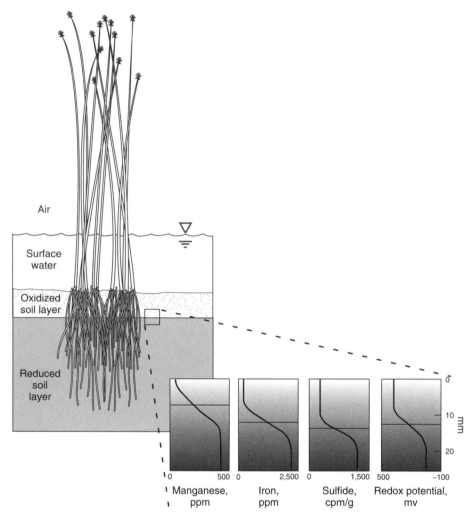

Figure 5.7 Characteristics of many wetland soils showing a shallow oxidized soil layer over a reduced soil layer. Also shown are soil profiles of reduced forms of manganese (sodium acetate–extractable manganese), iron (ferrous iron), and sulfur (sulfide), and redox potential. (After Patrick and Delaune, 1972; Gambrell and Patrick, 1978.)

at the surface of the soil at the soil–water interface (Fig. 5.7). The thickness of this oxidized layer is directly related to:

1. The rate of oxygen transport across the atmosphere–surface water interface
2. The small population of oxygen-consuming organisms present
3. Photosynthetic oxygen production by algae within the water column
4. Surface mixing by convection currents and wind action

Even though the deeper layers of the wetland soils remain reduced, this thin oxidized layer is often very important in the chemical transformations and nutrient cycling that occur in wetlands. Oxidized ions such as Fe^{3+}, Mn^{4+}, NO_3^-, and $SO_4^=$ are found in this microlayer, whereas the lower anaerobic soils are dominated by reduced forms such as ferrous and manganous salts, ammonia, and sulfides. Because of the presence of oxidized ferric iron (Fe^{3+}) in the oxidized layer, the soil surface often is a brown or brownish-red color, in contrast to the bluish-gray to greenish-gray color of the reduced gleyed sediments, dominated by ferrous iron (Fe^{2+}).

Redox potential, or oxidation–reduction potential, a measure of the electron pressure (or availability) in a solution, is often used to further quantify the degree of electrochemical reduction of wetland soils. *Oxidation* occurs not only during the uptake of oxygen but also when hydrogen is removed (e.g., $H_2S \rightarrow S^{2-} + 2H^+$) or, more generally, when a chemical gives up an electron (e.g., $Fe^{2+} \rightarrow Fe^{3+} + e^-$). *Reduction* is the opposite process of releasing oxygen, gaining hydrogen (hydrogenation), or gaining an electron.

Measuring Redox Potential

Redox potential can be measured in wetland soils and is a quantitative measure of the tendency of the soil to oxidize or reduce substances. When based on a hydrogen scale, redox potential is referred to as E_H and is related to the concentrations of oxidants {ox} and reductants {red} in a redox reaction by the *Nernst equation*:

$$E_H = E^0 + 2.3[RT/nF]\log[\{ox\}/\{red\}] \tag{5.2}$$

where

E^0 = potential of reference (mV)
R = gas constant = 81.987 cal deg^{-1} mol^{-1}
T = temperature (°K)
n = number of moles of electrons transferred
F = Faraday constant = 23,061 cal/mole-volt

Redox potential can be measured with a platinum electrode, which is easily constructed in the laboratory. Electric potential in units of millivolts (mV) is measured relative to a hydrogen electrode ($H^+ + e \longrightarrow H$) or to a calomel reference electrode. As long as free dissolved oxygen is present in a solution, the redox potential varies little (in the range of +400 to +700 mV). However, it becomes a sensitive measure of the degree of reduction of wetland soils after oxygen disappears, ranging from +400 mV down to –400 mV. Flooding and/or redox conditions in pond and wetland soils can be estimated by constructing platinum electrodes (Faulkner et al., 1989; Swerhone et al., 1999), using microplatinum electrodes (Meijer and Avnimelech, 1999), or using steel rods (Bridgham et al., 1991).

As organic substrates in a waterlogged soil are oxidized (donate electrons), the redox potential drops as a sequence of reductions (electron gains) takes place (Fig. 5.6). Because organic matter is one of the most reduced of substances, it can be oxidized when any number of terminal electron acceptors is available, including O_2, NO_3^-, Mn^{2+}, Fe^{3+}, or $SO_4^=$. Rates of organic decomposition are most rapid in the presence of oxygen and slower for electron acceptors such as nitrates and sulfates.

The oxidation of organic substrate is described by the following equation, which illustrates the organic substrate as an electron (e^-) donor (Fig. 5.6):

$$[CH_2O]n + nH_2O \rightarrow nCO_2 + 4ne^- + 4nH^+ \tag{5.3}$$

Various chemical and biological transformations take place as coupled oxidation (e^- donor)–reduction (e^- acceptor) reactions. Equations 5.3 and 5.4 make one such coupled reaction.

$$O_2 + 4e^- + 4H^+ \rightarrow 2H_2O \tag{5.4}$$

These transformations occur in a predictable sequence (Fig. 5.6), within predictable redox ranges to provide electron acceptors for this oxidation or decomposition (Table 5.2). The first and most common transformation is through aerobic oxidation when oxygen itself is the terminal electron acceptor (Eq. 5.4) at a redox potential of between 400 and 600 mV.

One of the first reactions that occur in wetland soils after they become anaerobic (i.e., the dissolved oxygen is depleted) is the reduction of NO_3^- (nitrate) first to NO_2^- (nitrite) and ultimately to N_2O or N_2; nitrate becomes an electron acceptor at

Table 5.2 Oxidized and reduced forms of several elements and approximate redox potentials for transformation

Element	Oxidized Form	Reduced Form	Approximate Redox Potential for Transformation (mV)
Nitrogen	NO_3^- (nitrate)	N_2O, N_2, NH_4^+	250
Manganese	Mn^{4+} (manganic)	Mn^{2+} (manganous)	225
Iron	Fe^{3+} (ferric)	Fe^{2+} (ferrous)	+100 to −100
Sulfur	$S_{O_4}^=$ (sulfate)	$S^=$ (sulfide)	−100 to −200
Carbon	CO_2 (carbon dioxide)	CH_4 (methane)	Below −200

a redox potential of approximately 250 mV:

$$2NO_3 + 10e^- + 12H^+ \rightarrow N_2 + 6H_2O \tag{5.5}$$

As the redox potential continues to decrease, manganese is transformed from manganic to manganous compounds at about 225 mV:

$$MnO_2 + 2e^- + 4H^+ \rightarrow Mn^{2+} + 2H_2O \tag{5.6}$$

Iron is transformed from ferric to ferrous form at about $+100$ to -100, while sulfates are reduced to sulfides at -100 to -200 mV:

$$Fe(OH)_3 + e^- + 3H^+ \rightarrow Fe^{2+} + 3H_2O \tag{5.7}$$
$$SO_4{}^= + 8e^- + 9H^+ \rightarrow HS^- + 4H_2O \tag{5.8}$$

Finally, under the most reduced conditions, the organic matter itself or carbon dioxide becomes the terminal electron acceptor below -200 mV, producing low-molecular-weight organic compounds and methane gas, as, for example,

$$CO_2 + 8e^- + 8H^+ \rightarrow CH_4 + 5H_2O \tag{5.9}$$

These redox potentials are not precise thresholds, because pH and temperature are also important factors in the rates of transformation. These major chemical transformations and others related to the nitrogen, sulfur, and carbon cycles are discussed in the next section.

The Nitrogen Cycle

The nitrogen cycle is one of the most important and studied chemical cycles in wetlands (Fig. 5.8). Nitrogen appears in a number of oxidation states in wetlands, several of which are important in a wetland's biogeochemistry. Nitrogen is often the most limiting nutrient in flooded soils, whether the flooded soils are in natural wetlands or on agricultural wetlands such as rice paddies. Nitrogen is considered one of the major limiting factors in coastal waters, making the nitrogen dynamics in coastal wetlands particularly significant, although this universal belief in nitrogen limitation in coastal wetlands has been challenged with studies that have suggested that a bacterial community in a pristine salt marsh in South Carolina was limited by phosphorus (Sundareshwar et al., 2005). Because of the presence of anoxic conditions in wetlands, microbial denitrification of nitrates to gaseous forms of nitrogen in wetlands and their subsequent release to the atmosphere remain one of the more significant ways in which nitrogen is lost from the lithosphere and hydrosphere to the atmosphere. Nitrates serve as one of the first terminal electron acceptors in wetland soils in this situation after the disappearance of oxygen (Table 5.2), making them an important chemical in the oxidation of organic matter in wetlands.

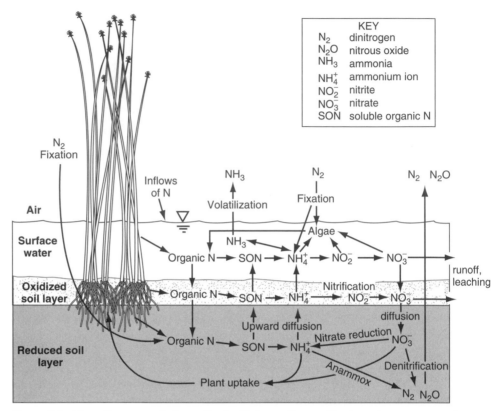

Figure 5.8 The nitrogen cycle in wetlands. Major pathways illustrated are nitrogen fixation, ammonia volatilization, nitrification, denitrification, plant uptake, dissimilatory nitrate reduction to ammonia (DNRA), and anammox (anaerobic ammonium oxidation).

Why the Nitrogen Cycle in Wetlands is Important

Humans have essentially doubled the amount of nitrogen that enters the land-based nitrogen cycle through fertilizer manufacturing, increased use of nitrogen-fixing crops, and fossil fuel burning (Vitousek et al., 1997; Galloway et al., 2003). Significant amounts of this excess nitrogen are transported as nitrate-nitrogen to rivers and streams, leading to eutrophication and episodic and persistent hypoxia (dissolved oxygen <2 mg/L) in coastal waters worldwide (NRC, 2000). For example, a hypoxic zone that often reaches 16,000 to 20,000 km^2 reappears annually in the Gulf of Mexico (Fig. 5.9; Rabalais et al., 2001), with the cause almost certainly being excessive nitrogen coming from the Mississippi-Ohio-Missouri (MOM) river basin. The extent of the hypoxia was half that area in the late 1990s. The ability of wetlands to serve as sinks for nitrogen is now being widely investigated as a

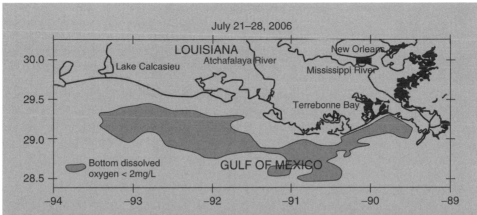

Figure 5.9   Extent of hypoxic conditions in Gulf of Mexico in summer 2006. Shaded area indicates where Gulf waters are less than 2 mg/L in dissolved oxygen. The hypoxia covered 17,300 km² that year and generally ranges from 15,000 to 20,000 km² in area. (From N. Rabalais, with permission.)

solution to this pollution problem in the American Midwest (see e.g., Mitsch et al., 2001; Mitsch and Day, 2006). The anaerobic process of denitrification is particularly important in this effort. More details are described in this chapter on nitrogen processes and in Chapter 12 on wetland restoration in the American Midwest to help solve the Gulf of Mexico hypoxia.

Nitrogen transformations in wetlands (Fig. 5.8) involve several microbiological processes, some of which make the nutrient less available for plant uptake. The ammonium ion (NH_4^+), with a nitrogen oxidation state of −3, is the primary form of mineralized nitrogen in most flooded wetland soils, although much nitrogen can be tied up in organic forms in highly organic soils. The presence of an oxidized zone over the anaerobic or reduced zone is critical for several of the pathways.

Nitrogen Mineralization

Nitrogen mineralization refers to a series of biological transformations that converts organically bound nitrogen to ammonium nitrogen as the organic matter is being decomposed and degraded. This pathway occurs under both anaerobic and aerobic conditions and is often referred to as *ammonification*. Typical formulas for the mineralization of a simple soluble organic nitrogen (SON) compound, urea, are given as

$$NH_2CONH_2 + H_2O \rightarrow 2NH_3 + CO_2 \tag{5.10}$$

$$NH_3 + H_2O \rightarrow NH_4 + OH^- \tag{5.11}$$

Ammonia Transformations and Nitrification

Once the ammonium ion (NH_4^+) is formed, it can take several possible pathways. It can be absorbed by plants through their root systems or by anaerobic microorganisms and converted back to organic matter. Under high-pH conditions (pH > 8), a common occurrence in marsh waters with excessive algal blooms, the ammonium ion can be converted to NH_3, which is then released to the atmosphere through *volatilization*. The ammonium ion can also be immobilized through ion exchange onto negatively charged soil particles. Because of the anaerobic conditions in wetland soils, ammonium would normally be restricted from further oxidation and would build up to excessive levels were it not for the thin oxidized layer at the surface of many wetland soils. The gradient between high concentrations of ammonium in the reduced soils and low concentrations in the oxidized layer causes an upward diffusion of ammonium, albeit very slowly, to the oxidized layer. In this aerobic environment, ammonium nitrogen can be oxidized through the process of *nitrification* in two steps by *Nitrosomonas* sp.:

$$2NH_4^+ + 3O_2 \rightarrow 2NO_2^- + 2H_2O + 4H^+ + \text{energy} \qquad (5.12)$$

and by *Nitrobacter* sp.:

$$2NO_2^- + O_2 \rightarrow 2NO_3^- + \text{energy} \qquad (5.13)$$

Nitrification can also occur in the oxidized rhizosphere of plants, where adequate oxygen is often available to convert the ammonium nitrogen to nitrate nitrogen.

Nitrate Transformations and Denitrification

Nitrate (NO_3^-), as a negative ion rather than the positive ammonium ion, is not subject to immobilization by negatively charged soil particles and is, thus, much more mobile in solution. If it is not assimilated immediately by plants or microbes (*assimilatory nitrate reduction*) or is lost through groundwater flow stemming from its rapid mobility, it has the potential to undergo *dissimilatory nitrogenous oxide reduction*, a term that refers to several pathways of nitrate reduction. It is called dissimilatory because the nitrogen is not assimilated into a biological cell. The most prevalent are reduction to ammonia and *denitrification*.

Denitrification, carried out by facultative bacteria under anaerobic conditions, with nitrate acting as a terminal electron acceptor, results in the loss of nitrogen as it is converted to gaseous nitrous oxide (N_2O) and molecular nitrogen (N_2):

$$C_6H_{12}O_6 + 4NO_3 \rightarrow 6CO_2 + 6H_2O + 2N_2 \qquad (5.14)$$

Denitrification is a significant path of nitrogen loss from most kinds of wetlands, including salt marshes, freshwater marshes, forested wetlands, and rice paddies. Denitrification is inhibited in acid soils and peat and is, therefore, thought to be of less consequence in northern peatlands. As illustrated in Figure 5.8, the entire process occurs after (1) ammonium nitrogen diffuses to the aerobic soil layer, (2) nitrification occurs, (3) nitrate nitrogen diffuses back to the anaerobic layer, and (4) denitrification, as described in Equation 5.14, occurs. The diffusion rates of the ammonium ion to the aerobic soil layer and the nitrate ion to the anaerobic layer are governed by the concentration gradients of the ions. There is generally a steep gradient of ammonium between the anaerobic and aerobic layers. Nevertheless, because nitrate diffusion rates in wetland soils are seven times faster than ammonium diffusion rates, ammonium diffusion and subsequent nitrification appear to limit the entire process of nitrogen loss by denitrification (Reddy and Patrick, 1984; Reddy and Graetz, 1988). Given an adequate supply of nitrate-nitrogen, the next most significant factor that affects denitrification appears to be soil temperature (Hernandez and Mitsch, 2007).

There are two gaseous products of denitrification—dinitrogen (N_2) and nitrous oxide (N_2O). The predominant gas that usually results from denitrification in most wetlands is N_2, and that is no environmental issue with an atmosphere already having 80 percent N_2. However, nitrous oxide is one of the so-called greenhouse gases that could cause climate change, so any attempt to design wetlands for nitrate removal should recognize this and understand conditions that minimize nitrous oxide production in favor of dinitrogen production. Hernandez and Mitsch (2006, 2007) found higher nitrous oxide fluxes in pulse-flooded high marsh areas compared to permanently flooded low marsh areas in midwestern United States freshwater wetlands (Fig. 5.10). The rate is probably limited by the lack of nitrates in the permanently flooded soils. Nitrous oxide production was highest when soil temperatures were greater than $20°$ C in the summer months both in pulsing and steady flow conditions. In addition, wetland plants did increase nitrous oxide emissions when sites were flooded but not when soils were exposed. Overall, the amount of nitrogen emitted as nitrous oxide in these riverine wetlands as a percentage of the amount of nitrate-nitrogen was quite small—about 0.3 percent of the total nitrate-nitrogen removed by the wetland. This study showed that nitrous oxide emissions and N_2O/N_2 ratios in denitrification are higher on the aerobic/anaerobic edges of wetlands than in the more anaerobic middle. It is reasonable to conclude then that if nitrate-nitrogen is denitrified in more aerobic farm fields, ditches, and even downstream coastal waters rather than in wetlands, higher nitrous oxide emissions would result from those systems than from the wetlands (Hernandez and Mitsch, 2006, 2007). Thus created and restored wetlands may not be the cause of additional nitrous oxide emissions; they may actually decrease the overall nitrous oxide emissions on a landscape scale.

Nitrogen Fixation

Nitrogen fixation results in the conversion of N_2 gas to organic nitrogen through the activity of certain organisms in the presence of the enzyme nitrogenase. It may be the source of significant nitrogen for some wetlands. Nitrogen fixation, which is carried out by certain aerobic and anaerobic bacteria and blue-green algae, is favored by low oxygen because nitrogenase activity is inhibited by high oxygen. In wetlands, nitrogen fixation can occur in overlying waters, in the aerobic soil layer, in the anaerobic soil

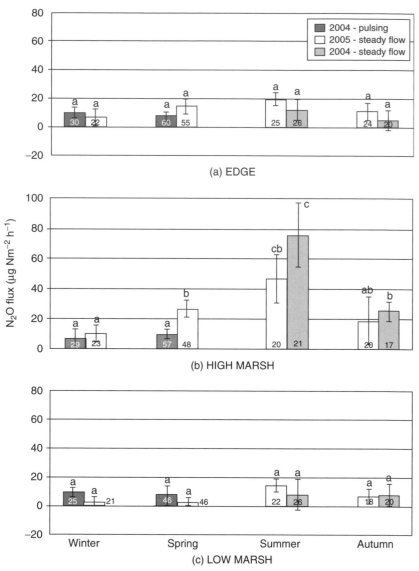

Figure 5.10 Seasonal nitrous oxide fluxes under different hydrologic conditions and along (a) dry edge, (b) high marsh (saturated soils with some standing water), and (c) low marsh (continuous standing water) in the freshwater experimental marshes at the Olentangy River Wetland Research Park in central Ohio. *(From Hernandez and Mitsch, 2006.)*

layer, in the oxidized rhizosphere of the plants, and on the leaf and stem surfaces of plants. Bacterial nitrogen fixation can be carried out by nonsymbiotic bacteria, by symbiotic bacteria of the genus *Rhizobium*, or by certain actinomycetes. Bacterial fixation is the most significant pathway for nitrogen fixation in salt marsh soils, while nitrogen-fixing bacteria are virtually absent from the low-pH peat of northern bogs. Cyanobacteria (blue-green algae) are common nitrogen fixers in wetlands, occurring in flooded delta soils in Louisiana, in northern bogs, and in rice cultures.

Dissimilatory Nitrate Reduction to Ammonia (DNRA)

Because conversion of nitrate-nitrogen to dinitrogen and nitrous oxide is considered to be the primary transformation of nitrates in anaerobic soils, an additional process whereby nitrate-nitrogen is transformed in anaerobic conditions is often overlooked (Megonigal et al., 2004). The process—called dissimilatory nitrate reduction to ammonia (DNRA)—occurs as follows, with mobile nitrates as the initial form of nitrogen and less-mobile ammonium as the product.

$$4NO_3^- + 4H_2 + 2H^+ \rightarrow 3H_2O + NH_4^+ \qquad (5.15)$$

The process yields energy to the many microorganisms capable of carrying out this process. The bacteria can be anaerobic, aerobic, or facultative. In some cases, nitrate reduction can be a more significant pathway than the other dissimilatory nitrate loss—denitrification. Many studies have supported the concept that high availability of organic carbon and/or low nitrate concentrations favors DNRA over denitrification (Megonigal et al., 2004).

Anammox

A newly discovered nitrogen transformation in anaerobic conditions is called *anammox* (for anaerobic ammonium oxidation). The process involves nitrite-nitrogen (rather than nitrate-nitrogen as originally thought) as the oxidant:

$$NO_2^- + NH_4^+ \rightarrow 2H_2O + N_2 \qquad (5.16)$$

Few studies have definitively determined the importance of anammox in the cycling of nitrogen in natural or created wetlands, but it does appear that nitrogen may be more important in wetlands where denitrification is limited by lack of organic carbon (Megonigal et al., 2004).

Iron and Manganese Transformations

Below the reduction of nitrate on the redox potential scale comes the reduction of manganese and iron (see Eqs. 5.6 and 5.7). Iron and manganese are among the most abundant minerals on the Earth, which are found in wetlands primarily in their reduced forms (ferrous and manganous, respectively; Table 5.2), and both are more soluble and more readily available to organisms in those forms. Manganese is reduced slightly before iron on the redox scale, but otherwise it behaves similarly to iron. The direct involvement of bacteria in the reduction of MnO_2 (Eq. 5.6) has been questioned by some researchers, although several experiments have shown the generation of energy by the bacterial reduction of oxidized manganese (Laanbroek, 1990).

Iron can be oxidized from the ferrous to the insoluble ferric form by chemosynthetic bacteria in the presence of oxygen:

$$4Fe^{2+} + O_2(aq) + 4H^+ \rightarrow 4Fe^{3+} + 2H_2O \qquad (5.17)$$

Although this reaction can occur nonbiologically at neutral or alkaline pH, microbial activity has been shown to accelerate ferrous iron oxidation by a factor of 10^6 in coal mine drainage water (Singer and Stumm, 1970). A similar type of bacterial process is believed to exist for manganese.

Iron bacteria are thought to be responsible for the oxidation to insoluble ferric compounds of soluble ferrous iron that originated in anaerobic groundwaters in northern peatland areas. These "bog-iron" deposits form the basis of the ore that has been used in the iron and steel industry. Iron in its reduced ferrous form causes a gray-green coloration (gleying) of mineral soils [$Fe(OH)_2$] instead of the normal red or brown color in oxidized conditions [$Fe(OH)_3$]. This appearance gives a relatively easy field check on the oxidized and reduced layers in a mineral soil profile (see Wetland Soils section earlier in this chapter).

Iron and manganese in their reduced forms can reach toxic concentrations in wetland soils. Ferrous iron, diffusing to the surface of the roots of wetland plants, can be oxidized by oxygen leaking from root cells, immobilizing phosphorus and coating roots with an iron oxide, and causing a barrier to nutrient uptake.

The Sulfur Cycle

Sulfur, as the 14th most abundant element in the Earth's surface, occurs in several different states of oxidation in wetlands, and like nitrogen, it is transformed through several pathways that are mediated by microorganisms (Fig. 5.11). Sulfur is rarely present in such low concentrations that it is limiting to plant or animal growth in wetlands. The release of the reduced form of sulfur, sulfide, when wetland sediments are disturbed, causes the odor familiar to those who carry out research in wetlands—the smell of rotten eggs. On the redox scale, sulfur compounds are the next major electron acceptors after nitrates, iron, and manganese, with reduction occurring at about –100 to –200 mV on the redox scale (see Table 5.2). The most common oxidation states (valences) for sulfur in wetlands are:

Form	Valence
S = (sulfide)	−2
S (elemental sulfur)	0
S_2O_3 (thiosulfate)	+2
$SO_4^=$ (sulfate)	+6

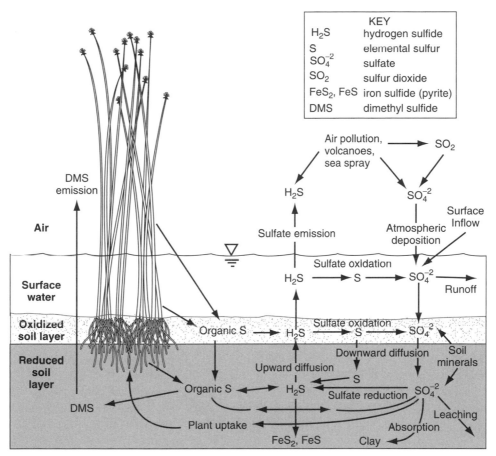

Figure 5.11   The sulfur cycle in wetlands. Major pathways illustrated are sulfur oxidation, sulfate reduction, iron sulfide production, sulfate absorption and leaching, and hydrogen sulfide emissions.

Sulfate Reduction

Sulfate reduction can take place as *assimilatory sulfate reduction* in which certain sulfur-reducing obligate anaerobes such as *Desulfovibrio* bacteria utilize the sulfates as terminal electron acceptors in anaerobic respiration:

$$4H_2 + SO_4^= \rightarrow H_2S + 2H_2O + 2OH^- \qquad (5.18)$$

This sulfate reduction can occur over a wide range of pH, with the highest rates prevalent near neutral pH.

There have been a few measurements of the rate at which hydrogen sulfide is produced in and released from wetlands, and those measurements have ranged over several orders of magnitude. It can be safely generalized that saltwater wetlands have higher rates of sulfide emission per unit area than do freshwater wetlands where sulfate ions are much less abundant ($\sim$2700 mg/L in sea water; $\sim$10 mg/L in fresh water). Sulfur can also be released to the atmosphere as organic sulfur compounds, especially as dimethyl sulfide (DMS), $(CH_3)_2S$; this flux is thought by some to be as important as or more important than H_2S emissions from some wetlands. The general consensus, however, is that most DMS comes from oceans as a product of decomposing phytoplankton cells and that the most important loss of sulfur from terrestrial freshwater wetland systems is H_2S.

Sulfide Oxidation

Sulfides can be oxidized by both chemoautotrophic and photosynthetic microorganisms to elemental sulfur and sulfates in the aerobic zones of some wetland soils. Certain species of *Thiobacillus*—and other bacteria collectively referred to as colorless sulfur bacteria (CSB)—obtain energy from the oxidation of hydrogen sulfide to sulfur, whereas other species in this genus can further oxidize elemental sulfur to sulfate. These reactions are summarized as follows:

$$2H_2S + O_2 \rightarrow 2S + 2H_2O + \text{energy} \tag{5.19}$$

and

$$2S + 3O_2 + 2H_2O \rightarrow 2H_2SO_4 + \text{energy} \tag{5.20}$$

Under anaerobic conditions, nitrate-nitrogen can be used as the terminal electron acceptor in oxidizing hydrogen sulfides.

Photosynthetic bacteria, such as the purple sulfur bacteria (PSB) found in salt marshes and mud flats, are capable of producing organic matter in the presence of light according to the following equation:

$$CO_2 + H_2S + \text{light} \rightarrow CH_2O + S \tag{5.21}$$

This reaction uses hydrogen sulfide as an electron donor rather than H_2O, but is otherwise similar to the more traditional photosynthesis equation. This reaction often takes place under anaerobic conditions where hydrogen sulfide is abundant, but at the surface of sediments where sunlight is also available.

Sulfide Toxicity

Hydrogen sulfide, which is characteristic of anaerobic wetland sediments, can be toxic to rooted higher plants and microbes, especially in saltwater wetlands where the concentration of sulfates is high. The negative effects of sulfides on higher plants include the following:

1. The direct toxicity of free sulfide as it comes in contact with plant roots;

2. The reduced availability of sulfur for plant growth because of its precipitation with trace metals; and

3. The immobilization of zinc and copper by sulfide precipitation.

In wetland soils that contain high concentrations of ferrous iron (Fe^{2+}), sulfides can combine with iron to form insoluble ferrous sulfides (FeS), thus reducing the toxicity of the free hydrogen sulfide. Ferrous sulfide gives the black color characteristic of many anaerobic wetland soils; one of its common mineral forms is pyrite, FeS_2, the form of sulfur commonly found in coal deposits.

The Carbon Cycle

The major processes of carbon transformation under aerobic and anaerobic conditions are shown in Figure 5.12. Photosynthesis and aerobic respiration dominate the aerobic

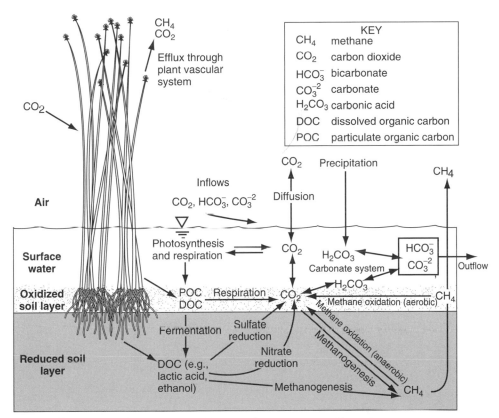

Figure 5.12 The carbon cycle in wetlands. Major pathways illustrated are photosynthesis, respiration, fermentation, methanogenesis, and methane oxidation (anaerobic and aerobic). Also indicated are the roles of sulfate and nitrate reduction in the carbon cycle.

horizons (aerial and aerobic water and soil), with H_2O as the major electron donor in photosynthesis and oxygen as the terminal electron acceptor in respiration:

$$6CO_2 + 12H_2O + \text{light} \rightarrow C_6H_{12}O_6 + 6O_2 + 6H_2O \tag{5.22}$$

$$C_6H_{12}O_6 + 6O_2 \rightarrow 6CO_2 + 6H_2O + 12e^- + \text{energy} \tag{5.23}$$

The degradation of organic matter by aerobic respiration (Eq. 5.23) is fairly efficient in terms of energy transfer. However, because of the anoxic nature of wetlands, anaerobic processes, less efficient in terms of energy transfer, occur in close proximity to aerobic processes. Two of the major anaerobic processes are fermentation and methanogenesis.

Fermentation

The *fermentation* of organic matter, also called *glycolysis* for the substrate involved, occurs when organic matter is the terminal electron acceptor in anaerobic respiration by microorganisms and forms various low-molecular-weight acids and alcohols and CO_2. Examples are lactic acid

$$\begin{array}{c} C_6H_{12}O_6 \rightarrow 2CH_3CH_2OCOOH \\ \text{(lactic acid)} \end{array} \tag{5.24}$$

and ethanol

$$\begin{array}{c} C_6H_{12}O_6 \rightarrow 2CH_3CH_2OH + 2CO_2 \\ \text{(ethanol)} \end{array} \tag{5.25}$$

Fermentation can be carried out in wetland soils by either facultative or obligate anaerobes. Although *in situ* studies of fermentation in wetlands are rare, it plays a central role in providing substrates for other anaerobes, such as methanogens in wetland sediments. Fermentation represents one of the major ways in which high-molecular-weight carbohydrates are broken down to low-molecular-weight organic compounds, usually as dissolved organic carbon, which are, in turn, available to other microbes.

Methanogenesis

Methanogenesis occurs when certain bacteria (*methanogens*) use CO_2 as an electron acceptor for the production of gaseous methane (CH_4), as shown in Equation 5.9, or, alternatively, use a low-molecular-weight organic compound such as one from a methyl group:

$$CH_3COO + 4H_2 \rightarrow 2CH_4 + 2H_2O \tag{5.26}$$

Methane, which can be released to the atmosphere when sediments are disturbed, is often referred to as *swamp gas* or *marsh gas*. Methane production requires extremely reduced conditions, with a redox potential below $-200\,\text{mV}$, after other terminal electron acceptors (O_2, NO_3, and $SO_4^=$) have been reduced. Methanogenesis is carried out by methanogens—a group of microbes called the *Archaea*, a group of prokaryotes that includes several obligate halophiles and thermophiles (Boon, 1999).

Methane Oxidation

Methane oxidation is carried out by obligate *methanotropic bacteria*, which are from a larger group of eubacteria; they convert methane gas in sequence to methanol, formaldehyde, and finally CO_2:

$$CH_4 \rightarrow CH_3OH \rightarrow HCHO- \rightarrow HCOOH \rightarrow CO_2 \qquad (5.27)$$

Nonflooded lands (e.g., forests, agricultural land, grasslands) are normally considered the major biological sinks of methane and are where most *methanotrophs* occur. But wetlands, which have stratified anoxic-oxic horizons, may have a lower anoxic zone dominated by methanogenesis and a surface oxygenated zone with methane oxidation (Fig. 5.12). Thus methane produced in the lower reaches of wetland soils may be "modulated" by methanotrophs that intercept methane from below and convert it to carbon dioxide. Methanotrophs are also able to tolerate extended periods of anoxia as with temporary flooding and can resume methane oxidation within a few hours of re-exposure to oxygen (Whalen, 2005). Methanogens, however, are extremely sensitive to oxygen; methane production does not continue very long once flooded soils are drained.

In addition to methanotrophs, the autotrophic nitrifier communities discussed previously are also able to carry out methane oxidation, because methane and ammonia molecules have a similar size and structure. As a result, the ammonium molecule can also essentially inhibit the methanotrophs from oxidizing CH_4 (Schimel, 2000), and CH_4 can substitute for NH_4^+ in nitrifiers and be co-oxidized.

Methane Emissions

Methane emissions, which are the net result of methanogenesis and methane oxidation, have a considerable range from both saltwater and freshwater wetlands, as well as from domestic wetlands such as rice paddies. Comparison of rates of methane production from different studies is difficult, because different methods are used and because the rates depend on both soil temperature (season) and hydroperiod. Methane emissions have clear seasonal patterns in temperate-zone wetlands (Fig. 5.13) and much less seasonality in tropical and subtropical wetlands (Fig. 5.14). As illustrated in Figure 5.14, summer rates can be highest in seasonal climates, but estimation of total methane generation requires year-long measurements, particularly in subtropical and tropical regions. The pattern also depends on the degree of flooding and the presence or absence of vegetation. Studies of methane fluxes in permanently and seasonally flooded marshes have shown that methane fluxes are much higher from

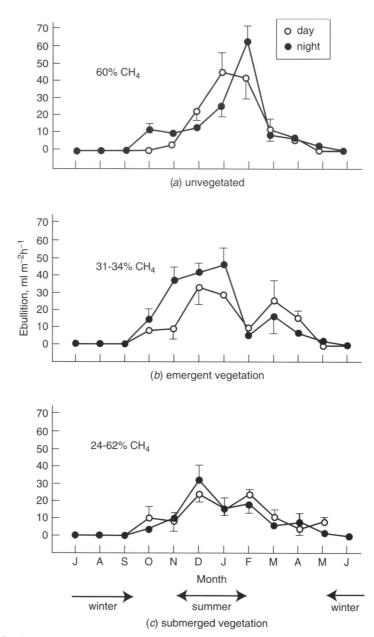

Figure 5.13 Seasonal patterns of gas ebullition (flux of methane-rich bubbles) from three different wetland community types in a floodplain lake (billabong) along the River Murray, New South Wales, Australia: (a) no vegetation; (b) beds of the emergent plant *Eleocharis sphaceiata*; and (c) beds of the submerged aquatic plant *Vallisneria gigantea*. Methane concentrations were 60 percent of the emissions from the bare area, 31 to 54 percent of the emergent plant site, and 24 to 62 percent of the submerged aquatic plant site. *(After Sorrell and Boon, 1992.)*

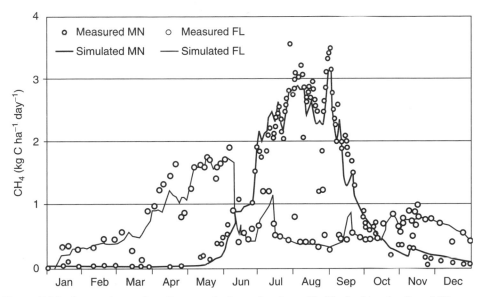

Figure 5.14 Comparison of methane emission rates from Florida (subtropical) and Minnesota (temperate with cold winters), and model results that attempted to simulate both conditions. (After Cui et al., 2005.)

permanently flooded parts of marshes than from intermittently exposed wetland areas (Altor and Mitsch, 2006), suggesting that seasonal pulsing rather than permanent flooding can minimize methane emissions. The lower rates of methane generation in the intermittently exposed marshes could result from lower methanogenesis or higher rates of methane oxidation.

Ebullition and Gaseous Transport in Plants

With the exception of CO_2 and O_2, gases emitted from wetlands (1) emanate from the sediment or soil surface through the water column by diffusive flux or diffusion, (2) bubble to the surface in a process called *ebullitive flux* or *ebullition* and then exit to the atmosphere, or (3) pass through the vascular system of emergent plants (Boon, 1999). Boon and Sorrell (1995) noted that there were substantially more methane fluxes during the day than during the night in chamber studies of Australian wetlands when wetland plants were included in the chambers. They also noted that there was a discrepancy in chambers between the total methane flux and the amount measured by inverted funnels (which capture the ebullitive flux). As a result, the pressures, flows, and gas concentrations were measured within a dominant wetland plant, *Eleocharis sphacelata*, in both "influx" culms, which could generate high pressures, and "efflux" culms, which could not. Sorrell and Boon (1994) and Sorrell et al. (1994) demonstrated that the methane concentration was three orders of magnitude greater in the efflux culms than in the influx culms (Fig. 5.15).

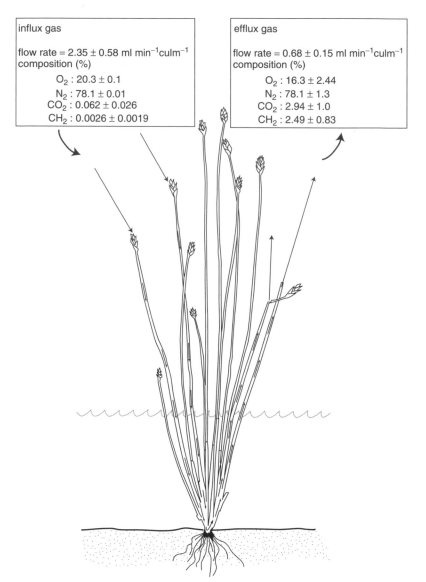

influx gas

flow rate = 2.35 ± 0.58 ml min^{-1}culm^{-1}
composition (%)

O_2 : 20.3 ± 0.1
N_2 : 78.1 ± 0.01
CO_2 : 0.062 ± 0.026
CH_2 : 0.0026 ± 0.0019

efflux gas

flow rate = 0.68 ± 0.15 ml min^{-1}culm^{-1}
composition (%)

O_2 : 16.3 ± 2.44
N_2 : 78.1 ± 1.3
CO_2 : 2.94 ± 1.0
CH_2 : 2.49 ± 0.83

Figure 5.15 Typical fluxes and concentrations of gases in "influx" and "efflux" of the emergent plant *Eleocharis sphacelata* as measured in Australian wetlands. Fluxes through emergent plants are caused by both simple diffusion along a gradient and convective flow of pressurized gas. (After Sorrell et al., 1994; Boon, 1999.)

Carbon dioxide concentrations, as expected, were 50 times higher, whereas dissolved oxygen concentrations decreased 20 percent. These studies and others suggest that between 50 and 90 percent of all methane generated from a vegetated wetland could come through the vascular system of emergent plants (Boon, 1999).

Carbon-Sulfur Interactions

The sulfur cycle is important in some wetlands for the oxidation of organic carbon. This is particularly true in most coastal wetlands where sulfur is abundant. In general, methane is emitted at low concentrations in reduced soils when sulfate concentrations are high. Possible reasons for this phenomenon include (1) competition for substrates that occurs between sulfur and methane bacteria, (2) the inhibitory effects of sulfate or sulfide on methane bacteria, (3) a possible dependence of methane bacteria on products of sulfur-reducing bacteria, and (4) a stable redox potential that does not drop low enough to reduce CO_2 because of an ample supply of sulfate. Other evidence suggests that methane may actually be oxidized to CO_2 by sulfate reducers.

Sulfur-reducing bacteria require an organic substrate, generally of low molecular weight, as a source of energy in converting sulfate to sulfide (Eq. 5.18). The process of fermentation described previously can conveniently supply these necessary low-molecular-weight organic compounds. such as lactate or ethanol (see Eqs. 5.24 and 5.25 and Fig. 5.12). The equations for sulfur reduction, also showing the oxidation of organic matter, are given as follows:

$$2CH_3CHOHCOO^- + SO_4^= + 3H^+ \text{ (lactate)} \tag{5.28}$$

and

$$\rightarrow 2CH_3COO_- + 2CO_2 + 2H_2O + HS^-$$

$$CH_3COO^- + SO_4^= \rightarrow 2CO_2 + 2H_2O + HS^- \text{ (acetate)} \tag{5.29}$$

This fermentation–sulfur reduction pathway is particularly important in the oxidation of organic carbon to CO_2 in saltwater wetlands, which have an excess of sulfates. Fully 54 percent of the carbon dioxide evolution from the salt marsh in New England was caused by the fermentation–sulfur reduction pathway, with aerobic respiration accounting for another 45 percent. By contrast, most of the carbon flux from freshwater systems is through the methane–methane oxidation pathway. In a freshwater billabong in Australia, Boon and Mitchell (1995) demonstrated that methanogenesis accounted for 30 to 50 percent of the total benthic carbon flux and that a major portion of the carbon fixed by plants leaves the wetland via methanogenesis. In general, the release of carbon by methane production is dominant in freshwater wetlands, whereas oxidation of organic carbon by sulfate reduction is dominant in saltwater wetlands.

The Phosphorus Cycle

Phosphorus (Fig. 5.16) is one of the most important limiting chemicals in ecosystems, and wetlands are no exception. It is a major limiting nutrient in northern bogs, freshwater marshes, and southern deepwater swamps. In other wetlands, such as agricultural wetlands and salt marshes, phosphorus is an important mineral, although

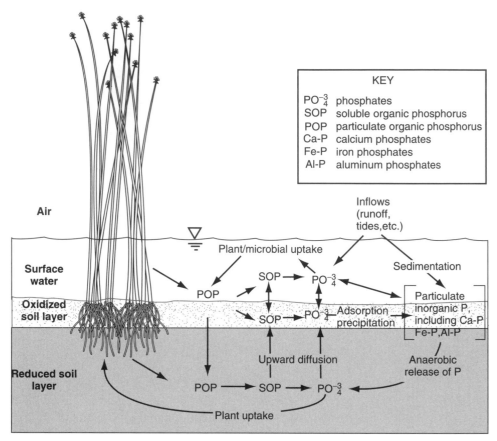

Figure 5.16 The phosphorus cycle in wetlands. Major pathways illustrated are plant/microbial uptake, mineralization, adsorption/precipitation, sedimentation, and anaerobic release.

it is not considered a limiting factor because of its relative abundance and biochemical stability. Phosphorus retention is considered one of the most important attributes of natural and constructed wetlands, particularly those that receive nonpoint source pollution or wastewater.

Phosphorus occurs as soluble and insoluble complexes in both organic and inorganic forms in wetland soils. Inorganic forms include the ions PO_4^{3-}, HPO_4^{2-}, and $H_2PO_4^-$ (collectively referred to as orthophosphates) with the predominant form depending on pH. Phosphorus also has an affinity for calcium, iron, and aluminum, forming complexes with those elements when they are readily available. Phosphorus occurs in a sedimentary cycle (Fig. 5.16) rather than in gaseous cycles such as the nitrogen, sulfur, and carbon cycles described earlier. At any one time, a major proportion of the phosphorus in wetlands is tied up in organic litter and peat and in inorganic sediments, with the former dominating peatlands and the latter dominating mineral soil wetlands.

The analytical measure of biologically available orthophosphates is sometimes called *soluble reactive phosphorus* (SRP), although the equivalence among soluble reactive phosphorus, exchangeable phosphorus, and orthophosphate is not exact. However, it is often used as indicators of the bioavailability of phosphorus. Dissolved organic phosphorus (DOP) and insoluble forms of organic and inorganic phosphorus are generally not biologically available until they are transformed into soluble inorganic forms.

Although phosphorus is not directly altered by changes in redox potential as are nitrogen, iron, manganese, and sulfur, it is indirectly affected in soils and sediments by its association with several elements, especially iron, that are so altered. Phosphorus is rendered relatively unavailable to plants and microconsumers by:

1. The precipitation of insoluble phosphates with ferric iron, calcium, and aluminium under aerobic conditions
2. The adsorption of phosphate onto clay particles, organic peat, and ferric and aluminum hydroxides and oxides
3. The binding of phosphorus in organic matter as a result of its incorporation into the living biomass of bacteria, algae, and vascular macrophytes

There are three general conclusions about the tendency of phosphorus to precipitate with selected ions: (1) Phosphorus is fixed as aluminum and iron phosphates in acid soils; (2) phosphorus is bound by calcium and magnesium in alkaline soils; and (3) phosphorus is most bioavailable at slightly acidic to neutral pH (Reddy et al., 1999). The precipitation of metal phosphates and the adsorption of phosphates onto ferric or aluminum hydroxides and oxides are believed to result from the same chemical forces, namely, those involved in the forming of complex ions and salts.

Co-precipitation of Phosphorus

In many surface water wetlands, high algal productivity can pull CO_2 out of the water, shift the whole carbonate equilibrium, and drive the pH as high as 9 or 10 on a diurnal basis. Under these conditions, co-precipitation of phosphorus as it adsorbs onto calcite and precipitates as calcium phosphate can be significant, just as precipitation of calcium carbonate is also accelerated. In a study of created marshes in central Ohio, calcite and dolomite were found in significant concentrations in algal mat biomass and wetland sediments but not in the river inflow, indicating that the precipitated calcite was produced within the wetlands in significant amounts. Phosphorus co-precipitating with calcite was up to 47 percent of the total phosphorus contained in the algal mat in these wetlands, suggesting that phosphorus co-precipitation essentially doubled the phosphorus removal capability of the algal mat (Liptak, 2000). Wetlands with high algal productivity can have two major pathways for phosphorus removal—the assimilation of phosphorus by algal cells and co-precipitation of phosphates caused by high pH created by the algal water column productivity.

Phosphates and Clay

The sorption of phosphorus onto clay particles is important in aquatic ecosystems. It is believed to involve both the chemical bonding of the negatively charged phosphates to the positively charged edges of the clay and the substitution of phosphates for silicate in the clay matrix. This clay–phosphorus complex is particularly important for many wetlands, including riparian wetlands and coastal salt marshes, because a considerable portion of the phosphorus brought into these systems by flooding rivers and tides is brought in sorbed to clay particles. Thus, phosphorus cycling in many mineral soil wetlands tends to follow the sediment pathways of sedimentation and resuspension. Because most wetland macrophytes obtain their phosphorus from the soil, sedimentation of phosphorus sorbed onto clay particles is an indirect way in which the phosphorus is made available to the biotic components of the wetland. In essence, the plants transform inorganic phosphorus to organic forms that are then stored in organic peat, mineralized by microbial activity, or exported from the wetland.

Phosphorus Release in Anaerobic Conditions

When soils are flooded and conditions become anaerobic, several changes in the availability of phosphorus result. A well-documented phenomenon in the hypolimnion of lakes is the increase in soluble phosphorus when the hypolimnion and the sediment–water interface become anoxic. In general, a similar phenomenon often occurs in wetlands on a compressed vertical scale. As ferric (Fe^{3+}) iron is reduced to more soluble ferrous (Fe^{2+}) compounds, phosphorus that is in a specific ferric phosphate analytically known as reductant-soluble phosphorus is released into solution. Other reactions that may be important in releasing phosphorus upon flooding are the hydrolysis of ferric and aluminium phosphates and the release of phosphorus sorbed to clays and hydrous oxides by the exchange of anions. Phosphorus can also be released from insoluble salts when the pH is changed either by the production of organic acids or by the production of nitric and sulfuric acids by chemosynthetic bacteria. Phosphorus sorption onto clay particles, however, is highest under acidic to slightly acidic conditions (Stumm and Morgan, 1996).

Chemical Budgets of Wetlands

The inputs of materials to wetlands occur through geologic, biologic, and hydrologic pathways. The geologic input from weathering of parent rock, although poorly understood, may be important in some wetlands. Biologic inputs include photosynthetic uptake of carbon, nitrogen fixation, and biotic transport of materials by mobile animals such as birds. Except for gaseous exchanges such as carbon fixation in photosynthesis and nitrogen fixation, however, elemental inputs to wetlands are generally dominated by hydrologic inputs.

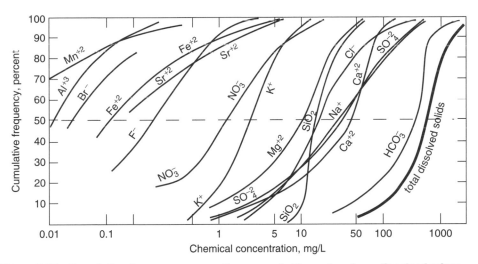

Figure 5.17 Cumulative frequency curves for concentrations of various dissolved minerals in surface waters. Horizontal dashed line indicates median concentrations, 90 percent indicates the 90th percentile concentration, etc. (After Davis and DeWiest, 1966.)

Streams, Rivers, and Groundwater

As precipitation reaches the ground in a watershed, it infiltrates into the ground, passes back to the atmosphere through evapotranspiration, or flows on the surface as runoff. When enough runoff comes together, sometimes combined with groundwater flow, in channelized streamflow, its mineral content is different from that of the original precipitation. There is not, however, a typical water quality for surface and subsurface flows. Figure 5.17 describes the cumulative frequency of the ionic composition of freshwater streams and rivers in the United States. It shows, for example, the average concentrations of the many ions at the 50 percent line. Average NO_3^- concentrations are about 1 mg/L, whereas the average for Mg^{2+} about 10 mg/L. The curves demonstrate the wide range over which these chemicals are found and the median values (50 percent line) of these ranges.

"Average" concentration of dissolved materials in the world's rivers is given in Table 5.3. The variability in concentrations of chemicals in runoff and streamflow that enter wetlands is caused by several factors:

1. *Groundwater influence.* The chemical characteristics of streams and rivers depend on the degree to which the water has previously come in contact with underground formations and on the types of minerals present in those formations. Soil and rock weathering, through dissolution and redox reactions, provides major dissolved ions to waters that enter the ground. The dissolved materials in surface water can range from a few milligrams per liter, found in precipitation, to 100 or even 1,000 mg/L. The ability of water to

Table 5.3   Average chemical concentrations (mg/L) of ocean water and river water

Chemical	Sea Water[a]	"Average" River[b]
Na^+	10,773	6.3
Mg^{2+}	1,294	4.1
Ca^{2+}	412	15
K^+	399	2.3
Cl^-	19,344	7.8
SO_4^{2-}	2,712	11.2
HCO_3^-/CO_3^{2-}	142	58.4
B	4.5[c]	0.01[c]
F	1.4[c]	0.1[c]
Fe	<0.01[c]	0.7
SiO_2	<0.1->10[c]	13.1[c]
N	0–0.5[c]	0.2[c]
P	0–0.07[c]	0.02[c]
Particulate Organic Carbon	0.01–10[c]	5–10[c]
Dissolved Organic Carbon	1–5[c]	10–20[c]

[a]Riley and Skirrow (1975).
[b]D. A. Livingston (1963).
[c]Burton and Liss (1976).

dissolve mineral rock depends, in part, on its nature as a weak carbonic acid. The rock being mineralized is also an important consideration. Minerals such as limestone and dolomite yield high levels of dissolved ions, whereas granite and sandstone formations are relatively resistant to dissolution.

2. *Climate.* Climate influences surface water quality through the balance of precipitation and evapotranspiration. Arid regions tend to have higher concentrations of salts in surface waters than do humid regions. Climate also has a considerable influence on the type and extent of vegetation on the land, and it therefore indirectly affects the physical, chemical, and biological characteristics of soils and the degree to which soils are eroded and transported to surface waters.

3. *Geographic effects.* The amounts of dissolved and suspended materials that enter streams, rivers, and wetlands also depend on the size of the watershed, the steepness or slope of the landscape, the soil texture, and the variety of topography. Surface waters that have high concentrations of suspended (insoluble) materials caused by erosion are often relatively low in dissolved substances. However, waters that have passed through groundwater systems often have high concentrations of dissolved materials and low levels of suspended materials. The presence of upstream wetlands also influences the quality of water entering downstream wetlands. Johnston et al. (2001) found in a comparison of two riverine wetland areas of different soils and geomorphology that there was nevertheless a seasonal convergence of surface

water chemistry caused by the wetlands that overrode the basin differences.

4. *Streamflow/ecosystem effects.* The water quality of surface runoff, streams, and rivers varies seasonally. There is generally an inverse correlation between streamflow and concentrations of dissolved materials (Fig. 5.18). During wet periods and storm events, the water is contributed primarily by recent precipitation that becomes streamflow very quickly without coming into contact with soil and subsurface minerals. During low flow, some or much of

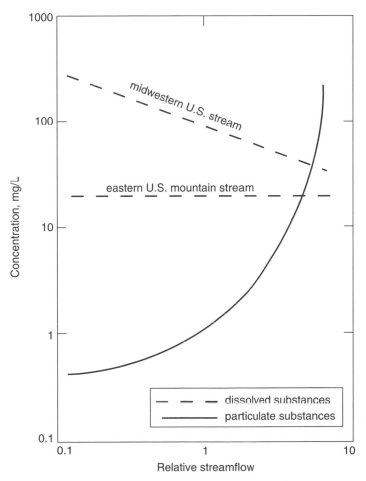

Figure 5.18   Generalized relationships between streamflow and concentrations of dissolved and particulate substances. Data for midwestern U.S. stream is generalization from several studies. Data for a small eastern U.S. watershed is Hubbard Brook, New Hampshire, from Likens et al. (1985).

the streamflow originates as groundwater and has higher concentrations of dissolved materials. This inverse relationship, however, is not always the case, as illustrated by the relationship found in several years of study at the Hubbard Brook, New Hampshire, watershed. There, the small watershed had remarkably similar concentrations of dissolved substances despite a wide range of streamflow, caused by the biotic and abiotic "regulation" of water quality in the small forested watershed and the stream (Likens et al., 1985). The relationship between particulate matter and streamflow is often the inverse (Fig. 5.18).

As with all generalizations, the increase in particulate matter concentration with flow is not always the case. For example, in many streams in the midwestern United States, bioturbation stemming from active fish populations (e.g., common carp, *Cyprinus carpio*) and summer algal blooms can actually cause sediment concentrations to be higher under the low-flow conditions typical of late summer.

5. *Human effects.* Water that has been modified by humans through, for example, sewage effluent, urbanization, and runoff from farms often drastically alters the chemical composition of streamflow and groundwater that reach wetlands. If drainage is from agricultural fields, higher concentrations of sediments and nutrients and some herbicides and pesticides might be expected. Urban and suburban drainage is often lower than that from farmland in those constituents, but it may have high concentrations of trace organics, oxygen-demanding substances, and some toxins.

Oceans and Estuaries

Wetlands such as salt marshes and mangrove swamps are continually exchanging tidal waters with adjacent estuaries and other coastal waters. The quality of these waters differs considerably from that of the rivers described previously. Although estuaries are places where rivers meet the sea, they are not simply places where sea water is diluted with fresh water (J. W. Day et al., 1989). Table 5.3 contrasts the chemical makeup of average river water with the average composition of sea water. The chemical characteristics of sea water are fairly constant worldwide compared with the relatively wide range of river water chemistry. Total salt concentrations typically range from 33 to 37 parts per thousand (ppt). Although sea water contains almost every element that can go into solution, 99.6 percent of the salinity is accounted for by 11 ions. In addition to seawater dilution, estuarine waters can also involve chemical reactions when sea and river waters meet, including the dissolution of particulate substances, flocculation, chemical precipitation, biological assimilation and mineralization, and adsorption and absorption of chemicals on and into particles of clay, organic matter, and silt. In most estuaries and coastal wetlands, biologically important chemicals such as nitrogen, phosphorus, silicon, and iron come from rivers, whereas other important chemicals such as sodium, potassium, magnesium, sulfates, and bicarbonates/carbonates come from ocean sources.

Nutrient Budgets of Wetlands

A quantitative description of the inputs, outputs, and internal cycling of materials in an ecosystem is called an *ecosystem mass balance*. If the material being measured is one of several elements such as phosphorus, nitrogen, or carbon that are essential for life, then the mass balance is called a *nutrient budget*. In wetlands, mass balances have been developed both to describe ecosystem function and to determine the importance of wetlands as sources, sinks, and transformers of chemicals.

A general mass balance for a wetland, as shown in Figure 5.19, illustrates the major categories of pathways and storages that are important in accounting for materials passing into and out of wetlands. Nutrients or chemicals that are brought into the system are called *inputs* or *inflows*. For wetlands, these inputs are primarily through hydrologic pathways (described in Chapter 4), such as precipitation, surface

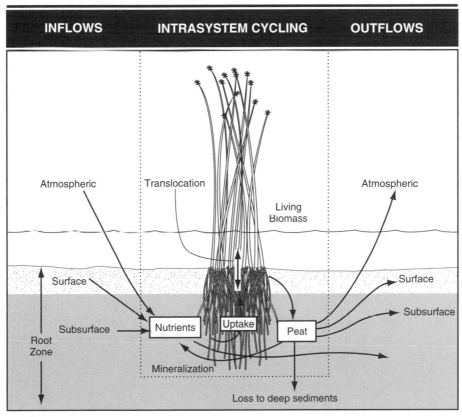

Figure 5.19 Components of a wetland nutrient budget, including inflows, outflows, and intrasystem cycling.

water and groundwater inflow, and tidal exchange. Biotic pathways of note that apply to the carbon and nitrogen budgets are the fixation of atmospheric carbon through photosynthesis and the capture of atmospheric nitrogen through nitrogen fixation.

Hydrologic *exports*, or *losses* or *outflows*, are by both surface water and groundwater, unless the wetland is an isolated basin that has no outflow, such as a northern ombrotrophic bog. The long-term burial of chemicals in the sediments is also considered a nutrient or chemical outflow, although the depth at which a chemical goes from internal cycling to permanent burial is an uncertain threshold. The depth of available chemicals is usually defined by the root zone of vegetation in the wetland. Biologically mediated exports to the atmosphere are also important in the nitrogen cycle (denitrification) and in the carbon cycle (respiratory loss of CO_2). The significance of other losses of elements to the atmosphere, such as ammonia volatilization and methane and sulfide releases, is not well understood, although they are potentially important pathways for individual wetlands, as well as for the global cycling of minerals (see Chemical Transformations in Wetlands earlier in this chapter).

Intrasystem cycling involves exchanges among various *pools*, or *standing stocks*, of chemicals within a wetland. This cycling includes pathways such as litter production, remineralization, and various chemical transformations discussed earlier. The *translocation* of nutrients from the roots through the stems and leaves of vegetation is another important intrasystem process that results in the physical movement of chemicals within a wetland.

Figure 5.20 illustrates in more detail the major pathways and storages that investigators should consider when developing chemical mass balances for wetlands. Major exchanges with the surroundings are shown as exchanges of particulate and dissolved material with adjacent bodies of water (pathways 1 and 2), exchange through groundwater (pathways 7 and 8), inputs from precipitation (pathways 9 and 10), and burial in sediments (pathway 28). Exchanges specific to a nitrogen mass balance, namely, nitrogen fixation (pathways 3–5), denitrification (pathway 6), and ammonia volatilization (pathway 26), are also shown in the diagram. Several intrasystem pathways such as stemflow (pathway 12), root sloughing (pathway 22), detritus–water exchanges (pathways 24 and 25), and sediment–water exchanges (pathway 27) can be very important in determining the fate of chemicals in wetlands but are extremely difficult to measure.

Few, if any, investigators have developed a complete mass balance for wetlands that includes measurement of all of the pathways shown in Figure 5.20, but the diagram remains a useful guide. A phosphorus budget developed for an alluvial river swamp in southern Illinois showed that 10 times more phosphorus was deposited with sediments during river flooding ($3.6 \text{ g P m}^{-2} \text{ yr}^{-1}$) than was returned from the swamp to the river during the rest of the year (Fig. 5.21). Thus, the swamp was a sink for a significant amount of phosphorus and sediments during that particular year of flooding, although the percentage of retention was low (3–4.5 percent) because a very large volume of phosphorus passed over the swamp ($80.2 \text{ g P m}^{-2} \text{ yr}^{-1}$) during flooding conditions.

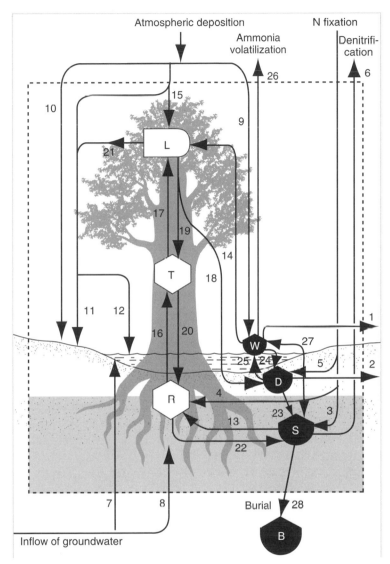

Figure 5.20 Model of major chemical storages and flows in a forested wetland. Storages: L, above-ground shoots or leaves; T, stems, branches, perennial above-ground storage; R, roots and rhizomes; W, surface water; D, litter and detritus; S, near-surface sediments; B, deep sediments essentially removed from internal cycling. Flows: 1 and 2 are exchanges of dissolved and particulate matter with adjacent waters; 3–5 are nitrogen fixation in sediments, rhizosphere microflora, and litter; 6 is denitrification; 7 and 8 are groundwater inputs; 9 and 10 are atmospheric inputs (e.g., precipitation); 11 and 12 are throughfall and stemflow; 13 is uptake by roots; 14 is foliar uptake from surface water; 15 is foliar uptake directly from precipitation; 16 and 17 are translocation from roots through stem to leaves; 18 is litterfall; 19 and 20 are translocation of materials from leaves back to stems and roots; 21 is leaching from leaves; 22 is death/decay of roots; 23 is incorporation of detritus into peat; 24 is adsorption from water to detritus; 25 is release from detritus to water; 26 is volatilization of ammonia; 27 is sediment–water exchange; and 28 is long-term burial of sediments. (After Nixon and Lee, 1986.)

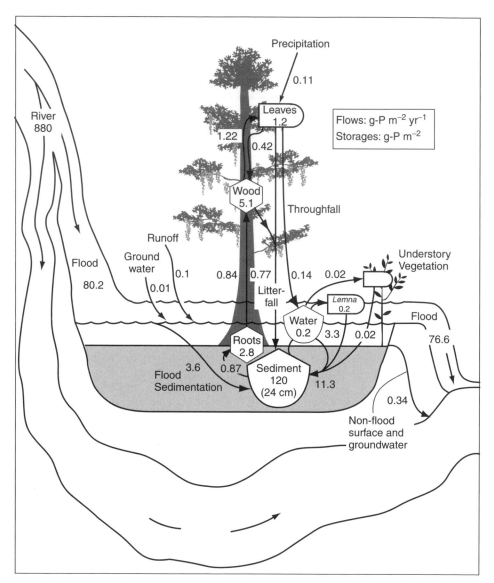

Figure 5.21 Annual phosphorus budget for alluvial cypress swamp in southern Illinois. (After Mitsch et al., 1979a.)

Chemical balances that have been developed for various wetlands are extremely variable, but a few generalizations have emerged from these studies:

1. Wetlands serve as sources, sinks, or transformers of chemicals (Fig. 5.22), depending on the wetland type, the hydrologic conditions, and the length of

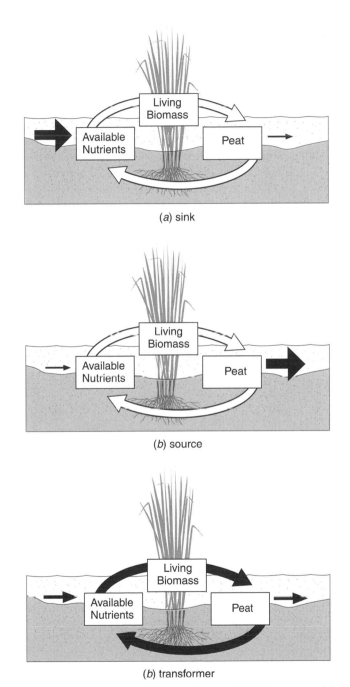

Figure 5.22 Wetlands as (a) inorganic nutrient sink; (b) source of total nutrients; and (c) transformer of inorganic nutrients to organic nutrients.

time the wetland has been subjected to chemical loadings. When wetlands serve as sinks for certain chemicals, the long-term sustainability of that situation depends on the hydrologic and geomorphic conditions, the spatial and temporal distribution of chemicals in the wetland, and the ecosystem succession. Wetlands can become saturated in certain chemicals after a number of years, particularly if loading rates are high.

2. Seasonal patterns of nutrient uptake and release are characteristic of many wetlands. In temperate climates, retention of certain chemicals such as nutrients is greatest during the growing season, primarily because of higher microbial activity in the water column and sediments and secondarily because of greater macrophyte productivity. For example, in cold temperate climates, distinct seasonal patterns of nitrate retention are evident in many cases, with greater retention during the summer months when warmer temperatures accelerate both denitrification microbial activity and algal and macrophyte growth.

3. Wetlands are frequently coupled to adjacent ecosystems through chemical exchanges that significantly affect both systems. Ecosystems upstream of wetlands are often significant sources of chemicals to wetlands, whereas downstream aquatic systems often benefit either from the ability of wetlands to retain certain chemicals or from the export of organic materials.

4. Contrary to popular opinion that all wetlands are highly productive, wetlands can be either highly productive ecosystems rich in nutrients or systems of low productivity caused by a scarce supply of nutrients.

5. Nutrient cycling in wetlands differs from both aquatic and terrestrial ecosystem cycling in temporal and spatial dimensions. For example, more nutrients are tied up in sediments and peat in wetlands than in most terrestrial systems, and deepwater aquatic systems have autotrophic activity more dependent on nutrients in the water column than on nutrients in the sediments.

6. Anthropogenic changes have led to considerable changes in chemical cycling in many wetlands. Although wetlands are quite resilient to many chemical inputs, the capacity of wetlands to assimilate anthropogenic wastes from the atmosphere or hydrosphere is not limitless.

Recommended Readings

Megonigal, J. P., M. E. Hines, and P. T. Visscher. 2004. "Anaerobic metabolism: Linkages to trace gases and aerobic processes." In W.H. Schlesinger, ed. *Biogeochemistry*. Elsevier-Pergamon, Oxford, UK, pp. 317–424.

Richardson, J. L., and M. J. Vepraskas, eds. 2001. *Wetland Soils: Genesis, Hydrology, Landscapes, and Classification*. Lewis Publishers, Boca Raton, FL.

Vepraskas, M. J. 1995. "Redoximorphic Features for Identifying Aquic Conditions." *Technical Bulletin 301*, North Carolina Agricultural Research Service, North Carolina State University, Raleigh, NC. 33 pp.

Biological Adaptations to the Wetland Environment

The wetland environment is, in many ways, physiologically harsh. Major stresses are anoxia and the wide salinity and water fluctuations characteristic of an environment that is neither terrestrial nor aquatic. Adaptations to this environment have an energy cost, either because an organism's cells operate less efficiently (conformer) or because the organism expends energy to protect its cells from the external stress (regulator). At the cell level, all organisms have similar adaptations, although unicellular organisms appear to show more novelty. Adaptations of these organisms include the ability to respire anaerobically, to detoxify end products of anaerobic metabolism, to use reduced organic compounds in the sediment as energy sources, and to use mineral elements in the sediment as alternative electron acceptors when oxygen is unavailable. Multicellular plants and animals have a wider range of responses available to them because of the flexibility afforded by the development of organ systems and division of labor within the body, mobility, and complex life-history strategies. To counter anoxia, one important structural adaptation in vascular plants is the development of pore space in the cortical tissues, which allows oxygen to diffuse from the aerial parts of the plant to the roots to supply root respiratory demands. Animals have developed both structural and physiological adaptations to reduced oxygen availability, such as specialized tissues or organ systems, mechanisms to increase the oxygen gradient into the body, better means of circulation, and more efficient respiratory pigment systems.

Salt stresses are met in plants and animals with specialized tissues or organs to regulate the internal salt concentration or to protect the rest of the body from the effects of salt (osmoregulators), or with increased metabolic and physiological tolerance to salt at high concentrations (osmoconformers). In motile organisms, behavioral adaptations, such as avoidance, are commonly found. Even in sessile seed-producing plants behavioral adaptations have evolved, which typically concern the timing of seed production, mechanisms of seed distribution, and advantageous germination patterns.

Wetland environments are characterized by stresses that most organisms are ill equipped to handle. Aquatic organisms are not adapted to deal with the periodic drying that occurs in many wetlands. Terrestrial organisms are stressed by long periods of flooding. Because of the shallow water, temperature extremes on the wetland surface are greater than would ordinarily be expected in aquatic environments. The most severe stress, however, is the absence of oxygen in flooded wetland soils, which prevents organisms from respiring through normal aerobic metabolic pathways. In the absence of oxygen, the supply of nutrients available to plants is also modified, and concentrations of certain elements and organic compounds can reach toxic levels. In coastal wetlands, salt is an additional stress to which organisms must respond. It is not surprising that those plants and animals regularly found in wetlands have evolved functional mechanisms to deal with these stresses.

Tolerators vs. Regulators

Adaptations by organisms to the harsh wetland environment can be broadly classified as those that enable the organism to tolerate stress and those that enable it to regulate stress. *Tolerators* (also called *resisters*) have functional modifications that enable them to survive and often to function efficiently in the presence of stress. *Regulators* (alternatively called *avoiders*) actively avoid stress or modify it to minimize its effects. The specific mechanisms for either tolerating or regulating are many and varied and are related to the type of organism, for example, bacteria/protests, plants, or animals (Table 6.1). In general, unicellular organisms show biochemical

Table 6.1 Types of adaptations of the wetland environment found in different kinds of organisms and communities of organisms

Type of Organisms	Level of Organization	Type of Adaptation
Bacteria/protists	Organelle	Biochemical
Plant	Cell Tissue	Physiological/structural
Animal	Organ Individual	Behavioral
Plant/Animal group	Community	Commensal/mutualistic

adaptations that are also characteristic of the range of cell-level adaptations found in more complex multicellular plants and animals. Vascular plants show both structural and physiological adaptations. Animals have the widest range of adaptations, not only through biochemical and structural means but also by behavioral responses, using to advantage their mobility and their life-history patterns.

Unicellular Organism Adaptations to Anoxia

For unicellular organisms that have little mobility, the range of adaptations is limited. Most adaptations of this group are metabolic. Because the metabolism of all living cells is similar, these adaptations are characteristic of cell-level adaptations in general, although some of the bacterial responses to anoxia are beyond anything found in multicellular organisms. When an organic wetland soil is flooded, the oxygen available in the soil and in the water is rapidly depleted through metabolism by organisms that normally use oxygen as the terminal electron acceptor for oxidation of organic molecules. The rate of diffusion of molecular oxygen through water is orders of magnitude slower than through air, and cannot supply the metabolic demand of submerged soil organisms under most circumstances. When the demand exceeds the supply, dissolved oxygen is depleted, the redox potential in the soil drops rapidly, and other ions (nitrate, manganese, iron, sulfate, and carbon dioxide) are progressively reduced (see Chapter 5).

Although some abiotic chemical reduction occurs in the soil, virtually all of these reductions are coupled to microbial respiration. When oxygen concentrations first become limiting, most cells, bacterial or otherwise, use internal organic compounds as electron acceptors. The glycolytic or fermentation pathway of sugar metabolism results in the anaerobic production of pyruvate, which is subsequently reduced to ethyl alcohol, lactic acid, or other reduced organic compounds, depending on the organism. Some bacteria also have the ability to couple their oxidative-respiratory reactions to the reduction of inorganic ions (other than molecular oxygen) in the surrounding medium, using them as electron acceptors. Many bacterial species are facultative anaerobes, capable of switching from aerobic to anaerobic respiration. Others have become so specialized, however, that they can grow only under anaerobic conditions and rely on specific electron acceptors other than oxygen in order to respire. *Desulfovibrio* is one such genus. It uses sulfate as its terminal electron acceptor, forming sulfides that give the marsh its characteristic rotten-egg odor.

Most bacteria require organic energy sources. In contrast, nonphotosynthetic autotrophic bacteria are adapted to use reduced inorganic compounds in wetland muds as an energy source for growth. The genus *Thiobacillus*, for example, captures the energy in the sulfide bonds formed by *Desulfovibrio* and, in the process, converts sulfide into elemental sulfur. Members of the genus *Nitrosomonas* oxidize ammonia to nitrite. *Siderocapsa* can capture the energy released in the oxidation of the ferrous ion to the ferric form. In this way, some bacteria not only survive in the anoxic environment of wetland soils but also require it and obtain their metabolic energy from it.

Vascular Plant Adaptations to Waterlogging and Flooding

Multicellular organization adds another layer of complexity to individuals compared to unicellular organization. This complexity has enabled plants and animals to develop a wider range of adaptations than bacteria to anoxia and to salt. At the same time, some adaptations found in unicellular organisms, such as the ability to use reduced inorganic compounds in the sediment as a source of energy, are not found in multicellular organisms. These adaptations typically develop in specialized tissue and organ systems.

In contrast to flood-sensitive plants, flood-tolerant species (*hydrophytes*) possess a range of adaptations that enable them either to tolerate stresses or to avoid them. There are several adaptations by hydrophytes that allow them to tolerate anoxia in wetland soils. These adaptations can be grouped into two main categories: structural or morphological adaptations and physiological adaptations (Table 6.2). Details of these adaptations are discussed as follows.

Table 6.2   Plant adaptations and responses to flooding and waterlogging

Structural (or morphological) adaptations

- a. Aerenchyma tissue in roots and stem
- b. Adventitious roots
- c. Stem hypertrophy (e.g., buttress trunks)
- d. Fluted trunks
- e. Rapid vertical growth/growth dormancy
- f. Shallow root systems/prop roots
- g. Lenticles
- h. Pneumatophores and cypress knees

Physiological adaptations

- a. Pressurized gas flow
- b. Rhizospheric oxygenation
- c. Decreased water uptake
- d. Altered nutrient absorption
- e. Sulfide avoidance
- f. Anaerobic respiration

Whole plant stragegies

- a. Timing of seed production
- b. Buoyant seeds and buoyant seedlings (viviparous seedlings)
- c. Persistent seed banks
- d. Resistant roots, tubers, and seeds

Plant Morphological Adaptations and Strategies

Aerenchyma

Virtually all hydrophytes have elaborate structural (or morphological) mechanisms to avoid root anoxia. These responses to flooding are mechanisms that increase the oxygen supply to the plant either by growth into aerobic environments or by enabling oxygen to penetrate more freely into the anoxic zone. The primary plant strategy in response to flooding is the development of air spaces (*aerenchyma*) in roots and stems, which allow the diffusion of oxygen from the aerial portions of the plant into the roots (Fig. 6.1). Aerenchyma development is not extensive in the absence of flooding and is characteristic of flood-tolerant plant species, not flood-sensitive ones. In plants with well-developed aerenchyma, the root cells no longer depend on the diffusion of oxygen from the surrounding soil, the main source of root oxygen to terrestrial plants. Unlike the plant porosity of normal mesophytes, which is usually a low 2 to 7 percent of volume, up to 60 percent of the volume of the roots of wetland species consists of pore space. Air spaces are formed either by cell separation during maturation of the root cortex or by cell breakdown. They result in a honeycomb structure. Air spaces are not necessarily continuous throughout the stem and roots. The thin lateral cellular partitions within the aerenchyma, however, are not likely to impede internal gas diffusion significantly. The same kind of cell lysis and air space development has been described in submerged stem tissue. Roots of flood-tolerant species, such as rice, form aerenchyma even in aerated apical cells.

Root porosity is the overriding factor governing internal root oxygen concentration. The effectiveness of aerenchyma in supplying oxygen to the roots has been demonstrated in several plant species. For example, the root respiration of flood-tolerant *Senecio aquaticus* was only 50 percent inhibited by root anoxia, whereas that of *S. jacobaea*, a flood-sensitive species, was almost completely inhibited. Greater root porosity in the tolerant species was the primary factor that contributed to the

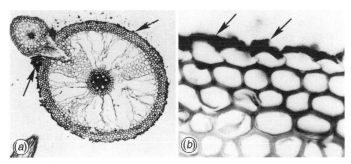

Figure 6.1   Light photomicrographs of *Spartina alterniflora* roots: (a) cross-section of a streamside root; arrows indicate the presence of red ferric deposits on the root epidermis. x192; (b) streamside root cross-section showing the presence of similar materials on the external walls of the epidermal cells. x1,143. Note the extensive pore space (aerenchyma) in the roots. (From Mendelssohn and Postek, 1982; ©1982 by the Botanical Society of America, reprinted by permission.)

difference. The most extensively studied flood-tolerant plant is rice. Rice plants grown under continuous flooding develop greater root porosity than unflooded plants, and this maintains the oxygen concentration in the root tissues. When deprived of oxygen, rice root mitochondria degraded in the same way as did flood-sensitive pumpkin plants, showing that the basis of resistance in flooded plants was by the avoidance of root anoxia, not by physiological changes in cell metabolism (Levitt, 1980).

Adventitious Roots

In addition to aerenchyma development, anaerobic conditions result in the formation of certain organs on wetland plants that assist the plant in getting oxygen to its root system. Hormonal changes, especially the concentration of ethylene in hypoxic tissues, initiate some of these structural adaptations (Fig. 6.2). Ethylene has been reported to stimulate the formation of *adventitious roots* in both flood-tolerant trees (e.g., *Salix* and *Alnus*) and flood-tolerant herbaceous species (e.g., *Phragmites*, *Ludwigia*, and *Lythrum salicaria*) and some flood-intolerant plants (e.g., tomato). These roots develop on the stem just above the anaerobic zone when these plants are flooded (Fig. 6.3a). They form as the original roots die and are able to function normally in an aerobic environment above the water line.

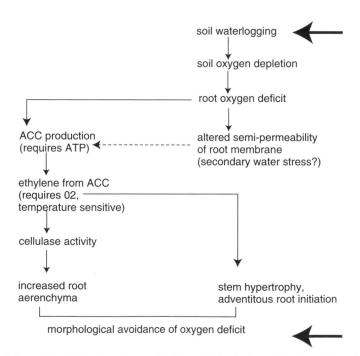

Figure 6.2 Schematic of the development of morphological acclimation characteristics in response to flooding stress, including the production of ethylene that, in turn, leads to both aerenchyma and stem hypertropy. (After McLeod et al., 1988.)

(a)

(d)

(b)

(e)

(c)

Figure 6.3 Illustrations of morphological adaptations to flooding and waterlogging by vascular plants: (a) adventitious roots; (b) stem hypertrophy or buttresses on cypress (*Taxodium*) trees in a deepwater swamp; (c) fluted trunk on pin oak tree (*Quercus palustris*) in a freshwater forested wetland; (d) prop roots extending from *Rhizophora* mangrove trees in Costa Rica; and (e) pneumatophores ("knees") of *Taxodium* in a freshwater swamp. *(Photographs b, c, d, and e by W. J. Mitsch; photo a by Ralph Tiner; reprinted with permission.)*

213

Stem Hypertrophy

Stem hypertrophy, a noticeable swelling of the lower stem of vascular plants, is another adaptation of many vascular plants to waterlogged conditions and hence serves as a good indicator or wetland conditions. When this hypertrophy occurs on a tree, it is called a *buttress* (Figure 6.3b). It is a characteristic of swamp trees such as bald and pond cypress (*Taxodium* spp.) and water and swamp black gum (*Nyssa* spp.). Hypertrophy is not caused by the formation of aerenchyma but rather by larger cells and lower density wood, also probably caused by ethylene production. A somewhat similar pattern of trees exhibiting flared or *fluted trunks* (Figure 6.3c) at the ground surface is common in wetlands with several tree species such as pin oak (*Quercus palustris*) and American elm (*Ulmus americana*).

Stem Elongation, Root Adaptations, and Lenticels

Another response stimulated by submergence is *rapid stem elongation* in such aquatic and semi-aquatic plants as the floating heart (*Nymphoides peltata*), rice (*Oryza sativa*), and bald cypress (*Taxodium distichum*), stimulated by rising water levels. Bald cypress seedlings have rapid vertical growth rates supposedly to get the photosynthetic organs out of harm's way before standing water levels increase. The formation of *shallow root systems* by wetland plants is another clear and common adaptation by vascular plants to avoid anaerobic conditions. Deep tap roots, common in upland forests, are almost never found in forested wetlands. Some species are facultative in the regard. Red maple (*Acer rubrum*) develops shallow root systems in wetlands but can have deep tap roots in upland forests.

The red mangrove (*Rhizophora* spp.) grows on arched *prop roots* in tropical and subtropical tidal swamps around the world (Fig. 6.3d). These prop roots have numerous small pores, termed *lenticels*, above the tide level, which terminate in long, spongy, air-filled, submerged roots. The oxygen concentration in these roots, embedded in anoxic mud, may remain as high as 15 to 18 percent continuously, but if the lenticels are blocked, this concentration can fall to 2 percent or less in two days. Lenticels are also in the stems of flood-tolerant species such as *Alnus glutinosa* and *Nyssa sylvatica* and serve as conduits to the aerenchymatous tissue in the stem.

Pneumatophores

Similarly, the black mangrove (*Avicennia* spp.) tree produces thousands of *pneumatophores* (air roots) about 20 to 30 cm high by 1 cm in diameter, spongy, and studded with lenticels. They protrude out of the mud from the main roots and are exposed during low tides. The oxygen concentration of the submerged main roots has a tidal pulse, rising during low tide and falling during submergence, reflecting the cycle of emergence of the air roots. These pneumatophores are often covered with lenticels that aid in root aeration. The "knees" of bald cypress (*Taxodium distichum*) (Fig. 6.3e) are pneumatophores that improve gas exchange to the root system. Cypress knees generally develop only when the trees are in waterlogged or flooded soils, and their height was often used as an indicator of high water levels in the wetlands.

Physiological Adaptations

Vascular emergent and floating-leaved wetland plants are sessile; only their roots are in an anoxic environment. Typically, if the roots of a flood-sensitive upland plant are inundated, the oxygen supply rapidly decreases. This shuts down the aerobic metabolism of the roots, impairs the energy status of the cells, and reduces nearly all metabolically mediated activities such as cell extension and division and nutrient absorption. Even when cell metabolism shifts to anaerobic glycolysis, adenosine triphosphate (ATP) production is reduced. Toxic metabolic end products of fermentation may accumulate, causing cytoplasmic acidosis and eventually death. Anoxia is soon followed by pathological changes in the mitochondrial structure. The complete destruction of mitochondria and other organelles occurs within 24 hours. Anoxia also changes the chemical environment of the root, increasing the availability of reduced forms of iron, manganese, and sulfur, which may accumulate to toxic levels in the root. Faced with these problems, wetland vascular plants have several physiological adaptations that attempt to solve the problem of anoxic conditions in the root system.

Pressurized Gas Flow

Dacey (1980, 1981) first described a particularly interesting adaptation that increases the oxygen supply to the roots of the floating-leaved water lily (*Nuphar luteum*). Since then, a similar adaptation of pressurized gas flow from the surface to the rhizosphere has been demonstrated for other floating-leaved species. Brix et al. (1992) tested 14 emergent plants in southwestern Australia and found significant gas flow ($0.2 \rightarrow 10$ cm^3 min^{-1} culm^{-1}) in 8 of the species, including *Baumea articulata*, *Cyperus involucratus*, *Eleocharis sphacelata*, *Schoenoplectus validus*, *Typha domingensis*, *T. orientalis*, *Phragmites australis*, and *Juncus ingens* (Table 6.3; Fig. 6.4). These results for such a wide variety of plants suggest that internal pressurization and pressurized gas flow may be common to many hydrophytes. They also investigated the diel variation in convective gas flow for *T. domingensis* and found a dramatic diel pattern related to air and subsequent leaf temperatures (Fig. 6.4). Gas flows of 0.1 to 0.2 cm^3 min^{-1} culm^{-1} occurred at night but increased to a rate as high as 3 cm^3 min^{-1} culm^{-1} during the afternoon (Fig. 6.4). The results were interpreted to suggest that humidity-induced pressurization was the dominant driving force for the gas flow.

The pressure produced by this range of plants matched very nicely the approximate depths at which these plants can potentially occur in wetlands (Table 6.3). *Phragmites*, *Eleocharis*, and the two *Typha* species can grow in water depths up to 2 m; *Schoenoplectus*, *Juncus*, and *Baumea* are found in water less than 1 m deep; the two *Cyperus* species, *Bolboschoenus*, and *Canna* grow in very shallow water to wet soils. Air moves into the internal gas spaces (the lacunar system) of aerial leaves and is forced down through the aerenchyma of the stem into the roots by a slight pressure ($\sim$200–1,300 Pa) generated by a gradient in temperature and water vapor pressure (Brix et al., 1992). Older leaves often lose their capacity to support pressure gradients, and so the return flow of gas from the roots is through the older leaves, which are rich in carbon dioxide and methane from root respiration. Brix (1989) did find that the gas exchange to the rhizosphere through dead culms of *P. australis* was sufficient

Table 6.3 Pressurized gas flow in culms or leaves of 13 wetland plants and 1 upland plant in Australia[a]

Water Depth Species	N	ΔP_s(Pa)	Flow Rate (cm^3 min^{-1} culm^{-1})
POTENTIALLY DEEPWATER PLANTS			
Phragmites australis	12	573 ± 54	5.3 ± 0.4[b]
Typha orientalis	8	1,070 ± 120	4.4 ± 0.3[b]
Typha domingensis	6	780 ± 140	3.4 ± 0.4[b]
MARGINAL DEPTH (<1 m) PLANTS			
Juncus ingens	11	222 ± 24	1.2 ± 0.1[c]
Eleocharis sphacelata	10	1,080 ± 86	0.85 ± 0.02
Schoenoplectus validus	9	1,310 ± 124	0.29 ± 0.05
Baumea articulata	16	494 ± 58	0.23 ± 0.06
VERY SHALLOW WATER OR MOIST-SOIL PLANTS			
Cyperus involucratus	11	903 ± 234	0.33 ± 0.09[c]
Canna sp.	5	27 ± 5	0.06 ± 0.01
Myriophyllum papillosum[d]	6	68 ± 12	0.04 ± 0.01
Cyperus eragrostis	8	111 ± 34	0.02 ± 0.01
Ludwigia pelloides[d]	5	57 ± 1	<0.01
Bolboschoenus medianus	15	2 ± 31	<0.01
NOT A TRUE WETLAND PLANT			
Arundo donax	6	1 ± 10	<0.01

[a]Water depths refer to the potential depths that these plants can grow based on other studies and plant size. ΔP_s refers to the static pressure differential in the plant stem. Plants are listed in order of decreasing gas flow rates. Numbers indicate averages ± standard deviations.
[b]Small specimens or leaves had to be removed to get flow rates within measuring range.
[c]Gas flow measured through detached culms.
[d]Creeping, floating plants that grow in shallow water.
Source: Brix et al. (1992).

to maintain aerobic respiration of the plant roots. The dead culms also provided an escape channel for excess CO_2 and CH_4 in the roots.

Grosse and others (Grosse and Schröder, 1984; Schröder, 1989; Grosse et al., 1992) described a similar process in swamp trees, specifically in common alder (*Alnus glutinosa*), the dominant tree species of European floodplain forests and riverine temperate forests. Seedlings and dormant (leafless) trees of flood-tolerant species show enhanced gas transport from aerial shoots to the roots when the shoots are heated by the sun or incandescent light, compared to plants in the dark. Grosse et al. (1998) called this phenomenon "pressurized gas flow" or "thermo-osmotic"

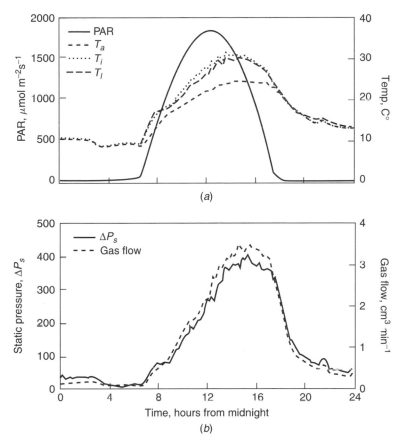

Figure 6.4 Diel variation of solar energy, temperature, internal pressure, and gas flow for *Typha domingensis*: **(a) photosynthetically available solar radiation (PAR), air temperature** (T_a)**, internal leaf temperature** (T_l)**, and leaf surface temperature** (T_l)**; (b) internal static pressure differential** (ΔP_s) **and gas flow.** *(After Brix et al., 1992.)*

gas flow. This phenomenon occurs when a temperature gradient is established between the exterior ambient air and the interior gas spaces in a plant's cortical tissue. A second requirement is a permeable partition between the exterior and interior with pore diameters "similar to or smaller than the 'mean free path length' of the gas molecules in the system (e.g., 70 nm at room temperature and standard barometric pressure" (Grosse et al., 1998). In alder, meristematic tissue in the lenticels forms such a partition.

When the surface of the stem is mildly heated by sunlight, the mean free path length of gas molecules in the intercellular spaces of the plant increases, preventing the molecules from moving out through the osmotic barrier in the lenticels. The cooler exterior molecules, however, can still diffuse into the plant. This sets up an internal pressure gradient that forces gas down through the plant stem to the roots.

This "thermal pump" is not as effective in moving oxygen to the roots in alder, as is extensive aerenchyma tissue. For example, alder seedlings grown in flooded soil for two months transported oxygen at eight times the rate of seedlings grown in aerated soil. The difference was because of aerenchyma and lenticel development under flooded conditions. By comparison, the thermo-osmotic effect (in the absence of flooding) led to a fourfold increase in the rate of gas transport. The thermal pump is also not as active in foliated trees as in dormant ones. Therefore, for trees, the adaptation appears to be most effective in enhancing root aeration during seedling establishment in saturated soils before aerenchyma development is accomplished, and in deciduous trees during the dormant season.

Rhizosphere Oxygenation

There are secondary effects of adaptations to root aeration that influence other parts of the plants or their environment. When anoxia is moderate, the magnitude of oxygen diffusion through many wetland plants into the roots is apparently large enough not only to supply the roots but also to diffuse out, oxidize the adjacent anoxic soil, and produce an oxidized rhizosphere (see Chapter 5). Through scanning electron microscopy coupled with X-ray microanalysis, Mendelssohn and Postek (1982) showed that the brown deposits found around the roots of *Spartina alterniflora* are composed of iron and manganese deposits formed when root oxygen comes in contact with reduced soil ferrous ions (Fig. 6.1). Oxygen diffusion from the roots is an important mechanism that moderates the toxic effects of soluble reduced ions such as manganese in anoxic soil and restores ion uptake and plant growth. These ions tend to be reoxidized and precipitated in the rhizosphere, which effectively detoxifies them. In a similar vein, McKee et al. (1988) determined that soil redox potentials were higher and that pore water sulfide concentrations were three to five times lower in the presence of the aerial prop roots of the red mangrove (*Rhizophora*) or the pneumatophores of the black mangrove (*Avicennia*) than in nearby bare mud soils, in all probability because of the diffusion of oxygen from the mangrove roots into the soil. An interesting possibility is that the root systems of these flood-tolerant plants may modify sediment anoxia enough to allow the survival of nearby nontolerant plants (Ernst, 1990).

The presence of *oxidized rhizospheres* (now called *oxidized pore linings* by soil scientists; see Chapter 5), which form as a result of root oxidation, is an important way in which wetlands can be identified. Long after the plant roots die, residual veins of red and orange, resulting from oxidized iron (Fe^{3+}) deposits, remain in many mineral soils, a telltale sign that hydrophytes had been living in the soil. They are used in wetland delineation practices as one indicator that hydric soils and, thus, wetlands are present.

Lower Water Uptake

Plants intolerant to anaerobic environments typically show decreased water uptake despite the abundance of water, probably as a response to an overall reduction of root metabolism. Decreased water uptake results in symptoms similar to those seen under drought conditions: closing of stomata, decreased carbon dioxide uptake, decreased transpiration, and wilting. The adaptive advantage of these responses is probably the same as for drought-stricken plants—to minimize water loss and accompanying damage to the cytoplasm. An accompanying depression of the photosynthetic machinery is generally seen as an unavoidable corollary.

Altered Nutrient Absorption

One of the earliest processes affected by anoxia is the absorption by plants of nutrients from the substrate. Anoxic conditions modify the availability of many nutrients in wetland soils. In general, flood-intolerant plant species lose the ability to control nutrient absorption because of the tissue energy deficit brought on by anoxia. Although some studies (e.g., Mendelssohn and Burdick, 1988) have concluded that nutrient uptake in most flood-tolerant species appears unchanged, perhaps because the plant can maintain near-normal metabolism, other studies (Grosse and Meyer, 1992) have illustrated that sufficient oxygen is necessary for nutrient uptake. This is illustrated in Figure 6.5, where hydroponic-grown alder (*Alnus glutinosa*) seedlings were subjected to light-induced pressurized gas flow. Under anoxic conditions, nitrate, potassium, and phosphate uptakes were reduced to about 20, 30, and 70 percent, respectively, of aerated conditions.

Wetland vascular plants have several adaptations related to the uptake, retention, and toxic avoidance of several major nutrients during soil anaerobiosis. In reduced soils, nitrates are replaced by ammonium, although most plants preferentially absorb the oxidized form (nitrate). Despite this change in the supply of available nitrogen, most wetland plants are able to maintain normal rates of nitrogen uptake due to two

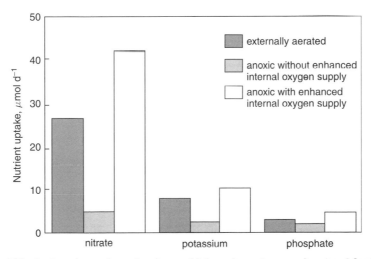

Figure 6.5 Effect of root anoxia and enhanced internal oxygen supply on nutrient uptake by one-year-old foliated *Alnus glutinosa* (common alder). Trees were grown hydroponically. (*After Grosse and Meyer, 1992; Grosse et al., 1998.*)

possibilities: (1) ammonium is oxidized to nitrate in the rhizosphere through radial oxygen loss from the roots; and (2) some wetland species are able to absorb the ammonium directly.

The availability of phosphorus generally increases in waterlogged soils. It is, however, precipitated by iron (which also increases in availability). Generally, studies have shown reduced uptake by flood-intolerant species, probably because of the energy requirement, but no effect or even enhanced uptake by flood-tolerant species. Because only small concentrations of iron and manganese are required by plants, they can reach toxic levels in many environments. Both elements are reduced and become much more available in flooded soils, where, because of their high concentrations, they escape every metabolic control and become concentrated in plant tissues. Wetland plants tolerate these elements by means of several adaptations: (1) the oxidized rhizosphere can precipitate and reduce the concentration that reaches the roots; (2) much of the minerals taken up into the tissues can be sequestered in cell vacuoles, in the shoot vascular tissue, or in senescing tissues where they do not influence the metabolism of healthy cytoplasm; and (3) many wetland plants appear to have a much higher than average metabolic tolerance for these ions (Ernst, 1990).

Sulfide Avoidance

Sulfur as sulfide is toxic to plant tissues. The element is reduced to sulfide in anaerobic soils and accumulates to toxic concentrations, especially in coastal wetlands (Goodman and Williams, 1961; Koch and Mendelssohn, 1989). Although sulfate uptake is metabolically controlled, sulfide can enter the plant without control and is found in elevated concentrations in many flood-adapted species under highly reduced conditions. In experiments with *Spartina alterniflora*, a salt marsh species, and *Panicum hemitomon*, a freshwater marsh species, Koch et al. (1990) reported that the activity of alcohol dehydrogenase (ADH), the enzyme that catalyzes the terminal step in alcohol fermentation, was significantly inhibited by hydrogen sulfide, and that this inhibition may help explain the physiological mechanism of sulfide phytotoxicity often seen in salt marshes. Sulfur tolerance in wetland plants varies widely, probably because of the variety of detoxification mechanisms available. These include the oxidation of sulfide to sulfate through root aeration of the rhizophere; the accumulation of sulfate in the vacuole; the conversion to gaseous hydrogen sulfide, carbon disulfide, and dimethylsulfide and their subsequent diffusive loss; and a metabolic tolerance to elevated sulfide concentrations (Ernst, 1990).

Anaerobic Respiration

Under conditions of oxygen deprivation, plant tissues respire anaerobically, as described for bacterial cells. In most plants, pyruvate, the end product of glycolysis, is decarboxylated to acetaldehyde, which is reduced to ethanol (see right side of Fig. 6.6). Both of these compounds are potentially toxic to root tissues. Flood-tolerant plants often have adaptations to minimize this toxicity. For example, under anaerobic conditions, *S. alterniflora* roots show much increased activity of ADH, the inducible enzyme that catalyzes the reduction of acetaldehyde to ethanol. The increase in the

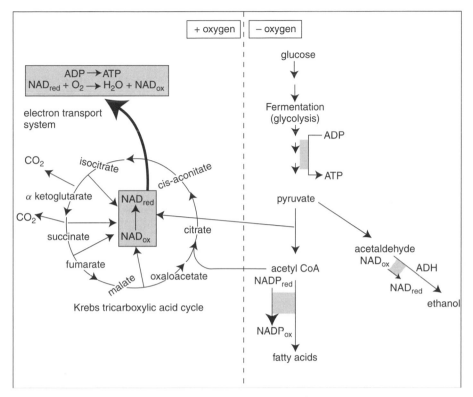

Figure 6.6 Schematic of metabolic respiration pathway in flood-tolerant plants. Left side of diagram is aerobic respiration; right side is anaerobic respiration (fermentation/glycolsis), which yields pyruvate, acetaldehyde, ethanol, and fatty acids such as malic acid. ADH, alcohol dehydrogenase; NAD, nicotinamide adenine dinucleotide; NADP, NAD phosphate; subscripts refer to oxidized (ox) and reduced (red) forms.

enzyme indicates a switch to anaerobic respiration, and it explains why acetaldehyde does not accumulate in the root tissue. Ethanol does not accumulate, either, although its production is apparently stimulated (Mendelssohn et al., 1982). It diffuses from rice roots during anaerobiosis (Bertani et al., 1980), thus preventing a toxic buildup, and the same probably occurs in other wetland plants. Another metabolic strategy reduces the production of alcohol by shifting the metabolism to accumulate nontoxic organic fatty acids instead (Fig. 6.6). At one time, it was suggested that malic acid (malate) accumulation may be a characteristic feature of wetland species. The accumulation of malate cannot easily be interpreted, however, in part because malate is an intermediate in several metabolic pathways.

The metabolic problem encountered by plants deprived of oxygen is the loss of the electron acceptor that enables normal energy metabolism through ATP formation and use. The metabolic bottleneck in this process is often the electron-accepting coenzyme nicotinamide adenine dinucleotide (NAD), which is reduced in the oxidative steps of carbohydrate metabolism and then reoxidized in the mitochondria by molecular oxygen to yield the biological energy currency ATP (see shaded boxes in left side of Fig. 6.6). In the absence of oxygen (right-hand side of Fig. 6.6), reduced NAD (NAD_{red}) accumulates and "jams" the metabolic system, blocking ATP generation. In the process of fermentation, acetaldehyde replaces oxygen, reoxidizing reduced NAD. Malate acts in the same way through the tricarboxylic acid cycle. Thus, glycolysis can occur as long as NAD is reoxidized to the oxidized form, NAD_{ox}.

Whole Plant Strategies

Many plant species have evolved avoidance or escape strategies by life-history adaptations. The most common of these strategies are (1) the timing of seed production in the non–flood season by either delayed or accelerated flowering (Blom et al., 1990); (2) the production of buoyant seeds that float until they lodge on high, unflooded ground; (3) the germination of seeds while the fruit is still attached to the tree (*vivipary*), as in the red mangrove; (4) the production of a large, persistent seed bank (Voesenek, 1990); and (5) the production of tubers, roots, and seeds that can survive long periods of submergence. In many riparian wetlands, flooding occurs primarily during the winter and early spring, when trees are dormant and much less susceptible to anoxia than they are during the active growing season. The *viviparous seedlings* that germinate live in the canopy of red mangrove (*Rhizophora*) trees fall in the water from the canopy after germination and are transported, sometimes great distances. The seedling rights itself to a vertical position, and develops roots if the water is shallow until it lodges in shallow sediments, allowing the seedling to then grow to a tree. The aquatic plant monocot *Sagittaria latifolia* has a similar way of distributing its seeds (the plant is sometimes called duck potato because of its seeds' resemblance to potatoes). The seed floats through a wetland until it lodges in a shallow area or amid other emergent macrophytes, after which it germinates. Seed banks are discussed in more detail in Chapter 7.

Animal Adaptations to Anoxia

Animals are exposed to the same range of environmental conditions in wetlands as unicellular organisms and plants, but because of their complexity, their adaptations are more varied. The adaptation may be as varied as a biochemical response at the cell level, a physiological response of the whole organism such as a modification of the circulatory system, or a behavioral response such as modified feeding habits (Table 6.4). Furthermore, although it is convenient to discuss the specific response mechanisms to individual kinds of environmental stresses, in reality an organism must respond simultaneously to a complex of environmental factors, and the success of this integrated response determines the organism's fate. For example, one possible response to stress is avoidance by moving out of the stress zone. In wetlands, however, that might mean moving from an anoxic zone within the soil to the surface, where temperature extremes and desiccation pose a different set of physiological problems.

Table 6.4 Adaptations of animals to control gas exchange in anoxic conditions

1. **Special organs**—Development or modification of specialized regions of the body for gaseous exchange
   (e.g., gills on fish and crustacea, parapodia on polychaetes)
2. **Improving oxygen conditions**—Mechanisms to improve the oxygen gradient across a diffusible membrane (e.g., by moving to oxygen-rich environments or by moving water across the gills by ciliary action)
3. **Internal structural changes**—such as increased vascularization, a better circulation system, or a stronger pump (the heart)
4. **Respitory pigments**—Modification of respiratory pigments to improve oxygen-carrying capacity
5. **Physiological adaptations**—including shifts in metabolic pathways and heart pumping rates
6. **Behavioral patterns**—such as decreased locomotor activity or closing a shell during low oxygen stress

Source: Vernberg and Vernberg (1972)

Thus, the organism's successful adaptations are often compromises that enable it to live with several competing environmental demands.

At the cell level, the metabolic responses of animals to anoxia are similar to those of bacteria. Vertebrates, however, tend to have less ability to adapt to anaerobic conditions than invertebrates. The vertebrates and many invertebrates are limited in anaerobic respiration to glycolysis or to the pentose monophosphate pathway whose dominant end product is lactate. In all chordates, the internal cell environment is closely regulated. As a result, most adaptations are organism-level ones to maintain the internal environment.

All but the behavioral adaptations involve modifications of tissue and organ systems and associated physiological changes. Adaptations in Table 6.4 are not mutually exclusive. Many animals combine several different types of adaptations. Examples of different kinds of adaptations are numerous. We give a few here to illustrate their diversity. Crabs inhabit a wide range of marine habitats. The number and the total volume of gills of crabs living on land are less per unit of body weight than those of aquatic species. In addition, the gills of some intertidal crabs have become highly sclerotized so that the gill leaves do not stick together when the crab is out of water. Tube-dwelling amphipods apparently can function efficiently with low oxygen supplies. At saturated oxygen tensions, they exhibit an intermittent rhythm of ventilation. At low tide, when the oxygen in their burrows drops to very low levels, they ventilate continuously but do not hyperventilate as free-swimming amphipods do. Because of the resistance of their tubes to water flow, hyperventilation would be energetically expensive for tube-dwelling amphipods. Many marine animals associated with anoxic wetland soils have high concentrations of respiratory pigments or pigments that have unusually high affinities for oxygen or both. These include the nematode (*Enoplus communis*), the Atlantic bloodworm (*Glycera dibranchiata*), the clam (*Mercenaria mercenaria*), and even the land crab (*Carooma quannumi*) (Vernberg and Coull, 1981).

Fiddler crabs (*Uca* spp.) illustrate the complex behavioral and physiological patterns to be found in the intertidal zone. These crabs are active during low tides, feeding daily when the marsh floor is exposed. (Incidentally, this pattern of activity is based on an innate lunar rhythm, not on a direct sensing of low water levels. When transported miles from the ocean, fiddlers continue to be active at the time that low tide would occur in their new location.) When the tide rises, they retreat to their burrows, where the oxygen concentration can become very low, because fiddler crabs apparently do not pump water in their burrows. Not only are these species relatively resistant to anoxia, but also their critical oxygen tension (i.e., the tension below which respiratory activity is reduced) is low, 0.01 to 0.03 atmospheres (atm) for inactive and 0.03 to 0.08 atm for active crabs. They can continue to consume oxygen down to a level of 0.004 atm (Vernberg and Vernberg, 1972). When oxygen levels get very low in the burrows, the crabs simply become inactive and consume very little oxygen. They may remain that way for several tidal cycles without harm.

Intertidal bivalves close their valves tightly or loosely when the tide recedes. Widdows et al. (1979) found that four different bivalves had lower respiration rates in air (valves closed) than in water. All could respire anaerobically, but the accumulation of end products of anaerobic respiration depended on how tightly their shells were closed and, thus, how much oxygen they received. The tolerance to anoxia may change during the life of an organism. The larvae of fiddler crabs, which are planktonic, are much more sensitive to low oxygen than are the burrowing adults. An interesting but rather unusual adaptation is that of a gastrotrich (*Thiodasys sterreri*), which is reported to be able to use sulfide as an energy source under extreme anaerobiosis (Maguire and Boaden, 1975).

Animal Feeding Adaptations

As with reproductive adaptations, the broad range of animal feeding responses closely reflects their habitats. Adaptations of feeding appendages, for example, seem to be more closely related to feeding habits than to taxonomic relationships. Many organisms that exist in marsh sediments are adapted for the direct absorption of dissolved organic compounds from their environment. For example, infaunal polychaetes can supply a major portion of their energy requirements from the rich supply of dissolved amino acids present in their environment, but epifaunal species are unable to take advantage of amino acids at concentrations typical of their environment (Vernberg, 1981). Many mud-dwelling organisms have one or more adaptations to selective feeding on microscopic particles by means of pseudopods, cilia, mucus, and setae, or they may ingest substrate unselectively. Sikora (1977) suggested that the appendages of many macrobenthic organisms (shrimp, crabs) are adapted to feeding on microscopic meiobenthic organisms and that these latter organisms are major intermediaries in the marsh/estuary food chain.

In addition to developing strategies for reproduction, many animals have adapted their lifestyles to take seasonal advantage of the wetland ecotone. For example, bottomland forests that flood during winter and drain in the spring are exceptionally high in productivity. While the forest floor is inundated in the early spring, many fish and shellfish species move up into the floodplain from the adjacent river to feed and to spawn. They move with the water's edge as floodwaters recede, feeding on abundant detritus and benthic invertebrates (Lambou, 1990). Terrestrial animals follow the receding water line for the same reasons. Deer, turkey, bear, migrating songbirds, and other species take advantage of the acorns and other hard seeds and nuts available in fall and early winter when surrounding uplands have lost much of their habitat value (Gosselink et al., 1990b).

Salinity Adaptations

Unicellular Adaptations to Salt

In a freshwater aquatic or soil environment, the osmotic concentration of the cytoplasm in living cells is higher than that of the surrounding medium. This enables the cells to develop turgor; that is, to absorb water until the turgor pressure of the cytoplasm is balanced by the resistance of their cell membranes and walls. In coastal wetlands, organisms must cope with high and variable external salt concentrations. The dangers of salts are twofold—osmotic and directly toxic. The immediate effect of an increase in salt concentration in a cell's environment is osmotic. If the osmotic potential surrounding the cell is higher than that of the cell cytoplasm, water flows out of the cell and the cytoplasm dehydrates. This rapid reaction can occur in a matter of minutes and may be lethal to the cell. Even the "tightest" membranes leak salts passively so that, in the absence of any active regulation by the cell, inorganic salts gradually diffuse into the cell. Although the absorption of inorganic ions such as Na^+ may relieve the osmotic gradient across the cell membrane, these ions at high concentrations in the cytoplasm are also toxic to most organisms, posing a second threat to survival.

Unicellular organisms have adapted in several ways to cope with these twin problems of osmotic shock and toxicity. There is no evidence that cells are able to retain water against an osmotic gradient. Instead, in order to maintain their water potential, the internal osmotic concentration of salt-adapted cells is usually slightly higher than the external concentration. Indeed, the high specific gravities of halophiles—literally "salt-loving" organisms—can be accounted for only by the presence of inorganic salts at high concentrations. Analyses of cell contents show, however, that the balance of specific ions is usually quite different from that of the external solution. For example, potassium is usually accumulated and sodium is usually diluted relative to external concentrations. Active transport mechanisms that accumulate or excrete ions across cell membranes are universal features of all living cells, and although they depend on a cellular supply of biological energy, there is no evidence that large energy expenditures are needed to maintain the gradients. They can be maintained in the cold, in cells apparently carrying out little metabolism.

Although inorganic ions seem to make up the bulk of the osmotically active cell solutes in most halophilic bacteria, in others the internal salt concentration can be substantially lower than the external concentration. Organic compounds supply the rest of the osmotic activity in these organisms. For example, the halophilic green alga *Dunaliella virigus* contains large amounts of glycerol, the concentration varying with the external salt concentration; and certain salt-tolerant yeasts regulate internal osmotic concentration with organic compounds such as glycerol and arabitol. The enzymes of these organisms seem to be salt sensitive, and it has been suggested that the organic compounds act like "compatible solutes" that raise the osmotic pressure without interfering with enzymatic activity.

Vascular Plant Adaptations to Salt

At the cell level, plants behave toward salt in much the same way as bacteria do, and their adaptive strategies are identical. Vascular plants, however, have also developed adaptations that take advantage of their structural complexity. These can be summarized into two separate plant strategies: (1) barriers or exclusions to prevent or control the entry of salts and (2) organs specialized to secrete salts. In both cases, specialized cells bear most of the burden of the adaptation, allowing the remaining cells to function in a less hostile environment. A third possible mechanism for plants dealing with the saline environment is through C_4 photosynthesis. All three of these strategies are discussed in the following sections.

Salt Exclusion

Roots of plants in high-salt environments often have much higher salt concentrations (and must also have higher salt tolerance) than the leaves do. Moon et al. (1986) reported that, in the mangrove *Avicennia marina*, the primary barrier to passive salt incursion into the plant is at the root periderm and exodermis, so that the root cortex is protected from high salt concentrations. Uptake of salt is restricted mainly to the terminal third- and fourth-order roots, and the large lateral roots serve primarily as the means of vascular transport and support. As a result of the filtering out of salt at the root apoplast, the sap of many halophytes is almost pure water. The mangroves *Rhizophora, Laguncularia*, and *Sonneratia* exclude salts almost completely. Their sap concentrations are only about 1 to 1.5 mg NaCl per mL (compared with about 35 mg/mL in sea water). *Avicennia* has a higher sap concentration of about 4 to 8 mg/mL, or about 10 percent of external concentration. When fluid is forced out of the leaves of these species by pressure, it is almost pure distilled water. Thus, both the root and the leaf cell membranes act like ultra-filters.

The leaf cytoplasm must have an osmotic potential higher than that of the sap in order to retain water, and in mangroves, 50 to 70 percent of this osmotic potential is obtained from sodium and chloride ions. Most of the remainder is presumably organic. In *Batis*, a succulent halophyte, NaCl alone makes up 90 percent of the total osmotic pressure.

Salt Secretion

Some plants that do not exclude salts at the root, or are "leaky" to salt, have organs that secrete salt. The leaves of many salt marsh grasses, for example, characteristically are covered with crystalline salt particles excreted through specialized salt glands embedded in the leaf. These glands do not function passively; instead, they selectively remove certain ions from the vascular tissues of the leaf. In *Spartina*, for example, the excretion is enriched in sodium, relative to potassium.

These two mechanisms, salt exclusion and salt secretion, protect the shoot and leaf cells of the plant from high concentrations of salt and presumably maintain an optimum ionic balance between mono- and divalent cations and between sodium and potassium. At the same time, the osmotic concentration of the cells of salt-tolerant plants must be maintained at a level high enough to allow the absorption of water from the root medium. Where the inorganic salt concentration is kept low, organic compounds make up the rest of the osmoticum in the cells.

Photosynthesis

One adaptation that many wetland plant species share with plants in other stressed environments, especially in hot dry environments, is the C_4 biochemical pathway of photosynthesis (formally called the Hatch–Slack–Kortschak pathway, after its discoverers). It gets its identity from the fact that the first product of CO_2 incorporation are four-carbon compounds, oxaloacetate and malate. The first compound resulting from CO_2 incorporation in C_3 plants is a three-carbon compound, phosphoglycerate. Although water is a universal feature of wetlands, plants in saline wetlands (salt marshes, mangrove swamps) have much the same problem of water availability as do plants in arid areas. In both cases, the water potential of the substrate is very low. In arid zones, it is low because the soil is dry; in saline wetlands, it is low because of the salt content.

Functions of C_3 plants are compared with C_4 plants in Table 6.5. Although C_3 plants are much more common, C_4 plants are fairly routine in the saline wetland environment. Among the common wetland angiosperms that have been shown to photosynthesize through the C_4 pathway are the salt marsh grasses *Spartina alterniflora*, *Spartina townsendii*, and *Spartina foliosa*, and brackish marsh reed grass *Phragmites australis*.

Plants that fix carbon by the C_4 pathway can use carbon dioxide more effectively than other plants (Table 6.5). The net rate of photosynthesis in C_4 plants can be two to three times higher than similar C_3 plants, mainly because photorespiration does not normally occur in C_4 plants. They can fix CO_2 in the dark, and are able to withdraw CO_2 from the atmosphere until its concentration falls below 20 parts per million (ppm), as compared with 30 to 70 ppm for C_3 plants. This is achieved by using phosphoenolpyruvate (PEP), which has a high affinity for CO_2, as the carbon dioxide acceptor instead of the ribulose diphosphate acceptor of the conventional pathway. These differences make C_4 plants more efficient than most C_3 plants, both in their rates of carbon fixation and in the amount of water used per unit of carbon fixed. Water conservation mechanisms in wetland plants have the additional function of reducing the rate at which soil toxins are drawn toward the root. This increases the probability of detoxifying them as they move through the oxidized rhizophere.

Table 6.5 Comparison of aspects of photosynthesis of herbaceous C_3 and C_4 plants

Photosynthetic Characteristics	C_3	C_4
Initial CO_2 fixation enzyme	RuBP carboxylase[a]	PEP carboxylase[a]
Theoretical energy requirement for net CO_2 fixation, CO_2: ATP : NADPH	1 : 3 : 2	1 : 5 : 2
CO_2 compensation concentration (ppm CO_2)	30–70	0–10
Transpiration ratio (g H_2O transpired/g dry weight)	450–950	250–350
Optimum day temperature for net CO_2 fixation ($^\circ$C)	15–25	30–47
Response of net photosynthesis to increasing light intensity	Saturation at 1/4 to 1/2 full sunlight	Proportional to or saturation at full sunlight
Maximum rate of net photosynthesis (mg CO_2/dm^2 leaf surface/hr)	15–40	40–80
Maximum growth rate (g $m^{-2}d^{-1}$)	19.5	30.3
Dry matter production (g m^{-2} yr^{-1})	2,200	3,860

[a]RuBP, ribulose-bis-phosphate; PEP, phosphoenolpyruvate.
Source: Based on data from Black (1973) and Fitter and Hay (1987).

Animal Adaptations to Salinity

Like their responses to oxygen stress, the major mechanisms of adaptation by animals to salt involve control of the body's internal environment. Most simple marine animals are *osmoconformers*; that is, their internal cell environment follows closely the osmotic concentration of the external medium. In animals that have greater body complexity, however, *osmoregulation* (i.e., control of internal osmotic concentration) is the rule. This is particularly true of animals that inhabit the upper intertidal zone, where they are exposed to widely varying salinities and to prolonged periods of desiccation. *Euryhaline* organisms can tolerate wide fluctuation in salinity. *Stenohaline* organisms, however, survive within fairly narrow osmotic limits. Most marsh organisms must be euryhaline, but they can be either osmoconformers or osmoregulators.

The difference between osmoregulators and osmoconformers is illustrated for marine animals in Figure 6.7. The blood concentration of a perfect osmoconformer would follow the solid line of isotonicity (internal concentration equal to ambient concentration) in Figure 6.7. In contrast, a perfect osmoregulator would have a constant internal concentration, which would be illustrated by a horizontal line on the graph. Organisms such as marsh crab (*Uca*) resemble osmoregulators, whereas the crabs *Hemigrapsus* and *Cancer* are almost perfect osmoconformers. *Uca* is from the high intertidal zone and is an excellent osmoregulator, possessing adaptations that are obviously useful in controlling the variable salinity and frequent desiccation of their habitat. Regulation in these species is controlled both by differences in exoskeleton permeability and by specialized excretory organs.

The brown shrimp (*Penaeus aztecus*) is intermediate between these two positions. At low external salt concentrations, the shrimp is hyper-osmotic, usually achieved

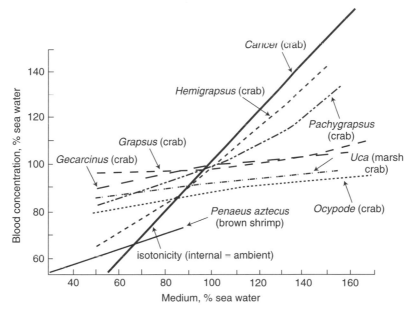

**Figure 6.7   Comparative osmoregulation of crabs and brown shrimp (*Penaeus aztecus*),
showing different degrees of adaptation to the marine environment. Line of perfect osmo-
conformity (isotonicity) is also shown for comparison. A horizontal line would indicate perfect osmoregulation. Of the crab species, Cancer is an aquatic species; *Hemigrapsus* and
Pachygrapsus are low intertidal zone species; others are terrestrial or high intertidal zone
inhabitants. (After Gross, 1964; Bishop et al., 1980.)**

by concentrating sodium and chloride ions. This condition indicates a water potential gradient into the organism. Regulation at high ambient salt concentration is
hypo-osmotic. In this case, the water potential gradient is directed out of the animal,
which means that dehydration would occur if the body covering were not, to some
extent, impervious to water movement.

Animals that possess the ability for hypo-osmotic regulation (blood salt is higher
than sea salt or conditions to the left of the isotonicity line in Figure 6.7) must have
some mechanism to lower the osmotic concentration of the body. This is accomplished
through special regulatory organs and organ systems, chiefly renal organs (kidneys or
more primitive nephridia, antennal glands), gills, salt-secretory, nasal, or rectal glands,
and the specialized excretory functions of the gut. These organs are able to move ions
across cell membranes against the concentration gradient (at an energy cost to the
animal), concentrating them in some excretory product such as urine.

Mutualism and Commensalism

The close interactions among members of an ecological community reflect the high
degree of adaptability of members of the community, not only to their physical

environment, but also to their biological environment. This chapter has documented adaptations to the physical and chemical wetland environment, but positive interactions among organisms are also predicted to play a significant role in ecosystem dynamics, especially in marginal or stressed environments such as wetlands. Two such possibilities of reactions between wetland populations are *mutualism*, when there are positive and obligatory benefits to both populations, and *commensalism*, where one population benefits and the other does not have either a positive or negative effect (i.e., it is neutral).

Documentations of these effects in wetland environments are relatively few. Many appear to involve nutrients that are limiting in these environments. For example, Grosse et al. (1990) reported a commensalism or mutualism between alder (*Alnus*) trees and fungi. Increased levels of nitrogen fixation by the symbiotic fungus *Frankia alni* presumably occurred because of thermal pumping of oxygen through and out of the root system of common alder into the rhizosphere. Ellison et al. (1996) reported a mutualistic interaction between root-fouling sponges (*Todania ignis* and *Haliclona implexiformis*) and the red mangroves (*Rhizophora mangle*) on which they grow. Fine, adventitious mangrove rootlets ramify throughout the sponges. They absorb dissolved ammonium from the sponges, which stimulates additional root growth. The sponges also protect the roots from isopod attack. Mangrove roots, in turn, provide the only hard substrate for sponges in this habitat, and they stimulate sponge growth by leaking carbon.

These examples of the positive interactions among wetland species point to an extremely interesting line of neglected research that may lead to important new insights into the complexity of mutualistic adaptations in wetland ecosystems, and their importance, not only to the organisms involved, but also to the energetic dynamics of the entire community.

Recommended Readings

Grosse, W., H. B. Büchel, and S. Lattermann. 1998. "Root aeration in wetland trees and its ecophysiological significance." In A. D. Laderman, ed. *Coastally Restricted Forests*. Oxford University Press, New York, pp. 293–305.

Larcher, W. 2003. *Physiological Plant Ecology*, 4th ed. Springer-Verlag, New York.

Tiner, R. W. 1999. *Wetland Indicators: A Guide to Wetland Identification, Delineation, Classification, and Mapping*. Lewis Publishers, Boca Raton, FL, p. 392.

Chapter 7

Wetland Ecosystem Development

Wetland development is another term for wetland succession. Wetland ecosystems have traditionally been considered transitional seres between open lakes and terrestrial forests. The accumulation of organic material from plant production was seen to build up the surface until it was no longer flooded and could support flood-tolerant terrestrial forest species (autogenic succession). An alternative theory is that the vegetation found at a wetland site consists of species adapted to the particular environmental conditions of that site (allogenic succession). Observed zonation patterns, in this view, reflect underlying environmental gradients rather than autogenic successional patterns. Present evidence seems to suggest that both allogenic and autogenic forces act to change wetland vegetation and that the idea of a regional terrestrial climax is inappropriate. Models used to describe wetland plant development include a functional guild model, an environmental sieve model, and a centrifugal organization concept.

If one looks at ecosystem attributes as indices of ecosystem maturity, wetlands appear to be mature in some respects and young in others. Generally, productivity is high, some production is exported, and mineral cycles are open, all indications of young systems. However, most wetlands accumulate much structural biomass in peat, all wetlands are detrital systems, spatial heterogeneity is generally high, and life cycles are complex. These properties indicate maturity. The strategy for ecosystem development in wetlands includes concepts such as pulse stability, whereby the wetland maintains itself because of and not despite natural pulses such as tides and floods, and self-organization or self-design, whereby the wetland develops according to the multitude of propagules that come its way. At landscape scales, observed patterns of wetlands, aquatic and upland habitats, reflect a complex and dynamic interaction of physical (allogenic) and biotic (autogenic) forces acting on the geomorphic template of the landscape.

Autogenic vs. Allogenic Succession

The beginning and subsequent development of a plant community is characterized by the initial conditions at the site and by subsequent events, including the availability of viable seeds or other propagules, appropriate environmental conditions for germination and subsequent growth, and replacement by plants of the same or different species as site conditions change in response to both abiotic and biotic factors. The concept of succession (i.e., the replacement of plant species in an orderly sequence of development), in particular, has exerted a strong influence on plant ecology for more than a century. Ecological theories of plant succession were advanced by H. C. Cowles in his classic work on plant succession based on the sequential exposure of sand dunes on the southern and eastern shores of Lake Michigan (Cowles, 1899). In that study, dunes left bare from a retreating Lake Michigan were shown through a series of successional ecosystems over thousands of years to go in an orderly primary succession to a climax beech-maple forests.

Autogenic succession was further enunciated by Clements (1916) and applied to wetlands by the English ecologist W. H. Pearsall in 1920 and by an American, L. R. Wilson, in 1935. E. P. Odum (1969) adapted and extended the ideas of those early ecologists to include ecosystem properties such as productivity, respiration, and diversity. This classical use of the term *succession* involves three fundamental concepts: (1) vegetation occurs in recognizable and characteristic *communities*; (2) community change through time is brought about by the biota (i.e., changes are *autogenic*); and (3) changes are linear and *directed* toward a mature, stable *climax* ecosystem. Although this classical concept of autogenic succession was a dominating paradigm of great importance in terrestrial ecology, the concept has been challenged and altered for almost a century. Gleason (1917) enunciated an *individualistic* hypothesis to explain the distribution of plant species. His ideas have developed into the *continuum* concept, which holds that the distribution of a species is governed by its response to its environment (*allogenic succession*). Because each species responds differently to its environment, no two occupy exactly the same zone. The observed invasion/replacement sequence is also influenced by the chance occurrence of propagules at a site. The result is a continuum of overlapping sets of species, each responding to subtly different environmental cues. In this view, no communities exist in the sense used by Clements, and although ecosystems change, there is little evidence that this is directed or that it leads to a particular climax.

A key issue in discussions of ecosystem development is whether biota determine their own future by modifying their own environment, or if the development of an ecosystem is simply a response to the external environment. In the classical view of succession, wetlands are considered transient stages in the *hydrarch development* of a terrestrial forested climax community from a shallow lake (Fig. 7.1). In this view, lakes and open water gradually fill in as organic material from dying

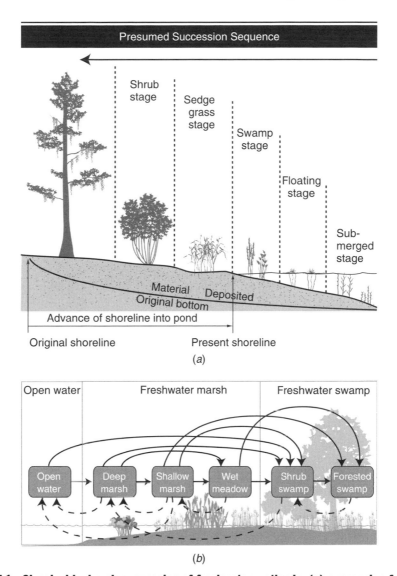

Figure 7.1 Classical hydrarch succession of freshwater wetlands: (a) succession from a pond to a terrestrial forest at the edge of a pond, and (b) general succession to mineral soil forested wetlands in glaciated regions of North America. (b after Golet et al., 1993.)

plants accumulates and minerals are carried in from upslope. At first, change is slow because the source of organic material is single-celled plankton. When the lake becomes shallow enough to support rooted aquatic plants, however, the pace of organic deposition increases. Eventually, the water becomes shallow enough to support emergent marsh vegetation, which continues to build a peat mat.

Shrubs and small trees appear. They continue to transform the site to a terrestrial one, not only by adding organic matter to the soil, but also by drying it through enhanced evapotranspiration. Eventually, a climax terrestrial forest occupies the site (Fig. 7.1a). The important point in this description of hydrarch succession is that most of the change is brought about by the plant community as opposed to externally caused environmental changes. A second important feature, shown in Figure 7.1b, is that the process can reverse if the environmental conditions, particularly the hydrology, change.

How realistic is this concept of succession? It is certainly well documented that forests do occur on the sites of former lakes, but the evidence that the successional sequence leading to these forests was autogenic is not clear. Because peat building is crucial to filling in a lake and its conversion to dry land, key questions involve the conditions for peat accumulation and the limits of that accumulation. Peat underlies many wetlands, often in beds 10 m or more deep. Some scientists (McCaffrey, 1977; Delaune et al., 1983b) have shown that, in coastal marshes, peat has accumulated and is still accumulating at rates varying from less than 1 to about 15 mm/yr. Most of this accumulation seems to be associated with rising sea levels (or submerging land). By contrast, northern inland bogs can accumulate peat at rates of 0.2 to 2 mm/yr.

In general, accumulation occurs only in anoxic sediment. When organic peats are drained, they rapidly oxidize and subside, as farmers who cultivate drained marshes have discovered. As the wetland surface accretes and approaches the water surface or at least the upper limit of the saturated zone, peat accretion in excess of subsidence must cease. It is difficult to see how this process can turn a wetland into a dry habitat that can support terrestrial vegetation unless there is a change in hydrologic conditions that lowers the water table. For example, Cushing (1963) used paleoecological techniques to show that most of the peatlands in the Lake Agassiz plain (Minnesota and south-central Canada) formed during the mid-Holocene (beginning about 4,000 years ago) during a moist climatic period when surface water levels rose about 4 m.

Wetlands are at the center of the dispute about the importance of autogenic versus allogenic processes because of their transitional nature. In addition to being *seres*, wetlands are often described as being *ecotones*—that is, transitional spatial gradients between adjacent aquatic and terrestrial environments. Thus, wetlands can be considered transitional in both space and time. As ecotones, wetlands usually interact strongly to varying (allogenic) forcing functions from both ends of the ecotone. These forces may push a wetland toward its terrestrial neighbor if, for example, regional water levels fall, or toward its aquatic neighbor if water levels rise.

Alternately, plant production of organic matter may raise the level of the wetland, resulting in a drier environment in which different species succeed. Because these environmental changes can be subtle, it is often difficult to determine whether the observed ecosystem response is autogenic or allogenic. Without careful measurements, the causes of the response are often obscure.

Revisiting the Lake Michigan Dunes

In the early 20th century, H. C. Cowles (1899, 1901, 1911) and Victor Shelford (1907, 1911, 1913) studied ponds of different ages in the Indiana dunes region along the southern shore of Lake Michigan. The ponds were of different ages and were thought to represent an autogenic successional sequence. Along this age gradient, the young ponds were deep and dominated by aquatic vegetation. Older ponds were shallower and supported emergent vascular plants along their borders. The oldest ponds were shallowest and contained the most "terrestrial" vegetation. This sequence was interpreted as evidence of classical autogenic succession.

Wilcox and Simonin (1987) and Jackson et al. (1988) revisited the Indiana dunes ponds. Using modern quantitative methods of ordination, they found the same progression of plant species from young to old ponds, supporting the sequence observed by earlier workers. In addition to the present vegetation, they also examined pollen and macrofossils in the sediments of a 3,000-year-old pond to determine whether the sediments support the presumed successional sequence found in the modern-day chronosequences. Pollen and macrofossil data older than 150 B.P. (before the present) consisted of a diverse assemblage of submersed, floating leaved, and emergent macrophyte groups. The data indicated a major and rapid vegetation change after 150 B.P., which the authors attributed to post-European settlement, such as railroad construction and forest clearing.

To further evaluate the historical changes in the Indiana dunes ponds, Singer et al. (1996) examined the sediment pollen and macrofossil record of aquatic and emergent plants in one of the old Indiana dunes pond sites and compared it with the regional terrestrial pollen record (of airborne pollen found in the same cores). The latter tracks long-term climate changes in the region. If aquatic and emergent paleotaxonomic remains showed changes in species dominance that mirrored the terrestrial pollen record, then the changes in the ponds could be attributed to regional climate change rather than to autogenic processes. From their 10,000-year record, Singer et al. (1996) determined that historic changes in pond vegetation did correspond to regional climate change (Fig. 7.2). Between 10,000 and 5700 B.P., the sampled area was a shallow lake; the regional climate was mesic (a pine/oak/elm terrestrial assemblage). A rapid increase in oak and hickory pollen around 5700 B.P. signaled a regional climate shift to a drier environment. At the same level in the sediment record, the pond macrofossil record showed a rapid shift to a peat-forming marsh environment. After about 3000 B.P., modest increases in beech and birch pollen suggested a trend toward a cooler, moister climate. The concomitant pond vegetation remained dominated by emergents, but transitions among several taxa suggest that water-level fluctuations and occasional fires were characteristic of the period.

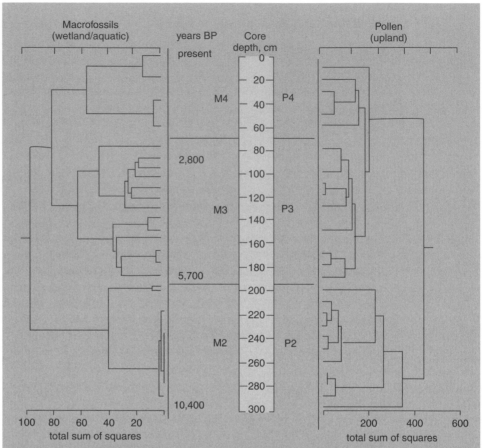

Figure 7.2 Stratigraphically constrained cluster diagrams for macrofossil and pollen data from an Indiana dunes pond in northern Indiana adjacent to Lake Michigan. Macrofossil zonation is based on the presence or absence of aquatic and wetland taxa; pollen zonation is based on percentages of selected upland pollen types. Close individual taxa indicate close occurrence in fossil record. Thus the macrofossil record indicates three different groups of organisms. *(After Singer et al., 1996.)*

These studies, taken together, provide a fuller, more complex picture of plant development than the autogenic succession process proposed in Cowles' and Shelford's earlier studies. The picture that emerges is one of an interaction between allogenic and autogenic processes, with allogenic forces driving the development of the biotic system, but modified by autogenic processes. Over the 10,000-year span of the fossil record, changes in the plant assemblage correlated well with regional climate change. However, during the same period, the lake was slowly filling with organic sediments, first 100 cm of gyttja, characteristic of open freshwater systems; then 200 cm of fibrous peat, characteristic of vascular aquatic plants. During the period of a slow climate shift to a less xeric environment after approximately 3000 B.P., the pond environment remained a marsh, although the species assemblage changed. This fits well with the idea that the organic sediments moderated the climatic influence on the local water levels. Finally, during the modern period after about 150 B.P., human activities, which probably altered water levels locally, resulted in rapid vegetation changes.

The Community Concept and the Continuum Idea

The Indiana dunes ponds example described previously is only a small part of the extensive literature concerning questions about plant and ecosystem development in wetlands. The idea of the community is particularly strong in wetland literature. Historic names for different kinds of wetlands—marshes, swamps, carrs, fens, bogs, reedswamps—often used with the name of a dominant plant (*Sphagnum* bog, leatherleaf bog, cypress swamp)—signify our recognition of distinctive associations of plants that are readily recognized and at least loosely comprise a community. One reason these associations are so clearly identified is that zonation patterns in wetlands often tend to be sharp, having abrupt boundaries that call attention to vegetation change and, by implication, the uniqueness of each zone. The plant community is central to the historic idea of succession because the mature climax resulting from succession was presumed to be a predictable group of plant species, with each group dependent on the regional climate.

The identification of a community is also, to some extent, a conceptual issue that is confused by the scale of perception. Field techniques are adequate to describe the vegetation in an area and its variability. However, its homogeneity—one index of community—may depend on size. For example, Louisiana coastal marshes have been classified into four zones, or communities, based on the dominant vegetation. If the size of the sampling area is large enough, any sample within one of these zones will always identify the same species. If smaller grids are used, however, differences appear within a zone. The intermediate marsh zone is dominated on a broad scale by *S. patens*, but aerial imagery shows patterns of vegetation within the zone, and intensive sampling and cluster analysis of the vegetation reveal five subassociations that are characteristic of intermediate marshes. Is the intermediate marsh a community? Are the subassociations communities? Or is the community concept a pragmatic device to reduce the bewildering array of plants and possible habitats to a manageable number of groups within which there are reasonable similarities of ecological structure and function?

Supporters of the continuum concept would argue that the scale dependence of plant associations illustrates that individual species are simply responding to subtle environmental cues, implying little, if anything, about communities, and that plant zonation simply indicates an environmental gradient to which individual species are responding. The reason zonation is so sharp in many wetlands, they argue, is that environmental gradients are "ecologically" steep, and groups of species have fairly similar tolerances that tend to group them on these gradients. Figure 7.3 shows the distribution of swamp trees and submersed aquatic vegetation along an ordination axis. Although the species overlap, the distribution of each seems to be distinct, leaving no reflection of a community. The idea that each species is found where the environment is optimal for it makes perfect sense to ecophysiologists and autecologists, who interpret the success of a species in terms of its environmental adaptation.

One major difference between classical community ecologists and proponents of the continuum idea is the greater emphasis put on allogenic processes by the latter. In some wetlands, abiotic environmental factors often seem to overwhelm biotic forces.

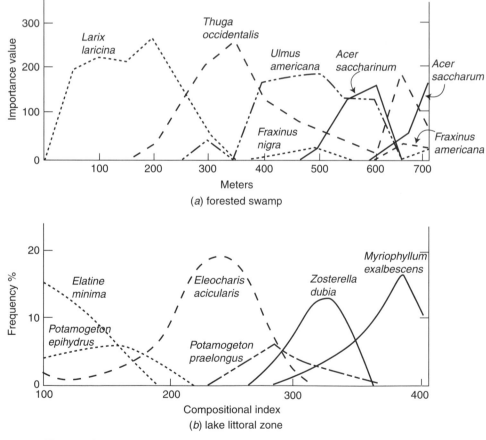

Figure 7.3 Examples of gradient analysis of wetlands in (a) a forested swamp and (b) submersed aquatic vegetation in the littoral zone of Wisconsin lakes. (*After van der Valk, 1982, based on original data from Beschel and Webber, 1962; Curtis, 1959.*)

In coastal areas, plants can do little to change the tidal pulse of water and salt. Tidal energy may be modified by vegetation as stems create friction that slows currents or as dead organic matter accumulates and changes the surface elevation. These effects are limited, however, by the overriding tides. These wetlands are often in dynamic equilibrium with the abiotic forces, an equilibrium that is sometimes called *pulse stability*.

In the low-energy environment of northern peatlands, in contrast to tidal marshes, hydrologic flows can be dramatically changed by biotic forces, resulting in distinctive patterned landscapes. Thus, changes in wetlands may be autogenic but are not necessarily directed toward a terrestrial climax. In fact, wetlands in dynamically stable environmental regimes seem to be extremely stable, contravening the central idea of succession. Pollen profiles were used to determine the successional sequence in British northern peatlands. Sequences were variable and there were reversals and skipped stages that may have been influenced by the dominant species first reaching a site (Fig. 7.4). A bog, not some type of terrestrial forest that hydrarch succession would have predicted, was the most common endpoint in most of the sequences described.

Linear Directed Change

If plant species development on a site is determined by allogenic processes and is, therefore, simply a response to environmental forcing processes, then the successional concept of linear directed change makes little sense. Although the scientific literature is replete with schematic diagrams showing the expected successional sequence from wetland to terrestrial forest, most of these are based on observed zonation patterns (or chronosequences), assuming that these spatial patterns presage the temporal pathway of change.

However, paleoecological analyses of soil profiles (such as those discussed earlier for the Indiana dunes ponds; Fig. 7.2) provide the best evidence to evaluate the concept. These records, mostly from northern peat bogs, suggest two generalizations: (1) In some sites, the present vegetation has existed for several thousands of years; and

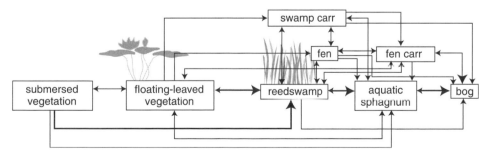

Figure 7.4 Successional sequences reconstructed from stratigraphic and palynological studies of postglacial British peatlands. Thicker lines indicate the more common transitions. (After Walker, 1970.)

(2) climatic change and glaciation had major impacts on plant species composition and distribution; generally, bogs expanded during warm, wet periods and contracted during cool, drier periods. Pollen sequences, however, are generally consistent across Europe and North America, indicating a response to similar global climate shifts (McIntosh, 1985). West (1964), as quoted in McIntosh (1985), wrote:

> We may conclude that our present plant communities have no long history in the Quaternary, but are merely temporary aggregations under given conditions of climate, other environmental factors, and historical factors.

Seed Banks

Seed banks, referred to as "buried reserves of viable seeds" (Keddy, 2000), are an important component of wetland succession. Many studies have documented the role of chance in the development of plant communities, especially in the early stages. The chance development can be the result of the availability of a seed bank and a changing environmental condition (e.g., the flooding of a site after years of dry conditions). In this respect, studies of seed banks and their role in the introduction and invasion of plant species have been important. If the development of plants on a site can be explained only in terms of the response of individual species to local conditions, then the previous history of the site is important because it determines what propagules are present for future invasion. This—the sediment *seed bank*—has been found to be extremely variable—both in space and in time. Pederson and Smith (1988) made the following generalizations about freshwater marsh seed banks:

1. Marshes with drawdowns produce the greatest number of seeds.
2. Seed banks are dominated by the seeds of annual plants and flood-intolerant species. Areas that contain emergent plants have greater seed densities than mud flats. Perennials generally produce fewer seeds that have shorter viability than annuals. They are more likely to reproduce by asexual means such as rhizomes.
3. Seed distribution decreases exponentially with the depth of the sediment.
4. Water is a major factor in seed banks. Seeds are concentrated along drift lines. The kinds of seeds produced depend on the flooding regime—by submergents when deep flooded, by emergents when periodically flooded, and by flood-intolerant annuals during drawdowns.
5. Saline zones produce few seeds. A salt marsh is an example of a perennial-dominated system in which most reproduction is asexual.

The germination of seedlings from a seed bank is similarly influenced by many factors that vary in space and time. Environmental factors such as flooding, temperature, soil chemistry, soil organic content, pathogens, nutrients, and allelopathy have been shown to influence recruitment. Water, in particular, is a critical variable, because most wetland plant seeds require moist but not flooded conditions for germination and early seedling growth. As a result of this restrictive moisture requirement, it is common to find even-aged stands of trees at low elevations in riparian wetlands, reflecting seed germination during relatively uncommon years when water levels were unusually low during the spring and summer.

Post-recruitment processes play a major role in the distribution of adult plants at a site, leading to plant assemblages that cannot be predicted from the seed bank alone. Thus, in coastal areas where the dominant plant, *Spartina alterniflora*, occurs in large monotypic stands, it is often the pioneer species and remains dominant throughout the life of the marsh. In contrast, in tidal and nontidal freshwater marshes, the seed bank is much larger and richer, and the first species to invade a site may later be replaced by other species.

Models of Wetland Community Development

Functional Guild Model

Historically, although the community concept has been of immense value in ecology, it has been criticized for being imprecise and not subject to accurate predictive models for ecological communities. Some ecologists have addressed this problem in different ways. One approach is to describe communities in terms of functional guilds that can be defined by measurable traits. A *guild* is defined as a group of functionally similar species in a community. The guild concept has at least three important advantages over the generalized community concept: (1) It collapses the large number of species in a community to a manageable subset; (2) it defines guilds in terms of measurable functional properties; and (3) it enables the prediction of what guilds will be found given specific environmental conditions. The guild concept is not new. Its use is well established, for example, in the study of birds and mammals (Simberloff and Dayan, 1991). Boutin and Keddy (1993) illustrated a functional classification of wetland plants, and Keddy (1992a) described how a similar approach could be used to predict the plant species (or guilds) in a specific habitat. In the Boutin–Keddy study, 27 functional traits (Table 7.1) of 43 wetland plant species from several different environments in the northeastern United States were measured and the results interpreted by cluster analyses. Figure 7.5 summarizes the results, which groups the species according to their traits into seven guilds, ranging from fast-growing obligate annuals that flowered in the first year to clonal species with deep roots that spread vegetatively. This exercise reduced 43 individual species to 7 groups or guilds within which the species are functionally similar. Most of the guilds appear to fall along a continuum of life histories adjusted to different light regimes, which is consistent with the results of other studies.

The traits of these guilds can then be used to predict their presence in defined habitats. For example, Keddy (1992a) described "assembly and response rules" by which a community of plants can be predicted from a list of species and their traits

Table 7.1 Traits measured on wetland plant species for functional guild classification

A. TRAITS MEASURED ON 1-YEAR-OLD PLANTS IN THE GARDEN

1	Life span:
	1 = annuals
	2 = facultative annuals (100% flowering)
	3 = partly facultative annuals ($>50<100$ % flowering)
	4 = perennials ($<50\%$ flowering)
2	Percentage flowering first year
3	Final height or highest height (cm)
4	Rate of shoot extension (cm/day): $\dfrac{\log_n \text{ height at day94} - \log_n \text{ height at day 36}}{\text{Day 94} - \text{Day 36}}$
5	Total biomass at harvest (g)
6	Above-ground biomass (g)
7	Below-ground biomass (g)
8	Ratio below-ground/above-ground biomass
9	Photosynthetic area (cm^2); includes leaves and green stems
10	Photosynthetic area/total biomass (cm^2/g)
11	Photosyntnetic area/total volume occupied by a plant (cm^2/ml) measured by displacement of water in graduated cylinder
12	Total biomass/total volume (g/ml)
13	Total number of tillers or shoots
14	Crown cover (cm^2): $((D_1 + D_2)/4)^2$
	where D_1 = first measure of crown diameter
	D_2 = second measurement at right angle to first
15	Stem diameter at ground level (cm)
16	Depth to belowground system (cm)
17	Diameter of belowground system, i.e., rhizome or main roots (cm)
18, 19	Shortest (18) and longest (19) distances between two shoots or tillers (measure of degree of clumping of aerial stems) (cm)

B. TRAITS MEASURED ON PLANTS IN NATURAL WETLANDS (ADULT TRAITS)

20	Total height (cm)
21	Total number of tillers or shoots
22	Stem diameter at ground level (cm)
23, 24	Shortest (23) and longest (24) distances between two shoots or tillers (cm)
25	Diameter of belowground system, i.e., rhizome or main roots (cm)
26	Depth to belowground system (cm)

C. TRAIT MEASURED UNDER GREENHOUSE CONDITIONS

27	Relative growth rate (RGR) (day^{-1}) between days 10 and 30

Source: Boutin and Keddy (1993).

and an environmental filter can be developed to delete those species that, because of their traits, will not be found in the community. The same procedure can be used to predict community species composition as a result of changes in the environmental filter. Figure 7.6 illustrates the procedure. From a total pool of freshwater wetland

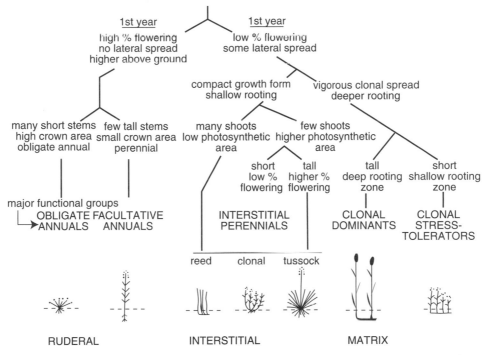

Figure 7.5   A functional classification of 43 species of plants from various wetland habitats in eastern North America, based on 27 plant traits displayed in Table 7.1. *(After Boutin and Keddy, 1993.)*

species ($s_1 \ldots s_n$ on left side of Fig. 7.6), those that cannot pass the first filter, for example, a shallow-water emergent environment, are deleted. If the marsh is later continually flooded, most of the adults will die from submergence or grazing by muskrats, as predicted from second deletion rules based on flood tolerance of adults. New vegetation then arises from buried propagules based on the regeneration requirements of species (addition rules). The result is a new set of species at time $t + 1$. This concept is quite similar to the environmental sieve model developed a decade earlier and described as follows.

Environmental Sieve Model

Van der Valk's (1981) environmental sieve model, also a Gleasonian model (Fig. 7.7), is similar to Keddy's model in several ways. The presence and the abundance of each species depend on its life history and its adaptation to the environment of a site. In van der Valk's model, all plant species are classified into life-history types, based on potential life span, propagule longevity, and propagule establishment requirements. Each life-history type has a unique set of characteristics and, thus, potential behavior in response to controlling environmental factors such as water-level

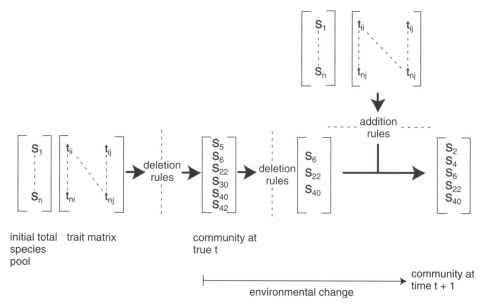

Figure 7.6 General procedure for response rules to delete and add species to a community after environmental perturbation. Species composition at time *t* is determined from an initial total list and trait matrix of all species, according to deletion rules specified by an environmental filter. Subsequent changes to the environmental filter result in deletion and/or addition to determine the plant composition at time *t* + 1. See discussion in text. (After Keddy, 1992a.)

changes. These environmental factors comprise the "environmental sieve" in van der Valk's model. As the environment changes, so does the sieve and, hence, the species present. L. M. Smith and Kadlec (1985) tested the model's ability to predict species composition in a fresh marsh after a fire, and they were satisfied with the qualitative results.

Centrifugal Organization Concept

Several other models of community change have been developed, although few have been applied to wetlands. Grime (1979) proposed that changes in species composition and richness of herbaceous plants was related to the gradients of disturbance and stress factors, which reduced biomass and determined which functional plant strategies would work best. Tilman (1982) suggested that competition among plants controlled community plant distribution, with each species limited by a different ratio of resources and spatial heterogeneity of the resources.

Wisheu and Keddy (1992) combined aspects of both Grime's and Tilman's models to propose a model of centrifugal organization of plant communities (Fig. 7.8a). Centrifugal organization describes the distribution of species and vegetation types

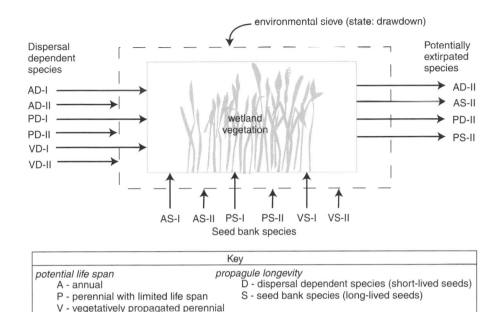

environmental sieve (state: drawdown)

Dispersal
dependent
species

AD-I
AD-II
PD-I
PD-II
VD-I
VD-II

wetland
vegetation

Potentially
extirpated
species

AD-II
AS-II
PD-II
PS-II

AS-I AS-II PS-I PS-II VS-I VS-II
Seed bank species

Key		
potential life span	*propagule longevity*	
A - annual	D - dispersal dependent species (short-lived seeds)	
P - perennial with limited life span	S - seed bank species (long-lived seeds)	
V - vegetatively propagated perennial		
propagule establishment requirement (e.g., hydrology)		
I - species only established in absence of standing water		
II - species can be established in standing water		

Figure 7.7 General sieve model of Gleasonian wetland (freshwater marsh) succession proposed by van der Valk. *(After van der Valk, 1981.)*

along standing-crop gradients caused by combinations of environmental constraints. Wisheu and Keddy (1992) summarize the concept as follows:

> Gradients radiate outwards from a single core habitat to many different peripheral habitats. The assumed mechanism is a competitive hierarchy where weaker competitors are restricted to the peripheral end of the gradient as a result of a trade-off between competitive ability and tolerance limits. The benign ends of the gradients comprise a core habitat which is dominated by the same species. At the peripheral end of each axis, species with specific adaptations to particular sources of adversity occur.

Wisheu and Keddy propose that the core habitat in wetlands has low disturbance and high fertility and is dominated by species that form dense canopies, such as *Typha* in the eastern United States (Fig. 7.8b). Peripheral habitats represent different kinds and combinations of stresses (infertility, disturbance) and support distinctive plant associations. The model allows one to predict how changes in gradients and, hence, peripheral habitats will change community composition. In the case of the *Typha*-core centrifugal model shown in Figure 7.8b, ice scouring, infertile sandy soils, flooding by beavers, and open shorelines are among the stresses that shift communities to less productive, albeit possibly more diverse, assemblages.

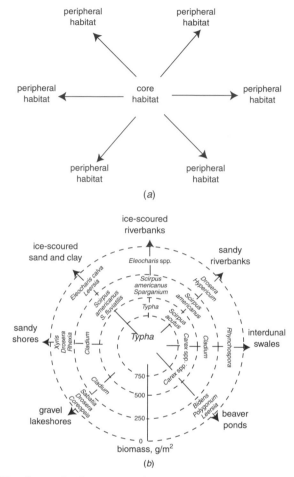

Figure 7.8 Centrifugal organization models illustrating (a) transitions from core habitat to peripheral habitats along resource or stress gradients (general model), and (b) freshwater wetland pattern for eastern North America, where large, leafy species such as cattail (*Typha spp.*) occupy the core habitat, while several different species and communities occupy peripheral habitats stressed by infertile sand, ice scouring, and beaver activity. *(After Wisheu and Keddy, 1992.)*

So far in this chapter, we have discussed vegetational changes in wetlands. We summarize this discussion with a statement by Niering (1989):

> Traditional successional concepts have limited usefulness when applied to wetland dynamics. Wetlands typically remain wet over time exhibiting a wetland aspect rather than succeeding to upland vegetation. Changes that occur may not necessarily be directional or orderly and are often not predictable on the long term. Fluctuating hydrologic conditions are the major factor controlling the vegetation pattern. The role of allogenic factors, including chance and coincidence, must be given new

emphasis. Cyclic changes should be expected as water levels fluctuate. Catastrophic events such as floods and droughts also play a significant role in both modifying yet perpetuating these systems.

Ecosystem Development

Eugene P. Odum (1969) described the maturation of ecosystems as a whole (as distinct from plants, communities and species) in an article entitled "The Strategy of Ecosystem Development." The concepts, in general, have withstood the test of time, and are republished with some update in Odum and Barrett (2005), published three years after Professor E. P. Odum's death. In ecosystem development, species composition in immature to mature (climax) stages are less important than are ecosystem functions, such as those described in Table 7.2. Immature ecosystems, Odum had observed, are characterized, in general, by high production to biomass ($P{:}B$) ratios; an excess of production over community respiration ($P{:}R$ ratio > 1); simple, linear, grazing food chains; low species diversity; small organisms; simple life cycles; and open mineral cycles. In contrast, mature ecosystems such as old-growth forests tend to use all of their production to maintain themselves and, therefore, have $P{:}R$ ratios about equal to one and little, if any, net community production. Production may be lower than in immature systems, but the quality is better, that is, plant production tends to be high in fruits, flowers, tubers, and other materials that are rich in protein. Because of the large structural biomass of trees in forested ecosystems, the $P{:}B$ ratio is small. Food chains are elaborate and detrital-based, species diversity is high, space is well organized into many different niches, organisms are larger than in immature systems, and life cycles tend to be long and complex. Nutrient cycles are closed; nutrients are efficiently stored and recycled within the ecosystem.

It is instructive to see how wetland ecosystems fit into this scheme of ecosystem development. Do their ecosystem-level characteristics fit the classical view that all wetlands are immature transitional seres? Or do they resemble the mature features of a terrestrial forest? The following conclusions can be made:

1. *Wetland ecosystems have properties of both immature and mature ecosystems.* For example, nearly all of the nonforested wetlands have $P{:}B$ ratios intermediate between developing and mature systems and $P{:}R$ ratios greater than one. Primary production tends to be very high compared with most terrestrial ecosystems. These attributes are characteristic of immature ecosystems. However, all of the ecosystems are detrital based, with complex food webs characteristic of mature systems.

2. *E. P. Odum (1971) used live biomass as an index of structure or "information" within an ecosystem.* Hence, a forested ecosystem is more mature in this respect than a grassland. This relationship is reflected in the high $P{:}B$ ratios (immature) of nonforested wetlands and the low $P{:}B$ ratios (mature) of forested wetlands. In a real sense, however, peat should be considered a structural element of wetlands because it is a primary autogenic factor

Table 7.2 Some attributes for ecosystem development in general[a]

| Ecosystem Type | Community Energetics | | | | Community Structure | | |
	P:R Ratio	P:B Ratio	Net Community Production	Food Chains	Total Biomass and Nonliving Organic Matter	Species Diversity	Organism Size
Developing	<1 or >1	High	High	Linear, grazing	Low	Increases Initially	Small
Mature (Climax)	1	Low	Low	Weblike, detrital	High	High or Declines	Large

| Ecosystem Type | Natural Selection | | Biogeochemical Cycles | | Regulation | |
	Growth Form	Life Cycle	Mineral Cycles	Internal Cycling	Resilience	Resistance
Developing	r-selection	Short, simple	Open	Unimportant	High	Low
Mature (Climax)	K-selection	Long, complex	Closed	Important	Low	High

[a]E. P. Odum (1969, 1971) as updated by Odum and Barrett (2005)

modifying the flooding characteristic of a wetland site. If peat is included in biomass, herbaceous wetlands would have the high biomass and low *P:B* ratios characteristic of more mature ecosystems. For example, a salt or fresh marsh has a live peak biomass of less than 2 kg/m^2. However, the organic content of a meter depth of peat (peats are often many meters deep) beneath the surface is on the order of 45 kg/m^2. This is comparable to the above-ground biomass of the most dense wetland or terrestrial forest. As a structural attribute of a marsh, peat is an indication of a maturity far greater than the live biomass alone would signify.

3. *Mineral cycles vary widely in wetlands.* ranging from extremely open riparian systems in which surface water (and nutrients) may be replaced thousands of times each year to bogs in which nutrients are derived from precipitation alone and are almost quantitatively retained. An open nutrient cycle is a juvenile characteristic of wetlands, directly related to the large flux of water through these ecosystems. Even in a system as open as a salt marsh that is flooded daily, however, about 80 percent of the nitrogen used by vegetation during a year is recycled from mineralized organic material.

4. *Spatial heterogeneity is generally well organized in wetlands along allogenic gradients.* The sharp, predictable zonation patterns and abundance of land–water interfaces are examples of this spatial organization. In forested wetlands, vertical heterogeneity is also well organized. This organization is an index of mature ecosystems. In most terrestrial ecosystems, however, the organization results from autogenic factors in ecosystem maturation. In wetlands, most of the organization seems to result from allogenic processes, specifically hydrologic and salinity gradients created by slight elevation

changes across a wetland. Thus, the "maturity" of a wetland's spatial organization consists of a high level of adaptation to prevailing microhabitat differences.

5. *Life cycles of wetland consumers are usually relatively short but are often exceedingly complex.* The short cycle is characteristic of immature systems, although the complexity is a mature attribute. Once again, the complexity of the life cycles of many wetland animals seems to be as much an adaptation to the physical pattern of the environment as to the biotic forces. Many animals use wetlands only seasonally or only during certain life stages. For example, small marsh fish and shellfish make daily excursions into wetlands during high tides, retiring to adjacent ponds during ebb tides. Many fish and shellfish species migrate from the ocean to coastal wetlands to spawn or for use as a nursery. Waterfowl use northern wetlands to nest and southern wetlands to overwinter, migrating thousands of miles between the two areas each year.

The Strategy of Wetland Ecosystem Development

In the previous sections, we showed that wetlands possess attributes of both immature and mature systems and that both allogenic and autogenic processes are important. Allogenic processes are important as forcing functions, which include factors such as hydrology and propagule introduction, change. Autogenic processes are important as the biota begin to control some of the physics and chemistry, as illustrated in Figure 4 in the hydrology chapter. In this section, we suggest that in all wetland ecosystems there is a common theme:

Development insulates the ecosystem from its environment.

At the level of individual species, this occurs through genetic (structural and physiological) adaptations to anoxic sediments and salt (see Chapter 6). At the ecosystem level, it occurs primarily through peat production, which tends to stabilize the flooding regime and shifts the main source of nutrients to recycled material within the ecosystem. In forests, shading is important in regeneration following disturbance.

Turnover Rates and Nutrient Influxes

The intensity of water flow over and through a wetland can be described by the water renewal rate (t^{-1}), the ratio of throughflow to the volume stored on the site (see Chapter 4). In wetlands, t^{-1} varies by four orders of magnitude (Fig. 7.9), ranging from about 1 per year in northern bogs to almost 10,000 per year in swamp forests. The nutrient input to a wetland follows closely the water renewal rate, because nutrients are carried to a site by water. The amount of nitrogen delivered to a wetland site, for example, also varies by five orders of magnitude, ranging from less than 1 g m^{-2} yr^{-1} in a northern bog to perhaps 10,000 g m^{-2} yr^{-1} in a riparian forested wetland (Fig. 7.9). Of course, not all of this nitrogen is available to plants in the

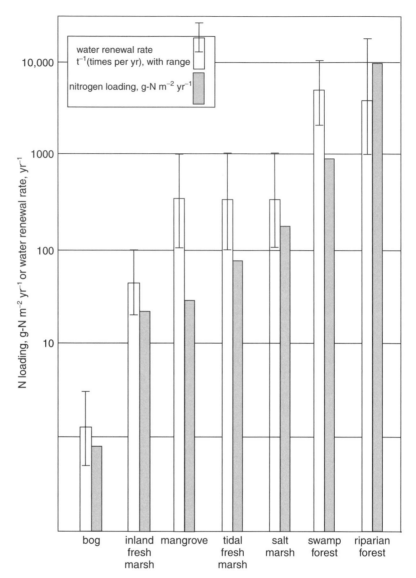

Figure 7.9 Renewal rates of water and nitrogen loading of major wetland types. This illustrates that wetlands have five orders of magnitude differences of hydrologic inflow and nutrient inflow.

ecosystem because, in extreme cases, it is flowing through much faster than it can be immobilized, but these figures indicate the potential nutrient supply to the ecosystem.

Despite the extreme variability in these outside (allogenic) forces, as illustrated in Figure 7.9, of hydrology and nutrient inflows varying over 10,000 times, wetland ecosystems are remarkably similar in many respects. Total stored biomass, including peat to 1 m depth, ranges from 40 to 60 kg m^{-2}—less than twofold. Soil nitrogen similarly varies only about threefold, from about 500 to 1,500 g m^{-2}. Net primary production, a key index of ecosystem function, varies by a factor of about five. Mean values for different wetland ecosystem types are usually in the range of 400 (peat bogs) to 2,000 (forested wetlands) g m^{-2} yr^{-1}.

Wetland Insularity

So allogenic forces vary by 10,000 times (4 orders of magnitude) but functions of wetlands vary by 2 to 5 times (not even one order of magnitude). Although several studies of individual species (e.g., *Spartina alterniflora*) or ecosystems (e.g., cypress swamps) have concluded that productivity is directly proportional to the water renewal rate (see Chapter 4), when different wetland ecosystems that constitute greatly different water regimes are compared, the relationship breaks down or at least is a logarithmic relationship. The apparent contradiction may be explained primarily by the role of stored nutrients within the ecosystem. As the large store of organic nutrients in the sediment mineralizes, it provides a steady source of inorganic fertilizer for plant growth. As a result, much of the nutrient demand is satisfied by recycling, even in systems as open as salt marshes and riparian wetlands. External nutrient inputs provide a subsidy to this basic supply. Therefore, growth is often apparently limited by the mineralization rate, which, in turn, is strongly temperature and hydroperiod dependent. Temperatures during the growing season are uniform enough to provide a similar nitrogen supply to plants in different wetland systems, except probably in northern bogs. There, low temperatures and short growing seasons limit mineralization and restrict nutrient input. The combination of the two factors limits productivity. Thus, as wetland ecosystems develop, they become increasingly insulated from the variability of the environment by storing nutrients. Often, the same process that stores nutrients (i.e., peat accumulation) also reduces the variability of flooding, further stabilizing the system. The surface of marshes, in general, is built up by the deposition of peats and waterborne inorganic sediments. As the elevation increases, flooding becomes less frequent and sediment input decreases. In the absence of overriding factors, coastal wetland marshes in time reach a stable elevation somewhere around local mean high water. The surfaces of riparian wetlands similarly rise until they become only infrequently flooded. Northern bogs grow by peat deposition above the water table, stabilizing at an elevation that maintains saturated peat by capillarity. Prairie potholes may be exceptions to these generalizations. They appear to be periodically "reset" by a combination of herbivore activity and long-term precipitation cycles and to achieve stability only in some cyclic sense.

Pulse Stability

In contrast to the lack of evidence for community succession to a stable set of species, which was summarized earlier, the concept of a progression toward a mature ecosystem (E. P. Odum, 1969) has greater merit. The attributes of a mature stable ecosystem place it in dynamic equilibrium with its environment, and although individual species may come and go, a mature ecosystem is stable in the sense that it has built-in mechanisms (species diversity, nutrient storage, and recycling) that resist short-term environmental fluctuations. In fact, W. E. Odum et al. (1995) suggested that natural processes pulse regularly, and the mature ecosystem responds in a pulsing steady state. In wetlands, examples of this phenomenon include salt marshes, tidal freshwater marshes, riverine forests, and seasonally flooded freshwater marshes, all of which are functionally similar despite marked differences in species composition, diversity, and community structure. Odum et al. (1995) suggest that natural pulses such as tides pump energy into ecosystems and enhance productivity. Biotic events are geared to and take advantage of these pulses, for example, the influx of small fish into flooded marshes to feed during high tide or the capturing of young fish in backwater oxbows and billabongs during flooding, with the captured fish serving as food for wading birds during periods of low water. This concept is referred to as *pulse stability*.

Self-Organization and Self-Design

Most wetland ecosystems are continually open to atmospheric, hydrologic, and biotic inputs of propagules of plants, animals, and microbes. *Self-organization*, as discussed by Howard T. Odum (1989), manifests itself in both microcosms and newly created ecosystems, "showing that after the first period of competitive colonization, the species prevailing are those that reinforce other species through nutrient cycles, aids to reproduction, control of spatial diversity, population regulation, and other means" (H.T. Odum, 1989). Self-organization is defined as "the process whereby complex systems consisting of many parts tend to organize to achieve some sort of stable, pulsing state in the absence of external interference" (E.P. Odum and Barrett, 2005).

Self-design, defined as "the application of self-organization in the design of ecosystems" (Mitsch and Jørgensen, 2004), relies on the self-organizing ability of ecosystems; natural processes (e.g., wind, rivers, tides, biotic inputs) contribute to species introduction; selection of those species that will dominate from this gene inflow is then nature's manifestation of ecosystem design (Mitsch and Wilson, 1996; Mitsch et al., 1998; Mitsch and Jørgensen, 2004). In self-design, the presence and survival of species resulting from the continuous introduction of them and their propagules is the essence of the successional and functional development of an ecosystem. This can be thought of as analogous to the continuous production of mutations necessary for evolution to proceed. In the context of ecosystem restoration and creation, self-design means that, if an ecosystem is open to allow "seeding," through human or natural means, of enough species' propagules, then the system will optimize its design by selecting for the assemblage of plants, microbes, and animals that is best adapted to the existing conditions. It is an important process to be investigated, particularly in view of the interest in restoring and creating wetlands.

In contrast to the self-design approach, the wetland restoration approach that is still used today involves the introduction of organisms (often plants), the survival of which becomes the measure of success of the restoration. This has sometimes been referred to as the "designer wetland" approach (Mitsch, 1998; van der Valk, 1998). This latter approach, while understandable because of the natural human tendency to control events, may be less sustainable than an approach that relies more on nature.

Wetland Primary Succession—A Wetland Experiment in Self-Design

In a multiyear, whole-ecosystem experiment in two created freshwater marsh basins at the Wilma H. Schiermeier Olentangy River Wetland Research Park at The Ohio State University in central Ohio (Fig. 7.10), Mitsch et al. (1998, 2005a,c) describe how 2,500 individual wetland plants representing 13 species were introduced to one 1-ha flow-through wetland basin while an adjacent identical basin remained an unplanted control, essentially testing the self-design capabilities of nature with and without human intervention. [NOTE: Other recent whole-ecosystem research at the Olentangy River Wetland Research Park on the importance of hydrologic pulsing is described in a box in Chapter 4.] Both basins have had identical inflows of river water and hydroperiods from 1994 through at least mid-2007. These experimental wetlands have allowed simultaneous long-term study of three different questions related to wetland development: (1) How important is wetland plant introduction on ecosystem function? (2) How long does it take for hydric soils and other wetland features to develop at a site where no hydric soils previously existed? and (3) What are the long-term patterns of biogeochemical changes of flow-through wetlands as they develop from open ponds of water to vegetated, hydric-soil marshes?

For the first six years of the wetland experiment, 17 different biotic and abiotic functional indicators of wetland function were measured, and similarities of the wetland basins were estimated from these indicators (Fig. 7.11). Indices were in six different categories, including macrophytes, algal communities, water quality changes, nutrient changes, benthic invertebrate diversity, and bird use. After only three years, there appeared to be a convergence of wetland function of the planted and unplanted basins, with a 71 percent similarity between the two basins after only one year of divergence. By three years, over 50 species of macrophytes, 130 genera of algae, over 30 taxa of aquatic invertebrates, and dozens of bird species found their way naturally to both wetlands to supplement the 13 introduced plant species (Mitsch et al., 1998). This convergence in year 3 followed the second year, where only 12 percent of the indicators were similar, probably because the planted wetland had macrophytes, but the unplanted wetland did not.

Typha dominated the naturally colonizing wetland (unplanted basin) but not in the planted wetland by the sixth year of the experiment (1999) (Fig. 7.12a), which caused a second divergence between the two basins in wetland function. During this year the similarity of the two wetlands dropped

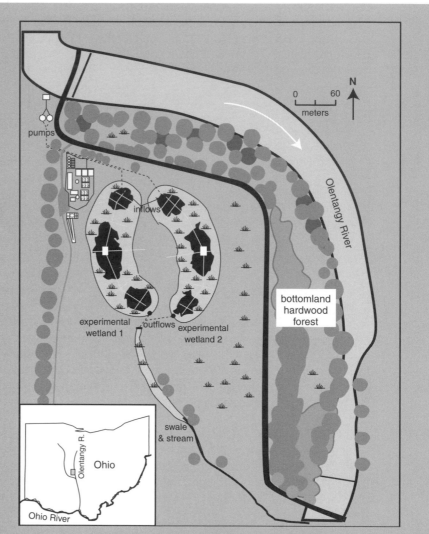

Figure 7.10 Wilma H. Schiermeier Olentangy River Wetland Research Park at The Ohio State University, Columbus, Ohio, showing two created 1-ha, flow-through wetland basins used to explore primary succession in created wetlands. One basin (experimental wetland 1) was planted with macrophytes in 1994, and the second basin (experimental wetland 2) remained an unplanted control. *(From Mitsch et al., 2005b.)*

to almost 40 percent. *Typha* was and continues to be held at bay by several planted species, most noteably *Sparganium eurycarpum* (bur reed), in the planted wetland. In general, the planted wetland has maintained a higher spatial macrophyte diversity throughout the study (Fig. 7.12b), but

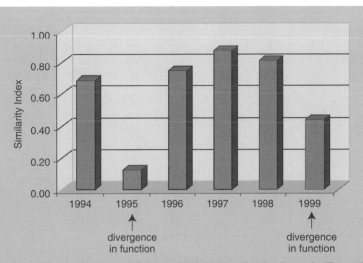

Figure 7.11 Similarity indices of two experimental wetlands shown in Figure 7.10 one to six growing seasons after experimental wetland 1 was planted with 2,500 individual plants representing 13 macrophytes in spring 1994, while experimental wetland 2 was not planted. Indices were in six different categories, including macrophytes, algal communities, water quality changes, nutrient changes, benthic invertebrate diversity, and bird use. Divergence in 1994 was due to plants only in one basin; divergence in 1999 was probably due to almost complete takeover of experimental wetland 2 (unplanted wetland) by *Typha* spp., while experimental wetland 1 remained much more diverse. (From Mitsch et al., 2005a.)

the naturally colonizing wetland has been more susceptible to disturbances such as muskrat herbivory and hydrologic pulses than has the more diverse planted wetland. The naturally colonizing wetland has had benthic invertebrate diversity and amphibian populations equal to or greater than the planted wetland for several of the years. Important to some, the naturally colonizing wetland was significantly greater in carbon sequestration through macrophyte primary productivity for several straight years when *Typha* dominated.

The continual introduction of species, whether introduced through flooding and other abiotic and biotic pathways, appeared to have a much longer-lasting effect in development of these ecosystems than the few species of plants that were introduced to one of the wetlands in the beginning. Studies done after 10 years of wetland development have shown that the planting had little long-term effect on functions such as nutrient retention (Mitsch et al., 2005c; see also Chapter 13), carbon accumulation in the soil (Anderson and Mitsch, 2006), methane emissions (Altor and Mitsch, 2006), nitrous oxide emissions (Hernandez and Mitsch, 2006), and denitrification (Hernandez and Mitsch, 2007). But the effect of planting remains observable in plant community diversity at least 12 years after planting (Fig. 7.12b).

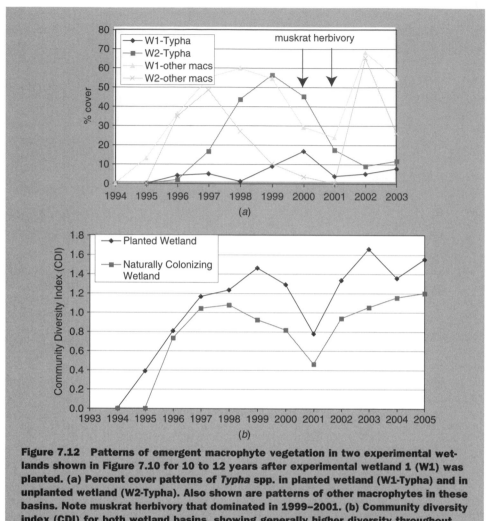

Figure 7.12 **Patterns of emergent macrophyte vegetation in two experimental wetlands shown in Figure 7.10 for 10 to 12 years after experimental wetland 1 (W1) was planted. (a) Percent cover patterns of *Typha* spp. in planted wetland (W1-Typha) and in unplanted wetland (W2-Typha). Also shown are patterns of other macrophytes in these basins. Note muskrat herbivory that dominated in 1999–2001. (b) Community diversity index (CDI) for both wetland basins, showing generally higher diversity throughout study for planted wetland compared to unplanted wetland. (a. from Mitsch et al., 2005c.)**

Ecosystem Engineers

A new way to describe the importance of autogenic successional processes involves the recently introduced term *ecosystem engineer*—a term that is used to describe organisms that have dramatic and important effects on an ecosystem (Jones et al., 1997; Alper, 1998). [This concept, also discussed in Chapter 4, should not be confused with the newly developing field of ecological engineering (Mitsch and Jørgensen, 2004).]

(a)

(b)

Figure 7.13   Landscape patterns in wetlands: (a) the physically controlled pattern of tidal creeks in a Louisiana salt marsh; (b) a muskrat eat-out in a brackish marsh on the Louisiana coast. Note the high density of muskrat houses.

In wetlands, examples of ecosystem engineers could be muskrats and beavers, both of which can have dramatic effects on vegetation cover and ecosystem hydrology in freshwater marshes. In these cases, the biota do show a dramatic feedback to the many features of the wetland (e.g., water levels or vegetation productivity). One could argue that these ecosystem engineers "set back" succession to an earlier stage; an alternative argument is that they are part of the ecosystem development and that their behavior and its effects should be both expected and appreciated as normal ecosystem behavior.

Landscape Patterns

Many large wetland landscapes develop predictable and often complex patterns of aquatic, wetland, and terrestrial habitats or ecosystems. In high-energy environments, these patterns appear to reflect abiotic forces, but they are largely controlled by biotic processes in low-energy environments. At the high-energy end of the spectrum, the microtopography and sediment characteristics of mature floodplains—complex mosaics of river channels, natural levees, back swamps, abandoned first and second terrace flats, and upland ridges—reflect the flooding pattern of the adjacent river. The vegetation responds to the physical topography and sediments with typical zonation patterns. Salt marshes similarly develop a characteristic pattern of tidal creeks, creekside levees, and interior flats that determine the zonation pattern and vigor of the vegetation (Fig. 7.13a).

At the low-energy end of the spectrum, the characteristic pattern of strings and flarks stretching for miles across northern peatlands appears to be primarily controlled by biotic processes. Similarly, in many freshwater marshes, herbivores can be major actors in the development of landscape patterns (Fig. 7.13b). In actuality, both physical (climatic, topographic, hydrologic) and biotic (production rates, root binding, herbivory, peat accumulation) processes combine in varying proportions and interact to produce observed wetland landscape patterns.

Recommended Readings

Keddy, P. A. 2000. *Wetland Ecology: Principles and Conservation*. Cambridge University Press, Cambridge, UK. 614 pp.

van der Valk, A.G. 2006. *The Biology of Freshwater Wetlands*. Oxford University Press, Oxford, UK, 173 pp.

Wetland Classification

Wetlands have been classified since the early 1900s, beginning with the peatland classifications of Europe and North America. The U.S. Fish and Wildlife Service has developed two major wetland classifications as the bases for wetland inventories. The early (1956) classification described 20 wetland types based on flooding depth, dominant forms of vegetation, and salinity regimes. "Classification of Wetlands and Deepwater Habitats of the United States," published in 1979, uses a hierarchical approach based on systems, subsystems, classes, subclasses, dominance types, and special modifiers to define wetlands and deepwater habitats precisely. Canadian and international wetland classification systems provide alternative systems that recognize 49 and 32 different wetland types, respectively. More recently, classifications based on wetland function have been developed, including a functionally based approach called the hydrogeomorphic (HGM) classification.

Wetland inventories are carried out at many different scales with several different imageries and with both aircraft and satellite platforms.

To deal realistically with wetlands on a regional scale, wetland scientists and managers have found it necessary both to categorize the different types of wetlands that exist and to determine their extent and distribution. The first of these activities is called *wetland classification*, and the second is called a *wetland inventory*. Some of the earliest efforts were undertaken to find wetlands that could be drained for human use; later classifications and inventories centered on the desire to compare different types of wetlands in a given region, often for their value to waterfowl. The protection of multiple ecological values of wetlands is the most recent purpose and now the most common reason for wetland classification and inventory. Recognition of wetland "value" has led some to now seek wetland classifications based on priorities for

Table 8.1   General wetland classification described in Mitsch and Gosselink (1986, 1993, 2000b), with estimated area in the United States, including Alaska

Type of Wetland	Area in United States (x 10^6 ha)
Coastal wetlands	
1. Tidal salt marshes	1.9
2. Tidal freshwater marshes	0.8
3. Mangrove wetlands	0.5
Inland wetlands	
4. Freshwater marshes	27
5. Peatlands	55
6. Freshwater swamps⎫	
7. Riparian systems ⎬	25
Total	111

protection, with highest protection afforded to those wetlands with the greatest value. As with other techniques, classifications and inventories are valuable only when the user is familiar with their scope and limitations.

One of the simplest classifications of wetlands was used in the previous three editions of our textbook (Mitsch and Gosselink, 1986, 1993, 2000b) and is still useful today. We described seven major types of wetlands divided into two major groups: (1) *coastal*—salt marsh, tidal freshwater marsh, mangrove, and (2) *inland*—freshwater marsh, peatland, freshwater swamp, and riparian ecosystems (Table 8.1). These classes of wetlands are generally recognizable ecosystems about which extensive research literature is available. Regulatory agencies also deal with these systems, and management strategies and regulations have been developed for these wetland types. Other types of wetlands, such as inland saline marshes, may fall between the cracks in this simple wetland classification, but the seven classes cover most wetlands currently found in North America. Collectively, these categories encompass about 111 million ha of wetlands in the United States (Table 8.1) and 127 million ha of wetlands in Canada.

Why Classify Wetlands?

Several attempts have been made to classify wetlands into categories that follow their structural and functional characteristics. These classifications depend on a well-understood general definition of wetlands (see Chapter 2), although a classification contains definitions of individual wetland types.

A primary goal of wetland classifications, according to Cowardin et al. (1979), "is to impose boundaries on natural ecosystems for the purposes of inventory, evaluation, and management." These authors identified four major objectives of a classification system:

1. To describe ecological units that have certain homogeneous natural attributes;

2. To arrange these units in a unified framework for the characterization and description of wetlands, that will aid decisions about resource management;

3. To identify classification units for inventory and mapping; and

4. To provide uniformity in concepts and terminology.

The first objective deals with the important task of grouping ecosystems that have similar characteristics in much the same way that taxonomists categorize species in taxonomic groupings. The wetland attributes that are frequently used to group and compare wetlands include the geomorphic and hydrologic regime, vegetation physiognomic type, and plant and/or animal species.

The second objective, to aid wetland managers, can be met in several ways when wetlands are classified. Classifications (which are definitions of different types of wetlands) enable the wetland manager to deal with wetland regulation and protection consistently from region to region and from one time to the next. Classifications also enable the wetland manager to pay selectively more attention to those types of wetlands that are most threatened or functionally the most valuable to a given region. Some classification systems [e.g., the U.S. Wetland Inventory system (Cowardin et al., 1979), the Canadian system (Warner and Rubec, 1997), the international Ramsar Convention system, and the hydrogeomorphic approach (Brinson, 1993a)] are scientific classifications based on natural properties, not evaluation systems developed for regulatory purposes. Thus, they do not focus on factors relating to environmental, social, or economic importance. Although these classification systems are useful planning tools, they are not structured by the requirements of management—that is, the need to make choices about relative social priorities and values.

The third and fourth objectives, to provide consistency in the formulation and use of inventories, mapping, concepts, and terminology, are also important in wetland management. The use of consistent terms to define particular types of wetlands is needed in the field of wetland science (see Chapter 2). These terms should then be applied uniformly to wetland inventories and mapping so that different regions can be compared and so that there will be a common understanding of wetland types among wetland scientists, wetland managers, and wetland owners.

Table 8.2 Early hydrologic classification of European peatlands

A. Rheophilous mire—Peatland influenced by groundwater derived from outside the immmediate watershed
   Type 1—Continuously flowing water that inundates the peatland surface
   Type 2—Continuously flowing water beneath a floating mat of vegetation
   Type 3—Intermittent flow inundating the mire surface
   Type 4—Intermittent flow of water beneath a floating mat of vegetation
B. Transition mire—Peatland influenced by groundwater derived solely from the immediate watershed
   Type 5—Continuous flow of water
   Type 6—Intermittent flow of water
C. Ombrophilous mire
   Type 7—Peatland never subject for flowing groundwater

Source: Bellamy (1968) and Moore and Bellamy (1974).

Wetland Classifications

Peatland Classifications

Many of the earliest wetland classifications were undertaken for the northern peatlands of Europe and North America. An early peatland classification in the United States, developed by Davis (1907), described Michigan bogs according to three criteria: (1) the landform on which the bog was established such as shallow lake basins or deltas of streams, (2) the method by which the bog was developed such as from the bottom up or from the shores inward, and (3) the surface vegetation such as tamarack or mosses. Based on the work of Weber (1907), Potonie (1908), Kulzynski (1949), and others in Europe, Moore and Bellamy (1974) described seven types of peatlands based on flow-through conditions (Table 8.2). Three general categories, called rheophilous, transition, and ombrophilous, describe the degree to which peatlands are influenced by outside drainage. The more modern terminology is minerotrophic, transition, and ombrotrophic peatlands. Most peatlands are limited to northern temperate climes and do not include all or even most types of wetlands in North America. These classifications, however, served as models for more inclusive classifications. They are significant because they combined the chemical and physical conditions of the wetland with the vegetation description to present a balanced approach to wetland classification.

Circular 39 Classification

In the early 1950s, the U.S. Fish and Wildlife Service recognized the need for a national wetlands inventory to determine "the distribution, extent, and quality of the remaining wetlands in relation to their value as wildlife habitat" (Shaw and Fredine, 1956). A classification was developed for that inventory (Martin et al., 1953), and the results of both the inventory and the classification scheme were published in U.S.

Fish and Wildlife Circular 39 (Shaw and Fredine, 1956). Twenty types of wetlands were described under four major categories:

1. Inland fresh areas
2. Inland saline areas
3. Coastal fresh areas
4. Coastal saline areas

In each of the four categories, the wetlands were arranged in order of increasing water depth or frequency of inundation. A brief description of the site characteristics of the 20 wetland types is given in Table 8.3.

Types 1 through 8 are freshwater wetlands that include bottomland hardwood forests (type 1), infrequently flooded meadows (type 2), freshwater nontidal marshes (types 3 and 4), open water less than 2 m deep (type 5), shrub-scrub swamps (type 6), forested swamps (type 7), and bogs (type 8). Types 9 through 11 are inland wetlands that have saline soils. They are defined according to the degree of flooding. Types 12 through 14 are wetlands that, although freshwater, are close enough to the coast to be influenced by tides. Types 15 through 20 are coastal saline wetlands that are influenced by both salt water and tidal action. These include salt flats and meadows (types 15 and 16), true salt marshes (types 17 and 18), open bays (type 19), and mangrove swamps (type 20).

This wetland classification was the most widely used in the United States until 1979, when the present National Wetlands Inventory classification was adopted. The earlier system is still referred to today by some wetland managers and is regarded by many as elegantly simple compared with its successor. It primarily used the physiognomy (life forms) of vegetation and the depth of flooding to identify the wetland type. Salinity was the only chemical parameter used, and although wetland soils were addressed in the Circular 39 publication, they were not used to define wetland types.

Coastal Wetland Classification

H. T. Odum et al. (1974) described coastal ecosystems by their major forcing functions (e.g., seasonal programming of sunlight and temperature) and stresses (e.g., ice) (Fig. 8.1). Coastal wetland types in this classification include salt marshes and mangrove swamps. Salt marshes, found in the type C category of natural temperate ecosystems with seasonal programming, have "light tidal regimes" and "winter cold" as forcing function and stress, respectively. Mangrove swamps are classified as type B (natural tropical ecosystems) because they have abundant light, show little stress, and reflect little seasonal programming. Three additional classes, type A (naturally stressed systems of wide latitudinal range), type D (natural arctic ecosystems with ice stress), and type E (emerging new systems associated with human activity), were included

Table 8.3 Early "Circular 39" wetland classification by U.S. Fish and Wildlife Service

Type Number	Wetland Type	Site Characteristics
INLAND FRESH AREAS		
1.	Seasonally flooded basins or flats	Soil covered with water or waterlogged during variable periods, but well drained during much of the growing season; in upland depressions and bottomlands
2.	Fresh meadows	Without standing water during growing season; waterlogged to within a few centimeters of surface
3.	Shallow fresh marshes	Soil waterlogged during growing season; often covered with 15 cm or more of water
4.	Deep fresh marshes	Soil covered with 15 cm to 1 m of water
5.	Open fresh water	Water less than 2 m deep
6.	Shrub swamps	Soil waterlogged; often covered with 15 cm or more of water
7.	Wooded swamps	Soil waterlogged; often covered with 30 cm of water; along sluggish streams, flat uplands, shallow lake basins
8.	Bogs	Soil waterlogged; spongy covering of mosses
INLAND SALINE AREAS		
9.	Saline flats	Flooded after periods of heavy precipitation; waterlogged within few centimeters of surface during the growing season
10.	Saline marshes	Soil waterlogged during growing season; often covered with 0.7 to 1 m of water; shallow lake basins
11.	Open saline water	Permanent areas of shallow saline water; depth variable
COASTAL FRESH AREAS		
12.	Shallow fresh marshes	Soil waterlogged during growing season; at high tide, as much as 15 cm of water; on landward side, deep marshes along tidal rivers, sounds, deltas
13.	Deep fresh marshes	At high tide, covered with 15 cm to 1 m water; along tidal rivers and bays
14.	Open fresh water	Shallow portions of open water along fresh tidal rivers and sounds
COASTAL SALINE AREAS		
15.	Salt flats	Soil waterlogged during growing season; sites occasionally to fairly regularly covered by high tide; landward sides or islands within salt meadows and marshes
16.	Salt meadows	Soil waterlogged during growing season; rarely covered with tide water; landward side of salt marshes
17.	Irregularly flooded salt marshes	Covered by wind tides at irregular intervals during the growing season; along shores of nearly enclosed bays, sounds, etc.
18.	Regularly flooded salt marshes	Covered at average high tide with 15 cm or more of wate; along open ocean and along sounds
19.	Sounds and bays	Portions of saltwater sounds and bays shallow enough to be diked and filled; all water landward from average lowtide line
20.	Mangrove swamps	Soil covered at average high tide with 15 cm to 1 m of water; along coast of southern Florida

Source: Shaw and Fredine (1956).

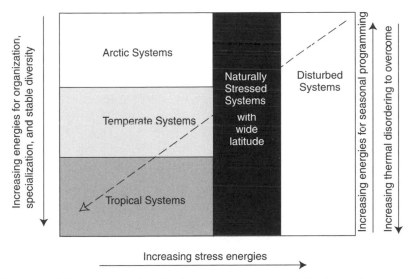

Figure 8.1 Coastal ecosystem classification system based on latitude (and, hence, solar energy) and major stresses. _(After H. T. Odum et al., 1974.)_

in this classification. The last class, which includes new systems formed by pollution such as pesticides and oil spills, is still an interesting concept that could be applied to other wetland classifications.

The United States Classification of Wetlands and Deepwater Habitats

The U.S. Fish and Wildlife Service began an inventory of the nation's wetlands in 1974. Because this inventory was designed to fulfill several scientific and management objectives, a new classification scheme, broader than the Circular 39 classification, was developed and finally published in 1979 as a "Classification of Wetlands and Deepwater Habitats of the United States" (Cowardin et al., 1979). Because wetlands were found to be continuous with deepwater ecosystems, both categories were addressed in this classification. It is thus a comprehensive classification of all continental aquatic and semi-aquatic ecosystems. As described in that publication in 1979:

> This classification, to be used in a new inventory of wetlands and deepwater habitats of the United States, is intended to describe ecological taxa, arrange them in a system useful to resource managers, furnish units for mapping, and provide uniformity of concepts and terms. Wetlands are defined by plants (hydrophytes), soils (hydric soils), and frequency of flooding. Ecologically related areas of deep water, traditionally not considered wetlands, are included in the classification as deepwater habitats.

This classification is based on a hierarchical approach analogous to taxonomic classifications used to identify plant and animal species. The first three levels of the classification hierarchy are given in Figure 8.2. The broadest level is _systems_: "a complex

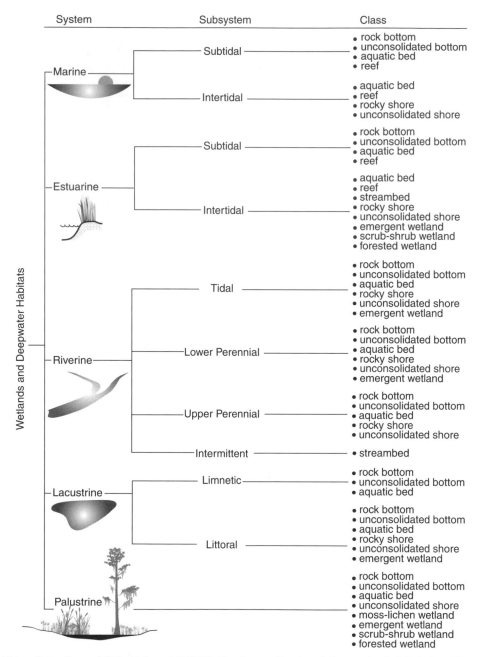

Figure 8.2   Current U.S. Fish and Wildlife Service wetland and deepwater habitat classification hierarchy showing five major systems, ten subsystems, and numerous classes. (*After Cowardin et al., 1979.*)

of wetlands and deepwater habitats that share the influence of similar hydrologic, geomorphologic, chemical, or biological factors." Thus, the systems, subsystems, and classes are based primarily on geologic and, to some extent, hydrologic considerations. Broad vegetation types are included primarily at the class level, and, even here, the vegetation types are generic (i.e., perennial, emergent, forested, scrub-shrub, or moss-lichen). Systems shown in Figure 8.2 include the following:

1. *Marine.* Open ocean overlying the continental shelf and its associated high-energy coastline.

2. *Estuarine.* Deepwater tidal habitats and adjacent tidal wetlands that are usually semi-enclosed by land but have open, partially obstructed, or sporadic access to the ocean and in which ocean water is at least occasionally diluted by freshwater runoff from the land.

3. *Riverine.* Wetlands and deepwater habitats contained within a channel with two exceptions: (1) wetlands dominated by trees, shrubs, persistent emergents, emergent mosses, or lichens; and (2) deepwater habitats with water containing ocean-derived salts in excess of 0.5 parts per thousand (ppt).

4. *Lacustrine.* Wetlands and deepwater habitats with all of the following characteristics: (1) situated in a topographic depression or a dammed river channel; (2) lacking trees, shrubs, persistent emergents, emergent mosses, or lichens with greater than 30 percent areal coverage; and (3) total area in excess of 8 ha. Similar wetland and deepwater habitats totaling less than 8 ha are also included in the lacustrine system when an active wave-formed or bedrock shoreline feature makes up all or part of the boundary or when the depth in the deepest part of the basin exceeds 2 m at low water.

5. *Palustrine.* All nontidal wetlands dominated by trees, shrubs, persistent emergents, emergent mosses, or lichens, and all such wetlands that occur in tidal areas where salinity stemming from ocean-derived salts is below 0.5 ppt. It also includes wetlands lacking such vegetation but with all of the following characteristics: (1) area less than 8 ha; (2) lack of active wave-formed or bedrock shoreline features; (3) water depth in the deepest part of the basin of less than 2 m at low water; and (4) salinity stemming from ocean-derived salts of less than 0.5 ppt.

Subsystems, as shown in Figure 8.2, give further definition to the systems. These include the following:

1. *Subtidal.* Substrate continuously submerged

2. *Intertidal.* Substrate exposed and flooded by tides, including the splash zone

3. *Tidal.* For riverine systems, gradient low and water velocity fluctuates under tidal influence

4. *Lower perennial.* Riverine systems with continuous flow, low gradient, and no tidal influence

5. *Upper perennial.* Riverine systems with continuous flow, high gradient, and no tidal influence

6. *Intermittent.* Riverine systems in which water does not flow for part of the year

7. *Limnetic.* All deepwater habitats in lakes

8. *Littoral.* Wetland habitats of a lacustrine system that extends from shore to a depth of 2 m below low water or to the maximum extent of nonpersistent emergent plants

The *class* of a particular wetland or deepwater habitat describes the general appearance of the ecosystem in terms of either the dominant vegetation life form or the substrate type. When more than 30 percent cover by vegetation is present, a vegetation class is used (e.g., shrub-scrub wetland). When less than 30 percent of the substrate is covered by vegetation, then a substrate class is used (e.g., unconsolidated bottom). The typical demarcation of many of the classes of the palustrine system is shown in Figure 8.3.

Most inland wetlands fall into the palustrine system, in the classes moss-lichen, emergent, scrub-shrub, or forested wetland. Coastal wetlands are classified in the same classes within the estuarine system and intertidal subsystem. Only nonpersistent emergent wetlands are classified into other systems.

Further descriptions of the wetlands and deepwater habitats are possible through the use of *subclasses, dominance types*, and *modifiers*. Subclasses such as "persistent"

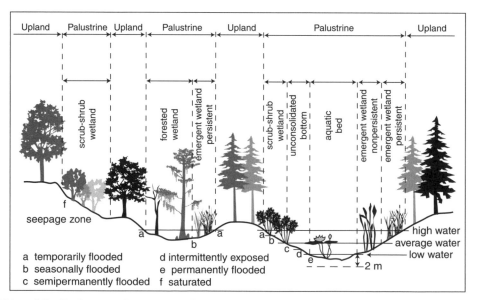

Figure 8.3   Features and examples of wetland classes and hydrologic modifiers in the palustrine system. (After Cowardin et al., 1979.)

and "nonpersistent" give further definition to a class such as emergent vegetation. Type refers to a particular dominant plant species (e.g., bald cypress, *Taxodium distichum*, for a needle-leaved deciduous forested wetland) or a dominant sedentary or sessile animal species (e.g., eastern oyster, *Crassostrea virginica*, for a mollusk reef). Modifiers (Table 8.4) are used after classes and subclasses to describe more precisely the water regime, the salinity, the pH, and the soil. For many wetlands, the description of the environmental modifiers adds a great deal of information about their physical and chemical characteristics. Unfortunately, those parameters are difficult to measure consistently in large-scale surveys such as inventories.

Canadian Wetlands Classification System

The Canadian Wetland Classification System (Warner and Rubec, 1997; Table 8.5) is designed to be practical as well as hierarchical. Its major features include:

1. *Classes*—based on natural features of the wetlands, rather than on interpretation for various uses. They have direct application to large wetland regions. Wetland classes are recognized on the basis of properties that reflect the overall "genetic origin" of the wetland system and the nature of the wetland environment. Division into classes allows ready identification in the field and delineation on maps. They are also convenient groupings for data storage, retrieval, and interpretation.

2. *Forms*—subdivisions of wetland classes based on surface morphology, water type, and morphology characteristics of the underlying mineral soil. Some forms are further subdivided into subforms. Forms are easily recognized features of the landscape and are the basic wetland-mapping unit.

3. *Types*—subdivisions of wetland forms and subforms based on physiognomic characteristics of the vegetation communities. They are comparable to the modifiers used in the U.S. Fish and Wildlife Service classification system. Types are most useful for evaluation of wetland values and benefits, management for wetland hydrology and wildlife habitat, and conservation and protection of rare and endangered species.

Currently, the system recognizes 5 wetland classes (bog, fen, swamp, marsh, shallow water marsh), 49 wetland forms, and 75 subforms (Table 8.5). Although geomorphological, hydrologic, and chemical characteristics do not appear in the classification, Figure 8.4 shows the ontogenetic development of wetland forms in Canada.

International Wetland Classification System

As a part of its mission to identify and conserve wetlands of international importance for biodiversity and wildlife, the International Union for the Conservation of Nature and Natural Resources (IUCN) developed a "Classification System for Wetland Types." As with both the U.S. Fish and Wildlife Service classification system and the Canadian Wetland Classification System, most of the wetland types are readily

Table 8.4   Modifiers used in current wetland and deepwater habitat classification by U.S. Fish and Wildlife Service

WATER REGIME MODIFIERS (TIDAL)

Subtidal—substrate permanently flooded with tidal water
Irregularly exposed—land suface exposed by tides less often than daily
Regularly flooded—alternately floods and exposes land surfaces at least daily
Irregularly flooded—land surface flooded less often than daily

WATER REGIME MODIFIERS (NONTIDAL)

Permanently flooded—water covers land surface throughout year in all years
Intermittently exposed—surface water present throughout year except in years of extreme drought
Semi-permanently flooded—surface water persists throughout growing season in most years; when surface water is absent, water table is at or near surface
Seasonally flooded—surface water is present for extended periods, especially in early growing season but is absent by the end of the season
Saturated—substrate is saturated for extended periods during growing season but surface water is seldom present
Temporarily flooded—surface water is present for brief periods during growing season but water table is otherwise well below the soil surface
Intermittently flooded—substrate is usually exposed but surface water is present for variable periods with no seasonal periodicity

SALINITY MODIFIERS

Marine and Estuarine	Riverine, Lacustrine, and Palustrine	Salinity (ppt)
Hyperhaline	Hypersaline	>40
Euhaline	Eusaline	30–40
Mixohaline (brackish)	Mixosaline	0.5–30
Polyhaline	Polysaline	18.0–30
Mesohaline	Mesosaline	5.0–18
Oligohaline	Oligosaline	0.5–5
Fresh	Fresh	<0.5

pH MODIFIERS

	Acid	pH less than 5.5
	Circumneutral	pH 5.5–7.4
	Alkaline	pH greater than 7.4

SOIL MATERIAL MODIFIERS

Mineral	(1) Less than 20% organic carbon and never saturated with water for more than a few days, or
	(2) Saturated or artificially drained and has
	(a) less than 18% organic carbon if 60% or more is clay
	(b) less than 12% organic carbon if no clay
	(c) a proportional content of organic carbon between 12 and 18% if clay content is between 0 and 60%
Organic	Other than mineral as described above

Source: Cowardin et al. (1979).

Table 8.5 Classes, forms, and subforms in the Canadian wetland classification system

Class/Form/Subform	Class/Form/Subform	Class/Form/Subform
Bog	**Swamp**	*Riparian meltwater channel marsh*
Palsa bog	Tidal swamp	Lacustrine marsh
Peat mound bog	*Tidal saltwater swamp*	*Lacustrine shore marsh*
Mound bog	*Tidal freshwater swamp*	*Lacustrine bay marsh*
Domed bog	Inland salt swamp	*Lacustrine lagoon marsh*
Polygonal peat plateau bog	Flat swamp	Basin marsh
Lowland polygon bog	*Basin swamp*	*Linked basin marsh*
Peat plateau bog	*Swale swamp*	*Isolated basin marsh*
Plateau bog	*Unconfined flat swamp*	*Discharge basin marsh*
Northern plateau bog	Riparian swamp	Hummock marsh
Atlantic plateau bog	*Lacustrine swamp*	Spring marsh
Collapse scar bog	*Riverine swamp*	Slope marsh
Riparian bog	*Floodplain swamp*	**Shallow water marsh**
Floating bog	*Channel swamp*	Tidal water
Shore bog	Slope swamp	*Tidal bay water*
Basin bog	*Unconfined slope swamp*	*Tidal shore water*
Flat bog	*Peat margin swamp*	*Tidal channel water*
String bog	*Lagg swamp*	*Tidal lagoon water*
Blanket bog	*Drainageway swamp*	*Tidal basin water*
Slope bog	Discharge swamp	Estuarine water
Veneer bog	*Spring swamp*	*Estuarine delta water*
Fen	*Seepage swamp*	*Estuarine bay water*
String fen	Mineral-rise swamp	*Estuarine shore water*
Northern ribbed fen	*Island swamp*	*Estuarine channel water*
Atlantic ribbed fen	*Levee swamp*	*Estuarine basin water*
Ladder fen	*Beach ridge swamp*	*Estuarine lagoon water*
Net fen	*Mound swamp*	Riparian water
Palsa fen	Raised peatland swamp	*Riparian stream water*
Snowpatch fen	**Marsh**	*Riparian meltwater channel water*
Spring fen	Tidal marsh	*Riparian floodplain water*
Feather fen	*Tidal bay Marsh*	*Riparian delta water*
Slope fen	*Tidal lagoon marsh*	Lacustrine water
Lowland polygon fen	*Tidal channel marsh*	*Lacustrine bay water*
Riparian fen	*Tidal basin marsh*	*Lacustrine shore water*
Floating fen	Estuarine marsh	*Lacustrine lagoon water*
Stream fen	*Estuarine delta marsh*	Basin water
Shore fen	*Estuarine bay marsh*	*Discharge basin water*
Collapse scar fen	*Estuarine lagoon marsh*	*Linked basin water*
Horizontal fen	*Estuarine shore marsh*	*Isolated basin water*
Channel fen	Riparian marsh	*Polygon basin water*
Basin fen	*Riparian stream marsh*	*Tundra basin water*
	Riparian floodplain marsh	*Thermokarst basin water*
	Riparian delta marsh	

Source: Warner and Rubec (1997).

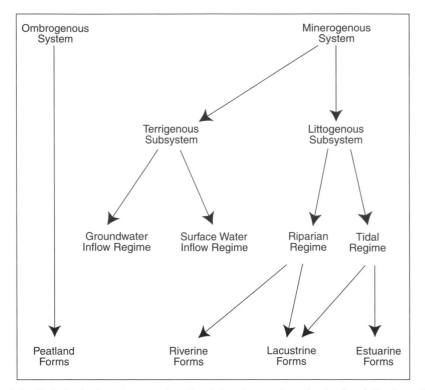

Figure 8.4 Hydrological systems and wetland development as the basis of the Canadian Wetland Classification System. (*After Warner and Rubec, 1997.*)

recognized by their location in the landscape combined with their vegetation life form. Table 8.6 compares the Canadian and current U.S. Fish and Wildlife Service systems to this Ramsar Convention system. The Ramsar system has 32 classes, divided into a marine/coastal group and an inland group. Because it attempts to be global, it has categories that neither the U.S. nor the Canadian system have, for example, underground karst systems and oases. The U.S. system, because it is hierarchical, has fewer classes at the system and subsystem level, but uses modifiers to identify specific wetland types. The Canadian system also has only five classes, but with forms and subforms reaches a total of over 70 different categories by which to classify wetlands.

Hydrogeomorphic Wetland Classification System

Brinson (1993a) described a classification system designed to be used for evaluation of wetland functions. It is being used increasingly as a means of assessing the physical, chemical, and biological functions of wetlands, and is extremely useful for comparing the level of functional integrity of wetlands within a functional class, or for evaluating the impact of proposed human activities on wetlands and mitigation alternatives

Table 8.6 Comparison of wetland types among the Ramsar Convention International Wetland classification system, the current U.S. Fish and Wildlife wetland and deepwater habitat classification, and the Canadian wetland classification system

Ramsar Convention Code Name		U.S. Fish and Wildlife System[a]	Canadian System[b]
MARINE/COASTAL WETLANDS			
A	Marine water <6 m	Marine Subtidal	Shallow (<2 m) water marsh
B	Marine subtidal aquatic beds	Marine subtidal aquatic bed	
C	Coral reefs	Marine subtidal reef	—
D	Rocky marine shores	Marine intertidal rock bottom	—
E	Sandy shore or dune	Marine intertidal unconsolidated	—
F	Estuarine waters	Estuarine subtidal	Estuarine marsh, water
G	Intertidal flats	Estuarine intertidal unconsolidated bottom	Estuarine water, tidal water
H	Intertidal marshes	Estuarine intertidal emergent wetland	Tidal marsh
I	Intertidal forested wetland	Estuarine intertidal forested wetland	Tidal swamp
J	Coastal saline lagoon	Estuarine subtidal unconsolidated, saline	Estuarine water
K	Coastal fresh lagoon	Estuarine subtidal unconsolidated, fresh	Estuarine water
Zk(a)	Marine/coastal karst	Estuarine subtidal rocky shore	—
INLAND WETLANDS			
L	Permanent inland deltas	Riverine perennial Estuarine delta marsh Shallow riparian delta water	Riparian delta marsh
M	Permanent rivers/streams	Riverine perennial swamp, marsh	Shallow riparian water
N	Intermittent rivers/streams	Riverine intermittent	—
O	Permanent fresh lakes Riparian water (oxbows)	Lacustrine littoral or limnetic	Shallow lacustrine water
P	Intermittent fresh lakes	Lacustrine or riparian littoral	—
Q	Permanent saline lakes	Lacustrine littoral unconsolidated, saline	—
R	Intermittent saline lakes	Lacustrine littoral intermittent, saline	—
Sp	Permanent saline marshes/pools	Palustrine emergent wetland or unconsolidated bottom, saline	Estuarine marsh, inland salt swamp
Ss	Intermittent saline marshes/pools	Palustrine emergent wetland or unconsolidated bottom, intermittent	Spring, slope, or basin marsh
Tp	Permanent fresh marsh/pools (<8 ha)	Palustrine emergent wetland or unconsolidated bottom, fresh	Shallow basin water, lacustrine marsh
Ts	Intermittent fresh marsh/pools, inorganic soils	Palustrine emergent wetland, intermittently flooded	Shallow basin water
U	Nonforested peatlands	Palustrine emergent wetland, persistent	Bogs, fens
Va	Alpine wetlands	Palustrine emergent wetland, persistent	—
Vt	Tundra wetlands	Palustrine emergent wetland, persistent	Bogs, fens, shallow basin water
W	Shrub-dominated wetlands	Palustrine scrub-shrub wetland	Riparian, flat, slope, discharge or mineral-rise swamp

(continued overleaf)

Table 8.6 (*continued*)

Ramsar Convention Code Name		U.S. Fish and Wildlife System	Canadian System
Xf	Fresh forested wetlands on inorganic soils	Palustrine forested wetland	Riparian swamp
Xp	Forested peatlands	Palustrine forested or scrub-shrub wetland	Flat bog, flat or raised peatland swamp
Y	Freshwater springs, oases	—	Hummock marsh
Zg	Geothermal wetlands	—	—
Zg(b)	Inland karst systems, underground	—	—

[a]Ramsar class can be further approximated with additional modifiers.
[b]Class and form indicated only; subforms can approximate Ramsar more closely. The term "shallow" in the wetland type indicates the shallow marsh class.

(Figure 8.5; Table 8.7). The classification received its inspiration from several earlier classification schemes that use hydrodynamics as a primary classifier: Gosselink and Turner (1978) for freshwater marshes; O'Brien and Motts (1980) and Hollands (1987) for glaciated northeastern U.S. wetlands; Novitzki (1979) for Wisconsin wetlands; Gilvear et al. (1989) for East Anglian fens; Prance (1979) for Amazon forests; National Wetlands Working Group (1988) for Canadian wetlands; Brinson (1989) for tidal wetlands; Winter (1977) for north-central U.S. lakes; and Leopold et al. (1964) and Rosgen (1985) for floodplains and riparian systems. Especially pertinent are the coastal classification of H. T. Odum et al. (1974) described earlier; Lugo and Snedaker's (1974) mangrove classification; and variations of the mangrove classification described by Cintrón et al. (1985) and Lugo et al. (1990a).

The classification is based primarily on hydrodynamic differences as they function within four geomorphic settings. Thus, the three core components of the classification system are geomorphology, water source, and hydrodynamics (Table 8.7). Geomorphic setting is the topographic location of a wetland in the surrounding landscape. Four geomorphic settings are identified: depressional, riverine, and fringe, and extensive peatlands. The first three are clearly related to the hydrologic setting. Extensive peatlands are different because the dominant influence on hydrology is biogenic accretion. Water sources are precipitation, surface or near-surface flow, and groundwater discharge (into a wetland). Hydrodynamics refers to the direction and strength of water movement within a wetland. The three core features are heavily interdependent, so it is difficult to describe any one without the other two. Taken as a group, the three core features may be pooled in 36 combinations, but because of the interdependence not all combinations are found in nature.

It becomes clear when examining Table 8.7 how interrelated the three core features are. The water entering a wetland is seldom from only one of the three sources—precipitation, groundwater discharge, and surface inflow. Figure 8.6 shows characteristic mixtures of water sources for some major wetland types. Ombrotrophic peat wetlands are typically dominated by precipitation; mineral fens and seep wetlands by groundwater discharge; and riverine and fringe wetlands by surface flows.

Table 8.7 Functional classification of wetlands by geomorphology, water source, and hydrodynamics

Core Component	Description	Example
GEOMORPHIC SETTING	Topographic location of a wetland in the surrounding landscape	
Depressional	Wetlands in depressions that typically receive most moisture from precipitation, hence often ombrotrophic; found in dry and moist climates	Kettles, potholes, vernal pools, Carolina bays, groundwater slope wetlands
Extensive peatlands	Peat substrate isolates wetland from mineral substrate; peat dominates movement and storage of water and chemicals	Blanket bogs, tussock tundra
Riverine	Linear strips in landscape; subject predominantly to unidirectional surface flow	Riparian wetlands along rivers, streams
Fringe	Estuarine and lacustrine wetlands with bidirectional surface flow	Estuarine tidal wetlands, lacustrine fringes subject to winds, waves, and seiches
WATER SOURCE	Relative importance of three main sources of water to a wetland	
Precipitation	Wetlands dominated by precipitation as the primary water source; water level may be variable because of evapotranspiration	Ombrotrophic bogs, pocosins
Groundwater discharge	Primary water source from regional or perched mineral groundwater sources	Fens, groundwater slope wetlands
Surface inflow	Water source dominated by surface inflow	Alluvial swamps, tidal wetlands, montane streamside wetlands
HYDRODYNAMICS	Motion of water and its capacity to do work	
Vertical fluctuation	Vertical fluctuation of the water table resulting from evapotranspiration and replacement by precipitation or groundwater discharge	Usually depressional wetlands, bogs (annual), prairie potholes (multiyear)
Unidirectional flow	Unidirectional surface or near-surface flow; velocity corresponds to gradient	Usually riverine wetlands
Bidirectional flow	Occurrence in wetlands dominated by tidal and wind-generated water-level fluctuations	Usually fringe wetlands

Source: Brinson (1993a).

The core component *hydrodynamics* is an expression of the fluvial energy that drives the system. This ranges from low-energy water table fluctuations typical of depressional wetlands to unidirectional flows found in riverine wetlands to bidirectional flows of tidal and high-energy lacustrine wetland systems. Uni- and bidirectional surface flows range widely in energy from hardly perceptible movement to strong erosive currents. Combined with the geographic setting, hydrodynamics can result in a range of different wetland types.

This classification system was designed to be independent of plant communities, because it depends on the geomorphic and hydrologic properties of the wetlands. In practice, however, vegetation often provides important clues to the hydrogeomorphic

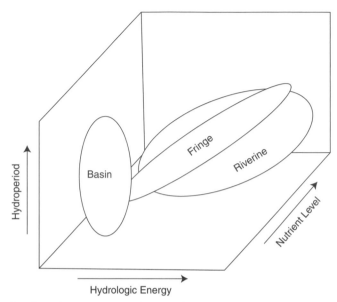

Figure 8.5 Basis of the hydrogeomorphic classification system. Geomorphic settings (basin, fringe, and riverine) are arranged around three core factors. (After Lugo et al., 1990; Brinson, 1993a.)

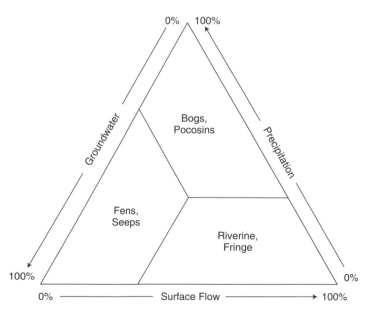

Figure 8.6 The relative contribution of a combination of three water sources—precipitation, groundwater discharge, and surface inflow—to wetlands. The location of major wetland types within the triangle shows the relative importance of different water sources. (After Brinson, 1987.)

forces at work, and because most modern classification systems developed for inventory purposes have some basis in hydrogeomorphology, their classes often give important clues as to function. The reverse is not true. Classification schemes based on functional processes, such as the hydrogeomorphic system, are difficult to delimit spatially, because the classes are not necessarily identified by different vegetation associations or other easily mapped features. Hence, the major classification schemes used for inventory purposes rely heavily on vegetation life forms, although all include some aspects of geomorphology and hydrology.

Wetland Classification for Value

Since the introduction of legislation aimed at wetland conservation and regulation, there has been considerable interest in classifying wetlands for their value to society in order to simplify the issuance of permits for wetlands activities; that is, high-value wetlands would presumably receive more protection than low-value wetlands. Figure 8.7 illustrates some of the difficulties of using this approach to classify a bottomland forest riparian system. We pose the question of whether one can classify this floodplain into high-value wetlands and wetlands of lesser value. This is not a trivial question because wetland status and, hence, the protection of the upper end of the bottomland zone, is an issue of serious concern to developers, farmers, and agency regulators. Floodplains are multiple-value resources; as Figure 8.7 shows, different ecological processes peak in different zones, and these peaks are not directly related to their value to humans. For example, although most floodwater is stored at low elevations in the floodplain, the highest zone is important because it helps moderate large, infrequent floods that do the most damage. Because tree growth rates are highest in the seasonally flooded zone, that zone might be considered most valuable for timber harvest. In fact, however, bald cypress, which is found primarily in a permanently flooded zone, is a more valuable timber crop because of its superior rot and insect resistance. The classification of wetlands by value involves trade-offs among competing social priorities that imply an earlier commitment to certain goals (e.g., to farming over conservation, or to hunting over fishing).

Wetland Remote Sensing and Inventory

One of the major objectives of wetland classification is to be able to inventory the location, extent, and type of wetlands in a region of concern. An inventory can be made of a small watershed, a county or parish, an entire state or province, or an entire nation. Whatever the size of the area to be surveyed, the inventory must be based on some previously defined classification and should be constructed to meet the needs of specific users of information on wetlands. Generally, inventories require not only information about the types and extent of wetlands, but also documentation of their geographic locations and boundaries. To accomplish this goal, *remote platforms*—aircraft and/or satellites—produce *imagery*—photographs or digital information that can be used to make images. This imagery must be *interpreted* to identify the locations, boundaries,

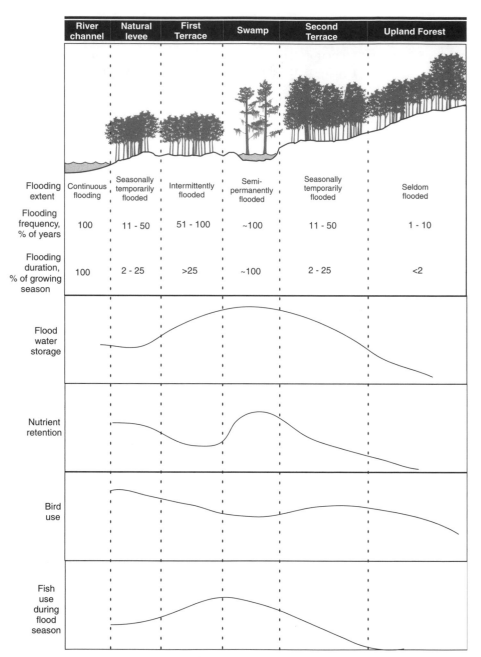

	River channel	Natural levee	First Terrace	Swamp	Second Terrace	Upland Forest
Flooding extent	Continuous flooding	Seasonally temporarily flooded	Intermittently flooded	Semi-permanently flooded	Seasonally temporarily flooded	Seldom flooded
Flooding frequency, % of years	100	11 - 50	51 - 100	~100	11 - 50	1 - 10
Flooding duration, % of growing season	100	2 - 25	>25	~100	2 - 25	<2
Flood water storage						
Nutrient retention						
Bird use						
Fish use during flood season						

Figure 8.7 A simplified representation of a section across a bottomland hardwood forest from stream to upland, showing how various functions of interest to humans change across the transect. In actuality, the area is a complex spatial pattern of intermixed zones, not a linear gradient. This is a multiple-value resource; different ecological processes peak in different zones, and these peaks are not directly related to their value to humans.

and types of wetlands on the image. The interpreted images are typically overlain on topographic maps, such as U.S. Geological Survey (USGS) quadrangle maps, to obtain wetland *maps* (paper or digital) of the scene.

Remote Sensing Platform

In the early days of wetland classification, wetlands were mapped from surveyors' records and from boats. Later, interpretation of aerial photographs combined with verification in the field made the process both faster and more accurate. Today, high-altitude imagery from aircraft such as that from the U-2 "spy plane" is used extensively, but satellite imagery is rapidly becoming the norm. Commonly used satellite systems include Landsat Multispectral Scanner (MSS), Landsat Thematic Mapper (TM), and Systéme Pour l'Observation de la Terre (SPOT). These remote platforms are an effective way of gathering data for large-scale wetland surveys. Satellites offer repeated coverage that allow seasonal monitoring of wetlands as well as providing data on the surrounding landscape readily translatable to a GIS format (Ozesmi and Bauer, 2002). The choice of which platform to use depends on the resolution required, the area to be covered, and the cost of the data collection. Low-altitude aircraft surveys offer a relatively inexpensive and fairly effective way to survey small areas. High-altitude aircraft offer much greater coverage in each image (photograph) and may be less expensive per unit area than low-altitude aircraft when costs of photo interpretation are included. The limitations of satellite remote sensing include the lack of ability to separate different wetland types or to even separate wetlands from upland forests or agricultural land.

Orbiting satellites have been providing data for Earth resources classification since the launching of the first of the Landsat satellites in 1972. Today, several highly effective satellites orbit the Earth. One problem of early satellites was the poor resolution (Landsat has a resolution of 30 m). Wetland scientists working on the National Wetlands Inventory (see BOX) found that Landsat 7 could not provide the desired level of detail without what appeared to be an excessive amount of collateral data, such as aerial photographs and field work. Nevertheless, the state of Ohio used satellite remote sensing as the basis for its wetland inventory (Figure 8.8). Today's satellites can resolve features of the Earth's surface to as little as 1 m, but high resolution is not an unmixed blessing. It requires the ability to transmit and process enormous amounts of data, because the data generated for a given surface area quadruples every time the resolution doubles.

Remote Sensing Imagery

In addition to choosing the remote-sensing platform, the wetland scientist or manager has the choice of several types of imagery from different types of sensors. Color photography and color–infrared photography have been popular for many years for wetland inventories from aircraft (see Shuman and Ambrose, 2003), although black-and-white photography has been used with some success. Color–infrared film

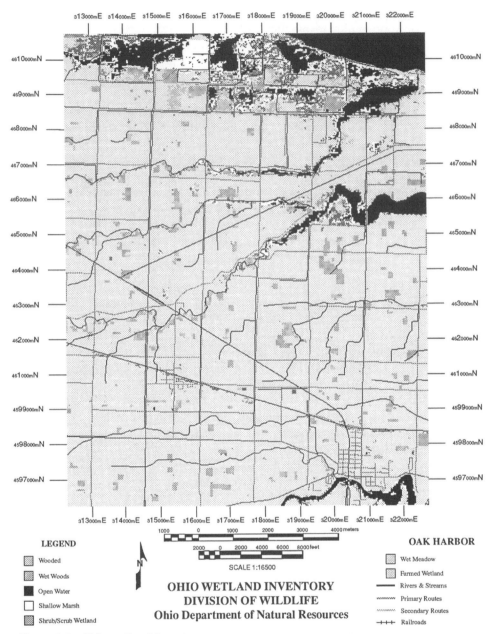

313000mE 314000mE 315000mE 316000mE 317000mE 318000mE 319000mE 320000mE 321000mE 322000mE

4610000mN   4610000mN
469000mN   469000mN
468000mN   468000mN
467000mN   467000mN
466000mN   466000mN
465000mN   465000mN
464000mN   464000mN
463000mN   463000mN
462000mN   462000mN
461000mN   461000mN
4599000mN   4599000mN
4598000mN   4598000mN
4597000mN   4597000mN

313000mE 314000mE 315000mE 316000mE 317000mE 318000mE 319000mE 320000mE 321000mE 322000mE

1000   0   1000   2000   3000   4000 meters

2000   0   2000   4000   6000   8000 feet

SCALE 1:16500

LEGEND

- ▨ Wooded
- ▨ Wet Woods
- ■ Open Water
- □ Shallow Marsh
- ▨ Shrub/Scrub Wetland

N

OHIO WETLAND INVENTORY
DIVISION OF WILDLIFE
Ohio Department of Natural Resources

OAK HARBOR

- ▨ Wet Meadow
- ▨ Farmed Wetland
- — Rivers & Streams
- ᠁ Primary Routes
- ᠁ Secondary Routes
- +++ Railroads

Figure 8.8   Ohio wetland inventory map prepared by the Ohio Department of Natural Resources, Division of Wildlife, from supervised classification of Landsat 5 Thematic Mapper (TM) satellite imagery (30-m resolution), hydric soil data, and limited on-site evaluation. *(Courtesy of Ohio Department of Natural Resources, Division of Wildlife.)*

provides good definition of plant communities and is the film of choice. Satellites, and some aircraft, gather digital data in one or more electromagnetic spectral bands. For example, *Landsat 7* has panchromatic and multispectral capability in seven color bands, including infrared. It can resolve the Earth's surface to 30 m in multispectral mode and 15 m in panchromatic mode. Although the U.S. National Wetland Inventory (see BOX) still relies primarily on manual photo or image interpretation, research suggests that computerized interpretation of satellite imagery could be superior for delineating wetland hydrology, particularly for agricultural areas (National Research Council, 1995).

Interpretation and Mapping

Interpretation of wetland areas from imagery is typically the most difficult and time consuming part of the mapping procedure, and the accuracy of the resulting maps is determined at this step. Much interpretation is still done visually by a photointerpreter, identifying and delimiting image colors correlated with field observation of the wetland types that correspond to different colors. The ability to program computers to recognize these colors (or their equivalent digital signal), and thus to interpret the imagery directly, is improving as computer power and image resolution increases. In particular, the use of several spectral bands at once allows for much improved resolution of the spectral signature of different surface features. Depending on the expected use of the classified map products, supervised computer interpretation is frequently used today, usually with satisfactory results. In addition, satellite data can be used in conjunction with other databases, such as hydric soil maps and on-site evaluation, as in the example of Figure 8.8. In this wetland inventory, completed for a region in Ohio primarily from satellite data, the following categories of wetlands were possible to map from satellite imagery combined with on-site evaluation and hydric soil maps:

1. Open water
2. Shallow marsh: emergent vegetation in water 1 m or less depth
3. Shrub-scrub wetland: emergent woody vegetation in water 1 m or less
4. Wet meadow: grassy vegetation in water 15 cm or less
5. Farmed wetland: wet meadow areas appearing on agricultural lands
6. Wet woods: forested wetlands with trees and hydric soils

The U.S. National Wetlands Inventory

The U.S. National Wetlands Inventory (NWI) is a good example of a major wetlands mapping project, and illustrates some of the problems encountered in any mapping enterprise. The Cowardin et al. (1979) classification scheme has provided the basic mapping units for the NWI being carried out by the U.S. Fish and Wildlife Service. For the NWI, aerial photography at scales ranging from 1:60,000 to 1:130,000 is the primary source of data, with color-infrared photography providing the best delineation of wetlands (Wilen and Pywell, 1981; Tiner and Wilen, 1983). In the 1970s, maps were mostly created from 1:80,000-scale black-and-white photography. Now 1:60,000 color-infrared photography is used for most of the mapping. Manual photointerpretation

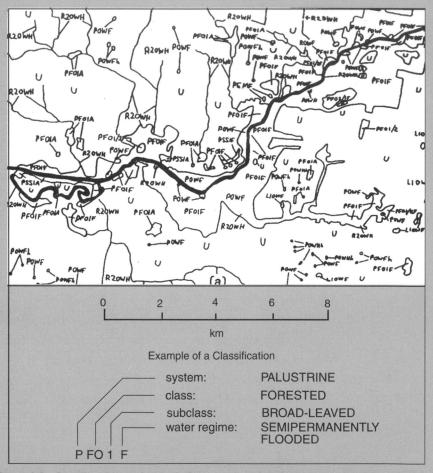

Figure 8.9 Sample of mapping technique used at 1:100,000 scale by National Wetlands Inventory, showing a portion of map (redrawn) and an example of a classification notation. *(After Dyersburg, Tennessee, 1:100,000 wetland map provided courtesy of National Wetlands Inventory, U.S. Fish and Wildlife Service.)*

and field reconnaissance are then used to define wetland boundaries according to the wetland classification system. The information is summarized on 1:24,000 and 1:100,000 maps, using an alphanumeric system based on the U.S. Fish and Wildlife Classification system as illustrated in Figure 8.9.

Wetland Maps

The product of image interpretation is a map delineating the locations, boundaries, and types of wetlands in the mapped scene. One such map, developed from satellite imagery and field studies, is shown in Figure 8.8. Another map, from the U.S. National Wetland Inventory (see BOX), is shown in Figure 8.9. These maps typically appear as an overlay on a standard topographic map or as a standalone map that has been geo-rectified using established coordinates and procedures. Maps are generally created for specific purposes, and none serves all management requirements equally well. Coarse-scale maps covering large areas are often useful for inventories addressing regional wetland types, areas, and changes through time (e.g., the Status and Trends Reports of the U.S. Fish and Wildlife Service). Finer-scale resolution maps generally cover much smaller areas because of the much greater data requirements per unit area, especially when computerized interpretation of multispectral, high-resolution satellite data is used. These maps tend to be much more accurate at the local level than regional maps.

Recommended Readings

Brinson, M. M. 1993. *A Hydrogeomorphic Classification for Wetlands. Wetlands Research Program Technical Report WRP-DE-4*. U.S. Army Corps of Engineers Waterways Experiment Station, Vicksburg, MS.

Cowardin, L. M., V. Carter, F. C. Golet, and E. T. LaRoe. 1979. *Classification of Wetlands and Deepwater Habitats of the United States*. U.S. Fish and Wildlife Service, FWS/OBS-79/31, Washington, DC. 103 pp.

Tiner, R. W. 1999. *Wetland Indicators: A Guide to Wetland Identification, Delineation, Classification, and Mapping*. CRC Press, Boca Raton, FL. 392 pp.

Part 3

Wetland Management

Human Impacts and Management of

Wetlands

Wetland impacts have included both wetland alteration and wetland destruction. In earlier times, wetland drainage was considered the only policy for managing wetlands. The most common alterations of wetlands have been draining, dredging, and filling of wetlands; modification of the hydrologic regime; highway construction; mining and mineral extraction; and water pollution. Peat resources, estimated to be 1.9 trillion tons in the world, are harvested in many countries as a source of fuel and horticultural materials. Wetlands can also be managed close to their natural state for certain objectives, such as fish and wildlife enhancement, agricultural and aquaculture production, water quality improvement, and flood control. Management of wetlands for coastal protection has now taken on more significance.

The concept of wetland management has had different meanings at different times to different disciplines and in different parts of the world. Until the middle of the 20th century, wetland management usually meant wetland drainage to many policy makers, except for a few resource managers who maintained wetlands for hunting, fishing, and waterfowl/wildlife protection. Landowners were encouraged through government programs to tile and drain wetlands to make the land suitable for agriculture and other uses. Countless coastal and inland wetlands were destroyed by dredging for navigation and filling for land development.

Until the 20th century, there was little understanding of and concern for the inherent values of wetlands. The value of wetlands as wildlife habitats, particularly for providing hunting grounds for waterfowl, was recognized in the first half of the 20th century by some fish and game managers. A whole science of "marsh management" developed around the idea of maintaining specific hydrologic conditions to optimize fish or waterfowl populations. Only since the 1970s have other values such as flood control, coastal protection, and water quality enhancement been recognized. It has taken disasters such as the 1993 Upper Mississippi River Basin flooding, the 2004 Indian Ocean tsunami, and the 2005 Hurricane Katrina disaster in New Orleans to cause societies to focus on the potential lives that could be saved and property damage minimized if wetland buffer systems were provided at our land–water margins.

Today, the management of wetlands usually means setting several objectives, depending on the priorities of the wetland managers, current environmental regulations, and wishes of a myriad of stakeholders who are usually involved. In some cases, objectives such as preventing pollution from reaching wetlands and using wetlands as sites of water quality improvement can be conflicting. Many floodplain wetlands are now managed and zoned to minimize human encroachment and maximize floodwater retention. Coastal wetlands are now included in coastal zone protection programs for storm protection and as sanctuaries and subsidies for estuarine fauna. In the meantime, wetlands continue to be altered or destroyed throughout the world by drainage, filling, conversion to agriculture, water pollution, and mineral extraction.

We are thankful to have witnessed a slowing of the destruction rate of wetlands, even since we wrote the first edition of this *Wetland* textbook (Mitsch and Gosselink, 1986), at least in the United States. We are not as certain that destruction of the world's wetlands is yet being slowed, but we are aware that there is a much greater international appreciation of wetlands than before. Vigilance is required, however, to make sure that wetland values continue to be protected. Wetland conservation and even wetland restoration and creation (see Chapter 12 for details on this type of wetland management) have accelerated, particularly in the developed world over the past quarter of a century. But there are few if any regulations or restrictions on wetland destruction or pollution in developing parts of the world. This may be the next frontier of wetland protection.

An Early History of Wetland Management

The early history of wetland management, a history that still influences many people today, was driven by the misconception that wetlands were wastelands that should be avoided or, if possible, drained and filled. Throughout the world, as long as there have been humans, there has been hydrologic alteration of the landscape. As described by Larson and Kusler (1979):

> For most of recorded history, wetlands were regarded as wastelands if not bogs of treachery, mires of despair, homes of pests, and refuges for outlaw and rebel. A good wetland was a drained wetland free of this mixture of dubious social factors.

In China, Europe, and regions of the Middle East, civilizations have both lived with but also significantly altered the natural hydrologic landscape for thousands of years. In the United States, this opinion of wetlands and shallow-water environments led to the destruction of more than half of the total wetlands in the lower 48 states over a 200-year period. In New Zealand, settlement by Europeans that began in earnest in the mid-1800s contributed significantly to a 90 percent loss of wetlands in a relatively short time. Preliminary estimates suggest that, over human history, about half of the world's wetlands have been lost (see Chapter 3).

With over 70 percent of the world's population living on or near coastlines, coastal wetlands have long been destroyed through a combination of excessive harvesting, hydrologic modification and seawall construction, coastal development, pollution, and other human activities. Likewise, inland wetlands have been continually affected, particularly through hydrologic modification and agricultural and urban development. Human activities such as agriculture, forestry, stream channelization, aquaculture, dam, dike, and seawall construction, mining, water pollution, and groundwater withdrawal all had impacts, some severe, on wetlands (Table 9.1). Wetlands are degraded and destroyed indirectly as well through alternation of sediment patterns in rivers, hydrologic alteration, highway construction, and land subsidence (Table 9.2). A third possibility is the loss of wetlands from natural causes (Table 9.3), although wetlands are normally resilient and can recover from natural events. Coastal wetlands that were devastated by the 2004 Indian Ocean tsunami or the 2005 hurricane that destroyed much of New Orleans are recovering.

It was estimated that, by 1985, 56 to 65 percent of wetlands in North America and Europe, 27 percent in Asia, 6 percent in South America, and 2 percent in Africa had been drained for intensive agriculture (Ramsar Convention Secretariat, 2004).

The propensity in the East was not to drain valuable wetlands entirely, as has been done in the West, but to work within the aquatic landscape, albeit in a heavily managed way. Dugan's (1993) interesting comparison between *hydraulic*

Table 9.1   Human actions that cause direct wetland losses and degradation[a]

Cause	Estuaries	Floodplains	Freshwater Marshes	Lakes/ Littoral Zone	Peatlands	Swamp Forest
Agriculture, forestry, mosquito control drainage	xx	xx	xx	x	xx	xx
Stream channelization and dredging; flood control	x		x			
Filling—solid-waste disposal; roads; development	xx	xx	xx	x		
Conversion to aquaculture/mariculture	xx					
Dikes, dams, seawall, levee construction	xx	x	x	x		
Water pollution— urban and agricultural	xx	xx	xx	xx		
Mining of wetlands of peat and other materials	x	x		xx	xx	xx
Groundwater withdrawal		x	xx			

[a]xx, common and important cause of wetland loss and degradation; x, present but not a major cause of wetland loss and degradation. Blank indicates that effect is generally not present except in exceptional situations
Source: Dugan (1993).

Table 9.2 Human activities that indirectly cause wetland losses and degradation[a]

Cause	Estuaries	Floodplains	Freshwater Marshes	Lakes/ Littoral Zone	Peatlands	Swamp Forest
Sediment retention by dams and other structures	xx	xx	xx			
Hydrologic alteration by roads, canals, etc.	xx	xx	xx	xx		
Land subsidence due to groundwater, resource extraction, and river alternations	xx	xx	xx			

[a]xx, common and important cause of wetland loss and degradation; x, present but not a major cause of wetland loss and degradation. Blank indicates that effect is generally not present except in exceptional situations
Source: Dugan (1993).

Table 9.3 Natural events that cause wetland losses and degradation[a]

Cause	Estuaries	Floodplains	Freshwater Marshes	Lakes/ Littoral Zone	Peatlands	Swamp Forest
Subsidence	x			x	x	x
Sea-level rise	xx					xx
Drought	xx	xx	xx	x	x	x
Hurricanes, tsunamis, and other storms	xx			x	x	
Erosion	xx	x			x	
Biotic effects		xx	xx	xx		

[a]xx, common and important cause of wetland loss and degradation; x, present but not a major cause of wetland loss and degradation. Blank indicates that effect is generally not present except in exceptional situations
Source: Dugan (1993).

civilizations (European in origin), which controlled water flow through the use of dikes, dams, pumps, and drainage tile, and *aquatic civilizations* (Asian in origin), which better adapted to their surroundings of water-abundant floodplains and deltas, is an interesting way to view humans' use of wetlands. The former approach of controlling nature rather than working it is becoming more dominant around the world today; that is why we continue to find such high losses of wetlands worldwide.

Wetland Drainage History in the United States

Had not politics intervened, George Washington may have succeeded in draining the Great Dismal Swamp in Virginia in the mid-18th century (see Chapter 3). The practice of draining swamps and other wetlands was an acceptable and even desired practice from the time Europeans first settled in North America. In the United States, public laws actually encouraged wetland drainage. Congress passed the Swamp Land

Act of 1849, which granted to Louisiana the control of all swamplands and overflow lands in the state for the general purpose of controlling floods in the Mississippi River basin. In the following year, the act was extended to the states of Alabama, Arkansas, California, Florida, Illinois, Indiana, Iowa, Michigan, Mississippi, Missouri, Ohio, and Wisconsin. Minnesota and Oregon were added in 1860. The act was designed to decrease federal involvement in flood control and drainage by transferring federally owned wetlands to the states, leaving to them the initiative of "reclaiming" wetlands through activities such as levee construction and drainage.

By 1954, an estimated 26 million ha of land had been ceded to those 15 states for reclamation. Ironically, although the federal government passed the Swamp Land Act to get out of the flood control business, the states sold those lands to individuals for pennies per acre, and the private owners subsequently successfully lobbied both national and state governments to protect these lands from floods. Further, governments are now paying enormous sums to buy the same lands back for conservation purposes. Although current government policies are generally in direct opposition to the Swamp Land Act and it is now disregarded, the act cast the initial wetland policy of the U.S. government in the direction of wetland elimination.

Other actions led to the rapid decline of the nation's wetlands. An estimated 23 million ha of wet farmland, including some wetlands, were drained under the U.S. Department of Agriculture's Agricultural Conservation Program between 1940 and 1977 (Office of Technology Assessment, 1984). An estimated 18.6 million ha of land, much of it wetlands, was drained in seven states in the upper Mississippi River basin alone (Table 9.4). Some of the wetland drainage activity was hastened by projects of groups such as the Depression-era Works Progress Administration (WPA), the Soil Conservation Service, and other federal agencies. Coastal marshes were eliminated or drained and ditched for intercoastal transportation, residential developments, mosquito control, and even for salt marsh hay production. Interior wetlands were converted primarily to provide land for urban development, road construction, and agriculture.

Table 9.4   Drainage statistics of selected states in the upper reaches of the Mississippi River basin

State	Total Area Drained (x 1,000 ha)	Percentage of All Land that is Drained	Percentage of Cropland that Is Drained
Illinois	3,965	30	35
Indiana	3,273	30	50
Iowa	3,154	20	25
Ohio	3,000	20	50
Minnesota	2,580	15	20
Missouri	1,720	10	25
Wisconsin	910	6	10
Total	18,602		

Source: USDA (1987), as cited in Zucker and Brown (1998).

Go South, Young Man?

Typical of the prevalent attitude toward wetlands in the mid-20th century is the following quote by Norgress (1947) discussing the "value" of Louisiana cypress swamps:

> With 1,628,915 acres of cutover cypress swamp lands in Louisiana at the present time, what use to make of these lands so that the ideal cypress areas will make a return on the investment for the landowner is a serious problem of the future. . . .
>
> The lumbermen are rapidly awakening to the fact that in cutting the timber from their land they have taken the first step toward putting it in position to perform its true function—agriculture. . . .
>
> It requires only a visit into this swamp territory to overcome such prejudices that reclamation is impracticable. Millions of dollars are being put into good roads. Everywhere one sees dredge boats eating their way through the soil, making channels for drainage.
>
> After harvesting the cypress timber crop, the Louisiana lumbermen are at last realizing that in reaping the crop sown by Nature ages ago, they have left a heritage to posterity of an asset of permanent value and service—land, the true basis for wealth.
>
> The day of the pioneer cypress lumberman is gone, but we need today in Louisiana another type of pioneer—the pioneer who can help bring under cultivation the enormous areas of cypress cutover lands suitable for agriculture. It is important to Louisiana, to the South, and the Nation as a whole, that this be done. Would that there were some latter-day Horace Greeleys to cry, in clarion tones, to the young farmers of today, "Go South, young man; go South!"

As an example of state action leading to wetland drainage, Illinois passed the Illinois Drainage Levee Act and the Farm Drainage Act in 1879, which allowed counties to organize into drainage districts to consolidate financial resources. This action accelerated draining to the point that 30 percent of Illinois and Indiana and 20 percent of Iowa and Ohio are now under some form of drainage (Table 9.4), and almost all of the original wetlands in these states (80 to 90 percent) have been destroyed. Chapter 3 described two very large wetlands in this region of the United States—The Great Kankakee Marsh in Indiana and the Great Black Swamp in Ohio—that essentially no longer exist. Drainage was absolute there.

Wetland Alteration

In a sense, wetland alteration or destruction is an extreme form of wetland management. One model of wetland alteration (Fig. 9.1) assumes that three main factors influence wetland ecosystem health: water level, nutrient status, and natural disturbances. Through human activity, the modification of any one of these factors can

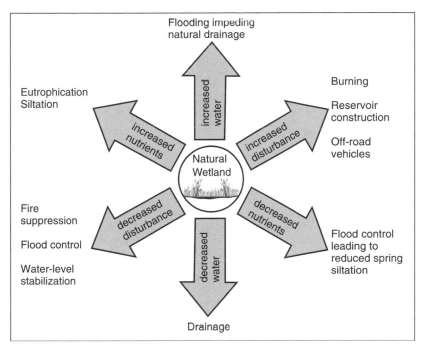

Figure 9.1 Model of human-induced impacts on wetlands, including effects on water level, nutrient status, and natural disturbance. By either increasing or decreasing any one of these factors, wetlands can be altered. (After Keddy, 1983.)

lead to wetland alteration, either directly or indirectly. For example, a wetland can be disturbed through decreased water levels, as in draining and filling, or through increased water levels, as in downstream drainage impediments. Nutrient status can be affected through upstream flood control that decreases the frequency of nutrient inputs or through increased nutrient loading from agricultural areas.

The most common alterations of wetlands have been (1) draining, dredging, and filling of wetlands; (2) modification of the hydrologic regime; (3) highway construction; (4) mining and mineral extraction; and (5) water pollution. These wetland modifications are described in more detail next.

Wetland Conversion: Draining, Dredging, and Filling

The major cause of wetland loss around the world continues to be conversion to agricultural use. Drainage for farms in the United States progressed at an average rate of 490,000 ha/yr since the early 1900s (slope of line in Fig. 9.2a). Little drainage occurred during the Depression and war years, and since 1985 little additional drainage has occurred, except in Minnesota and Ohio. This conversion was particularly

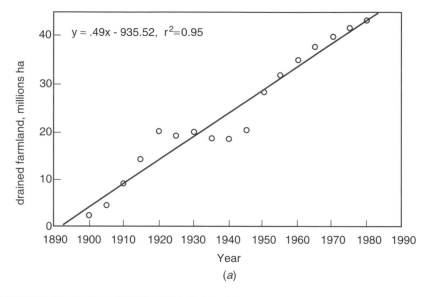

(a)

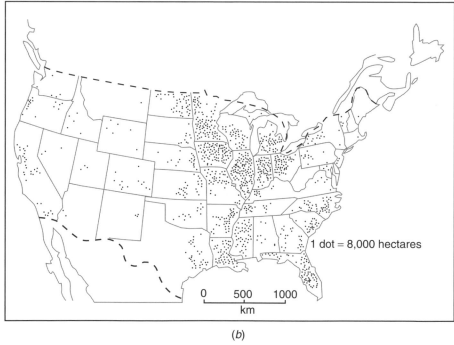

1 dot = 8,000 hectares

0   500   1000
    km

(b)

Figure 9.2 Artificially drained land in the United States: (a) Trend from 1900–1980. *(After Gosselink and Maltby, 1990, based on data from Office of Technology Assessment, 1984).* **(b) Extent and location as of 1985. Each dot represents 8,000 ha (20,000 acres), and total area drained is 43,500,000 ha.** *(After Dahl, 1990.)*

significant in the vast Midwestern United States "breadbasket," which has provided the bulk of the grain produced on the continent (Fig. 9.2b and Table 9.4). Some of the world's richest farming is in the former wetlands of Ohio, Indiana, Illinois, Iowa, and southern Minnesota. When drained and cultivated, the fertile soils of the prairie pothole marshes and east Texas *playas* also produce excellent crops. With ditching and modern farm equipment, it has been possible to farm former wetlands routinely (Fig. 9.3). The modern farm equipment of today and mass-produced reels of plastic drainage pipe also make it possible to drain much more area per day than was ever possible with earlier equipment and the use of clay tiles.

Some of the most rapid wetland losses have occurred in the bottomland hardwood forests of the lower Mississippi River alluvial floodplain (Fig. 9.4). As populations increased along the river, the floodplain was channeled and leveed so that it could be drained and inhabited. Since colonial times, the floodplain has provided excellent cropland, especially for cotton and sugarcane. Cultivation, however, was restricted to the relatively high elevation of the natural river levees, which flooded regularly after spring rains and upstream snowmelts but drained rapidly enough to enable farmers to plant their crops. Because river levees were naturally fertilized by spring floods, they required no additional fertilizers to grow productive crops. One of the results of drainage and flood protection is the additional cost of fertilization. The lower parts of the floodplain, which are too wet to cultivate, were left as forests but harvested for timber. As pressure for additional cropland increased, these agriculturally marginal forests were clear-cut at an unprecedented rate. This was feasible, in part, because of the development of soybean varieties that mature rapidly enough to be planted in June or even early July, after severe flooding has passed. Often, the land thus reclaimed was subsequently incorporated behind flood control levees, where it was kept dry by pumps. Clear-cutting of bottomland forests is still proceeding. Most of the available wetland has been converted in Arkansas and Tennessee; Mississippi and Louisiana are experiencing large losses.

Along the nation's coasts, especially the East and West coasts, the major cause of wetland loss is draining and filling for urban and industrial development or wetland loss due to subsidence. Compared to land converted to agricultural use, the area involved is rather small. Nevertheless, in some coastal states, notably California, almost all coastal wetlands have been lost. The rate of coastal wetland loss from 1954 to 1974 was closely tied to population density. This finding underscores two facts: (1) two-thirds of the world's population lives along coasts; and (2) population density puts great pressure on coastal wetlands as sites for expansion. The most rapid development of coastal wetlands occurred after World War II. In particular, several large airports were built in coastal marshes. Since the passage of federal legislation controlling wetland development, the rate of conversion has slowed.

Hydrologic Modifications

Ditching, draining, and levee building are hydrologic modifications of wetlands specifically designed to dry them out. Other hydrologic modifications destroy

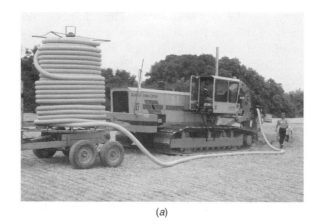

(a)

(b)

Figure 9.3   Modern drainage machinery such as that illustrated in these photos is able to drain dozens of hectares per day: (a) detail of the drainage machinery; (b) results of about 1 minute of drainage, showing new ditch and plastic pipe installed. *(Photographs by W. J. Mitsch.)*

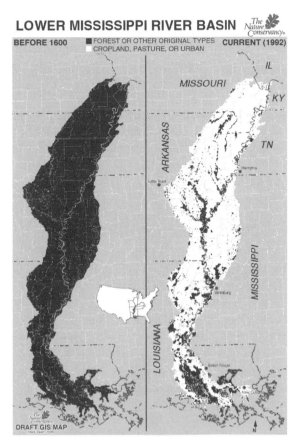

Figure 9.4 Historical and present distribution of bottomland wetland forests in the Mississippi River floodplain. *(After The Nature Conservancy.)*

or change the character of thousands of hectares of wetlands annually. Usually these hydrologic changes were made for some purpose that had nothing to do with wetlands; wetland destruction is an inadvertent result. Canals, ditches, and levees are created for three primary purposes:

1. *Flood control.* Most of the canals and levees associated with wetlands are for flood control. The canals have been designed to carry floodwaters off the adjacent uplands as rapidly as possible. Normal drainage through wetlands is slow surface sheet flow; straight, deep canals are more efficient. Ditching marshes and swamps to drain them for mosquito control or biomass harvesting is a special case designed to lower water levels in the wetlands. Along most of the nation's major rivers are systems of levees constructed to prevent overbank flooding of the adjacent floodplain. Most of these levees were built by the U.S. Army Corps of Engineers after Congress passed flood

control legislation following the disastrous floods of the 1920s and 1930s. (For a fascinating account of the great flood of 1927, the disruption it caused, and the social and political reverberations that led to flood control legislation, see Barry, 1997.) These levees, by separating the river from its floodplain, isolated wetlands so that they could be drained expeditiously. For example, along the lower Mississippi River, the construction of levees created a demand from farmers for additional floodplain drainage. The sequence of response and demand was so predictable that farmers bought and cleared floodplain forests in anticipation of the next round of flood control projects.

2. *Navigation and transportation.* Navigation canals tend to be larger than drainage canals. They traverse wetlands primarily to provide water transportation access to ports and to improve transport among ports. For example, the Intracoastal Waterway was dredged through hundreds of miles of wetlands in the northern Gulf Coast. In addition, when highways were built across wetlands, fill material for the roadbed was often obtained by dredging soil from along the right-of-way, thus forming a canal parallel to the highway.

3. *Industrial activity.* Many canals are dredged to obtain access to sites within a wetland to sink an oil well, build a surface mine, or other kinds of development. Usually pipelines that traverse wetlands are laid in canals that are not backfilled.

The result of all of these activities can be a wetland crisscrossed with canals, such as in the immense coastal wetlands of the northern Gulf Coast (Fig. 9.5). These canals modify wetlands in many ecological ways by changing normal hydrologic patterns. Straight, deep canals in shallow bays, lakes, and marshes capture flow, depriving the natural channels of water. Canals are hydrologically efficient, allowing the more rapid runoff of fresh water than the normal shallow, sinuous channels do. As a result, water levels fluctuate more rapidly than they do in unmodified marshes, and minimum levels are lowered, drying the marshes. The sheet flow of water across the marsh surface is reduced by the spoil banks that almost always line a canal and by road embankments that block sheet flow. Consequently, the sediment supply to the marsh is reduced, and the water on the marsh is more likely to stagnate than when freely flooded. In addition, when deep, straight channels connect low-salinity areas to high-salinity zones, as with many large navigation channels, tidal water, with its salt, intrudes farther upstream, changing freshwater wetlands to brackish. In extreme cases, salt-intolerant vegetation is killed and is not replaced before the marsh erodes into a shallow lake. On the Louisiana coast, the natural subsidence rate is high; wetlands go through a natural cycle of growth followed by decay to open bodies of water. There, canals accelerate the subsidence rate by depriving wetlands of natural sediment and nutrient subsidies.

Highway Construction

Highway construction can have a major effect on the hydrologic conditions of wetlands. Although few definitive studies have been able to document the extent of

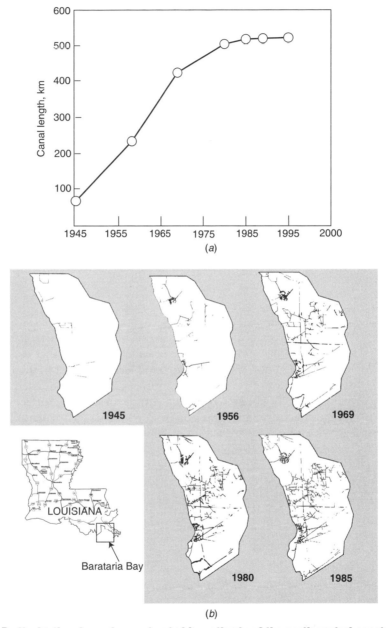

Figure 9.5 Navigational canals constructed in wetlands of the north-central coast of the Gulf of Mexico (Barataria Bay, Louisiana): (a) growth in the length of the canals from 1945–1995; (b) computer images of area for 1945–1985. The concentrated nodes of canals are sites of oil fields. Each short canal segment provides access to an oil well. Canal density has not increased since 1980 because of declining economically extractable oil reserves in the area, the low price of crude oil, and much stronger permit restrictions for wetland activities. *(After Sasser et al., 1986.)*

wetland damage caused by highways, the major effects of highways are alteration of the hydrologic regime, sediment loading, and direct wetland removal. In general, wetlands are more sensitive to highway construction than uplands are, particularly through the disruption of hydrologic conditions. Clewell et al. (1976) and Evink (1980) found that highway construction in Florida led to negative effects on coastal wetlands through hydrologic isolation. The authors of the former study discovered that isolated tidal marshes became less saline and began to fill with vegetation because of the construction of a filled roadway. Evink (1980) found that the decreased circulation that resulted from a causeway increased nutrient retention in the wetland and led to subsequent symptoms of eutrophication. Adamus (1983) concluded that the "best location for a highway that must cross a wetland is one which minimizes interference with the wetland ecosystem's most important driving forces." Other than solar energy and wind, the most important driving forces for wetlands are hydrologic, including tides, gradient currents (e.g., streamflow), runoff, and groundwater flow. The importance of protecting the hydrologic regime during highway construction is based on the contention presented in Chapter 4 that the hydrology of wetlands is the most important determinant of a wetland's structure and function.

Peat Mining

World resources of peat, principally in peatlands in the Northern Hemisphere, are estimated to be 1.9 trillion tons, of which countries that comprise the former Soviet Union have about 770 billion tons and Canada about 510 billion tons. In the United States, deposits of peat occur in most states, with estimated resources of about 310 billion tons, or about 16 percent of the world total. Surface peat mining has been a common activity in several European countries, particularly Ireland and countries in eastern and northeastern Europe, since the 18th century (Fig. 9.6). These countries account for over 19 million metric tons, or almost 75 percent of peat mining in the world (Table 9.5); some of this peat is used for heating as a fuel for electric power production. For centuries but no longer, turf (dried-out peat) was used for home heating in Ireland.

Since its inception, peat mining in North America has been primarily for horticultural and agricultural applications. The fibrous structure and porosity of peat promote a combination of water retention and drainage, which makes it useful for applications such as potting soils, lawn and garden soil amendments, and turf maintenance on golf courses. Peat is also used as a filtering medium to remove toxic materials and pathogens from wastewater, sewage effluent, and stormwater (Jasinski, 1999).

On an international basis, Canada ranks fifth in the global production of peat, after Finland, Ireland, Russia, and Germany (Table 9.5) and probably second behind Germany in the production of horticultural peat. In 1990, 749,000 metric tons of peat were sold by Canadian producers; by 1998, that production had increased significantly to 2,980,000 metric tons. Canadian peat, regarded as among the best-quality peats in the world, is sold to markets in the United States and Japan as well as Canada.

Figure 9.6   Peat mining near Tartu, Estonia. Peat is burned in power plant shown with smokestack in background. (Copyright by J. S. Aber; printed with permission.)

Peat produced in the United States was about 676,000 metric tons in 1998, ranking ninth in total production in the world. It is generally classified as reed-sedge peat, whereas the imports from Canada typically are a weakly decomposed *Sphagnum* peat, which has a higher market value per ton. The estimated value of marketable peat in the contiguous United States was about $16 million in 1995, with Alaskan peat output valued at $450,000. Geographically, about 85 percent of U.S. peat production was from the Great Lakes and Southeast regions, led by Florida, Michigan, and Minnesota, in order of importance. The remainder was produced in the Midwest, Northeast, and West. Approximately 95 percent of domestic peat is sold for horticulture/agriculture usage, including, in order of importance, general soil improvement, potting soils, earthworm culture, the nursery business, and golf course maintenance and construction.

Many people believe that peat production for horticultural purposes will continue to grow while its use as a fuel will continue to decrease, as it has since the political restructuring of Eastern Europe in the early 1990s. There is concern about the degradation of the peatlands as well as the carbon emissions that result when peat is combusted. Thus, peat usage may continue to decrease in Europe and Russia and increase in North America, especially Canada. The implications of this change in peat harvesting on both regional hydrology and climate change remain uncertain.

Mineral and Water Extraction

Surface mining activity for materials other than peat often affects major wetlands regions. Phosphate mining in central Florida is carried out over 120,000 ha and has had a significant impact on wetlands in the region (M. T. Brown, 2005). Thousands of

Table 9.5 World peat production by country for 1998

Country[a]	1998 Peat Production ($\_10^3$ tons/yr)		
	Fuel	Horticulture	Total
Finland	7,000	400	7,400
Ireland	4,500	300	4,800
Russia[b]	3,000		3,000
Germany	180	2,800	2,980
Canada		1,127	1,127
Sweden	800	250	1,050
Ukraine[b]	1,000		1,000
Estonia			1,000
United States		676	676
United Kingdom		500	500
Latvia			450
Belarus[b]	300		300
Netherlands		300	300
Denmark		205	205
France		200	200
Poland			200
Lithuania			195
Spain			60
Hungary		45	45
Norway	1	30	31
Australia		15	15
Argentina		5	5
Burundi		5	5
Grand total	16,800	6,900	25,500[c]

[a]In addition to the countries listed, Austria, Iceland, and Italy produced negligible amounts of peat.
[b]Production appears to be for fuel use.
[c]Includes production for which use is not known.
Source: Jasinski (1999).

hectares of wetlands may have been lost in central Florida because of this activity alone, although the reclamation of phosphate-mined sites for wetlands is now a common practice. H. T. Odum et al. (1981) argued that "managed ecological succession" on mined sites could be an economical alternative to current expensive reclamation techniques involving massive earth moving and reclamation planting.

Surface mining of coal has also affected wetlands in some parts of the United States (Brooks et al., 1985). Forty-six thousand hectares of wetlands in western Kentucky in the early 1980s, mostly bottomland hardwood forests, were or could have been affected by surface coal mining. The recognition of the potential benefits of including wetlands as part of the reclamation of coal mines has not been as widespread as one would have expected (Fig. 9.7), because of strict interpretation of measures regulating the return of the land to its original contours and because of liability questions. This is in contrast to the widespread acceptance of the reclamation of wetlands on phosphorus mine sites in Florida.

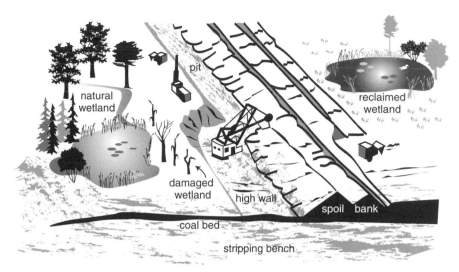

Figure 9.7   The impact of coal surface mining on wetlands and the possible use of wetlands in reclamation of coal surface mines for wildlife enhancement and control of mine drainage.

In some parts of the country, the withdrawal of water from aquifers or minerals from deep mines has resulted in accelerated subsidence rates that are lowering the elevations of marshes and built-up areas alike, sometimes dramatically. Land subsidence, which can also result in the creation of lakes and wetlands, is a geologically common phenomenon in Florida. Often, in karst deposits, when excessive amounts of water are removed from the ground, underground cave-ins occur, causing surface slumpage. Some believe that the cypress domes in north-central Florida are an indirect result of a similar natural process, whereby fissure and dissolutions of underground limestone cause slight surface slumpage and subsequent wetland development.

Water Pollution

Wetlands are altered by pollutants from upstream or local runoff and, in turn, change the quality of the water flowing out of them. The ability of wetlands to cleanse water has received much attention in research and development and is discussed elsewhere in this book. The effects of polluted water on wetlands has received less attention, although water quality standards for wetlands have been established in several regions of United States. Many coastal wetlands are nitrogen limited; one response to nitrogen as one of the pollutants is increased productivity of the vegetation and increased standing stocks of vegetation followed by increased rates of decay of the vegetation, at least initially, and higher community respiration rates.

Species composition may also change with eutrophication of wetlands. For example, increased agricultural runoff, laden with phosphorus, is believed to have caused a spread of *Typha domingensis* in conservation areas that are part of the original

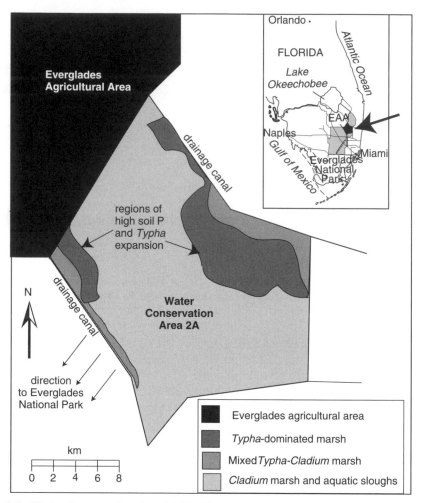

Figure 9.8 Water Conservation Area 2A (44,700 ha) in the south Florida Everglades, showing the area that has received high-nutrient surface overflow from agricultural land drainage since the 1960s. Excess nutrients from the Everglades Agricultural Area to the northwest have caused the spread of *Typha domingensis* and the loss of *Cladium jamaicense* over the 8,000-ha area shaded. (*After Koch and Reddy, 1992.*)

Everglades in Florida (Fig. 9.8). This, in turn, has increased fears that the phosphorus will eventually lead to invasion of *Typha* in the Everglades National Park, replacing the natural sawgrass (*Cladium jamaicense*).

When metals or toxic organic compounds are pollutants, effects on the wetland can be dramatic. In severe cases of water pollution, wetland vegetation can be killed, as occurred when oil was spilled on a coastal marsh (J. M. Baker, 1973) or sulfates were discharged into a forested wetland (J. Richardson et al., 1983). Acid drainage from active and abandoned coal mines has been shown to affect wetlands seriously. In a study of wetlands adjacent to coal surface mining in western Kentucky, Mitsch et al. (1983a–c) described the extensive ecological damage that could occur where waters with low pH and high iron and sulfur were discharged from the mines into or through wetlands.

In one of the most publicized and dramatic cases of water pollution of a wetland, selenium from farm runoff contaminated marshes in Kesterson National Wildlife Refuge in California's San Joaquin Valley (Ohlendorf et al., 1986, 1990; Presser and Ohlendorf, 1987; T. Harris, 1991). The selenium contamination led to excessive death and deformities of wildlife and to eventual "closing" of the contaminated marsh in the mid-1980s amid much controversy.

Wetland Management by Objective

Wetlands are managed for environmental protection, for recreation and aesthetics, and for the production of renewable resources. There can be several specific goals of wetland management that are applicable today:

1. Maintain water quality
2. Reduce erosion
3. Protect from floods and storm damage
4. Provide a natural system to process airborne pollutants
5. Provide a buffer between urban residential and industrial segments to ameliorate climate and physical impact such as noise
6. Maintain a gene pool of marsh plants and provide examples of complete natural communities
7. Provide aesthetic and psychological support for human beings
8. Produce wildlife
9. Control insect populations
10. Provide habitats for fish spawning and other food organisms
11. Produce food, fiber, and fodder (e.g., timber, cranberries, cattails for fiber)
12. Expedite scientific inquiry

One management approach is to fence in a wetland to preserve it. Although simple, this is an act of conservation of a valuable natural ecosystem involving no substantive changes in management practices. Often, however, management has one or more specific objectives that require positive manipulation of the environment. Efforts to maximize one objective may be incompatible with the attainment of others, although in recent years most management objectives have been broadly stated to enhance multiple objectives. Multipurpose management generally focuses on system-level support rather than individual species. This has often been achieved indirectly through plant species manipulation, because plants provide food and cover for the animals. In the management of many small wetland areas in proximity, the use of different practices or staggered management cycles so that the different areas are not all treated the same way at the same time, not only increases the diversity of the larger landscape but also attracts wildlife.

Waterfowl and Wildlife Management

The best wetland management practices are those that enhance the natural processes of the wetland ecosystem involved. One way to accomplish this is to maintain conditions as close as possible to the natural hydrology of the wetland, including hydrologic connections with adjacent rivers, lakes, and estuaries. Unfortunately, this cannot easily be accomplished in wetlands managed for wildlife; the vagaries of nature, especially in hydrologic conditions, make planning difficult. Hence, marsh management for wildlife, particularly waterfowl, has often meant water-level manipulation. Water-level control is achieved by dikes (impoundments), weirs (solid structures in marsh outflows that maintain a minimum water level), control gates, and pumps. In general, the results of the management activity depend on how well the water-level control is maintained, and control depends on the local rainfall and on the sophistication of the control structures. For example, weirs provide the poorest control; all they do is maintain a minimum water level. Pumps provide positive control of drainage or flooding depth at the desired time; and the management objectives can usually be met (Wicker et al., 1983), although the cost is much higher than fixed weirs.

Baldassarre and Bolen (2006) summarize several of the wetland management techniques used for waterfowl and waterbirds. They conclude that the practices fall into two general categories: *natural management* that takes advantage of natural attributes of wetlands such as seed banks, plant succession, water-level fluctuations, and herbivory; and *artificial management*, that includes practices such as planting, ditching, and island building. One of the most frequently used management techniques—perhaps a combination of the two above categories—is a water level *drawdown*. Drawdowns are carried out to recycle nutrients from otherwise undecomposed organic matter, to allow for "moist-soil management" to enhance vegetation regeneration from the wetland seedbank and sometimes to manage for a diversity of macroinvertebrate (important source of protein for ducks) communities. Quite often, there are trade-offs that occur during water level manipulations.

To illustrate the trade-offs in wetland management for wildlife enhancement, some generalizations about water-level manipulation of Lake Erie (Ohio) coastal marshes are shown in Figure 9.9. Maximum migratory wildlife use of the marshes occurs in moist soil conditions, but these conditions are also the best for the invasion of potentially undesirable plants and are generally least favorable for the overall abundance and diversity of resident plant and animal populations. Shallow-water (called *hemi* conditions by marsh managers; around 15 cm depth in summer) usually result in the highest plant species diversity and greatest fish and resident wildlife use but less migratory wildlife. Deepwater conditions (>30 cm) offer the least potential for both annual emergent plants and invading, undesirable plants, and desirable

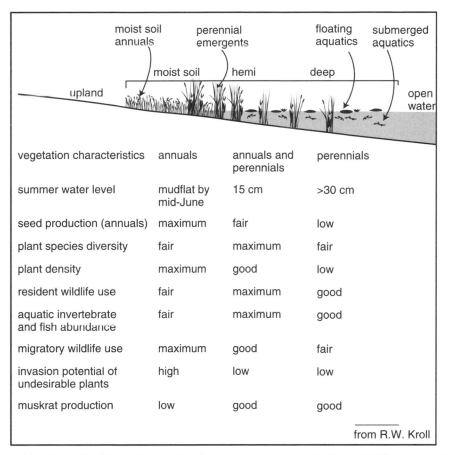

vegetation characteristics	annuals	annuals and perennials	perennials
summer water level	mudflat by mid-June	15 cm	>30 cm
seed production (annuals)	maximum	fair	low
plant species diversity	fair	maximum	fair
plant density	maximum	good	low
resident wildlife use	fair	maximum	good
aquatic invertebrate and fish abundance	fair	maximum	good
migratory wildlife use	maximum	good	fair
invasion potential of undesirable plants	high	low	low
muskrat production	low	good	good

from R.W. Kroll

Figure 9.9 Generalizations of water-level management for vegetation, wildlife use, and other characteristics as practiced on impounded marshes near Lake Erie in northern Ohio. (From Roy Kroll, Winous Point Shooting Club, Port Clinton, Ohio.)

migratory waterfowl use is only fair in deep water. Kroll et al. (1997) and Gottgens et al. (1998) point out that because the landward advance of marshes during high lake water times is restricted by human development, and because carp *(Cyprinus carpio)* are present in the lake, long-term above-average water levels probably mean that removal of dikes along the Great Lakes would lead to an irreversible loss of wetland vegetation fringing the lakes. Mitsch et al. (2000b) found that only 25 percent of the existing marshes encompassed by dikes would have had the right conditions to be emergent marshes more than 50 percent of the time during the last century.

The set of management recommendations by Weller (1978) for prairie pothole marshes in the north-central United States and south-central Canada is another example of multipurpose wildlife enhancement. Those recommendations mimic the natural cycle of marshes in the middle of North America. Although they may seem drastic, they are entirely natural in their consequences. In sequence, the practices are as follows:

1. When a pothole is in the open stage and there is little emergent vegetation, the cycle should be initiated by a spring drawdown. This stimulates the germination of seedlings on the exposed mud surfaces.
2. A slow increase in water level after the drawdown maintains the growth of flood-tolerant seedlings without shading them out in turbid water. Shallowly flooded areas attract dabbling ducks during the winter.
3. The drawdown cycle should be repeated for a second year to establish a good stand of emergents.
4. Low water levels should be maintained for several more seasons to encourage the growth of perennial emergents such as *Typha*.
5. Maintaining stable, moderate water depths for several years promotes the growth of rooted submerged perennial aquatic plants and associated benthic fauna that make excellent food for waterfowl. During that period, the emergent vegetation will gradually die out and will be replaced by shallow ponds. When that occurs, the cycle can be initiated again, as described previously in step 1.
6. Different wetland areas maintained in staggered cycles provide all stages of the marsh cycle at once, maximizing habitat diversity.

Weller (1994) makes the distinction between a complete drawdown of water levels, as described before, a management option when vegetation is completely lost because of high water levels, herbivory, winter kill, or plant disease; and a partial drawdown, which can be implemented when vegetation is reduced but not eliminated or when wildife use has declined but not disappeared.

Wildlife management in coastal salt marshes such as those found in Louisiana uses a similar strategy, although the short-term cycle is not as pronounced there. Drawdowns to encourage the growth of seedlings and perennials preferred by ducks are common practices, as is fall and winter flooding to attract dabbling ducks. As it happens, there is general agreement that stabilizing water levels is not good management, even though our society seems to feel intuitively that stability is a good thing. Wetlands thrive on cycles, especially flooding cycles, and practices that dampen these cycles also reduce wildlife productivity. Although the management practices described previously enhance waterfowl production, they are generally deleterious for wetland-dependent fisheries in coastal wetlands because free access between the wetlands and the adjacent estuary is restricted; the wetlands' role in regulating water quality is also often underutilized.

There is a tendency to want to control all external variables when we manage wetlands by objectives. This management tendency, although understandable, is particularly strong when herbivores such as geese, nutria, beavers, or muskrats "invade" a managed wetland. These animals can be discouraged and/or trapped to keep their influence on vegetation at a minimum, but one has to remember that these animals are not invaders at all, but are simply coming to a habitat that is generally well suited to their needs. In the ecosytem context, these animals are often nature's "ecosystem engineers" and provide many functions that, in the long term, may enhance marshes. Beavers cause water-level manipulations just as humans do. Muskrats and geese remove large areas of vegetation, but open up the system to allow for other vegetation to come into the wetland.

Whether management of ecosystem managers is a wise strategy is a complex issue. In coastal Louisiana, for example, muskrats and especially nutria (a South American immigrant) can "eat out" extensive marsh areas, which do not recover because of rising sea level and high marsh subsidence rates induced, in part, by human activities. Trapping used to keep the rodent populations in check, but the worldwide slump in fur sales (see Chapter 11) no longer makes trapping profitable, and rodent populations are rapidly escalating.

Six general principles have been presented by Baldassarre and Bolen (2006) that provide a useful set of rules for wetland managers. It is interesting to note that that many of the principles on wetland restoration described in Chapter 12 mirror many of these wetland management principles.

1. Protect wetland complexes that include a wide variety of wetland hydroperiods and wetland sizes;
2. Protect small wetlands, as these wetlands are most vulnerable to being lost along with their unique biota;
3. Consider all wetland-dependent wildlife when managing wetlands, not only one or two species;
4. Large wetlands need to be protected too for species with such requirements for large areas;
5. Recognize the importance of wetland complexes for species with complex life-history requirements;
6. Recognize that protected sites often require direct management intervention to protect wildlife values; and
7. Protect and restore upland habitats that are contiguous with wetlands.

Agriculture and Aquaculture

When wetlands are drained for agricultural use, they no longer function as wetlands. They are, as local farmers say, "fast lands" removed from the effects of periodic flooding, and they grow terrestrial, flood-intolerant crops. Some use is made of more or less undisturbed wetlands for agriculture, but it is minor. In New England, high salt marshes were harvested for "salt marsh hay," which was considered an excellent bedding and fodder for cattle. In fact, Russell (1976) stated that the proximity of fresh and salt hay marshes was a major factor in selecting the sites for the emergence of many towns in New England before 1650. Subsequently, marshes were ditched to allow the intrusion of tides to promote the growth of salt marsh hay *(Spartina patens)*, but the extent of this practice has not been well documented. On parts of the coast of the Gulf of Mexico where marshes are firm underfoot, they are still used extensively for cattle grazing. To improve access, small embankments or raised earthen paths are constructed in these marshes.

The ancient Mexican practice of *marceno* is unique. In the freshwater wetlands of the northern coast of Mexico, small areas were cleared and planted in corn during the dry season. These native varieties were tolerant enough to withstand considerable flooding. After harvest (or apparently sometimes before harvest), the marshes were naturally reflooded, and native grasses were reestablished until the next dry season. This practice is no longer followed, but there has been some interest in reviving it.

On a global scale, the production of rice in managed wetlands contributes a major proportion of the world's food supply. There are approximately 1.3 million km^2 of rice paddies in the world (Chapter 3), of which almost 90 percent are in Asia. In North America, especially in Minnesota, there are several commercial operations in the production of wild rice *(Zizania aquatica)* in wetlands and several other locations where Native American tribes have harvested wild rice in natural marshes for centuries.

Aquaculture, the farming of fish and shellfish, produces 21 million metric tons (mt), or about 20 percent of the annual worldwide total fish and shellfish harvest of 112 million mt (1995 production; World Resources Institute, Washington, DC). Most of this production occurs in Asia, with China by far the largest producer; however, in India, the second largest producer, shrimp farming is rapidly growing (Table 9.6). The United States is the major consumer of aquaculture products but accounts for only about 2 percent of worldwide production, mostly salmon and crayfish.

Fish farming practices vary. The most environmentally benign approach, similar to the Mexican *marceno* described previously, intercrops shellfish with a grain crop, usually rice. Typical is crayfish farming in the United States and Indian shrimp aquaculture in rotation with rice. The practice is described for crayfish in the southern United States. Crayfish are an edible delicacy in the southern United States and in

Table 9.6 Distribution of global aquaculture production

Country	Global Production (%)
China	57
India	9
Japan	4
Indonesia	4
Thailand	3
United States	2
Philippines	2
Korea, Republic of	2
Other countries	17

Source: Food and Agriculture Organization of the United Nations (1997).

many foreign countries. They live in burrows in shallow flooded areas such as swamp forests and rice fields, emerging with their young early in the year to forage for food. The young grow to edible size within a few weeks and are harvested in the spring. When floodwaters retreat, the crayfish construct burrows, where they remain until the next winter flood. In crayfish farms, this natural cycle is enhanced by controlling water levels. An area of swamp forest is impounded; it is flooded deep during the winter and spring and drained during the summer. This cycle is ideal for crayfish, which thrive. Fish predators are controlled within the impoundments to improve the harvest. The hydrologic cycle is also favorable for forest trees. It simulates the hydrologic cycle of a bottomland hardwood forest; forest tree productivity is high, and seedling recruitment is good because of the summer drawdown. Species composition tends toward species typical of bottomland hardwoods.

Some rice farmers have also found that they can take advantage of the annual flooding cycle typically used to grow rice to combine rice and crayfish production. Rice fields are drained during the summer and fall when the rice crop matures and is harvested. Then the fields are reflooded, allowing crayfish to emerge from their burrows in the rice field embankments and forage on the vegetation remaining after the rice harvest. The crayfish harvest ends when the fields are replanted with rice. When this rotation is practiced, extreme care has to be exercised in the use of pesticides.

Today, most aquaculture practices involve much more intensive aquatic farming than the intercropping method. In India, intercropped shrimp yields are about 200 to 500 kg/ha per crop (James, 1999). A slightly more intensive farming technique, which encloses natural ponds with mesh and uses weirs to control water levels and the influx and efflux of organisms, yields as much as 1 mt/ha per crop, with one crop per year. Typically, in this type of culture, recruitment of postlarval juveniles to the aquaculture site occurs naturally, after which the area is sealed off and the shrimp are allowed to grow. They are harvested as they emigrate over the weirs or by seining or trawling within the enclosure.

The most intensive aquacultural techniques control all aspects of production. Wetlands, salt flats, mangrove forests, and even high-quality farmland are dredged to form ponds in which water levels are controlled by pumps. "Seed" organisms, the young postlarvae, are raised in separate hatcheries. The young organisms are fed in the ponds on synthetic diets, often composed of fish bycatch from commercial fisheries. Water quality is monitored, the ponds are aerated, and, in the most sophisticated operations, wastes are treated. Yields from this kind of operation can be several metric tons per hectare per crop, and in tropical areas two crops per year are expected.

Whereas aquaculture farms in Asia have historically been small operations managed by local farmers, the worldwide boom in aquaculture, fueled by the high demand for fishery products, has led countries such as India to offer large incentives to initiate new fish farms and has drawn large corporations to invest in the industry. Aquaculture on this scale is a sophisticated enterprise. For the grower, one serious problem is that of diseases that infect the concentrated fish or shellfish populations. In India, for example, a viral disease nearly wiped out the shrimp crop in 1994 and 1995. More broadly, the concentration of fishponds in the coastal zone has serious environmental repercussions. Worldwide, 50,000 shrimp farms cover 9,800 km^2 of coastal lands. This has resulted in a serious loss of wetlands (coastal wetlands are required habitats for most commercial marine fish) and mangrove forests. These fish farms not only disrupt natural ecosystems but also bring in diseases, create enormous waste problems, deplete oxygen in shallow coastal waters, and reduce water quality. These disruptions have been cited as one reason for the decline in commercial fisheries in the areas where shrimp culture is concentrated.

Water Quality Enhancement

Several studies have shown natural wetlands to be sinks for certain chemicals, particularly sediments and nutrients. It is now common to cite the water quality role of natural wetlands in the landscape as one of the most important reasons for their protection. The idea of applying domestic, industrial, and agricultural wastewaters, sludges, and even urban and rural runoff to wetlands to take advantage of this nutrient sink capacity has also been explored in countless studies. The basic principles and practices of these so-called treatment wetlands are covered in detail in Chapter 13.

Flood Control and Stormwater Protection

Wetlands can be managed, often passively, for their role in the hydrologic cycle. Hydrologic values of wetlands include streamflow augmentation, groundwater recharge, water supply potential, and flood protection. It is not altogether clear how well wetlands carry out these functions, nor do all wetlands perform these functions equally well. It is known, for example, that wetlands do not necessarily always contribute to low flows or recharge groundwater. Some wetlands, however, should be and often are protected for their ability to hold water and slowly return it to surface water and groundwater systems during periods of low water. If wetlands are impounded to retain even more water from flooding downstream areas, considerable changes in vegetation will result as the systems adapt to the new hydrologic conditions. The values of wetlands for coastal protection and flood mitigation are discussed in more detail in Chapter 11.

Recommended Readings

Baldassarre, G.A. and E.G. Bolen. 2006. *Waterfowl Ecology and Management*, 2nd *ed.*, Krieger Publishing Company, Malabar, Florida, 567 pp.

Chapter 10

Climate Change and Wetlands

The earth's climate is changing, as witnessed by higher atmospheric temperatures, decreased snow and ice cover, and increasing sea levels in the 20th century and especially toward the end of that century. Wetlands emit 20 to 25 percent of global methane (CH_4) emissions to the earth's atmosphere, yet they also have the best capacity of any ecosystem to retain carbon through permanent burial (sequestration). Both processes have implications for climate change. Of the total storage of organic carbon in the earth's soils, 20 to 30 percent or more is stored in wetlands and is more vulnerable to loss back to the atmosphere if the climate warms or becomes drier. Preliminary estimates show that the world's wetlands are climate-change neutral; that is, the negative effects of methane emissions on climate are compensated for by carbon sequestration into peat or wetland soils. The effects of climate change on coastal wetlands could be significant if sea level rises, particularly in large river deltas where land subsidence is already occurring. For inland wetlands, change in precipitation patterns and warmer temperatures can likewise have detrimental effects on wetland function.

Wetlands have significant yet generally under-appreciated roles in the global carbon cycle. They are also positioned in the landscape where climate change could affect them more than most other ecosystems. So their roles both as players in and recipients of climate change are the subject of this chapter.

Climate Change

There is little doubt that something significant is happening to our climate. According to the consensus of hundreds of scientists who have been involved in the Intergovernmental Panel on Climate Change (IPCC), there are some major findings that should concern anyone interested in our planet and its future. The IPCC was established by the World Meteorological Organization (WMO) and the United Nations Environmental Programme (UNEP) to assess scientific, technical and socio-economic information relevant for understanding climate change, its potential impacts, and options for adaptation and mitigation. Some of the dominant conclusions of the panel, drafted in its two most recent mammoth multivolume reports and various summaries (IPCC, 2001, 2007a, b) are presented here. The collective picture of climate change has now been illustrated through the beginning of the 21st century by the following:

- *The global average surface temperature has increased over the 100 year period 1906–2005 by about 0.74° C.* This trend is illustrated in Figure 10.1. The temperate increase was about 0.14°C more than that estimated by the IPCC (2001) for the 20th century (0.6°C). It also determined that eleven of the twelve years 1995 – 2006 were the warmest since 1850 and 1998 was the warmest year since 1861. This temperature increase in the 20th century has also been determined to be the largest increase in the last 1,000 years.

- *Temperatures have risen during the last four decades of the 20th century in the lowest 8 km of the atmosphere.* These data are based on weather balloons, which were first used in the 1950s, and satellite records that started in 1979.

- *Snow cover and ice extent have decreased.* This includes retreat of many mountain glaciers in nonpolar regions and about a 10 percent decrease in snow cover since the 1960s.

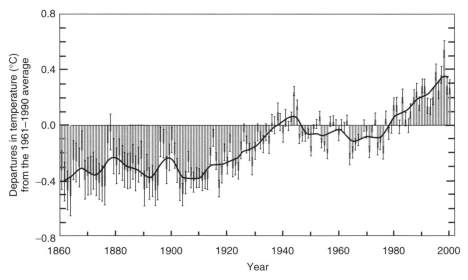

Figure 10.1 **Variations of the earth's surface temperature since 1860. Annual variations are shown in the gray bars (error bars indicate 95 percent confidence interval), and decadal average is shown by the black line.** *(From IPCC, 2001.)*

- *Global average sea level has risen, and ocean heat content has increased.* It is estimated that sea level has risen globally about 17 cm during the 20th century (1.7 mm per year) and at a much greater rate of 3.1 mm per year from 1993 to 2003, and that the heat content of the ocean has increased at least since the 1950s.

- *Arctic temperatures have increased and Arctic sea ice has shrunk.* Arctic temperatures have increased twice the global average over the past century and sea ice has decreased by 2.7% per decade since 1978.

It is important to note what has *not* happened over that period as well. There are a few areas in the Southern Hemisphere where temperatures have not increased, and there has been no trend in loss of ice in Antarctica. Also, although it has been tempting to make conclusions based on disastrous hurricanes and storms in the past few years, there is no conclusive evidence that tropical and extra-tropical storms, tornadoes, or hail events increased in the 20th century although there is evidence that there is increased cyclonic activity in the North Atlantic since about 1970.

Causes of Climate Change

The cause of climate change is the increasing concentration of the so-called greenhouse gases in the atmosphere, mostly caused by anthropocentric emissions. These gases adsorb several wavelengths of long-wave radiation, causing the earth to be a little warmer if the gas concentrations are increased. The primary greenhouse gas is carbon dioxide (CO_2), which is released through the burning of fossil fuels and also by cement production. Atmospheric carbon dioxide is estimated to have increased by over 30 percent since the mid-18th century (Fig. 10.2a), with concentrations of about

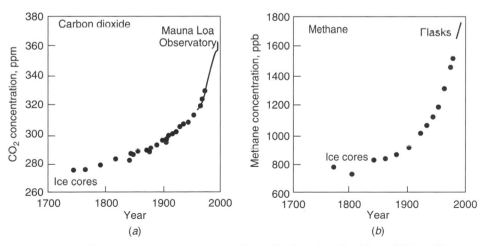

Figure 10.2   Estimated changes in concentrations of (a) carbon dioxide and (b) methane, since preindustrial times. *(From IPCC, 2001.)*

Table 10.1 Major greenhouse gas concentrations and fluxes that are influenced by humans and wetlands

	CO_2	CH_4	N_2O
Preindustrial concentration	~280 pm	~700 ppb	270 ppb
2005 concentration	379 ppm	1774 ppb	319 ppb
Rate of change	1.4 ppm/yr (1960 – 2005)	7.0 ppb/yr	0.8 ppb/yr
	1.9 ppm/yr (1995 – 2005)		
Atmospheric lifetime, yr	5–200	12	114

Source: Houghton et al. 2001; IPCC, 2007b

379 ppm by 2005 (IPCC, 2007b; Table 10.1). There has also been much discussion about sources of CO_2 besides fossil fuel burning, such as tropical forest deforestation and burning. In any event, as fossil fuel consumption continues to rise, the CO_2 has been increasing at a rate of 1.5 ppm per year over the past two decades (Table 10.1) and at a rate of 1.9 ppm per year from 1995 to 2005 (IPCC, 2007b). The second most important greenhouse gas is actually water vapor, but it is not known to have any trend or change. It is one of the most abundant gases in the troposphere.

The third most important greenhouse gas is methane (CH_4), which has been estimated to have more than doubled in concentration, from about 700 parts per billion (ppb) in preindustrial times to about 1,774 ppb in 2005 (Fig. 10.2b; Table 10.1). Before about 1980, methane was assumed to be a stable concentration in the atmosphere, but it increased by 13 percent between 1978 and 1999 alone (Whalen, 2005). Wetlands were described in Chapter 5 as being sources of methane gas, and that will be put in context with other sources later. What should be clear is that if we argue that we have lost half of the world's wetlands as a result of human activity over the same 250-year period shown in Figure 10.2b when methane is increasing, then the methane increase could not be due to wetlands. If wetlands were the major source of methane, we would have seen a decrease in methane in the atmosphere over the last 250 years.

A fourth important greenhouse gas, nitrous oxide (N_2O), also comes from wetlands as a result of nitrification and especially denitrification (see Chapter 5). While nitrous oxide is a normal product of denitrification, it is usually a small percentage of denitrification products, with most of nitrates converted to dinitrogen (N_2) gas. Nitrous oxide has increased by about 16 percent to about 319 ppb in the atmosphere since preindustrial times (Table 10.1).

Wetlands in the Global Carbon Cycle

Although soil carbon in wetland soils is recognized as an important component of global carbon budgets and future climate change scenarios, very little work has been done to consider the role of wetlands, particularly those in temperate and tropical regions of the world, in managing carbon sequestration. A carbon budget for the world, with wetlands included to show their relative contributions, is shown

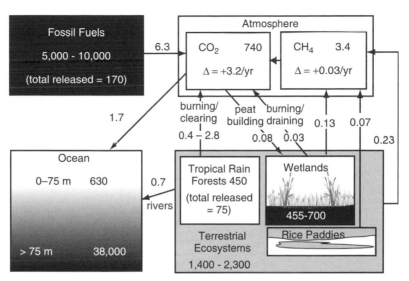

Figure 10.3   Estimated global carbon budget, with the relative role of wetlands superimposed. Storages are in Pg carbon; fluxes are in Pg carbon/yr.

in Figure 10.3. Following, we discuss the role of wetlands in this carbon budget in terms of carbon storage in peat, carbon sequestration through peat and organic soil development, and methane emissions from wetlands.

Peat Storage and a Global Carbon Budget

Peat deposits in the world's wetlands, particularly in boreal and tropical regions, are substantial storages of carbon (C) in the lithosphere. Of the total storage of C in the earth's soils of 1,400 to 2,300 Pg-C (Pg = 10^{15} g), anywhere from 20 to 30 percent is stored in wetlands (Mitsch and Wu, 1995; Roulet, 2000; Hadi et al., 2005; Fig. 10.3). Gorham (1991) estimated that 0.08 Pg-C yr^{-1} is sequestered in peatlands overall, a small fraction (1.3%) of the 6.3 Pg-C yr^{-1} that humans currently release to the atmosphere (Fig. 10.3). The sequestration is also a small percentage of the estimated global net primary productivity of wetlands of 4 to 9 Pg C/yr.

These peat deposits, if disturbed, however, could contribute significantly to worldwide atmospheric carbon dioxide levels, depending on the balance between draining and oxidation of the peat deposits and their formation in active wetlands. Gorham (1991) provided two other estimates of the role of wetlands in global carbon cycling. First, combustion of peat and oxidation of peat provide an estimated 0.026 Pg/yr of carbon back to the atmosphere. Second, drainage of wetlands is estimated to contribute 0.008 to 0.042 Pg C/yr back to the atmosphere (the low number is for long term; the high number for short term). Thus, wetland drainage and peat burning could be releasing back to the atmosphere from 45 to 89 percent of the carbon being sequestered.

Table 10.2   Carbon sequestration in wetlands

Wetland type	g-C m^{-2} yr^{-1}	Reference
General average for peatlands	12–25	Malmer (1975)
General range for wetlands	20–140	Mitra et al. (2005)
Peatlands (North America)	29	Gorham (1991)
Peatlands (Alaska and Canada)	8–61	Ovenden (1990)
Tropical wetland	56 (for 24,000 years)	Page et al. (2004)
	94 (for last 500 years)	
Boreal peatlands	15–26	Turunen et al. (2002)
Temperate peatlands	10–46	Turunen et al. (2002)
Thoreau's Bog, MA	90	Hemond (1980)
Created temperate marshes, OH	180–190	Anderson and Mitsch (2006)
Prairie pothole wetlands, North America		Euliss et al. (2006)
Restored (semi-permanently flooded)	305	
Reference wetlands	83	

Carbon Sequestration

Several studies have estimated carbon sequestration in wetlands per unit area (Table 10.2). The vertical accumulation rate of peat in bogs and fens is generally thought to be between 20 and 80 cm/1,000 yr in European bogs (Moore and Bellamy, 1974), although Cameron (1970) gave a range of 100 to 200 cm/1,000 yr for North American bogs, and Nichols (1983) reported an accumulation rate for peat of 150 to 200 cm/1,000 yr in warm, highly productive sites. Malmer (1975) described a vertical growth rate of 50 to 100 cm/1,000 yr as typical for western Europe. Assuming an average density of peat of 50 mg/mL, this rate is equivalent to a peat accumulation rate of 25 to 50 g dry wt m^{-2} yr^{-1} or about 12 to 25 g-C m^{-2} yr^{-1}. Hemond (1980) estimated a rapid accumulation rate of 430 cm/1,000 yr, equivalent to 180 g dry wt m^{-2} yr^{-1}, for Thoreau's Bog, Massachusetts, or about 90 g-C m^{-2} yr^{-1}.

Page et al. (2004) provide a useful comparison for tropical peatlands in Kalimantan, Indonesia. They investigated a 9.5-m core of peat that showed 26,000 years of history. They found an average carbon sequestration of the core of 56 g-C m^{-2} yr^{-1} with a rate of 94 g-C m^{-2} yr^{-1} for the past 500 years in the upper meter of the core (Table 10.2). These rates were higher than many rates published for temperate and boreal wetland systems. The accumulation of peat in tropical wetlands may be due more to the slow decomposition of recalcitrant lignin in roots and woody material under the constant high water rather than to the perceived high productivity of these systems (Chimner and Ewel, 2005).

Created and restored wetlands might be the best opportunity for carbon sequestration. Anderson and Mitsch (2006) found a total carbon sequestration rate of 180 to 190 g-C m^{-2} yr^{-1} for two created wetland basins in Ohio. About one-fourth of that carbon sequestration was as inorganic carbon, precipitated as calcite and calcium carbonate due to high productivities in the water column. Euliss et al. (2006)

compared several wetlands that had been restored over more than a decade in the prairie pothole wetlands of North America and found 305 g-C m^{-2} yr^{-1}, the highest number we report in Table 10.2. This is not surprising, because restoration in these cases meant reflooding agricultural land, allowing organic carbon to once again build up in the soil. For comparison, Euliss et al. (2006) estimated an accumulation rate in reference (natural) marshes in the region of 83 g-C m^{-2} yr^{-1} based on average sedimentation rates of 2 mm/yr.

Methane Emissions

In one of the most important connections between wetlands and climate change, wetlands are estimated to emit about 20 to 25 percent of current global methane emissions or about 115 to 145 Tg-CH$_4$ yr^{-1} (Tg $= 10^{12}$ g; Table 10.3). Rice paddies, which are essentially domestic wetlands, account for about 60 to 80 Tg-CH$_4$ yr^{-1}. Other anthropogenic sources account for most of the rest. Aselmann and Crutzen (1989) estimated a release of 30 to 120 Tg-CH$_4$ yr^{-1} (0.03 to 0.12 Pg C/yr) of methane from natural wetlands and another 40 to 100 Tg-CH$_4$ yr^{-1} (0.04 to 0.10 Pg C/yr) from rice paddies. Methane emissions are a concern because methane is estimated to be 21 times more effective as a greenhouse gas on a molecular basis than is carbon dioxide. Some researchers suggest that the factor 21 may not be justified because of the rapid degradation of methane in the atmosphere compared to carbon dioxide (Lashof and Ahuja, 1990; Gorham, 1991).

The rates of methane emissions from both saltwater and freshwater wetland have a considerable range (Table 10.4). Harriss et al. (1982) noted maximum methane production in a Virginia freshwater swamp between April and May and a net uptake of methane by the wetland during a drought when the wetland soil was exposed to

Table 10.3   Estimates of annual fluxes of methane from wetlands and other sources, Tg-CH$_4$/yr[a]

Sources	Megonigal et al. (2004)		Whalen (2005)
Natural wetlands		115	145
Tropics	65		
Northern latitude	40		
Others	10		
Other Natural Sources[b]		45	45
Anthropogenic			
Rice Paddies		60	80
Other[c]		315	330
TOTAL SOURCES		535	600

[a]Tg $= 10^{12}$ g
[b]Other natural sources include termites, ocean, freshwater, and geological sources
[c]Other anthropogenic sources include fossil fuels, landfills, domestic wastewater treatment, animal waste, enteric fermentation (ruminants), and biomass burning

Table 10.4   Ranges of mean methane emission rates (number of sites/treatments) for major wetland types

	CH₄ Emission Rates (mg C m⁻² day⁻¹)		
	Boreal	Temperate	Subtropical/Tropical
FRESHWATER WETLANDS			
Tundra	3.7–1,500 (12)	—	—
Bog	0.7–17 (5)	20–221 (7)	—
Fen	14–325 (11)	3–314 (8)	—
Freshwater marsh	23–80 (2)	0.1–498 (17)	29–443 (7)
Forest swamp	5–66 (2)	7.4–106 (6)	44–144 (7)
Rice paddy	—	10–880 (34)	47–486 (9)
SALTWATER WETLANDS			
Salt marsh	—	0–109 (17)	2.5 (1)
Mangrove	—	—	3–61 (3)

Source: Mitsch and Wu (1995).

the atmosphere. Wiebe et al. (1981) found methane production to generally peak in late summer in a Georgia salt marsh. Boon and Sorrell (1991) and Sorrell and Boon (1992), in a detailed study of gas production (mostly methane) from billabong sediments in Australia, found clear patterns of higher rates in the summer ($>25°C$) than in winter but no significant difference between rates during the night and during the day (see Chapter 5).

Most of the methane emission studies to date have been in peatlands (bogs and fens) and freshwater marshes. More studies have been undertaken in the temperate zone compared to boreal or tropical climes. Researchers in boreal wetlands have found beaver ponds to have much higher methane flux rates than other wetland types (Naiman et al., 1991; Roulet et al., 1992a,b) and neutral fens to have higher rates than acid fens and bogs (Crill et al., 1988). A comparison of methane emissions between freshwater wetlands (marshes and swamps) and marine wetlands (salt marshes and mangroves) shows that the rate of methane production is higher in the former (up to 500 mg C m⁻² day⁻¹) than in the latter (up to 100 mg C m⁻² day⁻¹). One of the reasons is because of the lower amounts of sulfate competition for oxidizable substrate in freshwater systems (see Carbon–Sulfur Interactions in Chapter 5).

The data in Table 10.4 can be compared with other generalizations that have been used to estimate global methane production from wetlands (Table 10.5). Gorham (1991) assumed an average of 77 mg C m⁻² day⁻¹ in estimating the global contributions of northern peatlands. Data in Table 10.4 suggest that the means of several studies range from 0.7 to 17 mg C m⁻² day⁻¹ for boreal bogs, which is much lower than the 14 to 325 mg C m⁻² day⁻¹ for the more mineral-rich fens. Thus, depending on the mix of bogs and fens in northern peatlands, Gorham's assumption could be low or high. Matthews and Fung (1987) used cross-latitude assumptions to

Table 10.5 Methane production means (mg C m^{-2} day^{-1}) used for global methane emission estimates

Wetland Type	Matthews and Fung (1987)	Aselmann and Crutzen (1989)	Gorham (1991)	LeMer and Roger (2001)	Whalen (2005)
Peatlands (bogs)	150	11	77	32	65–72
Fens	—	60			
Forested swamps	53	63			52–56
Marshes	90	190		54	
Riparian wetlands	23	75			
Rice paddies	—	230		75	
Tropical Wetlands					37–151

determine average emission rates of methane per wetland type and found that bogs have the highest rate of methane emissions. Aselmann and Crutzen (1989) found completely the opposite and suggested geometric mean emission rates of 11 to 230 mg C m^{-2} day^{-1} in increasing order for bogs, fens, swamps, marshes, and rice paddies (Table 10.5). Fens have much higher emissions in both boreal and temperate regions than do forested swamps and marshes.

Whalen's (2005) estimates (Table 10.5) suggest that tropical and subtropical wetlands may have higher rates of methane production than originally believed. Sorrell and Boon (1992) found high annual rates of methane emissions in Australian billabongs of about 16 to 30 g CH$_4$ m^{-2} yr^{-1} (32 to 60 mg C m^{-2} day^{-1}). Delaune and Pezeshki (2003) reported methane emissions from 3.6 to over 300 g CH$_4$ m^{-2} yr^{-1} (~7 to over 600 mg C m^{-2} day^{-1}) in Louisiana freshwater marshes, where the greatest methane emissions occurred in the summer months. Hadi et al. (2005) measured methane from tropical peatlands in Indonesia and found only 12 to 53 mg C m^{-2} day^{-1} from Indonesian peatlands, with the highest number from cultivated paddy fields.

Some of the preceding measurements may have led to integrated annual estimates of methane emission that are too high. Moore and Roulet (1995) suggest that most annual flux measurements in Canada are less than 10 g CH$_4$ m^{-2} yr^{-1} (~20 mg C m^{-2} day^{-1}), with the primary controlling mechanisms being soil temperature, water table position, or a combination of both. Roulet et al. (1992a), in a detailed field study of 12 minerotrophic wetlands and 3 beaver ponds in the low boreal region of central Canada, estimated a habitat-weighted average emission of 1.6 g CH$_4$ m^{-2} yr^{-1}, which translates to a daily rate of only 3.35 mg C m^{-2} day^{-1}. If their study is accurate, it suggests that emissions of methane from northern peatlands may be much less than originally estimated.

Studies of methane generation in one of the most common types of wetland in those regions—lotic (flowing) wetlands connected to rivers and streams—are almost nonexistent. Yet pulsing water fluxes and oscillating water levels of these wetlands are the norm rather than the exception and are keys to many biological processes in wetlands. Methane emissions from a seasonally pulsed wetland in Ohio were considerably lower ($36 \text{ mg-C m}^{-2} \text{ day}^{-1}$) in seasonally pulsed sites than in permanently flooded sites ($115 \text{ mg-C m}^{-2} \text{ day}^{-1}$; Altor and Mitsch, 2006).

Methanogenesis and Methane Oxidation

Methane emissions are actually the result of two competing processes going on at the same time by microbial communities—methanogenesis and methane oxidation (Fig. 10.4). The degradation of organic matter by aerobic respiration is fairly efficient in terms of energy transfer. Because of the anoxic nature of wetland soils, anaerobic processes, which are less efficient in terms of energy transfer, occur in close proximity to aerobic processes. *Methanogenesis* occurs when microbes called methanogens use CO_2 as an electron acceptor for the production of gaseous methane (CH_4) or, alternatively, use a low-weight organic compound such as one from a methyl group.

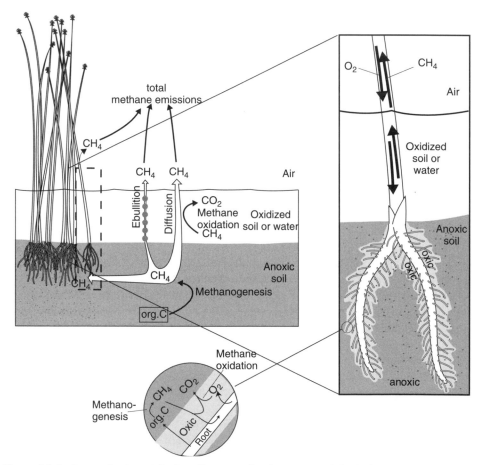

Figure 10.4 Conceptual model of methane cycling in a wetland, including methanogenesis and methane oxidation. *(From Conrad, 1993 and Whalen, 2005.)*

Methane production requires extremely reduced conditions, with a redox potential of less than -200 mv, after other terminal electron acceptors (O_2, NO_3, and $SO_4^=$) have been reduced.

Conversely, nonflooded upland soils (e.g., forests, grassland, arable land) are regarded as the major biological sink of atmospheric methane (the major sink overall is tropospheric photochemistry). Obligate aerobic methanotrophic bacteria use molecular oxygen to oxidize methane to CO_2 and cellular C. The consumption of atmospheric CH_4 is the result of two physiologically distinct microbial groups: (1) the methanotrophs, which have a membrane-bound enzyme system, and (2) an autotrophic nitrifier community. Methanotrophs are estimated to consume about 30 Tg CH_4 yr^{-1} (Whalen, 2005), an amount close to the net annual increase of atmospheric CH_4 (about 20 to 40 Tg CH_4 yr^{-1}) (IPCC, 2001).

Estimating the importance of methanotrophic activity in wetlands is only beginning. For example, a pulsing hydroperiod in wetlands could enhance methanotrophic activity. In the Ohio study described previously, Altor and Mitsch (2006) found net methane emissions in the pulsed part of the wetlands to be one-third the rate of methane emissions from a permanently flooded part of the same wetlands. This could be due to differences in redox potential, but it also could be due to methanotrophic activity in the wet-dry pulsing sites. It is probably due to both.

Comparing apples and oranges: The net balance of methane production and carbon sequestration of wetlands

There is a lot of confusion on the part of wetland conservationists, ecological engineers who are creating and restoring wetlands, and climatologists as to where wetlands fit into climate change. One the one hand wetlands are creating a greenhouse gas methane (and have been doing so for ages), but on the other hand, the vast peatlands of the world are sequestering carbon, some at significant rates. So are wetlands good or bad for climate change? Table 10.6 shows one set of calculations, albeit based on some enormously wide ranges of assumptions, that suggests that the effect of methane production is almost outweighed by the sequestration of carbon on a global scale. In this calculation, 1.5 t C ha^{-1} yr^{-1} is the CO_2 equivalent released as methane from the world's wetlands, while an estimated 0.2 to 1.4 t C ha^{-1} yr^{-1} is sequestered. This is after taking into account the magnification factor of 21 in the calculation.

In a specific case study in Ohio (described in a Wetland Creation and Restoration box on pages 413–415), a high number of spatially distinct samples were taken for both carbon sequestration (Anderson and Mitsch, 2006) and methane generation (Altor and Mitsch, 2006) in the same replicated flow-though wetlands. Using the measured rates of 180 to 190 gC m^{-2} yr^{-1} for carbon sequestration and a weighted average of 22 g CH_4-C m^{-2} yr^{-1} for methane production and the same conversions used in Table 10.6, carbon sequestration more than counterbalances the methane production on a net effect on climate change. These created wetlands were climate neutral or even climate positive.

Table 10.6 Net balance calculation between methane production and carbon sequestration in the world's wetlands

Assumptions

Average methane emission rates from the world's wetlands:	200 kg CH_4 ha^{-1} yr^{-1}
Carbon sequestered by the world's wetlands:	0.2–1.4 t C ha^{-1} yr^{-1}

One molecule of CH_4 is 21 times more effective in adsorbing radiant energy than one molecule of CO_2

Converting methane emissions to carbon dioxide equivalent:

200 kg CH_4 ha^{-1} yr^{-1} $\times$ 21 $\times$ 10^{-3} t/kg = 4.2 t CO_2 ha^{-1} yr^{-1}

Converting carbon dioxide to carbon:

4.2 t CO_2 ha^{-1} yr^{-1} $\times$ 12tC/44 t CO_2 = 1.5 t C ha^{-1} yr^{-1}

1.5 t C ha^{-1} yr^{-1} of carbon equivalent methane production is at the high end of the range of 0.2–1.4 t C ha^{-1} yr^{-1} of carbon being sequestered by wetlands suggesting the overall impact of wetlands on climate change in the carbon cycle is minimal.

Source: Mitra et al. (2005).

Climate Change Feedbacks

One of the interesting questions about the vast storages of peat in northern climes related to the potential positive feedback to climate change that could occur. Because there is significantly more carbon stored in the world's soils than in the atmosphere (see Fig. 10.3), there is the potential that if the climate were to warm and accelerate decomposition of peatlands, then these peatlands would become an additional major source of carbon, through aerobic respiration and possibly fires, to the atmosphere. Davidson and Janssens (2006) summarize the comparison of uplands, which have good drainage and aeration and are therefore less prone to having large releases of carbon dioxide in the event of warming, to peatlands, where drainage is poor and soils are anaerobic. They describe peatland soils as enormously vulnerable to climate change compared to upland soils (Table 10.7), even though peatland soils make up a relatively small percentage of the earth's landscape. The release of 100 Pg-C from peatlands by the year 2100 would mean that for several years, carbon would be released at rates comparable to those presently caused by fossil fuels. If peatland productivity were to increase with the increase in temperature, this could offset this positive feedback and even lead to a negative feedback, where more carbon is sequestered than released.

A possible positive feedback of increased temperatures expected from global change leading to increased emissions of greenhouse gases (CO_2 and CH_4) as a result

Table 10.7 Belowground carbon stocks in the world and their vulnerabilities to loss by 2100 due to global warming

Carbon Pool	Carbon Size, Pg-C	Potential Loss by 2100 from global Warming
Upland Soil Inventory (3 m depth)	2300	0 – 40
Peatlands (3 m depth)	450	100
Permafrost	400	100

Source: Davidson and Janssens (2006).

Table 10.8 Estimated changes in CO_2 and CH_4 emissions (10^{15} g/yr = Pg/yr) from northern wetlands with change in climatic conditions (assumes 5°C temperature rise)

	Warm/Wet		Warm/Dry	
	CO_2	CH_4	CO_2	CH_4
Tundra	+1.3	+0.1	+1.6	+0.1
Boreal peatland	—	+0.12	+0.83	+0.12

Source: Post (1990).

of increased metabolism is shown in Table 10.8. Comparing these figures with fluxes shown in Figure 10.3 also suggests that this feedback—climate change increasing the release of greenhouse gases from wetlands—may be significant. Christensen (1991) predicted that, as a result of a 5 percent global warming, the tundra would change from being a net sink of CO_2 to a net source of up to 1.25 Pg/yr carbon because of a combination of thermokarst erosion, deepening of the active layer in permafrost areas, lowering of the water table, and higher temperatures. Tarnocai (2006) was more direct and predicted severe degradation of the frozen peatlands in the subarctic and northern boreal Canada and severe drying in the southern boreal regions as well, but a scenario of 3 to 5°C increase in air temperature and 5 to 7°C increase over the oceans by the end of the 21st century. The affected area represents about 50 percent of all the organic carbon mass occurring in all Canadian wetlands.

In general, both the increase in temperature and the changes in water levels are important variables in the production of methane and carbon dioxide from wetlands, but their relative importance for methane generation is poorly understood (Moore and Knowles, 1989; Whalen and Reeburgh, 1990; Roulet et al., 1992b; Moore and Roulet, 1995; Updegraff et al., 2001). Using a model with inputs of a 3°C rise in temperature and a decrease in the water table between 14 and 22 cm for a subarctic fen, Roulet et al. (1992b) estimated that the increased temperature raised the methane flux between 5 and 40 percent, but the lowered water table decreased the methane flux by 74 to 81 percent. This decrease in methane flux in drier conditions was caused

Carbon Budgets

Carbon budgets for peatlands have drawn a great deal of interest, given the importance of these ecosystems in global carbon dynamics. It is accepted that boreal peatlands were once carbon sinks, but there is little consensus that they are contemporary sinks (Rivers et al., 1998). Carbon budgets have been developed for small peatlands (Carroll and Crill, 1997; Waddington and Roulet, 1997) and for substantial-sized peatland-dominated watersheds (Rivers et al., 1998). The latter, a 1,500-km^2 watershed in the Lake Agassiz peatlands in Minnesota, illustrated that the peat watershed had a net carbon storage of 12.7 g-C m^{-2} yr^{-1} but that there was a tenuous balance between the watershed being a source and a sink of carbon (Fig. 10.5). Inflows of carbon are groundwater, precipitation, and net community productivity, while outflows are groundwater and surface flow and outgassing of methane. It was estimated from a companion study (Glaser et al., 1997) that peat is accumulating at a rate of 1 mm/yr (100 cm/1000 yr). This budget illustrates the importance of accurate hydrologic measurements as well as biological productivity measurements in determining accurate carbon budgets.

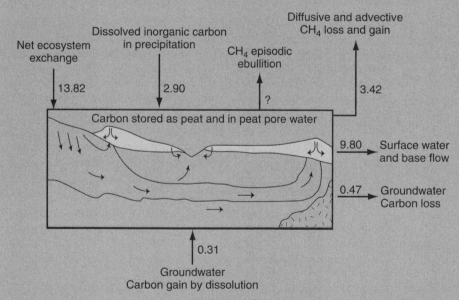

Figure 10.5 **Carbon dynamics of the 1,500-km^2 Rapid River watershed in the Lake Agassiz peatland basin of northern Minnesota. Fluxes are in g m^{-2} yr^{-1}.** *(After Rivers et al., 1998.)*

by a decrease in the zone of active methanogenesis and by an increase in methane oxidation in the aerobic layer. Thus, the influence of global temperature rise would depend locally on the temperature increase relative to the induced change in the moisture regime.

The Effects of Climate Change on Wetlands

Wetlands may be key ecosystems for mitigating the effects of fossil fuel emissions on climate. Conversely, climate change and resulting sea-level and temperature changes may have significant impacts on coastal and inland wetlands.

Coastal Wetlands

One of the major impacts of possible climate changes on wetlands is the effect that sea-level rise will have on coastal wetlands. Estimates of sea-level rise over the next century range from 50 to 200 cm. It has been estimated that if sea level were to rise by 100 cm, that half of the wetlands designated by the Ramsar Convention as wetlands of international importance would be threatened (Nicholls, 2004). The regions where wetlands are most at risk, even for a 44 cm rise in sea level by 2080, are shown in Figure 10.6. If the rise in sea level is not accompanied by equivalent vertical accretion of marsh sediments, then coastal marshes will gradually disintegrate as a result of increased inundation, erosion, and saltwater intrusion. Because much of the coastline of the world is developed, efforts to protect dry upland from inundation by the construction of bulkheads or dikes will exacerbate the problem. In essence,

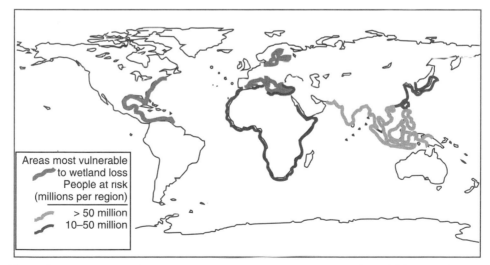

Figure 10.6 Coastal wetland areas most vulnerable to a sea-level rise of 44 cm by 2080. (From IPCC, 2001.)

Table 10.9   Estimated percentage coastal wetland loss in the United States with sea-level rise

	Sea-Level Rise		
	0.5 m	1 m	2 m
If no shores are protected	17–43%	26–66%	29–76%
If densely developed dry land is protected	20–45%	29–69%	33–80%
If all dry land is protected	38–61%	50–82%	66–90%

Source: Titus (1991).

the wetlands will be trapped between the rising sea and the protected dry land, a situation that has already occurred over the centuries in The Netherlands and China. This effect has been termed the *coastal squeeze* of sea-level rise (Nicholls, 2004). Even in the absence of bulkheads in most of our regions where coastal wetlands exist, "the slope above the wetland is steeper than that of the wetlands; so a rise in sea level causes a net loss of wetland acreage" (Titus, 1991).

Estimates of the loss of coastal wetlands in the United States vary, with much of the variability dependent on the assumed sea-level rise and the degree to which dry land is protected at all cost (Table 10.9). If there is no shoreline protection, a sea-level rise of 1 m could reduce coastal wetlands by 26 to 66 percent. If the policy were to protect all dry land, then the estimated loss of wetlands increases dramatically to 50 to 82 percent. How well these figures can be extrapolated to the rest of the world is unclear. In long-developed coastlines such as those of Europe and the Far East, the losses would probably be less.

Several studies have evaluated regional changes in coastal wetlands that would result in the event of dramatic sea-level rise. A spatial cell-based simulation model called SLAMM (Sea Level Affecting Marshes Model), used by Richard Park and colleagues (Park et al., 1991; Lee et al., 1991), illustrated the effects of sea-level rise on coastal regions. For example, Lee et al. (1991) predicted a 32 to 40 percent loss of wetlands in northeastern Florida with a rise of 1 to 1.25 m. Most of that loss is the low intertidal salt marsh.

The Mississippi River delta in Louisiana may be a model for seeing the effects of global sea-level rise on coastal wetlands (Day and Templet, 1989; McKee and Mendelssohn, 1989; Day et al., 2005). Here, the apparent sea-level rise is already 1 m/100 yr (1 cm/yr), primarily because of sediment subsidence rather than actual sea-level rise (see Chapter 12). In this delta marsh, vertical accretion is not keeping up with subsidence, in part because the Mississippi River is carrying only about 20 percent of the sediment load it did in 1850 (Kesel and Reed, 1995), and its flow is contained within levees, so riverborne sediments no longer reach the wetlands during spring floods. As a result, this region has the highest rate of wetland loss in the United States.

Day et al. (2005) describe the ramifications of global climate change on restoration efforts now underway in the delta. With a sea-level rise of 30 to 50 cm by 2100 possible, the relative sea-level rise will increase from 1 cm/yr (caused mostly by land subsidence) to 1.3 to 1.7 cm/yr, exacerbating an already difficult situation of wetland loss in the Louisiana Delta. In addition, Day et al. (2005) note that as a result of milder temperatures already, mangrove swamps were beginning to replace their temperate zone analog, the salt marsh, in several locations in the delta. This mangrove expansion is another effect that would be expected in subtropical regions that were previously dominated by salt marshes. Mangroves are valuable coastal ecosystems, as are salt marshes, but the overall effects of this substitution of ecosystems in unclear.

Patrick and Delaune (1990) measured the accretion and subsidence of sediments in the salt marshes of San Francisco Bay, California, and found that, because of an adequate supply of sediments, the salt marshes of south San Francisco Bay would probably have a net accretion rate despite sea-level rise. Lefeuvre (1990) presented a summary of the ecological effects of sea-level rise in coastal France near Mont St. Michel Bay. The salt marshes in that region are currently expanding at a rate of about 30 ha/yr because of a positive sediment influx of 2 cm/yr. He hypothesized that fragmentation would occur to the salt marshes and polders in the event of a maximum rise in sea level (1–2 m/100 yr), but that this fragmentation would enhance marine ecology by increasing marsh–tidal flat interfaces.

Management

The management possibilities for coastal wetlands in the face of sea-level rise offer some possibilities to save coastal wetlands. Figure 10.7 shows two future conditions. In Future 1, the house is protected with a bulkhead in the face of rising sea level and

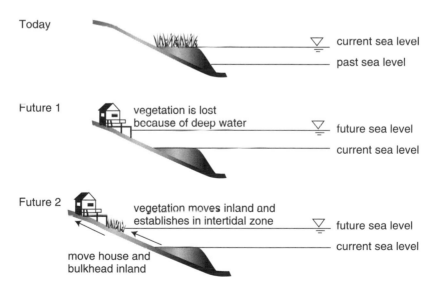

Figure 10.7   Coastal wetland management scenarios in the face of sea-level rise. Future 1 is without moving human habitation inland. Future 2 involves moving human activity inland to allow room for the wetland to move inland. (From Titus, 1991.)

the salt marsh is lost. In Future 2, the house is moved upland to accommodate the wetland, which would begin to form if a gentle slope and adequate sediment sources were available. This models the wetlands of the Great Lakes, which, for centuries, were "wetlands on skateboards," moving inland and lakeward with frequent (over periods of decades) water-level changes in the Great Lakes (Mitsch, 1992b). With stabilization of the coastline in the past century, diking the remaining wetlands along the Great Lakes has been necessary for their survival.

Day and Templet (1989) and Day et al. (2005) concluded, after extensive investigation of the apparent sea-level rise in coastal Louisiana, that we can manage coastal wetlands in periods of rising sea level through comprehensive, long-range planning and through the application of the principles of ecological engineering by using nature's energies such as upstream riverine sediments and fresh water, vegetation productivity, winds, currents, and tides as much as possible. Day and Templet (1989) cite the example of using vegetation to enhance sediment accretion, such as with brush fences developed by the Dutch.

Inland Wetlands

In addition to the effects of climate change on coastal wetlands through sea-level rise, the change in climate, particularly temperature (Fig. 10.8), will probably affect the function and distribution of inland wetlands. In the tundra, any melting of the permafrost would result in the loss of wetlands. In boreal and temperate areas, climate change would result in changing rainfall patterns, thus affecting runoff and groundwater inflows to wetlands. In general, a decrease in precipitation or an increase in evapotranspiration will result in less-frequent flooding of existing wetlands, although the types of wetlands may not change. Greater precipitation patterns would increase the length and depth of flooding of inland wetlands. Most susceptible to these effects are depressional wetlands that have very small watersheds and that are in regions between arid and mesic climates, such as the prairie potholes of North America.

The impact of climate change on the Prairie Pothole Region (PPR) of North America (see Chapter 3) was investigated by Johnson et al. (2005). These wetlands provide 50 to 80 percent of the continent's duck population and are exactly on the edge between areas to the east with abundant precipitation and arid climates to the west. By using a wetland simulation model, Johnson et al. (2005) were able to predict areas in the pothole region that would have highly favorable water conditions for three climate scenarios: (1) a 3°C temperature increase with no change in precipitation; (2) a 3°C temperature increase with a 20 percent increase in precipitation; and (3) a 3°C temperature increase with a 20 percent decrease in precipitation (Fig. 10.9). Basically any temperature increase coupled with precipitation decrease shifted the area favorable for ducks to the east. Overall, the climate change would "diminish the benefits of wetland conservation in the central and western PPR. Simulations further indicate that restoration of wetlands along wetter fringes of the PPR may be necessary to ameliorate potential impacts of climate change on waterfowl populations" (Johnson et al., 2005).

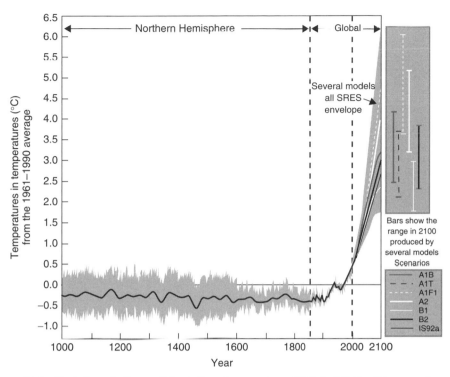

Figure 10.8 Variations in the earth's surface temperature, 1000 to 2000, with several model predictions of temperature departures from 1961–1990 average through the year 2100. (From IPPC, 2001.)

Management

Limited experimentation, especially in rice paddies, suggests some management alternatives that might be appropriate for inland wetlands, especially to reduce methane emissions. Sass et al. (1992) measured the effects on methane emissions of four different water management methods in some rice fields in Texas and found that temporary drainage (midseason drainage and multiple aeration) decreased methane emission caused by both increased CH_4 consumption in the aerobic layer and decreased CH_4 production. Such management may only be practical in flat systems with sufficient control of water levels.

Nutrient and compost management may also offer opportunities for reducing methane emissions. There may be a relationship between the C:N ratio of the organic matter in wetlands and CH_4 emissions, although the trends are not clear. Yagi and Minami (1990) found, in rice paddies in Japan, that compost with a low C:N ratio (enriched in nitrogen) causes lower emissions of methane than uncomposted rice straw with a high C:N ratio. Conversely, Schutz et al. (1989) found high emissions in fields applied with compost. Because of its competition with methanogenesis, enhancing sulfate reduction is often suggested as a management alternative to reduce methane

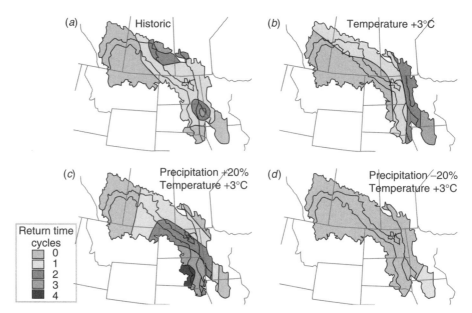

Figure 10.9 Simulation results for locations of highly favorable water and cover conditions in the Prairie Pothole Region of North America for waterfowl breeding under options (a) historic, (b) a 3°C temperature increase with no change in precipitation; (c) a 3°C temperature increase with a 20 percent increase in precipitation; and (d) a 3°C temperature increase with a 20 percent decrease in precipitation. (From Johnson et al., 2005, reprinted with permission.)

emissions (see Chapter 6). This has long been known as one of the primary reasons that methanogenesis is lower in saltwater wetlands than in freshwater wetlands.

One of the easiest management approaches for minimizing methane emissions from freshwater wetlands is to allow the wetlands to have their natural fluctuating hydroperiods and, in some cases, a pulsing hydrology. Studies by Altor and Mitsch (2006) showed that a pulsing hydrology had methane emissions that were 69 percent lower than those from permanently flooded sites.

We cannot estimate, at present, with much certainty whether wetlands are significant global carbon sources or sinks. Nevertheless, the opportunities cited previously for managing carbon dioxide and methane emissions in wetlands are not generally on a scale large enough to make much difference to the global carbon balance.

Recommended Readings

IPPC. 2007. *Climate Change 2007, 4th Assessment Report*. Published for the Intergovernmental Panel on Climate Change. Cambridge University Press, UK.

Whalen, S. C. 2005. "Biogeochemistry of methane exchange between natural wetlands and the atmosphere." *Environmental Engineering Science* 22:73–94.

Chapter **11**

Values and Valuation of Wetlands

Wetlands provide many services and commodities to humanity. At the population level, wetland-dependent fish, shellfish, fur animals, waterfowl, and timber provide important and valuable harvests and millions of days of recreational fishing and hunting. At the ecosystem level, wetlands moderate the effects of floods, improve water quality, protect coastlines from storms, hurricanes, and tsunamis, and have aesthetic and heritage value. At the global level, they contribute to the stability of global levels of available nitrogen, atmospheric sulfur, carbon dioxide, and methane.

 The valuation of these services and commodities is complicated by (1) the difficulty of comparing by common denominator the various values of wetlands against human economic systems, (2) the conflict between a private owner's interest in the wetlands and the values that accrue to the public at large, and (3) the need to consider the value of a wetland as part of an integrated landscape. Valuation techniques include nonmonetary scaling and weighting approaches for comparing different wetlands or different management options for the same wetland, and common-denominator approaches that reduce the various values to some common term such as dollars, embodied energy, or emergy. These common-denominator methodologies can include willingness to pay, replacement value, energy analysis, and emergy analysis. None of these approaches is without problems, and no universal agreement about their use has been reached. But when compared to other ecosystems or use of the landscape, sustainable values of wetlands are often among the highest of any ecosystems.

The term *value* imposes an anthropocentric orientation on a discussion of wetlands. The term is often used in an ecological sense to refer to functional processes, as, for example, when we speak of the "value" of primary production in providing the food energy that drives the ecosystem. However, in ordinary parlance, the word connotes something worthy, desirable, or useful to humans. The reasons that wetlands are often legally protected have to do with their value to society, not with the abstruse ecological processes that occur in wetlands; this is the sense in which the word *value* is used in this chapter. Perceived values arise from the functional ecological processes described in previous chapters but are determined also by human perceptions, the location of a particular wetland, the human population pressures on it, and the extent of the resource.

Regional wetlands are integral parts of larger landscapes—drainage basins, estuaries. Their functions and their values to people in these landscapes depend on both their extent and their location. Thus, the value of a forested wetland varies. If it lies along a river, it probably plays a greater functional role in stream water quality and downstream flooding than if it were isolated from the stream. If situated at the headwaters of a stream, a wetland functions differently from a wetland located near the stream's mouth. The fauna it supports depend on the size of the wetland relative to the home range of the animal. Thus, to some extent, each wetland is ecologically unique. This complicates the measurement of its value.

Wetland Values

Wetland values can conveniently be considered from the perspective of three hierarchical levels—population, ecosystem, and global.

Population Values

The easiest wetland values to identify are the populations that depend on wetland habitats for their survival.

Animals Harvested for Pelts

Fur-bearing mammals, and even alligators and crocodiles, are harvested for their pelts throughout the world. In contrast to most other commercially important wetland species, these animals typically have a limited range and spend their lives within a short distance of their birthplaces. The most abundant furbearer historically harvested in wetlands in the United States is the muskrat (*Ondatra zibethicus*). Muskrats (Fig. 11.1a) are found in wetlands throughout the United States except, strangely, the South Atlantic Coast. They prefer fresh inland marshes, but along the northern Gulf Coast are more abundant in brackish marshes. About 50 percent of the nation's harvest is from the Midwest and 25 percent from along the northern Gulf of Mexico, mostly Louisiana.

The nutria (*Myocastor coypus*), an ecological analog of the muskrat, is the next most abundant species. It is very much like a muskrat but is larger and more vigorous (Fig. 11.1b). This species was imported from South America to Louisiana and escaped from captivity in 1938, spreading rapidly through the state's coastal marshes. In the

Figure 11.1   Five fur-bearing animals found in wetlands that have been historically harvested for their pelts: (a) muskrat, *Ondatra zibethicus*, (b) nutria, *Myocastor coypus*, (c) beaver, *Castor canadensis*, (d) mink, and (e) otter. (*a, d, e courtesy of National Park Service.*)

1940s, the animal was promoted by state agents for controlling aquatic weeds, particularly water hyacinth (*Eichhornia crassipes*). It is now abundant in freshwater swamps and in coastal freshwater marshes, from which it may have displaced muskrats to more brackish locations, and is spreading up the coastal Atlantic states well beyond Louisiana. Most of the U.S. nutria harvest occurs in Louisiana (Fig. 11.2). In order of decreasing abundance in the United States, other harvested fur animals are beaver

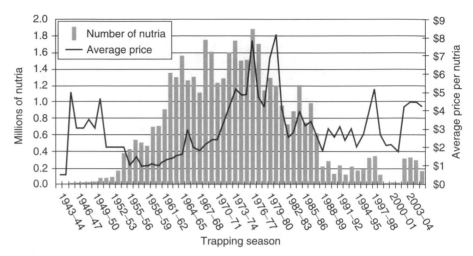

Figure 11.2 Annual harvest of nutria in Louisiana and average price for a pelt, 1943–2004. From 2002–03 and three subsequent seasons, a $4 incentive payment was made available to hunters. *(From Louisiana Department of Wildlife and Fisheries.)*

(Fig. 11.1c), mink (Fig. 11.1d), and otter (Fig. 11.1e). Beavers are associated with forested wetlands, especially in the Midwest. Minnesota harvests a high percentage of the nation's beaver catch.

Waterfowl and Other Birds

Birds, as our only remaining evolutionary link to the dinosaurs that once roamed the Earth, may have actually survived the dinosaur die-off precisely because of wetlands (Gibbons, 1997; Weller, 1999). Although not all present bird species require wetlands as their primary habitat, a great many do. Eighty percent of America's breeding bird population and more than 50 percent of the 800 species of protected migratory birds rely on wetlands (Figs. 11.3 to 11.5). Wetlands, which are probably known best for their waterfowl abundance, also support a large and valuable recreational hunting industry. We use the term "industry" because hunters spend large sums of money in the local economy for guns, ammunition, hunting clothes, travel to hunting spots, food, and lodging.

Most of the birds hunted are hatched in marshes in the far North, sometimes above the Arctic Circle, but are shot during their winter migrations to the southern United States and Central America. There are exceptions—the Wood Duck (*Aix sponsa*) breeds locally throughout the continent—but the generalization holds for most species. Different groups of geese and ducks have different habitat preferences, and these preferences change with the maturity of the duck and the season.

A broad diversity of wetland habitat types is important for waterfowl success. The freshwater prairie potholes of North America are the primary breeding place for waterfowl in North America. There, an estimated 50 to 80 percent of the continent's main game species are produced. Wood Ducks prefer forested wetlands. During the

(a)

(b)

Figure 11.3   Two wetland waterfowl known around the world: (a) Mallard, *Anas platyrhynchos*, and (b) Canada Goose, *Branta canadensis*. (Photographs courtesy of Alan and Elaine Wilson.)

winter, diving ducks (*Aythya* spp. and *Oxyura* spp.) are found in brackish marshes, preferably adjacent to fairly deep ponds and lakes. Dabbling ducks (*Anas* spp.) prefer freshwater marshes and often graze heavily in adjacent rice fields and in very shallow marsh ponds. Gadwalls (*Anas strepera*) like shallow ponds with submerged vegetation.

Figure 11.4   Herons are consummate symbols of wetlands throughout the world. Different species that dominate this wading niche in parts of the world include: (a) Great Blue Heron, *Ardea herodias*, North America; (b) White-necked Heron, *Ardea cocol*, South America; (c) Black-headed Heron, *Ardea melanocephala, eastern Africa*; (d) White-faced heron, *Ardea novaehollandiae*, Australia/New Zealand; (e) Gray Heron, *Ardea cinerea*, Europe and Africa. *(Photograph a by T. Daniel, Ohio Department of Natural Resources; b, c by W. J. Mitsch; d by B. Harcourt, courtesy of New Zealand Department of Conservation; e by P. Marion; reprinted by permission.)*

Figure 11.5   Selected waterfowl from North American wetlands: (a) Gadwall, *Anas strepera*; (b) Green-winged teal, *Anas crecca*; (c) Blue-winged teal, *Anas discors*; and (d) Northern pintail, *Anas aouta*. (Photograph c. from U.S. Fish and Wildlife Service; others from Alan and Elaine Wilson.)

The waterfowl value of wetlands such as the prairie pothole region of North America (see Chapter 3) is unmistakable. When waterfowl census data for the prairie pothole region over the 30-year period were compared to the number of potholes flooded in May of each year, there was a clear positive correlation (Fig. 11.6), indicating the importance of wetland hydrology in the breeding success of waterfowl. On average, there were almost 22 million waterfowl (dabbling and diving ducks) in the region, dominated by the Mallard (average 3.7 million; Fig. 11.3a).

Generally, the duck population of North America has shown a 10- to 20-year cycle of increase and decline, with low points in the early 1960s and 1990s and highs in the mid-1950s, mid-1970s, and late 1990s (Table 11.1 and Fig. 11.6). Populations of nine of the ten duck species listed in Table 11.1 were lower than historical averages after the dry years 1987–1991, while populations of seven of the same ten duck species were higher than historical averages after the wet years 1995–1998. Over that period, from dry period to wet period, the total number of ducks increased by 60 percent. The trends of below-average populations during dry periods and above-average populations during wet periods are particularly apparent for

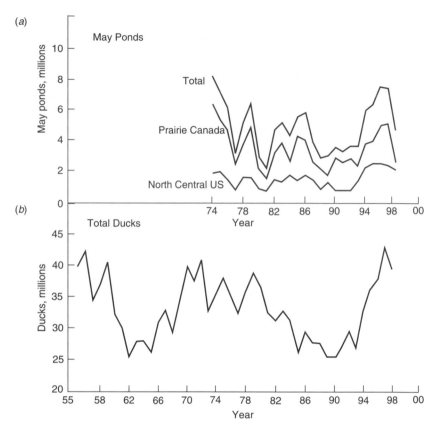

Figure 11.6   Connection between waterfowl populations and how wet conditions are in the prairie pothole region of North America: (a) estimated number of ponds in May in the prairie pothole region; (b) duck breeding populations in North America, 1955–1998, excluding scooters, elders, mergansers, and oldsquaws. *(From U.S. Fish and Wildlife Service, Washington, DC.)*

Mallards, Green-winged and Blue-winged Teals, Northern Shovelers, and Canvasback (Table 11.1).

Habitat degradation and loss, both in the northern breeding grounds and in the wintering areas, are certainly factors in population declines. Northern Pintail and Scaup have showed a steady decline. However, climatic changes that influence the number of ponds from year to year in the breeding grounds appear to be the major cause of year-to-year fluctuations.

Hunting is closely regulated and tailored to the local region. The Mallard makes up about one-third of the U.S. total of harvested ducks. About 50 percent are shot in wetlands; most of the rest are shot in agricultural fields. In Louisiana, one-third of the Mallard population is killed each year. The percentage is lower for other species—about 8 to 13 percent. The vast flocks of geese that used to be so abundant

Table 11.1 Population estimates of the ten most common species of breeding ducks and four species of goose in North America for a dry year (1991) and a wet year (1998) in the prairie pothole region, with percentage change in 1991 and 1998 compared to 1995–1990 and 1955–1997 averages, respectively.

Species	Population (× 1,000)		Percentage Change	
	1991(Dry Year)	1998(Wet Year)	1991[a]	1998[b]
All species	24,200	39,100		20
Mallard (*Anas platyrhynchos*)	5,353 ± 188	9,640 ± 302	−27	+32
Gadwall (*Anas strepera*)	1,573 ± 94	3,742 ± 206	+22	+149
American Wigeon (*Anas americana*)	2,328 ± 135	2,858 ± 145	−14	−5
Green-winged Teal (*Anas crecca*)	1,601 ± 88	2,087 ± 139	−4	+16
Blue-winged Teal (*Anas discors*)	3,779 ± 245	6,399 ± 332	−10	+36
Northern Shoveler (*Anas clypeata*)	1,663 ± 84	4,120 ± 194	−8	+106
Northern Pintail (*Anas acuta*)	1,794 ± 199	3,558 ± 194	−62	−36
Redhead (*Aythya americana*)	437 ± 37	918 ± 77	−26	+48
Canvasback (*Aythya valisneria*)	463 ± 57	689 ± 57	−16	+28
Scaup (*Aythya* spp.)	5,247 ± 333	4,122 ± 234	−7	−35
Average of 10 duck species			−15	+34
Canada Goose (*Branta canadensis*)	3,750	4,683		
Snow Goose (*Chen caerulsecens*)	2,440	3,776		
White-fronted Goose (*Anser albifrons*)	492	941		
Brant (*Branta bernicla*)	275	276		

[a]Compared to average for 1955–1990
[b]Compared to average for 1955–1997
Source: U.S. Fish and Wildlife Service. Duck surveys on summer breeding grounds: goose surveys during summer, fall, and winter.

along the eastern seaboard and the Gulf Coast are smaller now but are still abundant and are considered to be important as hunted species in some areas.

Fish and Shellfish

A direct relationship between shrimp and fish harvests and wetland area has been illustrated for many fisheries around the world, including marine, freshwater, and pond-raised (Fig. 11.7).

Over 95 percent of the fish and shellfish species that are harvested commercially in the United States are wetland dependent (Feierabend and Zelazny, 1987). The degree of dependence on wetlands varies widely with species and with the type of wetland. Some important species are permanent residents; others are merely transients that feed in wetlands when the opportunity arises. Some shallow wetlands, which may exhibit several other wetland values, may be virtually devoid of fish, whereas other types of deepwater and coastal wetlands may serve as important nursery and feeding areas. Table 11.2 lists some of the major nektonic species associated with wetlands, while Table 11.3 lists the economic value of coastal fisheries.

Virtually all of the freshwater species are dependent, to some degree, on wetlands, often spawning in marshes bordering lakes or in riparian forests during spring flooding.

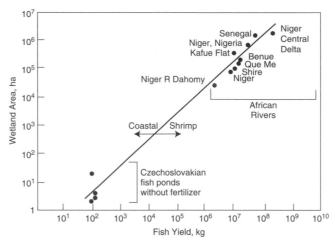

Figure 11.7   Relationship between wetland area and fish harvests. The linear slope describes the line of about 60 kg/ha yield. Pond fisheries are unfertilized, managed ponds in Czechoslovakia. African floodplain river fisheries after Welcomme, 1976. Wetland-dependent coastal fisheries yields per ha (adapted from Turner, 1977) are generally 10 times higher than inland ecosystems. *(After Turner, 1982.)*

These species are primarily recreational, although some small local commercial fisheries exploit them. The saltwater species tend to spawn offshore, move into the coastal marsh "nursery" during their juvenile stages, and then emigrate offshore as they mature. They are often important for both commercial and recreational fisheries. The menhaden is caught only commercially, but competition between commercial and sport fishermen for shrimp, blue crab, oyster, catfish, seatrout, and striped bass can be intensive and acrimonious. Anadromous fish probably use wetlands less than the other two groups. However, young anadromous fish fry sometimes linger in estuaries and adjacent marshes on their migrations to the ocean from the freshwater stream in which they were spawned.

Analyses of fishery harvests from wetlands show the importance of recreational fishing. Although the commercial harvest is usually much better documented, several studies have shown that the recreational catch far outweighs the commercial catch for certain species. Furthermore, the value to the economy of recreational fishing is usually far greater than the value of the commercial catch, because sports fishermen spend more money per fish caught (they are less efficient) than their commercial counterparts.

Timber and Other Vegetation Harvest

Wetlands often provide an abundance of building materials and foodstuffs for local economies. Timber from forested wetlands was one of the staples of the economy of southeastern United States. The antebellum homes of the South were often supported by giant trusses of cypress trees harvested from nearby swamps. The Mississippi River

Table 11.2   Dominant commercial and recreational fish and shellfish associated with wetlands

Species Common Name	Scientific Name	Commercial Harvest (Metric Tons)[a]
FRESHWATER		
Catfish and bullhead	*Ictalurus* sp.	16,800
Carp	*Cyprinus carpio*	11,800
Buffalo	*Ictiobus* sp.	11,300
Crayfish	*Procambarus clarkia*	11,300
Perch	*Perca* sp., *Istizostedion* sp.	—
Pickerel	*Esox* sp.	—
Sunfish	*Lepomis* sp., *Micropterus* sp., *Pomoxis* sp.	—
Trout	*Salmo* sp., *Salvelinus* sp.	—
ANADROMOUS		
Salmon	*Oncorhynchus* sp.	396,600
Shad and alewife	*Alosa* sp.	4,400
Striped bass	*Morone saxatills*	3,600
SALTWATER		
Menhaden	*Brevoortia* sp.	901,300
Shrimp	*Penaeus* sp.	135,200
Blue crab	*Callinectes sapidus*	102,100
Oyster	*Crassostrea* sp.	17,200
Mullet	*Mugil* sp.	11,000
Atlantic croaker	*Micropogonias undualtus*	7,700
Hard clam	*Mercenaria* sp.	5,800
Bluefish	*Pomatomus saltatrix*	4,300
Seatrout	*Cynoscion* sp.	4,000
Spot	*Leiostomus xanthurus*	3,300
Drum	*Pogonias cromis, Sciaenops ocellatus*	2,200
Soft clam	*Mya arenaria*	1,300

[a]Landings are 1993–1997 averages, except 1971–1975 landings for freshwater fish.
Source: National Marine Fisheries Service, U.S. Commercial Landings.

alluvial floodplain and the floodplains of rivers entering the South Atlantic are mostly deciduous wetlands, whereas the forested wetlands along the northern tier of states are primarily evergreen. The former are more extensive and potentially more valuable commercially because of the much faster growth rates in the South.

In addition to the timber harvest, the production of herbaceous vegetation in marshes is a potential source of energy, fiber, and other commodities. These prospects have not been explored widely in North America but are viable options elsewhere. For example, many commercial products are harvested from restored and natural salt marshes and freshwater marshes in China. The productivity of many wetland species (e.g., *Spartina alterniflora, Phragmites Australis, Typha angustifolia, Eichhornia crassipes, Cyperus papyrus*) is as great as our most vigorous agricultural crops.

Table 11.3 Value of U.S. landings of commercial fish and shellfish species, 1998 (About one-half of the value of the fish catch is from estuarine-dependent species, nearly all of which use salt and brackish marshes for food and shelter during some part of their lives.)

Species	Landings (metric tons)	Value (×$1,000)	Percentage of Total
Estuarine-dependent speices			
Crab, blue	101,675	184,250	
Flounder—all	17,622	58,748	
Menhaden—Atlantic	774,631	103,950	
Oyster—eastern	10,807	66,297	
Shrimp—brown, pink, white	116,313	505,860	
Other—striped bass, Atlantic croaker, drum, mullet, seatrout	20,226	28,340	
Subtotal		947,451	50
Marine species			
Lobster—>90% American	38,800	275,600	
Scallops	5,500	75,100	
Squid, longfin	18,880	32,141	
Tuna—all	4,267	27,085	
Subtotal		409,926	22
Other		521,540	28
Total		$1,878,917	

Source: National Marine Fisheries Service.

Peat Harvesting

In addition to the annual production of living vegetation in wetlands, great reservoirs of buried peat exist around the world. Peat harvesting was described in detail in Chapter 9. This buried peat is a nonrenewable energy source that destroys the wetland habitat when it is mined. In the United States and Canada, peat is mined primarily for horticultural peat production, but in other parts of the world—for example, several republics of the former Soviet Union and in Finland—it has been used as a fuel source for hundreds of years. It is used to generate electricity, formed into briquettes for home use, and gasified or liquified to produce methanol and industrial fuels.

Endangered and Threatened Species

Wetland habitats are necessary for the survival of a disproportionately high percentage of endangered and threatened species. Table 11.4 summarizes the statistics but imparts no information about the particular species involved, their location, wetland habitat requirements, degree of wetland dependence, and factors contributing to their demise. Although wetlands occupy only about 3.5 percent of the land area of the United States, of the 209 animal species listed as endangered, about 50 percent depend on wetlands for survival and viability. Almost one-third of native North American freshwater fish species are endangered, threatened, or of special concern. Almost all of these were adversely affected by habitat loss. Sixty-three species of plants

Table 11.4 Threatened and endangered species associated with wetlands

Taxon	Number of Species Endangered	Number of Species Threatened	Percentage of U.S. Total Threatened or Endangered
Plants	17	12	28
Mammals	7	—	20
Birds	16	1	68
Reptiles	6	1	63
Amphibians	5	1	75
Mussels	20	—	66
Fish	26	6	48
Insects	1	4	38
Total	98	25	

Source: Niering (1988).

and 34 species of animals that are considered endangered, threatened, or candidates for listing occupy southern U.S. forested wetlands. Of these, amphibians and many reptiles are especially linked to wetlands. In Florida, where the number of amphibian and reptile species is about equal to the number of mammal and breeding bird species, 18 percent of all amphibians and 35 percent of all reptiles are considered threatened or endangered or their status is unknown (Harris and Gosselink, 1990).

The fate of several endangered species is discussed here to illustrate the ecological complexity of species endangerment. Whooping Cranes nest in wetlands in the Northwest Territories of Canada, in water 0.3 to 0.6 m deep, during the spring and summer. In the fall, they migrate to the Aransas National Wildlife Refuge, Texas, stopping off in riverine marshes along the migration route. In Texas, they winter in tidal marshes. All three types of wetlands are important for their survival. The decline in the once-abundant species has been attributed both to hunting and to habitat loss. The last Whooping Crane nest in the United States was seen in 1889. In 1941, the flock consisted of 13 adults and 2 young. Since then, the flock has been gradually built up to about 75 birds.

The American Alligator: From Endangered to Plentiful

The American alligator (*Alligator mississippiensis*; Fig. 11.8) represents a dramatic success story of the return from the edge of extinction to a healthy U.S. population. Alligators are abundant in fresh and slightly brackish lakes and streams and build nests in adjacent marshes and swamps in the southwestern United States, especially in Florida and Louisiana. Alligators have an interesting role in wetlands—they depend on them, and, in return, the character of the wetland is shaped by the alligator, at least in the south Florida Everglades. They are another example of an ecosystem engineer. As the annual dry season approaches, alligators dig "gator holes." The material

Figure 11.8 The American alligator (*Alligator mississippiensis*). (Photograph by W. J. Mitsch.)

thrown out around the holes forms a berm high enough to support trees and shrubs in an otherwise treeless prairie. The trees provide cover and breeding grounds for insects, birds, turtles, and snakes. The hole is a place where the alligator can wait out the dry period until the winter rains. It also provides a refuge for dense populations of fish and shellfish (up to $1,600/m^2$). These organisms, in turn, attract top carnivores, and so the gator holes are sites of concentrated biological activity that may be important for the survival of many species.

American alligator populations were reduced by hunters and poachers to such low levels that the species was declared endangered in the 1970s. The species was threatened by severe hunting pressure, not by habitat loss. When that pressure was removed, its numbers increased rapidly. The animal is now harvested under close regulation and grown commercially in both Louisiana and Florida. About 250,000 alligators are harvested in the wild and in farms annually in Louisiana, and yet the population remains constant or is slightly increasing. Alligator hunting and farming in Louisiana has increased dramatically; it was worth $16 million in 1992 and $26 million in 2004 for both wild and farm-raised animals.

In Florida, where limited hunting is permitted, the harvest in the wild and on farms is considerably less that that in Louisiana, but the compatibility of alligators and a rapidly increasing human population is constantly being challenged. It is probably extraordinary that there have been fewer than 20 confirmed fatal alligator attacks on humans in the last 50 years in Florida, given the high number of both alligators and people in Florida.

In addition to the harvest of alligators for their meat, alligator skins from both Florida and Louisiana are sold worldwide, particularly for high-end luxury handbags, wallets, belts, and boots. Apparently with the increased interest in wetlands and wildlife, the fashion world has gone reptile-chic.

Ecosystem Values

At the level of the whole ecosystem, wetlands have value to the public for flood mitigation, storm abatement, aquifer recharge, water quality improvement, aesthetics, and general subsistence. Some of the ecosystem values of wetlands vary from year to year or from season to season. For example, Figure 11.9 illustrates several of the potential ecosystem values of riparian forested wetlands during flooding (spring) and dry (summer) seasons.

Flood Mitigation

Chapter 4 dealt with the importance of hydrology in determining the character of wetlands. Conversely, wetlands influence regional water flow regimes. One way they do this is to intercept storm runoff and to store storm waters, thereby changing sharp runoff peaks to slower discharges over longer periods of time (Fig. 11.10). Because it is usually the peak flows that produce flood damage, the effect of the wetland area is to reduce the danger of flooding. Riverine wetlands are especially valuable in this regard. In a classic study on the Charles River in Massachusetts, the floodplain wetlands were deemed so effective for flood control by the U.S. Army Corps of Engineers that it purchased them rather than build expensive flood control structures to protect Boston (U.S. Army Corps of Engineers, 1972). The study on which the Corps' decision was based demonstrated that if the 3,400 ha of wetlands in the Charles River basin were drained and leveed off from the river, flood damages would increase by $17 million per year.

Bottomland hardwood forests along the Mississippi River before European settlement stored floodwater equivalent to about 60 days of river discharge. Storage capacity has been reduced to only about 12 days as a result of leveeing the river and draining the floodplain. The consequences—the confinement of the river to a narrow channel and the loss of storage capacity—are major reasons that flooding is increasing along the lower Mississippi River.

Novitzki (1985) analyzed the relationship between flood peaks and the percentage of basin area in lakes and wetlands. In the Chesapeake Bay drainage basin, where the wetland area was 4 percent, flood flow was only about 50 percent of that in basins containing no wetland storage. However, in Wisconsin river basins that contained 40 percent lakes and wetlands, spring streamflow was as much as 140 percent of that in basins that do not contain storage. This apparent anomaly is probably related to a

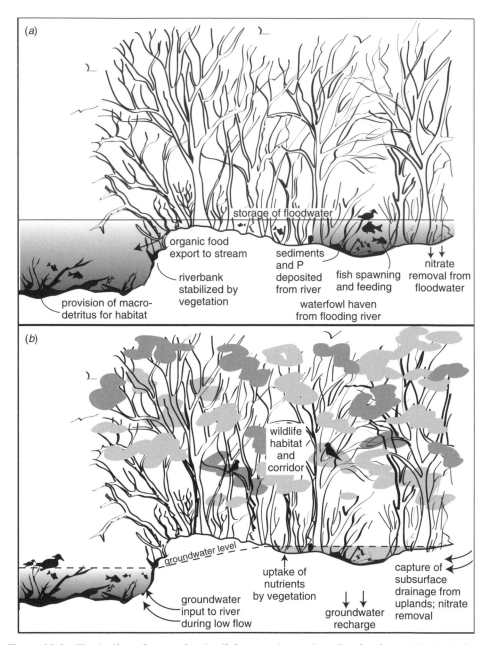

Figure 11.9   Illustration of several potential ecosystem values for riparian wetlands during (a) flood season and (b) dry season.

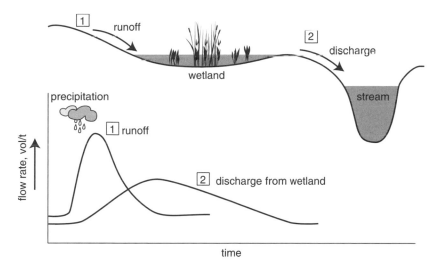

Figure 11.10   The general effect of wetlands on streamflow and stormwater runoff.

reduction in the proportion of precipitation that can infiltrate the soil and to a lack of additional storage capacity in lakes and wetlands that are already at full capacity during spring floods. Thus, the location of wetlands in the river basin can complicate the response downstream. For example, detained water in a downstream wetland of one tributary can combine with flows from another tributary to increase the flood peak rather than to desynchronize flows.

Ogawa and Male (1983, 1986) used a hydrologic simulation model to investigate the relationship between upstream wetland removal and downstream flooding. Their study found that for rare floods—that is, those predicted to occur only once in 100 or more years—the increase in peak stream flow was significant for all sizes of streams when wetlands were removed. The authors concluded that the usefulness of wetlands in reducing downstream flooding increases with (1) an increase in wetland area, (2) the distance that the wetland is downstream, (3) the size of the flood, (4) the closeness to an upstream wetland, and (5) the lack of other upstream storage areas such as reservoirs.

Storm Abatement and Coastal Protection

Coastal wetlands absorb the first fury of ocean storms as they come ashore (Fig. 11.11). Salt marshes and mangrove wetlands act as giant storm buffers. This value can be seen in the context of marsh conservation versus development. Natural marshes, which sustain little permanent damage from these storms, can shelter inland developed areas. Buildings and other structures on the coast are vulnerable to storms, and hurricane and typhoon damage is increasing almost every year in the world. Inevitably, the public pays much of the cost of this damage through taxes for public assistance, rebuilding public services such as roads and utilities, and federally guaranteed insurance.

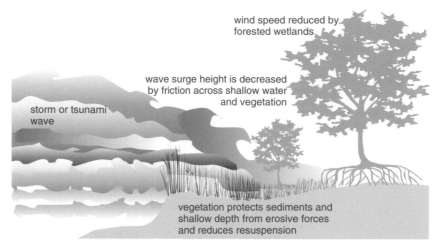

Figure 11.11 The general effects that coastal wetlands provide to buffer coastlines from tidal surges caused by hurricanes, typhoons, and tsunamis.

Two coastal disasters in the first decade of the 21st century poignantly illustrate in hindsight the value of coastal wetlands for coastal protection (see boxes). In both of these cases, as the memory of the disaster fades, there will be the tendency to go back to the ways things were done in the past.

Mangrove Swamps and the Indian Ocean Tsunami of December 2004

On December 26, 2004, an earthquake-caused tsunami produced unprecented damage and loss of life (estimated that 230,000 were killed or missing) around the entire Indian Ocean. The earthquake activity center was off the west coast of Sumatra, Indonesia, and so the greatest devastation occurred in that region. While no coastal defense system is capable of buffering areas that were hit with a 10-m-high wall of water, it is clear that the destruction of mangrove wetlands for shrimp farms and tourist meccas and the habitation of these areas by humans are at least partially responsible for the carnage. The mangrove swamps suffered significant temporary destruction as well (Fig. 11.12), but they have evolved to survive a violent seascape and will certainly restore themselves. The same cannot be said for human settlements that were built in former mangrove swamps.

One year prior to the Indian Ocean tsunami event, simulation models had illustrated that a wide (100 m) belt of dense mangrove trees (referred to as a ''greenbelt'') could reduce a tsunami pressure flow by more than 90 percent (Hiraishi and Harada, 2003). That information was not made public quickly, and the Indian Ocean tsunami happened with little to no warning. In the five

Figure 11.12 Mangrove wetlands and tidal creek in Phra Thong Island, Phang Nga, Thailand, before and after the December 26, 2004 Indian Ocean tsunami. The mangrove forests were significantly impacted, but their woody biomass had a role in significantly reducing the energy of the tidal tsunami that was 6 m high or more. *(Photographs by C. Conti, Naucratis Collection, permission provided courtesy of Monica Aureggi.)*

countries hit hardest by the Indian Ocean tsunami, at least 1.5 million ha of mangrove wetlands, or 26 percent of the mangrove cover, were destroyed between 1980 and 2000 (Check, 2005).

The protective role that mangrove wetlands provided during the Indian Ocean tsunami was illustrated in hindsight for a region along the southeast coastline in Tamil Nadu, India (Danielsen et al., 2005). In an area without mangroves and coastal *Casuarina* plantations, a sand spit was totally removed and parts of the local village was destroyed; there were "significantly less damaged" areas where mangroves and plantations were present. Danielsen et al. (2005) concluded that "conserving or replanting coastal mangroves and greenbelts should buffer communities from future tsunami events." There is hope that such a tsunami disaster will never occur again, but restoring mangrove swamps for coastal protection now has the attention of all countries that surround the Indian Ocean.

Hurricane Katrina of 2005 and New Orleans' Wetland Wet Suit

Hurricane Katrina struck the Louisiana coastland and the city of New Orleans, Louisiana, in late August 2005 with devastating results to lives and property (Fig. 11.13). One of the reasons for the extensive destruction is the fact that New Orleans and Louisiana are losing their deltaic wetlands due to land subsidence caused by natural and human effects (see discussions in Chapters 3 and 12). Studies over 50 years in Louisiana led to the conclusion that "New Orleans was becoming a more vulnerable city with each passing year" (Costanza et al., 2006). The formerly extensive salt marshes and other wetlands that used to surround New Orleans could have provided some coastal protection from the 6-m storm surge that overwhelmed New Orleans' levee system during Hurricane Katrina. But the wetlands have been lost at a rate of 65 km^2 per year since the beginning of the 20th century, after 6,000 years of gradual land building. Almost 4,800 km^2 of coastal wetlands have been lost since the 1930s alone (Day et al., 2005).

Since marsh plants hold and accrete sediments (Cahoon et al., 1995), often reduce sediment resuspension (Harter and Mitsch, 2003), and consequently maintain shallow water depths, the presence of vegetation contributes in two ways: (1) by actually decreasing surges and waves, and (2) by maintaining the shallow depths that also accomplish the same. Because wetlands indicate shallow water, the presence of wetland vegetation is also an "indicator" of the degree to which New Orleans and other human settlements are protected. While few experimental studies or modeling efforts have specifically addressed the effect of coastal marshes on storm surges, anecdotal data accumulated after Hurricane Andrew in 1992 in Louisiana suggested that the storm surge from that hurricane was reduced about 4.7 cm per km of marsh that it traveled over (Louisiana Coastal Wetlands Conservation Task Force and Wetlands Conservation and Restoration Authority, 1998). Extrapolating

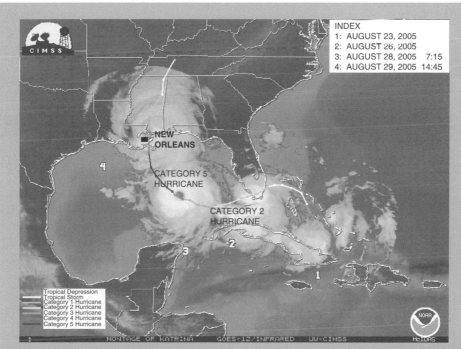

Figure 11.13   Path of Hurricane Katrina across Florida, the Gulf of Mexico, Louisiana, and Mississippi in August 2005. The hurricane crossed to the east of New Orleans on August 29, 2005, and a tidal wave caused by the hurricane caused extensive damage and loss of life in New Orleans and surrounding parishes. A more robust system of coastal wetlands and barrier beaches, many of which have been lost in the last century, would have provided more protection for the city. *(From the National Oceanic and Atmospheric Administration.)*

from this number, a storm tracking from the south of New Orleans through existing coastal marshes could have its surge reduced by 3.7 m if it crossed 80 km of marsh before reaching the city. It is not inappropriate to refer to the disappearing marshes around New Orleans as that city's wet suit. Without its wet suit, New Orleans will not survive.

Aquifer Recharge

Another value of wetlands related to hydrology is groundwater recharge. This function has received too little attention, and the magnitude of the phenomenon has not been well documented. Some hydrologists believe that, although some wetlands recharge groundwater systems, most wetlands do not. The reason for the absence of recharge is that soils under most wetlands are impermeable. In the few studies available, recharge occurred primarily around the edges of wetlands and was related to the edge:

volume ratio of the wetland. Thus, recharge appears to be relatively more important in small wetlands such as prairie potholes than in large ones. These small wetlands can contribute significantly to recharge of regional groundwater.

Water Quality

Wetlands, under favorable conditions, have been shown to remove organic and inorganic nutrients and toxic materials from water that flows across them. The concept of wetlands as sinks for chemicals was discussed in Chapter 5, and the practice of using wetlands for wastewater treatment and water quality improvement is discussed in detail in Chapter 13. Wetlands have several attributes that influence the chemicals that flow through them, whether the chemicals are naturally added or artificially applied:

1. A reduction in water velocity as streams enter wetlands, causing sediments and chemicals sorbed to sediments to drop out of the water column

2. A variety of anaerobic and aerobic processes in close proximity, promoting denitrification, chemical precipitation, and other chemical reactions that remove certain chemicals from the water

3. A high rate of productivity in many wetlands that can lead to high rates of mineral uptake by vegetation and subsequent burial in sediments when the plants die

4. A diversity of decomposers and decomposition processes in wetland sediments

5. A large contact surface of water with sediments because of the shallow water, leading to significant sediment–water exchange

6. An accumulation of organic peat in many wetlands that causes the permanent burial of chemicals

Aesthetics

A real but difficult aspect of a wetland to capture is its aesthetic value, often hidden under the dry term "nonconsumptive use values," which simply means that people enjoy being out in wetlands. There are many aspects of this kind of wetland use. Wetlands are excellent "biological laboratories," where students in elementary, secondary, and higher education can learn natural history first hand. They are visually and educationally rich environments because of their ecological diversity. Their complexity makes them excellent sites for research. Many visitors to wetlands use hunting and fishing as excuses to experience wildness and solitude, expressing that frontier pioneering instinct that may lurk in all of us. In addition, wetlands are a rich source of information about our cultural heritage. The remains of prehistoric Native American villages and mounds of shells or middens have contributed to our understanding of Native American cultures and of the history of the use of our wetlands. Many artists—the Georgia poet Sidney Lanier, the painters John Constable and John Singer Sargent, and many others who paint and photograph wetlands—have been drawn to them. Two artists—one a photographer and the other a painter—took a one-year excursion through the wetlands of the Louisiana delta in 2004 and 2005 (Lockwood and Gary, 2005). Their works, shown as exquisite photographs and paintings, have been shown in several museums throughout the United States.

Subsistence Use

In many regions of the world, the subsistence use of wetlands is extensive. There, wetlands provide the primary resources on which village economies are based. These societies have adapted to the local ecosystems over many generations and are integrated into them. Some of these cultures, including the Camarguais in France, the Louisiana Cajuns in the United States, and the Marsh Arabs in Iraq, are described in Chapter 1.

Regional and Global Values

The wetlands function of maintaining water and air quality influences a much broader scale than that of the wetland ecosystem. Wetlands may be significant factors in the global cycles of nitrogen, sulfur, and carbon. The natural supply of ecologically useful nitrogen comes from the fixation of atmospheric nitrogen gas (N_2) by a small group of plants and microorganisms that can convert it into organic form. Currently, ammonia is manufactured from N_2 for fertilizers, at more than double the rate of all natural fixation. Wetlands may be important in returning a part of this "excess" nitrogen to the atmosphere through denitrification. Denitrification requires the proximity of an aerobic and a reducing environment, such as the surface of a marsh, as well as a source of organic carbon, something abundant in most wetlands. Because most temperate wetlands are the receivers of fertilizer-enriched agricultural runoff and are ideal environments for denitrification, it is likely that they are important to the world's available nitrogen balance. The phenomenon of nitrogen enrichment of coastal waters causing "dead zones" or hypoxia (dissolved oxygen <2.0 mg/L) in the hypolimnion has already begun to occur worldwide. Wetlands have been recommended as a key ecosystem in providing a solution to this eutrophication (Mitsch et al., 2001; Mitsch and Day, 2006).

Sulfur is another element whose cycle has been modified by humans. The atmospheric load of sulfate has been greatly increased by fossil fuel burning. It is almost equally split between anthropogenic sources—chiefly caused by fossil fuel burning—and natural biogenic sources, of which salt marshes account for about 25 percent. When sulfates are washed out of the atmosphere by rain, they acidify oligotrophic lakes and streams. When sulfates are washed into marshes, however, the intensely reducing environment of the sediment reduces them to sulfides. Some of the reduced sulfide is recycled to the atmosphere as hydrogen, methyl, and dimethyl sulfides, but most of it forms insoluble complexes with phosphate and metal ions. These complexes can be more or less permanently removed from circulation in the sulfur cycle.

As discussed in Chapter 10, the global carbon cycle and wetlands are tightly linked. Wetlands, particularly northern peatlands, have stored enormous quantities of carbon in the peat. When these peatlands are protected and their water table is not affected, this carbon remains essentially in storage forever. When this peat is oxidized, whether by burning directly as a fuel or indirectly by altering the hydrology and causing drying and oxidation of the peat, the peatlands could become important sources of carbon dioxide to the atmosphere. Wetlands can be significant sinks of carbon if they are still building peat or accumulating carbon in their soil. This could be a significant advantage for tropical wetlands and for created and restored wetlands that are still building carbon storage in their soils compared to terrestrial systems that accumulate organic carbon in the soil slowly.

Quantifying Wetland Values

Efforts have been made to quantify the "free services" and amenities that wetlands provide to society for more than 30 years. Starting with the publication of "The Southern River Swamp—A Multiple-Use Environment" (Wharton, 1970) and "The Value of the Tidal Marsh" (Gosselink et al., 1974), a significant literature now exists on ascribing values to wetlands for the services they provide. Costanza et al. (1997) took these types of calculations one step further by estimating the public service functions of all the earth's ecosystems, including wetlands. That study and others have generated a new vocabulary on ecosystem values with terms such as *public service function*, *natural capital*, *environmental services*, and *ecosystem goods and services*. All of these terms mean essentially the same thing. Nature, including wetlands, provides values to humans, and these need to be recognized whenever wetlands are either threatened or conserved (Söderquist et al., 2000).

Just as with the hierarchical system described previously for wetland values, the valuation of wetlands can be described at three levels of ecological hierarchy—population, ecosystem, and biosphere (Fig. 11.14). This value also accrues to different segments of the economy. Ecological populations, generally harvested for food or fiber, are the easiest values to estimate and agree on, and they generally accrue to the landowners and local population. At the ecosystem scale, wetlands provide flood control, drought prevention, and water quality protection. These ecosystem values are real, but their quantification is difficult, and the benefits are generally regional and less specific to individual landowners. At the highest level of ecosystem worth, the biosphere, we know the least about how to estimate the values, and benefits accrue to the entire world.

Several approaches to the valuation of wetlands have been advanced. Because of the complexities described previously, there is no universal agreement about which approach is preferable. In part, the choice depends on the circumstances. Valuations fall broadly into two classes, ecological (or functional) evaluation and economic (or monetary) evaluation. The former are generally necessary before attempting the latter, because the valued ecological functions determine the monetary value.

Ecological Valuation

The ecological-valuation approach has been widely used as a means of forming a rational basis for deciding on different management options. Probably the best-developed

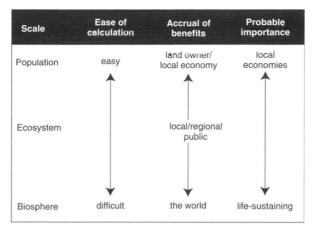

Scale	Ease of calculation	Accrual of benefits	Probable importance
Population	easy	land owner/ local economy	local economies
Ecosystem		local/regional public	
Biosphere	difficult	the world	life-sustaining

Figure 11.14   The ease of quantification, accrual, and probable importance of values for population, ecosystem, and biosphere. (After Mitsch and Gosselink, 2000a.)

procedures assess the relative value of wildlife habitats. E. P. Odum (1979b) suggested a general procedure, as follows:

a. Make a list of all the values that a knowledgeable person or panel can apply to the situation in question, and assign a numerical value of "1" to each.

b. Scale each factor in terms of a maximum level; for example, if 200 ducks per acre could be supported by a first-class marsh but only 100 are supported by the marsh in question, then the scaled factor is 0.5, or 50 percent of the maximum value for that item.

c. Weigh each scaled factor in proportion to its relative importance; for example, if the value 2 is considered 10 times more important to the region than the value 1, then multiply the scale value of 2 by 10.

d. Add the scaled and weighted values to obtain a value index. Because the numbers are only arbitrary and comparative, the index is most useful in comparing different wetlands or the same wetland under different management plans. It is desirable that each value judgment reflect the consensus of several "experts," for example, determined by the "Delphi method."

Habitat Evaluation Procedures

Table 11.5 shows an example from the Habitat Evaluation Procedure (HEP) of the U.S. Fish and Wildlife Service of the application of this technique to evaluate different development plans for a cypress–gum swamp ecosystem. The present value of the swamp for a representative group of terrestrial and aquatic animals was evaluated (baseline condition) using a habitat suitability index (HSI) based on a range of 0 to 1 for the optimum habitat for the species in question. The evaluation resulted in a

Table 11.5 Comparison of the impact of two management plans and a no-management control in a cypress–gum swamp[a]

Species	Baseline Condition	Future with Project Plan A[b]		Future with Project Plan B[c]		Future Without Project	
		50 Years	100 Years	50 Years	100 Years	50 Years	100 Years
TERRESTRIAL							
Raccoon	0.7	0.5	0.6	0.8	0.8	0.7	0.9
Beaver	0.7	0.2	0.2	0.4	0.3	0.6	0.4
Swamp rabbit	0.7	0.2	0.2	0.8	0.8	0.7	0.4
Green heron	0.9	0.2	0.1	0.8	0.9	0.9	1.0
Mallard	0.8	0.3	0.2	1.0	0.9	0.9	1.0
Wood duck	0.8	0.3	0.2	0.9	1.0	1.0	1.0
Prothonotary warbler	0.8	0.3	0.1	0.6	0.7	0.8	0.9
Snapping turtle	0.8	0.4	0.3	0.8	0.7	0.8	0.9
Bullfrog	0.9	0.3	0.2	0.8	0.9	1.0	1.0
Total terrestrial HSI	7.1	2.7	2.1	6.9	7.0	7.4	7.5
Mean terrestrial HSI	0.8	0.3	0.2	0.8	0.8	0.8	0.8
AQUATIC							
Channel catfish	0.3	0.3	0.4	0.4	0.4	0.4	0.4
Largemouth bass	0.4	0.2	0.3	0.7	0.8	0.4	0.4
Total aquatic HSI	0.7	0.5	0.7	1.1	1.2	0.8	0.8
Mean aquatic HSI	0.4	0.3	0.4	0.6	0.6	0.4	0.4

[a]Numbers in the tables are habitat suitability index (HSI) values, which have a maximum value of 1 for an optimal habitat.
[b]Channelization of water and clearing of swamp for agricultural development with a loss of 324 ha of wetland.
[c]Construction of levees around swamp for flood control with no loss of wetland area.
Source: Schamberger et al. (1979).

mean terrestrial HSI of 0.8 and a mean aquatic HSI of 0.4. This baseline condition was compared with the projected habitat condition in 50 and 100 years under three projected scenarios—Plan A, Plan B, and a no-project projection. The results suggest that Plan A would be detrimental to the environment, whereas Plan B would have no effect on terrestrial habitat values and would improve aquatic ones. Whether to proceed with either of these plans is a decision that requires weighing the projected environmental effects against the projected economic benefits of the project.

One often-neglected feature of the analysis is the effect of aggregating HSIs for different species. Although, overall, Plan B appears to be about equivalent environmentally to the no-project option, scrutiny of Table 11.5 shows that Plan B is expected to improve the habitat for swamp rabbits and large-mouthed bass but decrease its value for warblers and turtles. This kind of detailed scrutiny may be important because it indicates a change in the quality of the environment, but it is often neglected when the "apples and oranges" are combined into "fruit."

Wetland Evaluation Technique

Several wetland evaluation procedures have been developed that attempt to deal with two shortcomings of the habitat evaluation discussed previously by (1) evaluating all relevant goods and services (not just biotic ones) derived from the site and (2) incorporating a landscape focus. The Wetland Evaluation Technique (WET), once in favor with the U.S. Army Corps of Engineers but now not used, rates a broad range of functional attributes on a scale of high, moderate, and low. The result is a list of functions, each involving a quality rating for three attributes: (1) *social significance* assesses the value of a wetland to society in terms of its economic value, strategic location (e.g., upstream from an urban area that requires flood protection), or any special designations it carries (e.g., habitat for an endangered species); (2) *effectiveness* is the site's capacity to carry out a function because of its physical, chemical, or biological characteristics (e.g., to store floodwaters); and (3) *opportunity* refers to the opportunity of a wetland to perform a function to its level of capability (e.g., whether the upstream watershed is capable of producing floodwaters). The evaluator is charged with the task of weighing each function to get an integrated evaluation.

Hydrogeomorphic Analysis

The hydrogeomorphic (HGM) classification (Brinson, 1993a) also allows a quantification of the functions of wetlands. Its uniqueness lies in its quantification of natural wetland functions without regard to their significance to society. This is done by comparing the wetland of interest to a reference site that is characteristic of the same hydrogeomorphic class. Brinson et al. (1994) summarized the assessment procedure:

1. Group wetlands into HGM classes with shared properties (the classification is discussed in Chapter 8).

2. Define the relationship between hydrogeomorphic properties and the functions of wetlands. The goal is to select functions that are linked clearly and logically to wetland HGM properties and that have hydrologic, geomorphic, and ecological significance. This step represents the scientific basis for the presence of the function.

3. Develop functional profiles for each wetland class. These can range from descriptive narratives to multivariate data sets covering numerous sites.

4. Develop a scale for expressing functions within each wetland class, by using indicators and profiles from the reference wetlands of that class. These scales serve as benchmarks for each wetland class. Reference wetlands should include the full range of natural and human-induced variations due to stress and disturbance.

5. Develop the assessment methodology. The assessment relies on indicators to reveal the likelihood that the functions being evaluated are present in the wetland and depends on reference populations to scale the assessment. The reference wetlands are also used to set goals for compensatory mitigation.

Evaluating Alternatives with the HGM Technique—An Illustration from North Carolina

In an illustration of the method to estimate the impact of a project or restoration on wetland functions, Rheinhardt et al. (1997) apply the HGM method to evaluate mitigation strategies in mineral soil forested pine (*Pinus palustris*) flats in North Carolina. Fourteen variables were used to estimate the function of both study and reference wetlands (Table 11.6). Absolute values of some of the variables (e.g., tree density) are then translated into indices on a scale of 0.0 to 1.0 by comparing those functions to a reference wetland site. Such indices, in turn, are applied to model functions such as "maintain hydrologic regime" as in Table 11.7, and comparisons of human impact on wetlands can be assessed. Table 11.7 shows a hypothetical case in which an airport is destroying a wetland (with an overall loss index of 0.71 when compared to a nearby reference, which, by definition, has an index of 1.0), and two restoration alternatives are being considered. The analysis shows that restoration of a cropland back to a wetland (Restoration Alternative 1) would be a good alternative because the cropland has 0.0 value in maintaining hydrologic regime presently. Thus, the restoration is estimated to require only 1 ha of that cropland (gain =+0.71) for every hectare of wetland lost due to

Table 11.6 Field parameters used to estimate ecosystem function in a hydrogeomorphic assessment of forested wetlands in southeastern North Carolina

Variable	Description
HYDROLOGY/TOPOGRAPHY	
V_{DITC}	Lack of ditches nearby (<50 m)
V_{MICR}	Microtopographic complexity
HERBACEOUS VEGETATION	
V_{GRAM}	Percentage cover of graminoids
V_{FORB}	Percentage cover of forbs
CANOPY VEGETATION	
V_{TREE}	Total basal area for trees (m^2/ha; >10 cm DBH)
V_{TDEN}	Density of canopy trees (stems/ha; >10 cm DBH)
V_{TDIA}	Average tree diameter (m)
V_{CVEG}	Sorensen similarity index of canopy importance value
SUBCANOPY VEGETATION	
V_{SUBC}	Density of subcanopy (stems/ha)
V_{SDLG}	Percentage cover of trees and shrubs <1 m tall
V_{SVEG}	Sorensen similarity index of subcanopy importance value
LITTER/STANDING DEAD	
V_{LTR}	Litter depth (cm)
V_{SNAG}	Density of standing dead stems (stems/ha)
V_{CWD}	Volume of coarse woody debris (cm^3/ha)

Source: Rheinhardt et al. (1997).

Table 11.7 Predicted changes in hydrologic regime function resulting from a hypothetical airport construction on one wetland site and the comparison of the mitigation required for two different wetland restoration alternatives (variables are defined in Table 11.6)

	Reference Wetland		Wetland Being Destroyed — Now		Wetland Being Destroyed — After Airport		Restoration Alternative 1[a] — Now		Restoration Alternative 1[a] — After Restoration		Restoration Alternative 2[b] — Now		Restoration Alternative 2[b] — After Restoration	
Variable	Raw	Index	Raw	Index	Raw	Index	Raw	Index	Raw	Index	Raw	Index	Raw	Index
V_{TREE}	14.7	1.0	14.6	1.0	—	0.0	0.0	0.0	0.0	0.0	15.3	1.0	10.0	0.7
V_{SUBC}	12,550	1.0	13,314	1.0	—	0.0	0.0	0.0	6,963	0.5	18,402	0.5	9,800	0.8
V_{MICR}	2.5	1.0	2.5	1.0	—	0.0	0.0	0.0	2	1.0	4.2	1.0	4.2	1.0
V_{DITC}	1.0	1.0	0.5	0.5	—	0.0	0.0	0.0	1.0	1.0	0.5	0.5	1.0	1.0
Functional index[c]		1.0		0.71		0.0		0.0		0.71		0.64		0.91
Relative impact					−0.71				+0.71				+0.27	
Mitigation ratio[d]							0.71/0.71 = 1:1				0.71/0.27 = 2.6:1			

[a]Restoration of an agricultural field (former wetland) to a forested wetland.
[b]Restoration of a pine plantation to a forested wetland.
[c]Hydrologic functional index $= [((V_{TREE} + V_{SUBC} + V_{MICR})/3) \pm V_{DITC}]^{1/2}$.
[d]Ratio of wetland must be restored to area of wetland destroyed to achieve functional equivalent hydrologic regime.
Source: Rheinhardt et al. (1997).

the airport (loss = −0.71). This is a 1:1 mitigation ratio (the ratio of area of wetland restored to wetland lost).

Restoration of an existing pine plantation to a natural pine wetland, on the other hand, would probably be easier but, functionally, the plantation already has some of the desired values of wetlands (it rates a functional index of 0.64 before any restoration takes place and would rate an index of 0.91 after restoration, a net change of +0.27). So that restoration strategy would require 2.6 ha (0.71/0.27) of pine plantation to be restored for every hectare of wetland lost for the airport (mitigation ratio = 2.6:1).

Economic Evaluation

Evaluation systems that seek to compare natural wetlands to human economic systems usually reduce all values to monetary terms (thus losing sight of the apples and oranges). Conventional economic theory assumes that in a free economy, the economic benefit of a commodity is the dollar amount that the public is willing to pay for the good or service rather than be without it.

Although this characterization of value is reasonable under most conventional economic conditions, it leads to real problems in monetizing nonmarket commodities such as pure water and air and in pricing wetlands whose value in the marketplace is determined by their value as real estate, not by their "free services" to society. Consequently, attempts to monetize wetland values have generally emphasized the commercial crops from wetlands: fish, shellfish, furs, and recreational fishing and hunting, for which pricing methodologies are available. As E. P. Odum (1979b) and many others have pointed out, this kind of pricing ignores ecosystem- and global-level values related to clean air and water and other life-support functions. Even in the cases of market commodities from wetlands, available data are seldom adequate to develop reliable demand curves.

Economists recognize four more or less independent aspects of "value" that contribute to the total. These aspects are (1) *use value*—the most tangible portion of total value derived from identifiable direct benefits to the individual; hunting, harvesting fish, and nature study are examples; (2) *social value*—those amenities that accrue to a societal group rather than an individual; examples are improved water quality, flood protection, and the maintenance of the global sulfur balance; (3) *option value*—the value that exists for the conservation of perceived benefits for future use; and (4) *existence value*—the benefits deriving from the simple knowledge that the valued resource exists—irrespective of whether it is ever used. For example, the capacity of an extant wetland to conserve biological diversity is an existence value. As we have seen, use value is the easiest to estimate. The other three values, which are more difficult to quantify and also generally reflect longer-term viewpoints, have been addressed by economists using alternative methods, as illustrated next.

Willingness-to-Pay Methods

In the absence of a well-developed free-market alternative, pricing methodologies have been applied. One of these, "willingness-to-pay," establishes a more or less hypothetical (contingency) market for nonmarket goods or services. *Willingness-to-pay* or, more accurately, *net* willingness-to-pay is "the amount society would be willing to pay to produce and/or use a good beyond that which it actually does pay" (Scodari, 1990). The principle is illustrated as follows: Suppose a fisherman were willing to pay $30 a day to use a particular fishing site but had to spend only $20 per day in travel and associated costs. The net benefit, or economic value, to the fisherman of a fishing day at the site is not the $20 expenditure but the $10 difference between what he was willing to spend and what he had to spend. If the fishing opportunity at the site were eliminated, the fisherman would lose $10 worth of satisfaction fishing; the $20 cost that he would have incurred would be available to spend elsewhere. In the case of commercial goods such as harvested fish, the total value of a wetland is the sum of the net benefit to the consumer plus the net benefit to the producer (the fisherman).

The evaluation of willingness-to-pay has been carried out in several ways, including:

1. *Gross expenditure.* This approach evaluates the total expenditures for a specific activity (e.g., the willingness of hunters to spend dollars to travel to a site, buy equipment, and rent hunting easements) as a measure of the value of

the wetland site. Aside from the fact that this method measures only one aspect of the total value of a wetland, it is beset with serious methodological difficulties.

2. *Travel cost.* In this method, the costs of travel from different locations to a common property resource such as a wetland site are used to create a demand curve for the goods and services of that site. The total net value of the site is represented by the integral of the demand function. This approach has been used most effectively for recreational activity. It is not effective in estimating the demand for off-site services such as downstream water quality enhancement or flood mediation.

3. *Imputed willingness-to-pay.* In circumstances in which goods or services that depend on wetlands are produced and are recognized in the marketplace (e.g., commercial fish harvests), the values of these goods can be interpreted as measures of society's willingness-to-pay for the productictivity of the wetlands and, hence, for the wetland itself (Farber and Costanza, 1987). This method is limited to specific products or services and does not cover the entire range of wetland values.

4. *Direct willingness-to-pay (contingent value).* This method uses direct surveys of consumers to generate willingness-to-pay values for the entire range of potential wetland benefits. This method can separate public from private benefits, derive marginal values rather than average ones, and deal with individual sites or the entire resource base (Bardecki, 1987). However, an individual's willingness-to-pay for a wetland "service" is probably strongly influenced by his or her understanding of the ecology and functions of wetlands. Thus, the contingent-value method, like the other approaches, has both strengths and weaknesses.

Opportunity Costs

A second approach to resource evaluation in the absence of a free-market model is the *opportunity cost* approach. In general terms, the opportunity cost associated with a resource is the net worth of that resource in its best alternative use. For example, "the opportunity cost of conserving a wetland area is the net benefit which might have been derived from the best alternate use of the area which must be foregone in order to preserve it in its natural state" (Bardecki, 1987). Because determining the opportunity cost associated with wetland conservation would require the evaluation of each wetland service as well as the identification and valuation of the best alternative use, in practice, a comprehensive evaluation of the opportunity cost of wetland conservation is far from possible. Nevertheless, it may represent a useful approach to the valuation of specific wetland functions.

Replacement Value

If one could calculate the cheapest way of replacing various services performed by a wetland and could make the case that those services would have to be replaced if the

Table 11.8   Some replacement technologies for societal support values provided by wetlands

Societal Support	Replacement Technologies
PEAT ACCUMULATION	
Accumulating and storing organic matter (peat)	Artificial fertilizers
	Artificial flooding
HYDROLOGIC FUNCTIONS	
Maintaining drinking water quality	Water transport
	Pipeline to distant source
Maintaining groundwater level	Well-drilling
	Saltwater filtering
Maintaining surface water level	Dams for irrigation
	Pumping water to dam
	Irrigation pipes and machines
	Water transport for domestic animals
Moderation of water flows	Regulating gate
	Pumping water to stream
BIOGEOCHEMICAL FUNCTIONS	
Processing sewage; cleansing nutrients and chemicals	Mechanical sewage treatment
	Sewage transport
	Sewage treatment plant
	Clear-cutting ditches and stream
Maintaining drinking water quality	Water quality inspections
	Water purification plant
	Silos for manure from domestic animals
	Nitrogen filtering
	Water transport
Filter to coastal waters	Nitrogen reduction in sewage treatment plants
FOOD CHAIN FUNCTIONS	
Providing food for humans and domestic animals	Agriculture production
	Import of food
Providing cover	Roofing materials
Sustaining anadromous trout populations	Releases of hatchery-raised trout
	Farmed salmon
Sustaining other fish species and wetland-dependent flora and fauna	Work by nonprofit organizations
Species diversity; storehouse for genetic material	Replacement not possible
Bird watching, sport fishing, boating, and other recreational values	Replacement not possible
Aesthetic and spiritual values	Replacement not possible

Source: Folke (1991).

wetland were destroyed, then the figure arrived at would be the "replacement value." Some of the replacement technologies that might be necessary to replace services provided by wetland processes are listed in Table 11.8. A sample calculation of the replacement cost method is shown in Table 11.9. In this example, a fish hatchery is used to calculate fishery production, a flood reservoir to calculate flood and drought

Table 11.9   Estimated value of 770-ha riparian wetlands along the Kankakee River, northeastern Illinois, estimated by replacement value approach and by energy analysis

REPLACEMENT COST APPROACH

		Total Value
Ecosystem Function (Replacement Technology)	$/Year	
Fish productivity (fish hatchery)	$91,000	
Flood control/drought prevention (flood control reservoir)	$691,000	
Sediment control (sediment dredging)	$100,000	
Water quality enhancement (wastewater treatment)	$57,000	
Total replacement cost	$939,000	
Value/area $939,000 yr^{-1}/770 ha =		US$1,219 ha^{-1} yr^{-1}

ENERGY FLOW APPROACH

	Number	Total Value
Energy Flow Parameter		
Ecosystem gross primary productivity (kcal m^{-2} yr^{-1})	20,000	
Energy quality conversion, (kcal GPP/kcal fossil fuel)	20	
Energy conversion in U.S. economy (kcal fossil fuel /U.S.$)	14,000	
Value/area =		
$\dfrac{20{,}000\ \text{kcalGPPm}^{-2}\text{yr}^{-1} \times 10{,}000\ \text{m}^{-2}\ \text{ha}^{-1}}{20\ \text{kcal GPP/kcal ff} \times 14{,}000\ \text{kcal ff/\$}} =$		US$714 ha^{-1} yr^{-1}

Source: Mitsch et al. (1979c).

control, sediment dredging to estimate sediment retention, and wastewater treatment to estimate water quality enhancement.

This approach has the merit of being accepted by some conventional economists. For certain functions, it gives very high values compared with those of other valuation approaches discussed in this section. For example, the tertiary treatment of wastewater is extremely expensive, as is the cost of replacing the nursery function of marshes for juvenile fish and shellfish. Serious questions, however, have been raised about whether these functions would be replaced by treatment plants and fish nurseries if the wetlands were destroyed. Some ecologists and economists argue that in the long run, either the services of wetlands would have to be replaced or the quality of human life would deteriorate. Other individuals argue that this assertion cannot be supported in any convincing manner.

Energy Analysis

A completely different approach uses the idea of energy flow through an ecosystem or the similar concept of "embodied energy." The concepts of embodied energy (Costanza, 1980), and *emergy* (= energy memory; H. T. Odum, 1988, 1989, 1996), both attempt to estimate the total energy required to produce something and then translating the energy analysis into economic terms. It is assumed to be a valid index of the totality of ecosystem functions and is applicable to human systems as well. In this way, both natural and human systems can be evaluated on the basis of one common currency—energy. Because there is a clear relationship between energy and money in our society, energy flow can be translated to the more familiar currency of dollars at the end of the evaluation.

A simple calculation using the annual energy flow of a bottomland forested wetland in Illinois is illustrated in Table 11.9. Here, an estimated ecosystem energy flow (gross primary productivity = GPP) of 20,000 kcal m^{-2} yr^{-1} yielded an estimated value of \$714 ha^{-1} yr^{-1}. The energy analysis method gave a number about 60 percent of the replacement value. The concept of energy "quality" was used in this calculation to differentiate between energy flow in the ecosystem (based on gross primary productivity) and energy flow in the human-based fossil fuel economy. This is a precursor to the current approach of emergy discussed below.

In a comparison of economic evaluation approaches in Sweden, the replacement cost of a 2.5-km^2 peatland lake on the island of Gotland was estimated to be about \$1,600 ha^{-1} yr^{-1}, with most of the cost involved in replacing the biogeochemical processes of the wetland (52–82 percent of the cost) and less involved in replacing the hydrologic processes (7–40 percent) and food chain functions (8–11 percent) (see Table 11.8). When the energy cost of the economic replacements was compared with the energy lost when the wetland was lost, the results were remarkably similar. If the 2.5-km^2 wetland was lost, the economic replacement cost in energy terms would range from 3.5 to 12 $\times$ 10^9 kcal yr^{-1}; the ecosystem loss calculation ranged from 13 to 18 $\times$ 10^9 kcal yr^{-1}. In this case, the energy analysis method gave a slightly higher estimate of the energy (and, hence, the money) cost of wetland loss than did the replacement cost method.

Louisiana Coastal Wetlands: Comparing Energy and Economic Analyses

Costanza (1984) and Costanza et al. (1989) showed that the economist's willingness-to-pay approach and energy analysis converge to a surprising degree for coastal marshes in Louisiana, although both methods result in a great deal of uncertainty (Table 11.10). The energy analysis approach yielded higher wetland values, but the ranges overlap. The sensitivity of both conventional and energy analysis methods to the choice of a discount rate, which has been vital for decades in the outcome of cost–benefit studies, is also demonstrated in this comparison. The energy analysis method is based on using the total amount of energy captured by natural ecosystems as a measure of their ability to do useful work (for nature and, hence, for society). The gross primary productivity (GPP) of representative coastal marsh systems, which ranges from 48,000 to 70,000 kcal m^{-2} yr^{-1}, is converted to monetary units by multiplying by a conversion factor of 0.05 units fossil fuel energy/unit GPP energy and dividing by the energy/money ratio for the economy (15,000 kcal fossil fuel/1983 \$). These calculations resulted in an estimate of annual coastal wetland value of about \$1,560 ha^{-1} yr^{-1}, which, when converted to present value for an infinte series of payments, yields the

Table 11.10   Estimates of wetland values in $/ha of Louisiana coastal marshes based on willingness-to-pay and energy analysis at two discount rates

Method	Discount Rate	
	3 percent	8 percent
Willingness-to-pay		
Commercial fishery	$2,090	$783
Fur trapping (muskrat and nutria)	991	373
Recreation	447	114
Storm Protection	18,653	4,732
Total willingness-to-pay value	$22,181	$6,002
Energy analysis	$42,000–$70,000	$16,000–$26,000
Best estimate	$22,000–$42,000	$6,000–$16,000

Source: Costanza et al. (1989).

range of capitalized values of $16,000 to $70,000 ha^{-1} for the discount rates used in Table 11.10.

In comparison, the willingness-to-pay estimates reflect the assessment that a reasonable range of wetland value for coastal Louisiana is between $6,000 and $22,000 ha^{-1}, depending on the discount rate applied to determine the present value. Costanza et al. (1989) used this range from the willingness-to-pay and energy analysis approaches to suggest that the annual loss of Louisiana coastal wetlands is costing society from $77 million to $544 million per year.

Emergy analysis is a variation on the energy analysis (both were pioneered by H. T. Odum at the University of Florida in the 1970s and 1980s). The key to emergy analysis is the determination of transformities, or ratios that allow the conversion of one form of energy to another, as was done previously for gross primary productivity and fossil-fuel energy (see Table 11.9). These ratios are usually expressed in terms of solar emjoules (sej) per joule (or similar unit) of base energy or ecosystem flow. An example of an emergy flow analysis used for wetlands is illustrated in the box below.

Emergy Analysis of Wetlands in Florida

A comparison was made among three types of wetlands in Florida—a forested wetland, a shrub-scrub wetland, and a marsh (Bardi and Brown, 2001)—to compare their ecosystem services. The services considered were not only gross primary productivity, but also infiltration of water to the groundwater (groundwater recharge) and transpiration. In addition, the storages of natural

Table 11.11 Results of emergy analysis comparing the economic value of three types of wetlands in Florida for their environmental services and natural capital (Values are US$/ha)

Ecosystem Type	Environmental Services[a]	Natural Capital[b]	Total value
Forested Wetland	$231,880	$1,322,723	$1,554,603
Shrub/Scrub Wetland	$31,831	$1,075,536	$1,107,366
Freshwater Marsh	$13,173	$626,645	$639,817

[a]environmental services include gross primary productivity, infiltration, and transpiration
[b]natural capital includes live biomass, peat, water, and basin structure (formed by geological processes)
Source: Bardi and Brown (2001).

capital (stored water, biomass, and basin structure) were added. When all of the environmental services and natural capital are first converted to solar emjoules (sej) and then to dollars, (Table 11.11) the data suggest that a 1-ha forested wetland is approximately 2.4 times more valuable than a similar sized marsh. Furthermore, the analysis points out that the wetland values range from $640,000 to $1.5 million per ha. At the time, the going rate for buying wetland mitigation credit in Florida was $187,000/ha. Thus the rate being paid for mitigation credit was one-third to one-eighth that of the values calculated for these wetlands. Wetlands were being sold to destruction at too low a price.

Energy and emergy analyses, although imprecise because of the several estimates used, are more satisfying to many scientists than conventional cost-accounting methods, because they are based on the inherent function of the ecosystem, not on perceived values that may change from generation to generation and from location to location.

Valuing Ecosystem Goods and Services

Costanza et al. (1997) wrote a seminal paper on the value of the goods and services from ecosystems. That study used ecosystem unit estimators that showed that wetlands are 75 percent more valuable than lakes and rivers, 15 times more valuable than forests, and 64 times more valuable than grasslands and rangelands (Table 11.12). Only coastal estuaries (which include salt marshes and mangrove swamps) had higher unit values than wetlands. The overall value of ecosystems of the world, updated from the Costanza paper to year 2000 U.S. dollars, was estimated to be $18 trillion to $61 trillion, with a rough average of $38 trillion (Balmford et al., 2002).

Balmford et al. (2002) argued that the *net marginal benefits* of ecosystems should be estimated rather than the aggregated numbers developed by Costanza

Table 11.12   Estimated unit values of ecosystems

Ecosystem	Unit Value (U.S.\$ ha^{-1} yr^{-1})
Estuaries	22,832
Wetlands	14,785
Lakes/rivers	8,498
Forests	969
Grasslands	232

Source: Costanza et al. (1997).

et al. (1997), which often were simple replacement values. The net marginal benefit is the difference between values of relatively intact ecosystems and the values to humans of the same ecosystems converted to human use. After investigating over 300 case studies, Balmford et al. (2002) came up with only five studies worldwide where economic estimates were available for both conditions—intact ecosystems and the same landscape heavily managed. Two of those five case studies were of wetlands (Fig. 11.15). An economic analysis of a mangrove swamp in Thailand showed that conversion of a swamp to aquaculture made economic sense in the short term, but in the long term, the total economic value of an intact mangrove swamp was \$60,400, about 3.6 times that of the value of converting the swamp to shrimp aquaculture. The values provided by the natural mangrove swamp included timber, charcoal, nontimber forest products, offshore fisheries, and storm protection.

In a similar comparison, a freshwater marsh in Canada was found to have a total economic value of \$8,800/ha, about 2.4 times the value realized by converting the wetland to intensive agriculture. Here, the major values of the natural marsh were for sustainable hunting, fishing, and trapping.

Problems and Paradoxes of Quantifying Wetland Values

Regardless of which kind of ecosystem evaluation is used, several generic problems and paradoxes to quantifying wetland values need to be recognized and addressed.

1. *The term "value" is anthropocentric; hence, assigning values to different natural processes usually reflects human perceptions and needs rather than intrinsic ecological processes.* Because wetlands are multiple-value systems, the evaluator is often faced with the problem of comparing and weighing different commodities. For example, a fresh marsh area is more valuable for waterfowl than a salt marsh area, but the salt marsh may be much more valuable as a fish habitat. Which is rated higher depends on the value judgment made by the evaluator, a judgement that has nothing to do with the intrinsic ecological viability of either area. This is the old "apples versus oranges" problem. The decision, to some extent, reflects a matter of preference. Furthermore, in most wetland evaluations, evaluators are not concerned with single commodities. Instead, they wish to approximate the

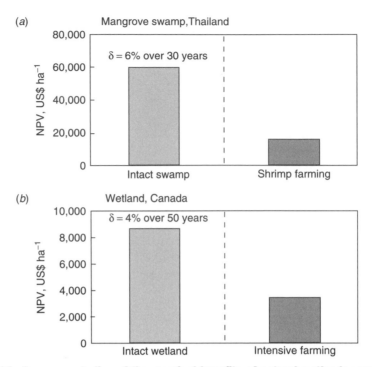

Figure 11.15 Two case studies of the marginal benefits of natural wetlands versus conversion of the wetland to intensive human industry: (a) mangrove system in Surat Thani, southern Thailand, and (b) freshwater marshes in Canada. ∂ indicates discount rates, and NPV indicates net present value in year 2000 US$/ha. *(From Balmford et al., 2002.)*

overall value of an area—that is, the value of the whole fruit basket, rather than the apples, oranges, and pears. Complexity is added when the concern is to compare the value of a natural wetland with the same piece of real estate proposed to be used for economic development—a dammed lake, a parking lot, or an oil well. In that case, the comparison is not between apples and oranges but between apples and computer chips or a fruit basket and electrical energy. Conventional economics solves the comparison problem by reducing all commodities to a single index of value—dollars. This is difficult when some of the commodities are natural products of wetlands that do not compete in the marketplace.

2. *The most valuable products of wetlands are public amenities that have no commercial value for the private wetland owner.* The wetland owner, for example, has no control over the harvest of marsh-dependent fish that are caught in the adjacent estuary or even offshore in the ocean. The owner does not control and usually cannot capitalize on the ability of the wetland to purify wastewater, and certainly cannot control its value for the global nitrogen balance. Thus, there is often a strong conflict between what a

private wetland owner perceives as his or her best interests and the public's best interests. In coastal Louisiana, a marsh owner can earn revenues of perhaps $50 per hectare annually from the renewable resource of his or her marsh by leasing it for hunting and fur trapping. In contrast, depending on where it is situated, a wetland may be worth $100,000 per hectare as a housing development or as a site for an oil well. Riparian wetlands in the Midwest and lower Mississippi River basin yield little economic benefit to the owner for the flood mitigation and water quality maintenance that they provide for downstream populations. Yet if cleared and planted in corn or soybeans, the land will provide economic benefits to the owner. Many of the current regulations that govern wetlands were initiated to protect the public's interest in privately owned wetlands.

3. *The ecological value, but not necessarily the economic value, of a wetland depends on its context in the landscape.* This applies to both scale issues and location within the landscape. Wetland values are different, accrue to different stakeholders, and probably have different importance, depending on the spatial scale on which we base our estimations, and their location in the watershed. Small wetlands are often important locally for their support of populations of ducks, geese, migratory birds, and fish and shellfish. This is true of prairie pothole wetlands, for example. Linear riparian wetlands that lie along rivers and streams, although narrow, may be extremely important because they remove nutrients and soil from upland runoff, thus helping maintain stream-water quality. Large areas of wetland—for example, the great expanses of peatlands in northern latitudes—may have global value as regulators of global climate.

4. *The relationships among wetland area, surrounding human population, and marginal value are complex.* Conventional economic theory states that the less there is of a commodity, the more valuable the remaining stocks. This generalization is complicated in nature because different natural processes operate on different scales. This is an important consideration in parklands and wildlife preserves, for example. Large mammals require large ranges in which to live. Small plots below a certain size will not support *any* large mammals. Thus, the marginal-value generalization falls apart because the marginal value ceases to increase below a certain plot size. In fact, it becomes zero. In wetlands, the situation is even more complex, because they are open ecosystems that maintain strong ties to adjacent ecosystems.
Therefore, two factors that govern the ecological value (and, hence, the value to society) of the wetland are (1) its *interspersion* with other ecosystems—that is, its place in the total regional landscape—and (2) the *degree of linkage* with other ecosystems. A small wetland area that supports few endemic organisms may be extremely important during critical times of the day or during certain seasons for migratory species that spend only a day or a week in the area. A narrow strip of riparian wetland along a stream that

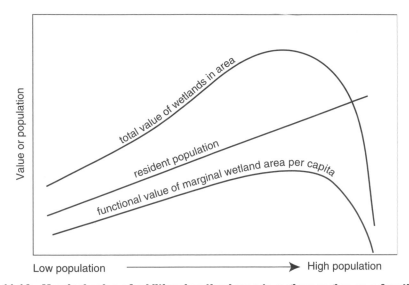

Figure 11.16 Marginal value of additional wetland area to a given region as a function of nearby human population (from Mitsch and Gosselink, 2000a, based on King and Herbert, 1997). Total value is a product of population times functional value per capita. Functional marginal value of additional wetland per capita does initially increase as population increases and wetlands are becoming rare. At some point of population density, however, these functions become taxed with pollution, lost corridors, and so on, and value drops precipitously for additional population increase.

amounts to very little acreage may efficiently filter nutrients out of runoff from adjacent farmland, protecting the quality of the stream water. Its value is related to its interspersion in the landscape, not to its size. A typical curve of marginal wetland value versus surrounding human population (Fig. 11.16) shows that, as demand increases for the goods and services of a wetland area, or supply decreases as the wetland is encroached upon, the marginal value at first increases. However, as wetland functions are compromised, either by overuse (e.g., overfishing or hunting, tree harvesting, increased fertilizer runoff, water-level stabilization) or by reduction in wetland area (which may, for example, eliminate wide-ranging predators or reduce water treatment capacity), the marginal value declines because the wetland can no longer produce the services that make it useful ecologically and economically. These kinds of considerations have only recently begun to be addressed in a quantitative way, and the methodology is not well developed.

5. *Commercial values are finite, whereas wetlands provide values in perpetuity.* Wetland development is often irreversible. The time frame for most human projects is 10 to 30 years. Private entrepreneurs expect to recoup their investments and profits in projects within this time frame and seldom consider longer-term implications. Even large public-works projects

such as dams for energy generation seldom are seen in terms longer than 50 to 100 years. The destruction of natural areas, however, removes their public services forever. Often, especially for wetlands, the decision to develop is irreversible. If an upland field is abandoned, it will gradually revert to a forest; but once a wetland is drained and developed, it is usually lost forever because of associated changes in the hydrologic regime of the area. For example, in Louisiana and elsewhere, marsh areas were diked and drained for agriculture early in the 20th century. Many of these developments have subsequently been abandoned. They did not revert to wetlands, but are now large, shallow lakes, distinguishable by their straight edges.

6. *A comparison of economic short-term gains with wetland value in the long term is often not appropriate.* Wetland values, even when multiple functions are quantified, often cannot compete with short-term economic calculations of high-economic-yield projects, such as commercial development or intensive agriculture. This is especially true because economic analyses typically discount the value of future amenities. The issue of wetland conservation versus development has an intergenerational component. Future generations do not compete in the marketplace, and decisions that will affect the public resources they inherit are often made without regard to their interests.

7. *Estimates of value, by their nature, are colored by the biases of individuals and society and by the economic system.* There are good reasons why we wish to protect nature; developed countries, having taken care of the basic needs of their citizens, are particularly involved in protecting ecosystems, including wetlands, for their aesthetic as well as more functional attributes, not all of which translate into direct economic benefit. Other cultures, where the basic needs of food and shelter cannot be taken for granted, have different views of the economics of wetlands. Many cultures do live in and among wetlands and use them for daily subsistence—the production of food and fiber. Yet they generally leave the normal wetland functions intact. The values that we ascribe to wetlands are not separate from the institutions and cultures from which we come.

8. *A landscape view of wetlands is required to make intelligent decisions about the values of created and managed wetlands.* A paradox of assigning values to ecosystems is that, unless we take a landscape view, it can be argued that we should replace a less valuable system (e.g., a grassland) with another more valuable one (e.g., a wetland). A straightforward economic analysis from the data shown in Table 11.12 would thus argue for the replacement of forests and prairies with wetlands. While this physical substitution is not possible in most instances because climatic and hydrologic variables determine what ecosystem occurs in a particular landscape, on a microscale it is not only possible to substitute wetlands for grasslands and upland forests, but it is frequently done to meet regulatory requirements for wetland mitigation in the United States. Many question whether the created wetland can achieve

the same functional and, hence, economic value as did the original ecosystem at that site. Some argue that these created ecosystems are doomed to failure, whereas others are more optimistic that these systems do provide real, measurable value that might even exceed what was at the site previously.

If one ignores the technical problems of functional ecosystem substitution, the idea attracts many people because of the common perception among economists that any commodity can be replaced. As scarcity of one product drives the price up, the creativity of the free market will surely result in the development of a cheaper substitute. This is not true of ecosystems, however. Much of the value of an ecosystem, especially an open system such as a wetland, depends on its landscape context and the strong interactions among the parts of the landscape. Thus, the value of a riparian forest depends on its ecological links to the adjacent stream on one side and the upland fields or forest on the other.

Similarly, a coastal wetland derives much of its value from the movement of upstream and coastal ocean organisms, sediments, and chemicals into and out of the wetland. In this context, a landscape is like a tile mosaic: The individual tiles have little value in themselves. Rather, their value lies in their precise location and color within the entire mosaic. In the same way, the value of a wetland or a pond or a stream is in its location and ecological properties as they interact with other ecosystems to form a functional landscape.

A Faustian Bargain

Because of the many problems documented in this chapter related to valuation of natural ecosystems, many ecologists oppose their economic valuation, because it implies that natural systems can be equated in the marketplace to other market products. Attempts to place dollar values on natural ecosystems, however, such as those cited in this chapter, have raised public awareness of the high value of the goods and services of nature, and in this way helped in efforts to conserve natural resources. Thus, ecologists are caught in a "Faustian bargain with the devil," trying to make the case for natural resource values in the common currency of our civilization, while clearly documenting the reasons why natural ecosystem conservation should not depend on the operation of free-market forces. There is no easy answer to this dilemma.

Recommended Readings

Barbier, E. B., J. C. Burgess, and C. Folke. 1994. *Paradise Lost? The Ecological Economics of Biodiversity.* Earthscan Publications, London. 267 pp.

Costanza, R., R. d'Arge, R. de Groot, S. Farber, M. Grasso, B. Hannon, K. Limburg, S. Naeem, R. V. O'Neill, J. Paruelo, R. G. Raskin, P. Sutton, and M. van den Belt. 1997. "The value of the world's ecosystem services and natural capital." *Nature* 387:253–260.

Odum, H. T. 1996. *Environmental Accounting: Energy and Environmental Decision Making*. John Wiley & Sons, New York. 370 pp.

Söderquist, T., W. J. Mitsch, and R. K. Turner, eds. 2000. "The Values of Wetlands: Landscape and Institutional Perspectives." Special Issue of *Ecological Economics* 35:1–132.

Chapter **12**

Wetland Creation and Restoration

Loss rates of wetlands around the world and the subsequent recognition of wetland values have stimulated restoration and creation of these systems. Policies such as "no net loss" of wetlands in the United States have made wetland creation and restoration a veritable industry in that country. Wetland restoration involves returning a wetland to its original or previous wetland state, whereas wetland creation involves conversion of uplands or shallow open-water systems to vegetated wetlands. Wetland restoration and creation can occur for replacement of habitat, for coastal restoration, and for restoration of mined peatlands. Wetland mitigation banks may overcome many of the limitations of current approaches to replacing lost wetlands. Generally, wetland restoration and creation first involve establishment or reestablishment of appropriate natural hydrologic conditions, followed by establishment of appropriate vegetation communities. Although many of these created and restored wetlands have become functional, there have been some cases of "failure" of created or restored wetlands generally caused by a lack of proper hydrology. Creating and restoring wetlands should be based on the concept of self-design, whereby any number of native propagules can be introduced, but the ecosystem adapts and changes itself according to its physical constraints, and success should not solely be determined by specific plant and animal presence. Giving these systems sufficient time to carry out their self-design is another factor that is generally overlooked.

There are two general starting points for anyone interested in getting involved in wetland restoration and creation:

1. Learn and understand wetland science and its principles first.
2. Broaden your horizons beyond the field that you were trained in so that you resist the ever-present temptation to over-engineer, over-botanize, or over-zoologize the wetlands that you create and restore.

The principles and practices of wetland creation and restoration are based on wetland science (hydrology, biogeochemistry, adaptations, and succession). Our advice if you are interested in creating and restoring wetlands is to first become an expert in wetland science. Know how the real wetlands work first. That was the intent of the first eight chapters in this book. Only after you understand the function and structure of natural wetlands are you qualified to create and restore wetlands.

The second point is one that needs to be emphasized to all professions. Most of us have been taught in our lives and professions that we can improve upon nature. Indeed, human civilization is based on that premise. But when we are attempting to create or re-create natural ecosystems, Mother Nature is in control. In all situations of wetland creation and restoration, human contribution to the design of wetlands should be kept simple and should strive to stay within the bounds established by the natural landscape. As stated by Boulé (1988): "simple systems tend to be self-regulating and self-maintaining."

The literature on wetland creation and restoration has exploded like no other wetland topic since our last edition, and it is impossible for us to include all of the possible principles, case studies, and techniques in this chapter. Streever (1999) provides regional overviews and case studies of wetlands from around the world. A critique of the policies and techniques of wetland creation and restoration in the United States was published as NRC (2001). Some notable new papers discuss specific wetland creation and restoration projects for salt marshes (Alphin and Posey, 2000; Craft et al., 2002; Edwards and Proffitt, 2003; Callaway and Zedler, 2004; Peterson et al., 2005), mangrove swamps (Lewis, 2005), freshwater marshes (Atkinson et al., 2005; Anderson et al., 2005; Mitsch et al., 2005a,c; Anderson and Mitsch, 2006), peatlands (Gorham and Rochefort, 2003), and forested wetlands (Rodgers et al., 2004). Some details not included here are in books entitled *Ecological Engineering and Ecosystem Restoration* (Mitsch and Jørgensen, 2004) and *Wetland Creation, Restoration and Conservation: The State of the Science* (Mitsch, 2006).

Definitions

Several terms are frequently used in connection with the creation and restoration of wetlands. Precise definitions are important, and confusion about the exact meaning of wetland creation, restoration, and related terms is common (Lewis, 1990a). Bradshaw (1996) concurs that "we must be clear in what is being discussed." *Wetland restoration* refers to the return of a wetland from a disturbed or altered condition caused by human activity to a previously existing condition. The wetland may have been degraded or hydrologically altered, and restoration then may involve reestablishing hydrologic conditions to reestablish previous vegetation communities. *Wetland creation* refers to the conversion of a persistent upland or shallow water area into a wetland by human activity. *Wetland enhancement* refers to a human activity that increases one or more functions of an existing wetland. One type of created wetland, a *constructed wetland,* refers to a wetland that has been developed for the primary purpose of contaminant or pollution removal from wastewater or runoff. This last type of wetland is also referred to as a *treatment wetland* and is discussed in Chapter 13.

Significant efforts now focus on the voluntary restoration and creation of wetlands. Part of the interest in wetland creation and restoration stems from the fact that we are losing or have lost so much of this valuable habitat (see Chapter 3). Often interest is less voluntary and more in response to government policies, such as "no net loss" in the United States, that require the replacement of wetlands for those unavoidably lost. New Zealand, which has lost 90 percent of its wetlands, has major efforts underway to restore marshes and other wetlands in the Waikato River Basin on North Island and in the vicinity of Christchurch on South Island. In southeastern Australia, restoration of the Murray-Darling watersheds, particularly the riverine billabongs, has become a major undertaking, while coastal plain wetland restoration and creation are occurring in southwestern Australia.

There are concerted efforts to restore mangrove forests in the Mekong Delta of Vietnam, along South American coastlines where shrimp farming has destroyed thousands of hectares of mangroves, and around the Indian Ocean to provide tsunami and typhoon protection for coastal areas. Tidal marshes have been created along much of China's eastern coastline, and restoration is now occurring in the Yangtze Delta in Shanghai. Wetland restoration and creation are being proposed or implemented on very large scales to prevent more deterioration of existing wetlands (Everglades in Florida), to mitigate the loss of fisheries (Delaware Bay in Eastern United States), to reduce land loss and provide protection from hurricanes (Mississippi Delta in Louisiana), to stabilize a watershed and provide water quality improvement (Skjern River, Denmark), and to solve serious cases of overenrichment of coastal waters (Baltic Sea in Scandinavia; Gulf of Mexico in United States).

Mitigating Wetland Habitat Loss

Wetland protection regulations in the United States and now elsewhere have led to the practice of requiring that wetlands be created, restored, or enhanced to replace wetlands lost in developments such as highway construction, coastal drainage and filling, or commercial development. This is referred to as the process of "mitigating" the original loss, and these "new" wetlands are often called *mitigation wetlands* [NOTE: To *mitigate* means to make less harsh or harmful. The term "mitigation wetland" or "wetland mitigation" is therefore poor use of English. We should rather refer to "mitigating the loss of a wetland."] Perhaps it might be more appropriate to refer to these wetlands as *replacement wetlands*.

Figure 12.1 illustrates conceptually how success should be measured for replacement wetlands. *Legal success* involves a comparison of the lost wetland function and area with that which is gained in the replacement wetland. *Ecological success* should involve a comparison of the replacement wetland with a reference wetland (natural wetlands of the same type that may occur in the same setting or generally accepted "standards" of regional wetland function) (Wilson and Mitsch, 1996). Overall success would then be gauged by a combination of the legal and ecological comparisons.

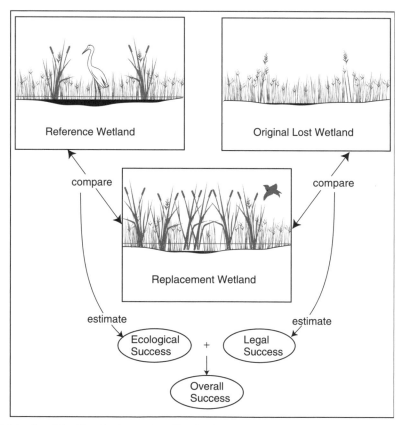

Figure 12.1   In mitigating the loss of wetlands through creation and restoration of wetlands, the proper procedure should involve comparison with both what has been lost (legal success) and with regional reference wetlands (ecological success). *(From Wilson and Mitsch, 1996.)*

While this model represents an ideal, the comparison involving both standards is rarely done.

In fact, the usual decision is based on the size of the wetland lost and little else. Replacement wetlands are designed to be at least the same size as the lost wetlands, but more often a *mitigation ratio* is applied so that more wetlands are created and/or restored than are lost. For example, a mitigation ratio of 2:1 means that two hectares of wetlands will be restored or created for every hectare of wetland lost to development. Considerable controversy exists, for example, in the United States, on the question as to whether wetland loss can be mitigated successfully or if it is essentially impossible (NRC, 2001). Robb (2002) reviewed several years' efforts on mitigating wetland loss in Indiana and suggested, based on failure rates of various wetland types, that there should be the following mitigation ratios: 3.5:1 for forested wetlands, 7.6:1 for wet meadows, 1.2:1 for freshwater marshes, and 1:1 for open-water systems.

On paper, the U.S. Army Corps of Engineer's implementation of the U.S. policy of "no net loss" of wetlands (see Chapter 14) appears to be working (Fig. 12.2). An estimated net gain of 10,000 ha/yr of wetlands and associated uplands in the United States over the most recent decade, from 1996 through 2005, were seen as a result of enforcement of the Clean Water Act through mitigation of wetland loss. This number is the result of the issuing of permits for the destruction of 10,000 ha/yr of wetlands and the creation, restoration, enhancement, or preservation of approximately 20,000 ha/yr of wetlands and associated uplands. Overall, the 13 years of record shown in Figure 12.2 shows that the United States has lost 122,000 ha of wetlands and "gained" 237,000 ha of mitigation credit, hopefully most of that as created or restored wetlands. There are two reasons why one should not be so euphoric about this "net gain" of wetlands. First, it is impossible to tell from these general numbers

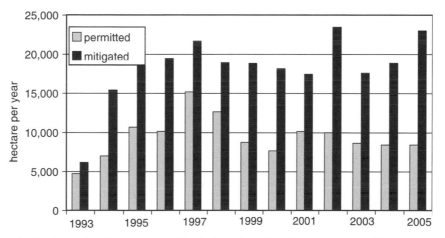

Figure 12.2 Approximate area of wetlands lost (permitted) and gained (mitigated) through U.S. Army Corps of Engineers' Section 404 dredge-and-fill permit program, fiscal (October–September) years 1993–2005 in the United States. *(U.S. Army Corps of Engineers.)*

just how "successful" this wetland trading has been, because few statistics exist on what functions were lost versus what functions were gained (NRC, 2001). Second, the estimated gain of 115,000 ha over 13 years does not make much of an impact on the loss of 47,000,000 ha of wetlands that occurred from presettlement time up to the 1980s in the United States.

Mitigation Banks

One of the more interesting strategies that the private sector and government agencies have developed to deal with the piecemeal approach to mitigation of wetland loss is the concept of a *mitigation bank*. A mitigation bank is defined as "a site where wetlands and/or other aquatic resources are restored, created, enhanced, or in exceptional circumstances, preserved expressly for the purpose of providing compensatory mitigation in advance of authorized impacts to similar resources" (*Federal Register*, November 28, 1995, "Federal Guidance for the Establishment, Use, and Operation of Mitigation Banks"). In this approach, wetlands are built in advance of development activities that cause wetland loss, and credits of wetland area can be sold to those who are in need of mitigation for wetland loss. Banks are seen as a way of streamlining the process of mitigating wetland loss and, in many cases, providing a large, fully functional wetland rather than small, questionable wetlands near the site of wetland loss. The mitigation bank can be set up with bonds ensuring compliance. Arrangements are easier for wetland mitigation banks to be managed "in perpetuity" through conservation easements or transfer of titles to resource agencies. Financial resources can be arranged ahead of time for proper monitoring of the wetland bank. Mitigation banks can be publicly or privately owned, although there is a potential conflict of interest if public agencies run mitigation banks. Public agencies could be involved in enforcing regulations on mitigating wetland loss and then steer permitees toward their own banks rather than to private banks.

By 2002, there were 219 mitigation banks, both private and public, covering 50,000 ha in 29 states in the United States (Spieles, 2005). At one time, there were 62 formal mitigation banks (proposed and operating) and hundreds of quasi-mitigation banks in Florida alone (Ann Redmond, personal communication). It appears that, if mitigation of wetland loss continues to be the nation's policy and if regulation of mitigation banks can be developed that is fair and uncomplicated, the use of mitigation banks to solve this "wetland trading" issue will continue to increase well through the 21st century.

Agricultural Land Restoration

For the past several decades in the United States, farm pond creation has been encouraged as a way of providing drinking water for domestic animals and other functions on the farm. Although individually quite small (about 0.2 ha), the total number of constructed ponds is large. Several years ago, ponds were being constructed at a rate of about 50,000 ponds per year. Marshes often develop around the perimeter of many of these ponds, while others have converted to marshes. Many of these ponds were built with large, shallow areas to attract waterfowl, and these shallow zones have become typical pothole marshes. Dahl (2006) estimated that between 1998 and 2004, wetland pond areas increased by 280,000 ha in the United States, a 12.6 percent increase. Most of this gain (141,000 ha) resulted on nonagricultural upland, while 29,000 ha were constructed on farmland. Many of the nonagricultural ponds are built in housing and commercial developments, especially in states like Florida, as storm-water runoff ponds. There are those who question the ecological value of these ponds: for example, some regulators do not like ponds because they have fish and therefore cannot support amphibians.

Conservation programs are now in place to encourage individual farmers in the United States to restore wetlands on their land. Both the Conservation Reserve Program (CRP) and the Wetlands Reserve Program (WRP) under the U.S. Department of Agriculture have led to significant areas of wetlands being restored or protected. CRP guidelines, announced in 1997, give increased emphasis to the enrollment and restoration of *cropped wetlands*—that is, wetlands that produce crops but serve wetland functions when crops are not being grown. The CRP also encourages wetland restoration, particularly through hydrologic restoration. In the CRP, participants voluntarily enter into contracts with the U.S. Department of Agriculture to enroll erosion-prone and other environmentally sensitive land in long-term contracts for 10 to 15 years. In exchange, participants receive annual rental payments and a payment of up to 50 percent of the cost of establishing conservation practices.

The Wetland Reserve Program (WRP) is another voluntary program established in 1990 and is specific for wetland restoration; it offers landowners the opportunity to protect, restore, and enhance wetlands on their property and provides funds for the farmer to do so. The USDA Natural Resources Conservation Service (NRCS) provides technical and financial support to help landowners. The WRP options to protect, restore, and enhance wetlands and associated uplands include permanent easements, 30-year easements, or 10-year restoration cost-share agreements. As of 2004, approximately 660,000 ha of wetlands and adjacent uplands have been enrolled in the WRP in the United States, with the most intense activity in the lower Mississippi River basin and Florida (Fig. 12.3). Most of the enrolled lands are flood-prone, and each project averages about 70 ha in size.

Forested Wetland Restoration

There is less experience with forested wetland restoration and creation compared to herbaceous marshes, although these wetlands have been lost at alarming rates, particularly in the southeastern United States. Forested wetland creation and restoration are different from marsh creation and restoration, because forest regeneration takes decades rather than years to complete, and there is more uncertainty about the results.

Much riparian forest restoration in the United States has centered on the lower Mississippi River alluvial valley, where more than 78,000 ha were reforested

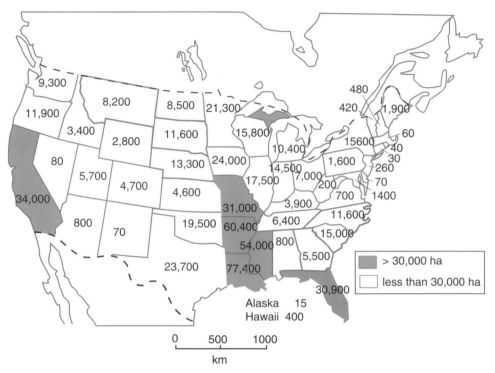

Figure 12.3 Area of wetlands and adjacent uplands voluntarily enrolled by farmers in Wetlands Reserve Program (WRP) in the conterminous United States through 2003. Numbers are in hectares for each state; shading represents states with more than 30,000 ha enrolled in the WRP. National total as of that year was almost 600,000 ha. *(From U.S. Department of Agriculture, Natural Resources Conservation Service.)*

by federal agencies over the 10-year period from 1988 to 1997 (King and Keeland, 1999), primarily with bottomland hardwood species and, to a lesser extent, deepwater swamp species. This is a small contribution to the restoration of this alluvial floodplain, where 7,200,000 ha of bottomland hardwood forest were estimated to have been lost (Hefner and Brown, 1985).

Hydrologic and Water Quality Restoration with Wetlands

Lines often blur between wetlands created and restored for habitat restoration and those restored for water quality and hydrology improvement. In fact, most wetlands that are restored or created are done so for both habitat and water quality values. One of the largest freshwater wetland restorations in the world is being carried out in the Florida Everglades to restore, at least to some degree, the natural hydrologic conditions, at least in the Everglades that are left (see Case Study 1). Another example of a proposed watershed restoration is a large-scale wetland and riparian forest restoration and creation, on the order of millions of hectares, being proposed to help solve a major coastal pollution problem in the Gulf of Mexico (Mitsch et al., 2001; Mitsch and Day, 2006), see Case Study 2.

Restoration done for water quality improvement also has the major advantages of providing habitat restoration and flood mitigation in addition to water quality improvement. A third example of a hydrologic restoration that is well underway is the restoration of the Mesopotamian Marshlands of Iraq. There, the hydrology was purposefully disrupted by the regime of Saddam Hussein and indirectly through upstream river management by Iraq's upstream neighbors. The Iraqi people, with some international assistance, are undertaking a hydrologic restoration of this historically and culturally important wetland (see Case Study 3).

CASE STUDY 1: Restoring the Florida Everglades

The restoration of the Florida Everglades, the largest wetland area in the United States, is actually several separate initiatives being carried out in the 4.6-million-ha Kissimmee–Okeechobee–Everglades (KOE) region in the southern third of Florida (Fig. 12.4). The basic plan involves restoring something closer to the original hydrology of the KOE region, by sending less of the water from the upper watershed to the Caloosahatchee River to the west and the St. Lucie Canal to the east (see Fig. 12.4 middle diagram) and directing more of the water to the Everglades south of Lake Okeechobee (see Fig. 12.4 right).

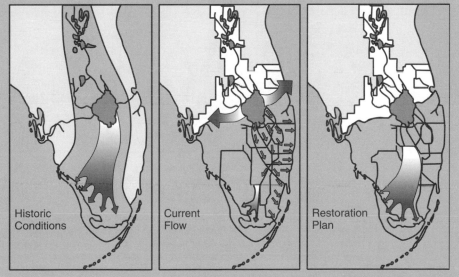

Figure 12.4 Overall plan of the Everglades restoration as planned by federal and state agencies. *(From Mitsch and Jørgensen, 2004.)*

Specific problems in the Everglades have developed because of (1) excessive nutrient loading to Lake Okeechobee and to the Everglades, primarily from agricultural runoff, (2) loss and fragmentation of habitat caused by urban and agricultural development, (3) spread of *Typha* and other invasives and exotics to the Everglades, replacing native vegetation, and (4) hydrologic alteration due to an extensive canal and straightened river system built by the U.S. Army Corps of Engineers and others for flood protection, and maintained by several water management districts.

One major restoration project in the KOE region that has received a lot of attention is the restoration of the Kissimmee River (Fig. 12.5). As a result of the channelization of the river in the 1960s, a 166-km-long river was transformed into a 90-km-long, 100-m-wide canal, and the extent of wetlands along the river decreased by 65 percent (Table 12.1). The restoration of the Kissimmee River is a major undertaking to reintroduce the sinuosity to the artificially straightened river. The river restoration work, expected to be completed in stages over the next several decades, will return some portion

Figure 12.5   Part of the Everglades area restoration has involved the "re-meandering" and restoration of the Kissimmee River that flows from central Florida to Lake Okeechobee. The river is being partially "restored" by refilling the dredged flood control canal and reconnecting the old meanders of the river channel. *(Courtesy of Lou Toth, South Florida Water Management District; photograph by Paul Whalen, reprinted with permission.)*

Table 12.1 Wetland changes due to channelization of the Kissimmee River in south Florida (Channelization took place between 1962 and 1971 and transformed a 166-km meandering river into a 90-km-long, 10-m-deep, 100-m-wide canal.)

Wetland Type	Prechannelization (ha)	Postchannelization (ha)	Percentage Change
Marsh	8,892	1,238	−86
Wet prairie	4,126	2,128	−48
Scrub–shrub wetland	2,068	1,003	−51
Forested wetland	150	243	+62
Other	533	919	+72
Total	15,769	5,531	−65

Source: Toth et al. (1995).

of lost wetland habitat to the riparian zone and will also provide sinks for nutrients that are otherwise causing increased eutrophication in downstream Lake Okeechobee.

Everglades restoration also involves halting the spread of high-nutrient cattail (*Typha domingensis*) through the low-nutrient sawgrass (*Cladium jamaicense*) communities that presently dominate the Everglades (see Chapter 9 for general description of the water pollution problem in the Everglades and Chapter 13 for a description of the treatment wetlands being used to solve this problem). Overall, the Everglades restoration, as now planned by the U.S. Army Corps of Engineers, will cost over $8 billion and will be carried out over the next 20 years or more.

CASE STUDY 2: Creating and Restoring Wetlands to Solve the Gulf of Mexico Hypoxia

A hypoxic zone has developed off the shore of Louisiana in the Gulf of Mexico, where hypolimnetic waters with dissolved oxygen less than 2 mg/L O_2 now extend over an area of 1.6 to 2.0 million ha (Rabalais et al., 1996, 1998, personal communication). Nitrogen, particularly nitrate-nitrogen, is the most probable cause; 80 percent of the nitrogen input is from the 3-million-km^2 Mississippi River basin (Fig. 12.6). The basin represents 41 percent of the lower 48 states of the United States. The control of this hypoxia is important in the Gulf of Mexico because the continental shelf fishery in the Gulf is approximately 25 percent of the U.S. total.

Several options were investigated for controlling nutrient flow into the Gulf (Mitsch et al., 2001). Among the options are modifying agricultural practices (e.g., reduced fertilizer use or alternate cropping techniques); tertiary

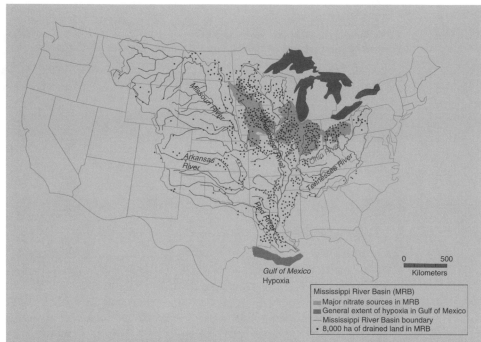

Figure 12.6 The Mississippi–Ohio–Missouri (MOM) River Basin showing the location of Gulf of Mexico hypoxia (low-dissolved-oxygen zone) and the entire Mississippi River watershed that contributes to the problem. The area of major nitrate-nitrogen runoff from the Midwest is illustrated. Each dot represents 8,000 ha of drained land, which also contributes to the problem in the Gulf by leading to more rapid transport of nitrate-nitrogen through the watershed. (From Mitsch et al., 2001, 2005b.)

treatment (biological, chemical, physical) of point sources; landscape restoration (e.g., riparian buffers and wetland creation to control non–point source pollution from farmland); stream and delta restoration; and atmospheric controls of NO_x. In the end, there are three general approaches (Fig. 12.7) that involve either revision of agronomic approaches or wetland creation and riparian restoration that make the most sense. Two million ha of restored and created wetlands and restored riparian buffers would be necessary to provide enough denitrification to substantially reduce the nitrogen entering the Gulf of Mexico (Mitsch et al., 2001, 2005b; Mitsch and Day, 2006). That area is less than 1 percent of the Mississippi River Basin. Interestingly, Hey and Phillipi (1995) found that a similar scale of wetland restoration would be required in the Upper Mississippi River Basin to mitigate the effects of very large and costly floods such as the one that occurred in the summer of 1993 in the Upper Mississippi River Basin.

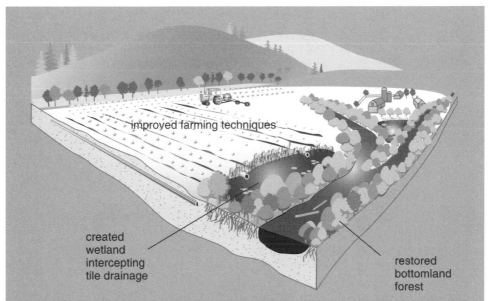

Figure 12.7 **Wetland restoration in an agricultural setting for water quality improvement, particularly for nitrate nitrogen, should involve created and restored wetlands that intercept surface drainage, riparian buffers that intercept subsurface drainage and river flooding, and improved farming techniques for more efficient use of fertilizers.** *(From Mitsch et al., 2001.)*

CASE STUDY 3: Restoration of the Mesopotamian Marshlands

The Mesopotamian Marshlands of southern Iraq and Iran were described in Chapter 3. These wetlands, found at the confluence of the historic Tigris and Euphrates rivers, were 15,000 to 20,000 km^2 in area as recently as the early 1970s (Fig. 12.8a) but were drained and diked, especially in the 1990s, to less than 10 percent of that extent by 2000 (Fig. 12.8b). Among the main causes are upstream dams and drainage systems constructed in the 1980s and 1990s that altered the river flows and eliminated the flood pulses that sustained the wetlands.

Since the overthrow of Saddam Hussein's dictatorship in 2003 in Iraq, there has been a concerted effort by the Iraqis and then the international community at restoring the marshlands (Richardson et al., 2005). The restoration has often occurred with local residents breaking dikes or removing impediments to flooding. Remote sensing images showed that at least 37 percent of the wetlands were restored by 2005. *Frontiers in Ecology and the Environment* (Vol. 3, No. 8, Oct. 2005) reported in 2005 that "at least 74 species of migratory waterfowl and many endemic birds have been sighted in a survey of Iraq's marshland.'' It was also reported that as many as 90,000 Marsh Arabs have returned to the wetlands already (Azzam Alwash, personal communication, 2006).

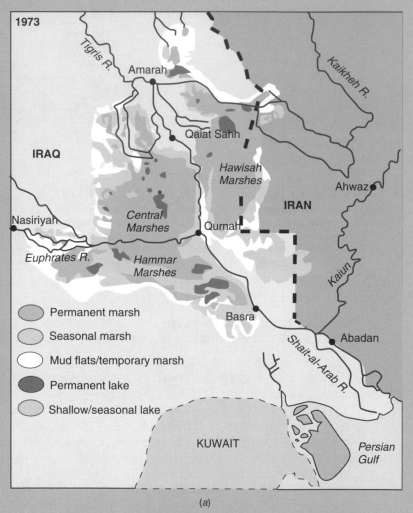

(a)

Figure 12.8 The Mesopotamian Marshlands of Iraq, with shading indicating extent of the marshlands: (a) in 1970 before extensive drainage; (b) in 2000 after extensive drainage; and (c) as expected in the future with 75 percent restoration of the marshland. *(a, b from UNEP, 2001; c from illustration by Azzam Alwash.)*

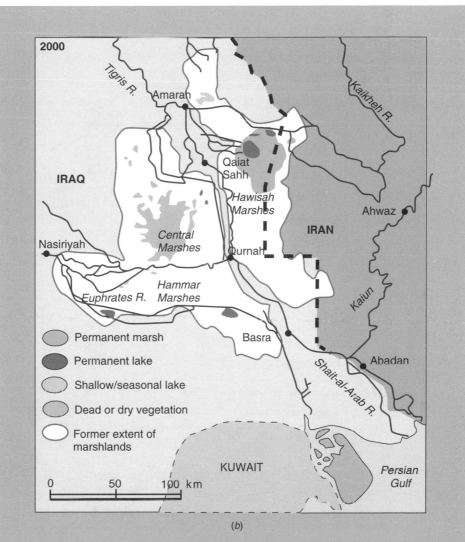

(b)

Figure 12.8 (*continued*)

Alwash, the Director of the Eden Again effort, has suggested that perhaps as much as 75 percent of the marshlands will be restored (Fig. 12.8c). Several questions still remain unanswered about whether full restoration can occur, including whether adequate water supplies exist given the competition from Turkey, Syria, and Iran, and within Iraq itself, and whether landscape connectivity of the marshes can be reestablished (Richardson and Hussain, 2006).

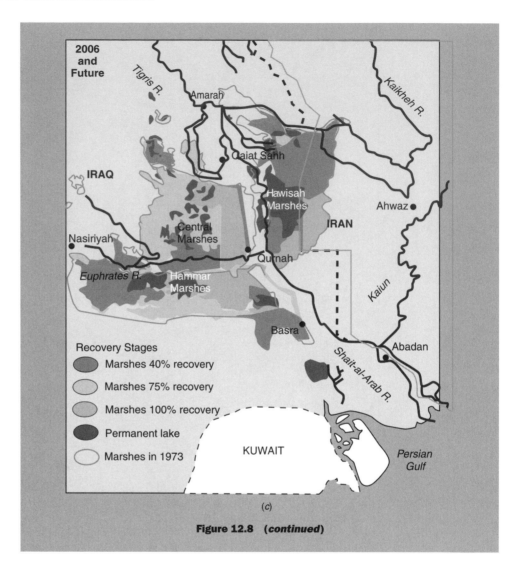

(c)

Figure 12.8 (*continued*)

Peatland Restoration

Peatland restoration is a relatively new type of wetland restoration compared to other types and potentially could be the most difficult (Gorham and Rochefort, 2003). Early attempts with peatlands occurred in Europe, specifically in Finland, Germany, the United Kingdom, and The Netherlands. Increased peat mining in Canada and elsewhere has led to increased interest in understanding if and how mined peatlands can be restored. When peat surface mines are abandoned without restoration, the area rarely returns through secondary succession to the original moss-dominated system (Quinty and Rochefort, 1997). There is promise that restoration can be successful

(Lavoie and Rochefort, 1996; Wind-Mulder et al., 1996; Quinty and Rochefort, 1997; Price et al., 1998), but because surface mining causes major changes in local hydrology and peat accumulates at an exceedingly slow rate, restoration progress will be measured in decades rather than years.

In the 1960s and 1970s, block harvesting of peat was replaced by vacuum harvesting in southern Quebec and in New Brunswick, necessitating the development of different restoration techniques. While traditional block-cutting of peat left a variable landscape of high ground and trenches, vacuum harvesting leaves relatively flat surfaces bordered by drainage ditches. Abandoned block-cut sites appear to revegetate with peatland species more easily than do vacuum-harvested sites, and the latter can remain bare for a decade or more after mining (Rochefort and Campeau, 1997).

Coastal Wetland Restoration

Salt Marsh Restoration

There is a great deal of interest in coastline restoration. Early pioneering work on salt marsh restoration was done in Europe (Lambert, 1964; Ranwell, 1967), China (Chung, 1982, 1989), and in the United States on the North Carolina coastline (Woodhouse, 1979; Broome et al., 1988), in the Chesapeake Bay area (Garbisch et al., 1975; Garbisch, 1977, 2005), and along the coastlines of Florida, Puerto Rico (Lewis, 1990b,c), and California (Zedler, 1988, 2000b; Josselyn et al., 1990). Some of this coastal wetland restoration has been undertaken for habitat development as mitigation for coastal development projects (Zedler, 1988, 1996).

For coastal salt marshes in the Eastern United States, the cordgrass *Spartina alterniflora* is the primary choice for coastal marsh restoration, but the same species is considered an invasive and unwanted plant on the West Coast of North America. Both *Spartina townsendii* and S. *anglica* have been used to restore salt marshes in Europe and in China, although the latter species is considered an invasive species in parts of the world such as New Zealand. Salt marsh grasses tend to distribute easily through seed dispersal, and the spread of these grasses can be quite rapid once the reintroduction has begun, as long as the area being revegetated is intertidal—that is, the elevation is between ordinary high tide and low tide.

The details of successful coastal wetland creation are site specific, but these generalizations seem to be valid in most situations:

1. Sediment elevation is the most critical factor determining the successful establishment of vegetation and the plant species that will survive. The site must be intertidal.

2. In general, the upper half of the intertidal zone is more rapidly vegetated than lower elevations.

3. Sediment composition does not seem to be a critical factor in colonization by plants unless the deposits are almost pure sand that is subject to rapid desiccation at the upper elevations.

4. The site needs to be protected from high wave energy. It is difficult or impossible to establish vegetation at high-energy sites.

5. Most sites revegetate naturally from seeds if the elevation is appropriate and the wave energy is moderate. Sprigging live plants has been accomplished successfully in some cases, and seeding also has been successful in the upper half of the intertidal zone.

6. Good stands can be established during the first season of growth, although sediment stabilization does not occur until after two seasons. Within four years, successfully planted sites are often indistinguishable superficially from natural marshes.

Several early studies emphasized the importance of restoring tidal conditions, including salinity, to marsh areas that had become more "freshwater" because of isolation from the sea. In cases such as this, the restoration is simple: remove whatever impediment is blocking tidal exchange. Case Study 4 describes a salt marsh restoration where that is exactly the case—restore the natural tidal hydrology, and the vegetation and aquatic species will follow.

CASE STUDY 4: Delaware Bay Salt Marsh Restoration

A large coastal wetland restoration project in the Eastern United States involves the restoration, enhancement, and preservation of 5,000 ha of coastal salt marshes on Delaware Bay in New Jersey and Delaware (Figure 12.9). This estuary enhancement, being carried out by New Jersey's electric utility (PSEG), with advice from a team of scientists and consultants, was undertaken as mitigation for the potential impacts of once-through cooling from a nuclear power plant operated by PSEG on the bay. The reasoning was that the impact of once-through cooling on fin fish, through entrainment and impingement, could be offset by increased fisheries production from restored salt marshes. Because of uncertainties involved in this kind of ecological trading, the area of restoration was estimated as the salt marshes that would be necessary to compensate for the impacts of the power plant on fin fish, times a safety factor of four. Three distinct approaches are being utilized in this project to restore the Delaware Bay coastline:

1. *Reintroduce flooding.* The most important type of restoration involves the reintroduction of tidal inundation to about 1,800 ha of former diked salt hay farms. Many marshes along Delaware Bay have been isolated by dikes from the bay, sometimes for centuries, and put into the commercial production of "salt-hay" (*Spartina patens*). Hydrologic restoration was accomplished by excavating breaches in the dikes and, in most cases, connecting these new inlets to a system of recreated tidal creeks and existing canal systems.

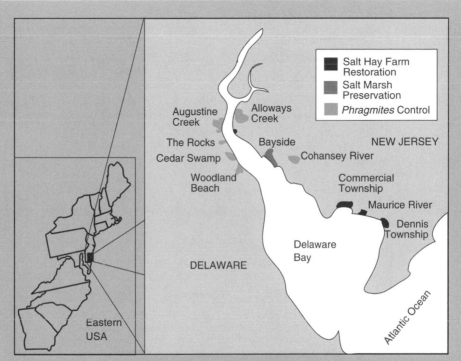

Figure 12.9   Delaware Bay between New Jersey and Delaware, showing locations of 5,800-ha salt marsh restoration/enhancement that is being carried out in lieu of cooling towers at the Salem Nuclear Generating Station on the bay, to mitigate the loss of fin fish due to entrainment and impingement caused by once-through cooling. Wetlands are being preserved, restored from salt-hay farms by reintroducing flooding, and enhanced by removal of *Phragmites australis*. (From PSEG Service Corporation, Salem, NJ.)

2. *Reexcavate tidal marshes.* Additional restoration involves enhancing drainage by reexcavating higher-order tidal creeks in these newly flooded salt marshes, thereby increasing tidal circulation. This is particularly important in marshes that were formerly diked, because the isolation from the sea has led to the filling of former tidal creeks. After initial tidal creeks were established, it was expected that the system would self-design more tidal channels and increase the channel density.

3. *Reduce Phragmites domination.* In another set of restoration sites in Delaware and New Jersey, restoration involves the reduction in cover of the aggressive and invasive *Phragmites australis* in 2,100 ha of nonimpounded coastal wetlands. Alternatives that were investigated

include hydrological modifications such as channel excavation, breaching remnant dikes, microtopographic changes, mowing, planting, and herbicide application.

Results of this study were reported in many presentations and reports and several early journal articles (e.g., Weinstein et al., 1997, 2001; Teal and Weinstein, 2002). More recent results are reported in a special issue of *Ecological Engineering* (Peterson et al., 2005).

From a hydrodynamic perspective, in those marshes where tidal exchange was restored, the development of an intricate tidal creek density from the originally constructed tidal creeks has been impressive. Figure 12.10 illustrates the development of a stream network at one of the newly restored marsh sites—Dennis Township. The "order" of the stream channels increased from 5 or less to well over 20 from 1996 through 2004. The number of small tributaries increased from "dozens" to "hundreds" at all three salt-hay farm sites that were reopened to tidal flushing. For the first 3 years, there was a rapid increase in the growth of channel orders 3 through 9; in the next 3 years, there was a rapid increase in the channel orders 10 through 16.

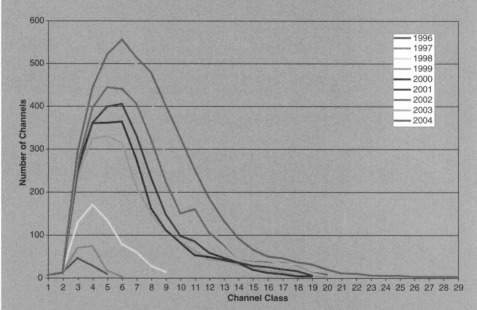

Figure 12.10   Total number of stream channels by channel class, 1996–2004, at Dennis Township salt marsh, which was restored from a former salt-hay farm in 1996. Note the physical self-design of the landscape that developed smaller-order tidal channels on its own. (Provided with permission, Kenneth A. Strait, PSEG Service Corporation, Salem, NJ.)

[*Note*: This definition uses channel order is opposite to the normal method on stream order; here the largest channels are designated as channel order 1.] Hydrologic design did occur in a self-design fashion after only initial cuts by construction of the first-order channels.

For the salt-hay farms that were flooded, typical goals include a high percent cover of desirable vegetation such as *Spartina alterniflora*, a relatively low percent of open water, and the absence of the invasive reed grass *Phragmites australis*. The success of this coastal restoration project, subject to a combination of legal, hydrologic, and ecological constraints, is also being estimated through comparison of restored sites to natural reference marshes. Results of this part of the project after almost a decade are encouraging (Figs. 12.11 and 12.12). At the formerly diked salt-hay farms, reestablishment of *Spartina alterniflora* and other favorable vegetation has been rapid and extensive.

At Dennis Township, approximately 70 percent of the site was dominated by *Spartina alterniflora* after only two growing seasons and almost 80 percent by the fifth year after construction (Figure 12.12a). Tidal restoration was completed at the Maurice River site, which is twice the size of Dennis Township, in early 1998. Major revegetation by *Spartina alterniflora* and some *Salicornia* has already occurred, with 71 percent of the site showing desirable vegetation after four growing seasons (Figure 12.12b). At the third and the largest salt-hay farm restoration site, Commercial Township, which is five times larger than Dennis Township site, revegetation is occurring rapidly from the bayside, and 25 percent of the site was in *Spartina* after four growing seasons (Figure 12.12c).

The study has shown that the speed with which restoration takes place is dependent on three main factors:

a. The degree to which the tidal "circulatory system" works its way through the marsh

b. The size of the site being restored

c. The initial presence of *Spartina* and other desirable species

No planting was necessary on these sites as *Spartina* seeds arrive by tidal fluxes, but the design of the sites to allow that tidal connectivity (and hence the importance of appropriate site elevations relative to tides) was critical. Self-design works when the proper conditions for propagule disbursement are provided. Extensive ponding in some areas of the marshes, especially in Commercial Township, which has the highest ratio of area to edge, has impeded the reestablishment of *Spartina* in some locations (Teal and Weinstein, 2002). Creating additional streams or waiting for the tidal forces to cause the same effect eventually allows these areas to develop tidal cycles and *Spartina* to establish itself.

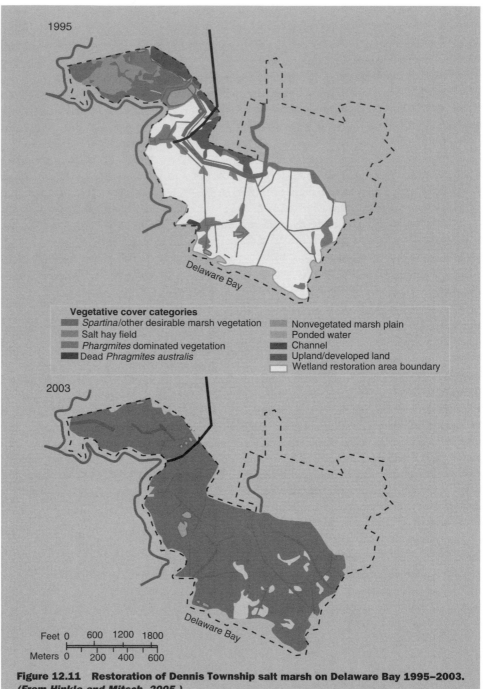

1995

Vegetative cover categories

Spartina/other desirable marsh vegetation
Salt hay field
Phargmites dominated vegetation
Dead Phragmites australis

Nonvegetated marsh plain
Ponded water
Channel
Upland/developed land
Wetland restoration area boundary

Delaware Bay

2003

Delaware Bay

Feet 0   600 1200 1800
Meters 0   200   400   600

Figure 12.11   Restoration of Dennis Township salt marsh on Delaware Bay 1995–2003.
(From Hinkle and Mitsch, 2005.)

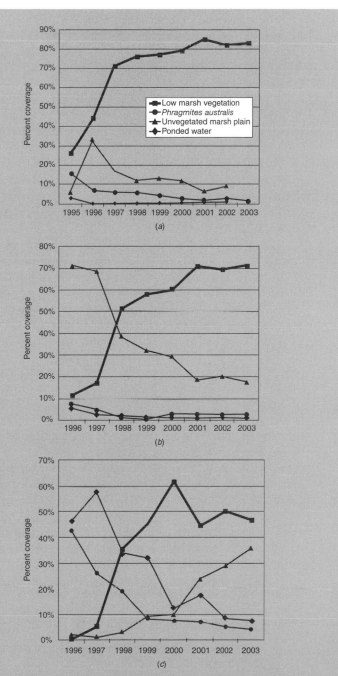

Figure 12.12 Low marsh (mostly *Spartina alterniflora* salt marsh cord grass), *Phragmites*, nonvegetated marsh plain, and ponded water cover for restored salt mashes at: (a) Dennis Township; (b) Maurice River Township; and (c) Commercial Township sites, Delaware Bay, New Jersey 1996–2003. *(From Hinkle and Mitsch, 2005.)*

Mangrove Restoration

Restoring mangrove swamps in tropical regions of the world has some similar characteristics to restoring salt marshes in that the establishment of vegetation in its proper intertidal zone is the key to success. But that is generally where the similarities end. Mangrove restoration is more cosmopolitan in that it has been attempted throughout the tropical and subtropical world (Lewis, 2005); salt marsh restoration has been attempted primarily on the Eastern North American and Chinese coastlines, and to some extent in Europe and the West Coast of North America. Salt marsh restoration can often rely on waterborne seeds distributing through an intertidal zone; mangrove restoration often involves the physical planting of trees. In countries such as Viet Nam, mangrove declines have been attributed to the spraying of herbicides during the Viet Nam war, and immigration of people to the coastal regions, leading to cutting of lumber for timber, fuel, wood, and charcoal.

Mangroves are being cleared for construction of aquaculture ponds at unprecedented rates in Viet Nam and many other tropical coastlines of the world (Benthem et al., 1999). Most of the edible shrimp sold in the United States and Japan are produced in artificial ponds constructed in mangrove wetlands in Thailand, Indonesia, and Viet Nam. Sold in the United States and Japan at very low prices, these products are the result of massive destruction of mangrove forests. More than 100,000 ha of abandoned ponds located in former mangrove swamps currently exist in these countries (R. Lewis, personal communication). In Viet Nam, mangroves are being restored and protected to provide coastal protection and coastal fisheries support. In the Philippines, despite a presidential proclamation in 1981 prohibiting the cutting of mangroves, it is estimated that the country still was losing 3,000 ha/yr in the late 1990s (2.4 percent/yr; deLeon and White, 1999). But the insatiable appetites in the United States, Japan, and several other developed countries for shrimp continue to cause mangroves to be destroyed. The shrimp ponds last only about five to six years before they develop toxic levels of sulfur and are then abandoned and more mangroves are destroyed. These abandoned ponds present a challenge for mangrove restoration.

Lewis 2005 argues that common ecological engineering approaches would work best in restoring mangrove swamps and that more of an analytic approach and less of a "gardening" approach should be taken. He recommends seven principles to correctly restore mangroves:

1. Get the hydrology right (see Case Study 5).
2. Do not build a nursery, but grow mangroves and plant only areas totally devoid of mangroves.
3. See if the conditions that prevent natural colonization can be corrected. If they cannot, pick another site.
4. Examine normal hydrology in reference mangrove swamps as your model for restored sites.

5. Remember that mangrove swamps do not have flat floors, but have subtle topographic patterns.

6. Construct tidal creeks to facilitate flooding and drainage of tide waters.

7. Evaluate the costs of restoration early in the project design to make the project as cost effective as possible.

CASE STUDY 5: A Costly Mangrove Restoration

Lewis (2000) describes a case in southeastern Florida where inattention to the fundamental ecology of how mangrove swamps function led to costly delays and confusion in a mangrove creation/restoration project. The project involved the widening of Route 1 between Homestead and Key Largo at the entrance to the Florida Keys. Because of impact on mangrove wetlands by the highway project, the Florida Department of Transportation (FDOT) was required to mitigate the pending wetland loss by restoring additional wetlands. The project chosen was mangrove restoration on a site that had been filled during previous construction activity but near the site affected.

The project was designed and reviewed over a two-year-period by consultants, agency personnel, and regulatory agencies and found to be acceptable. In this case, the fill was to be removed, with the restoration depending on natural recruitment of mangrove seedlings. The success criterion was 80 percent total coverage by mangroves two years after excavation, and one of the primary goals was restoring habitat for the American crocodile (*Crocodylus acutus*), a species that required open water. The success criterion was probably doable, but it did not fit the goal of crocodile habitat—tidal streams connecting to deepwater sites. Also, no one thought to determine the exact grade that the land needed to be for mangrove success. With adjacent mangrove communities, that could be easily done.

Figure 12.13 illustrates the final revised plan where a specific grade above national geodetic vertical datum (NGVD) is specified and is based on the elevation range in which natural mangroves are found adjacent to the project. Two issues were important here. First, there was so much attention paid to unrealistic goals of 80 percent vegetation cover that caused overdesign on plant introduction techniques, that no one thought to review the elevation "design" of the project to see if any mangroves were going to be intertidal. Second, there was a disconnect between biologists who wanted crocodiles and engineers who designed the hydrology. Ecological engineers, with a real knowledge of ecology, might have designed the appropriate system and recognized that 80 percent vegetation cover in two years is neither practical nor necessary. Nature should be trusted when she takes her time.

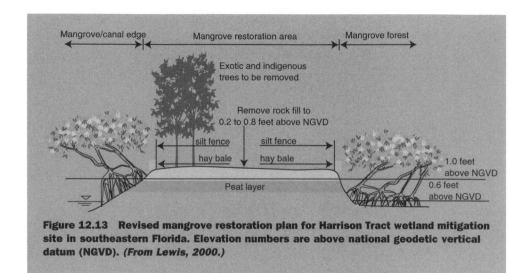

Figure 12.13 Revised mangrove restoration plan for Harrison Tract wetland mitigation site in southeastern Florida. Elevation numbers are above national geodetic vertical datum (NGVD). *(From Lewis, 2000.)*

Delta Restoration

As large rivers connect to the sea, multitributary deltas tend to develop, allowing the river to discharge to the sea in many channels. Many of these rich-soil deltas are among the most important ecological and economic regions of the world, from the ancient Nile delta in Egypt to the modern-day Mississippi River delta in Louisiana. There should be two major ecological resource goals of delta areas: (1) protecting and restoring the functioning of the deltaic ecosystems in the context of a geologically dynamic framework; and (2) controlling pollution from entering the downstream lakes, oceans, gulfs, and bays. Delta restoration should have this dual emphasis where possible—ecosystem enhancement of the delta itself and improvement of coastal water quality downstream. The best strategy for delta restoration when "land building" is a necessary prerequisite (see Case Study 6) is to restore the ability of the river to "spread out its sediments" in deltaic form through as wide an area as possible, particularly during flood events and by not discouraging (or encouraging and even creating) river distributaries. When river distributaries are not possible on a large scale because of navigation requirements or population locations, then restoring and creating riverine wetlands and constructing river diversions to divert river water to adjacent lands may be the best alternatives to maximize nutrient retention and sediment retention. In some cases this involves the conversion of agricultural lands back to wetlands; in other cases, the dikes that "protect" wildlife protection ponds or retain rivers in their channels only need to be carefully breached to allow lateral flow of rivers during flood season.

CASE STUDY 6: Louisiana Delta Restoration

Louisiana is one of the most wetland-rich regions of the world, with 36,000 km^2 of marshes, swamps, and shallow lakes. Yet Louisiana is suffering a rate of coastal wetland loss of 6,600 to 10,000 ha/yr as it converts to open-water areas on the coastline, due to natural (land subsidence) and human causes, such as river levee construction, oil and gas exploration, urban development, sediment diversion, and possibly climate change. Since the early 1990s, there has been a major interest in reversing this rate of loss and even gaining coastal areas, particularly freshwater marshes and salt marshes, the loss of which are the major symptom of this "land loss." Clearly since the disaster in Louisiana and New Orleans caused by Hurricane Katrina in 2005 (see Chapter 11), there is intense interest in restoring the Louisiana delta. But the enormous cost of reestablishing human settlements and putting back levees that were breeched during the hurricane has led some to doubt if there are enough resources to carry out the needed wetland restoration as well (Costanza et al., 2006). Yet the wetlands are vital to the long-term survival of New Orleans.

Initially, the solution to the loss of wetlands and uplands in Louisiana was thought to be the passage by the U.S. Congress of the Coastal Wetland Planning, Protection and Restoration Act (CWPPRA; pronounced "quipra") in 1990. This act initiated comprehensive planning in Louisiana aimed at the protection, restoration, and creation of millions of hectares of coastal wetlands. The plans called for diverting the water and sediment of the Mississippi River to build new deltas and smaller *splays* to mimic spring floods and restore subsiding marshes; restoring barrier islands; and instituting measures to protect many smaller wetlands with dikes, plantings, and disposing of dredge spoil.

Some wetland enhancement and creation projects have created *crevasses* or "diversions" in the levee of the lower Mississippi River. These crevasses allow river water and sediments to flow into shallow estuaries and create crevasse splays or mini-deltas that rapidly become vegetated marshes. Their extent and life span can seldom be predicted, and they function in a natural manner because they mimic the natural geomorphic processes of the river. The CWPPRA led to the Louisiana Coastal Wetland Restoration Plan in 1993 and was spending about $40 million annually on individual projects (LCWCR Task Force, 1998).

A more ambitious project—first called Coast 2050 and later referred to as the Louisiana Coastal Area (LCA) project was then developed in Louisiana to reengineer the coastline to curtail the land loss. That project,

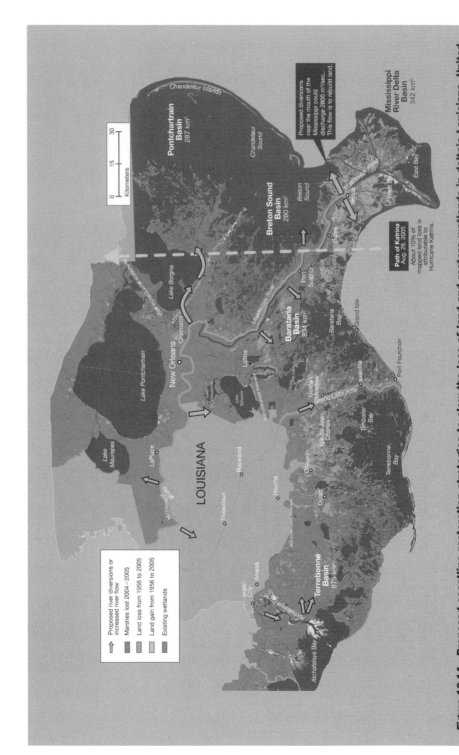

Figure 12.14 Proposed coastline restoration strategies to slow the loss of land and restore wetlands in deltaic Louisiana, United States. Square kilometer numbers next to basin names are land (including wetland) area lost from each drainage basin since 1956. (From Coastal Protection and Restoration Authority of Louisiana, 2007, USGS, Open-File Report 2006-1274 by John A. Barras, and Washington Post illustration May 1, 2007). Composite graphic prepared by Anne Mischo.

estimated to cost $14 billion, would have been the largest wetland restoration in the world if undertaken. As a result of the devastating hurricanes in Louisiana in 2005—Katrina and Rita—the state of Louisiana refocused its efforts into an even more ambitious plan for coastal restoration and hurricane protection—Louisiana's Comprehensive Master Plan for a Sustainable Coast (Coastal Protection and Restoration Authority of Louisiana, 2007). The plan (Fig. 12.14) calls for both coastal restoration and coastal protection and is expected to cost more than twice as much as the LCA. It includes river diversions, closing of the Mississippi River Gulf Outlet, increased utilization of the Atchafalaya River, and hurricane protection with levees and non-structural alternatives for several Louisiana parishes.

River diversions (Figure 12.15) are a significant part of the delta restoration in Louisiana. The deltaic wetlands in Louisiana cover more than 20,000

Figure 12.15   Caernarvon Diversion on Mississippi River near New Orleans and downstream Breton Sound, southeastern Louisiana, illustrating conditions before river diversion (left) and during river diversion (right). Note extensive sediment plume traveling at least 50 km through the sound when the diversion is open. *(provided by J. W. Day, Jr. with permission.)*

km^2 and are critical to building land and wetlands from open water in the delta and may be critical nutrient removal sites for abatement of Gulf hypoxia [see Case Study 2]. The Caernarvon freshwater diversion is one of the largest of several diversions currently in operation on the Mississippi River in Louisiana. The diversion structure is on the east bank of the river below New Orleans, 131 km upstream of the Gulf of Mexico (Fig. 12.15). The structure is a five-box culvert with vertical lift gates with a maximum flow of 226 m^3 sec^{-1}. Freshwater discharge began in August 1991, and discharge from then until December 1993 averaged 21 m^3 sec^{-1}; current minimum and maximum flows are 14 and 114 m^3 sec^{-1}, respectively, with summer flow rates generally near the minimum and winter flow rates of 50 to 80 percent of the maximum (Lane et al., 1999). The Caernarvon diversion delivers water to the Breton Sound estuary, a 1,100-km^2 area of fresh, brackish, and saline wetlands.

Wetland Creation and Restoration Techniques

General Principles

Some general principles of ecological engineering that apply to the creation and restoration of wetlands are outlined as follows (Mitsch and Jørgensen, 2004):

1. Design the system for minimum maintenance and a general reliance on self-design.
2. Design a system that utilizes natural energies, such as the potential energy of streams, as natural subsidies to the system.
3. Design the system with the hydrologic and ecological landscape and climate.
4. Design the system to fulfill multiple goals, but identify at least one major objective and several secondary objectives.
5. Give the system time.
6. Design the system for function, not form.
7. Do not overengineer wetland design with rectangular basins, rigid structures and channels, and regular morphology.

Many other principles can be invoked, but these are good starting points. For example, Zedler (2000a) had suggested the following ecological principles that should be applied to wetland restoration:

1. *Landscape context and position are crucial to wetland restoration.* See the number 3 design principle in the previous list. Wetlands are always a function of the watershed and ecological setting in which they are placed.

2. *Natural habitat types are the appropriate reference systems.* This suggests that while we may know how to build ponds, for example, are those the natural habitats of the area, even if they do increase waterfowl?

3. *The specific hydrologic regime is crucial to restoring biodiversity and function.* See the number 2 and 3 previous design principles. In many cases, such as the Florida Everglades, the restoration is being done in the face of a massive change in the hydrologic character of the landscape.

4. *Ecosystem attributes develop at different paces.* Give the system time; see number 5 in the previous list. Hydrology develops quickly, vegetation over several years, and soils over decades. Yet we are quick to review and criticize created and restored wetlands after a couple of years.

5. *Nutrient supply rates affect biodiversity recovery.* There are low-nutrient and high-nutrient wetland systems (see Chapter 7). Low-nutrient wetlands are often more difficult to create or restore. We live in a eutrophic landscape with very few exceptions (the Okavango Delta in Botswana might be one). High-nutrient inflows cause wetlands to go for power, often instead of diversity. In other cases, insufficient nutrients are the cause of wetland failure, as was the case in southern California salt marshes (Zedler, 2000b).

6. *Specific disturbance regimes can increase species richness.* This can clearly be the case if we allow the word "disturbance" to include flood pulses, fire, and even tropical storms.

7. *Lack of seed banks and dispersal can limit recovery of plant species richness.* This is why restoring wetlands with seed banks can be so important. Another solution is to have a hydrologically or biologically "open" system with a multitude of inputs of propagules (plants, animals, microbes) more likely.

8. *Environmental conditions and life-history traits must be considered when restoring biodiversity.* See our discussion on self-design versus designer wetlands on pages 419–421.

9. *Predicting wetland restoration begins with succession theory.* Again, design principle number 5 says we need to give the system time. Ecological succession cannot be accelerated without other consequences. This also supports our contention that one must understand wetland science first before attempting to create and restore wetlands.

10. *Genotypes influence ecosystem structure and function.* This is an important but often overlooked principle about wetland restoration. Species are not the same everywhere. This has been shown in common garden experiments on *Spartina alterniflora* (Seliskar, 1995) and freshwater rush *Juncus effusus* (Weihe and Mitsch, 2000). A brackish/freshwater wetland plant with several genotypes that has invaded many natural and restored wetlands in the United States is *Phragmites australis*. For example, a new subspecies *Phragmites australis* subsp. *americanus* has been identified and has been shown to be distinctly different from the introduced and Gulf Coast lineages of *P. australis*

(Saltonstall et al., 2004). This creates a difficult problem in wetland creation and restoration, because wetland managers must be able to distinguish between the invasive "bad" *Phragmites* and the native "good" *Phragmites*.

These principles provide useful guides for wetland restoration, and they are focused mostly on the development of wetland structure, not function, but they mesh well with the ecological engineering principles given previously.

Defining Goals

The design of an appropriate wetland or series of wetlands, whether for habitat recreation, the control of non–point source pollution, or for wastewater treatment, should start with forming the overall objectives of the wetland. One view is that wetlands should be designed to maximize ecosystem longevity and efficiency and minimize cost. The goal, or a series of goals, should be determined before a specific site is chosen or a wetland is designed. If several goals are identified, one must be chosen as primary.

Placing Wetlands in the Landscape

In some cases, particularly when sites are being chosen for habitat replacement, many choices are available of sites in the landscape to locate a restored or created wetland. The natural design for a riparian wetland fed primarily by a flooding stream or river (Fig. 12.16a) allows for flood events of a river to deposit sediments and chemicals on a seasonal basis in the wetland, and for excess water to drain back to the stream or river. Because there are natural and also often constructed levees along major sections of streams, it is often possible to create such a wetland with minimal construction. The wetland could be designed to capture flooding water and sediments and slowly release the water back to the river after the flood passes, or to receive flooding water and retain it through the use of flap gates.

Wetlands can be designed as instream systems by adding control structures to the streams or by impounding a distributary of the stream (Fig. 12.16). Blocking an entire stream is a reasonable alternative only in headwater streams, and it is not generally cost-effective or ecologically advisable. This design is particularly vulnerable during flooding, and its stability might be unpredictable, but it has the advantage of potentially "treating" a significant portion of the water that passes that point in the stream. The maintenance of the control structure and the distributary might mean making significant management commitments to this design.

A riparian wetland fed by a pump (Fig. 12.16c) creates the most predictable hydrologic conditions for the wetland but at an obvious extensive cost for equipment and maintenance. If it is anticipated that the primary objective of a constructed wetland is the development of a research program to determine design parameters for future wetland construction in the basin, then wetlands fed by pumps is a good design. Two examples of wetlands of this type constructed primarily for research and education are the Des Plaines River wetlands in northeastern Illinois (Sanville and Mitsch, 1994) and the Olentangy River wetlands in central Ohio

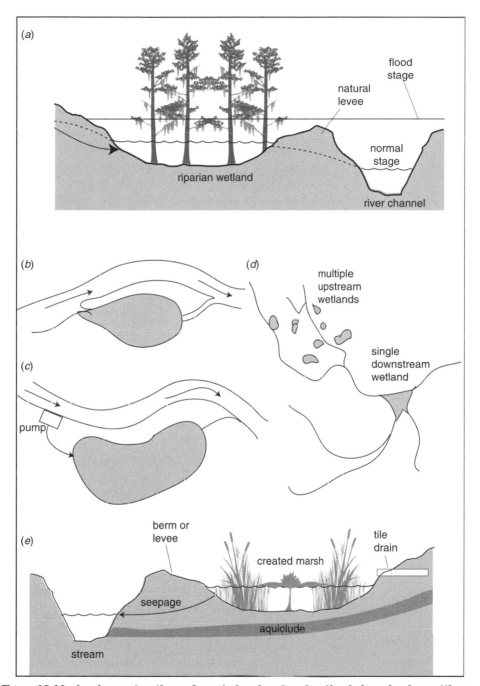

Figure 12.16 Landscape locations of created and restored wetlands in a riverine setting: (a) riparian wetland that both intercepts groundwater from uplands but also receives annual flood pulse from adjacent river; (b) riparian wetland with natural flooding; (c) riparian wetland with pump; (d) multiple upstream wetlands versus single downstream wetland; and (e) lateral wetland intercepting groundwater carried by tile drains.

(Mitsch et al., 1998, 2005c). If other objectives are more important, then the use of large pumps is usually not appropriate unless the wetland is constrained in an urban setting with no recourse.

Locating several small wetlands on small streams or intercepting ditches in the upper reaches of a watershed (but not in the streams themselves), rather than creating fewer larger wetlands in the lower reaches, should be considered (Fig. 12.16d). The usefulness of wetlands in decreasing flooding increases with the distance the wetland is downstream.

Figure 12.16e shows a design involving the creation of a wetland along a stream to intercept tile drains from agricultural fields. The stream is not diverted, but the wetlands receive their water, sediments, and nutrients from small tributaries, swales, and especially tile drains that otherwise would empty straight to the stream. If tile drains can be located and broken or blocked upstream to prevent their discharge into tributaries, they can be rerouted to make effective conduits to supply adequate water to constructed wetlands. Because tile drains are often the sources of the highest concentrations of chemicals, such as nitrates from agricultural fields, the lateral wetlands would be an efficient means of controlling certain types of non–point source pollution while creating a needed habitat in an agricultural setting.

Site Selection

Several important factors ultimately determine site selection. When the objective is defined, the appropriate site should allow for the maximum probability that the objective can be met, that construction can be done at a reasonable cost, that the system will perform in a generally predictable way, and that the long-term maintenance costs of the system are not excessive. These factors are elaborated as follows:

1. Wetland restoration is generally more feasible than wetland creation.
2. Take into account the surrounding land use and the future plans for the land.
3. Undertake a detailed hydrologic study of the site, including a determination of the potential interaction of groundwater with the proposed wetland.
4. Find a site where natural inundation is frequent.
5. Inspect and characterize the soils in some detail to determine their permeability, texture, and stratigraphy.
6. Determine the chemistry of the soils, groundwater, surface flows, flooding streams and rivers, and tides that may influence the site water quality.
7. Evaluate on-site and nearby seed banks to ascertain their viability and response to hydrologic conditions.
8. Ascertain the availability of necessary fill material, seed, and plant stocks and access to infrastructure (e.g., roads, electricity).
9. Determine the ownership of the land, and hence the price.
10. For wildlife and fisheries enhancement, determine if the wetland site is along ecological corridors such as migratory flyways or spawning runs.

11. Assess site access.
12. Ensure that an adequate amount of land is available to meet the objectives.

Creating and Maintaining the Proper Hydrology

The key to restoring and creating wetlands is to develop appropriate hydrologic conditions. Groundwater inflow is often desired because this offers a more predictable and less seasonal water source. Surface flooding by rivers gives wetlands a seasonal pattern of flooding, but such wetlands can be dry for extended periods in flood-absent periods. Depending on surface runoff and flow from low-ordered streams can be the least predictable. Often wetlands developed in these conditions are isolated pools and potential mosquito havens for a good part of the growing season; their design should be carefully considered. It is generally considered to be optimum to build wetlands where they used to be and where hydrology is still in place for the wetland to survive. But tile drainage, ditches, and river downcutting have often changed local hydrology from prior conditions. Most biologists have difficulty estimating hydrologic conditions, while engineers often overengineer control structures that need substantial maintenance and are not sustainable.

Wetland basins are constructed either by establishing levees around a basin that may be partially excavated in the landscape or excavating a depression without constructing any levee. Construction engineers will note that if they use excavated soil for a levee, they can save large sums of money because excavation is often the largest cost of wetland restoration or creation. This is usually not a good idea, because levees are bound to have problems with leakage and, in many parts of the world, burrowing animals like muskrats (*Ondatra zibethicus*).

Some sort of control structure is often needed at the outflow of the wetland basin, whether there is a levee or not. The control devices are the outflow of the wetland. Three such control devices are shown in Figure 12.17:

1. Drop pipes
2. Flashboard risers
3. Full-round riser (combination of drop pipe and flashboard riser)

Each has its own advantages and disadvantages. Drop pipes are the least flexible because they do not allow water-level manipulation. A flashboard riser is more flexible but can be easily vandalized. Full-round risers are a little more secure and can be designed for control of beavers, but they are a little more expensive. In the last two cases, the outflow risers include removable "stoplogs" that allow manual changes in water level. This option is desirable when the exact hydrology is not known for the wetland basin because it allows flexibility.

But these types of control devices have several disadvantages. They require occasional maintenance, if only for removing accumulation of plant debris and resetting stoplogs. Also, stoplog removal is a favorite pastime of vandals. Control devices such as risers are also favorite locations for nature's ecological engineer—the

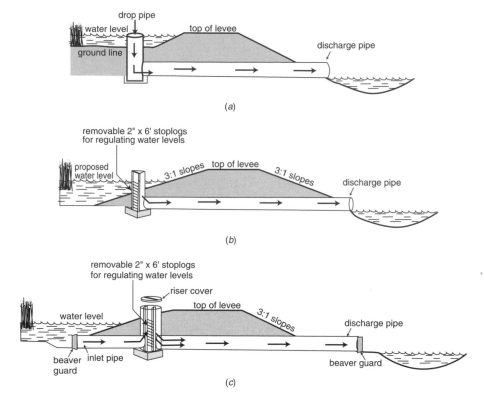

Figure 12.17 Designs for control systems for created and restored wetlands including (a) drop pipe, (b) flashboard riser, and (c) full-round riser. (From Massey, 2000.)

beaver *Castor canadensis*—to provide its idea of water management, usually creating blockages that can raise the water level by a meter or more, changing the vegetation patterns dramatically.

The best design situation is when the local topography allows the wetland to be naturally flooded without control devices, but this opportunity is rarely available. For a reliable source of water, groundwater is generally less sensitive to seasonal highs and lows than is surface water. Also, a wetland fed by groundwater invariably has better water quality and generally fewer sediments that will eventually fill the wetlands.

Soils

Choice of the site for wetland creation and restoration is often limited by property ownership. If a choice exists, a wetland that is restored on former wetland (hydric) soils is much preferred over one constructed on upland soils. Hydric soils develop certain color and chemical patterns, because they have spent long periods flooded and thus under anaerobic conditions. The soil color is mostly black in mineral hydric soils, because iron and manganese minerals have been converted to reduced soluble forms and have leached out of the soil (see Chapter 5). In most cases, developing wetlands on hydric soils has the following three advantages:

1. Hydric soils indicate that the site may still have or can be restored to appropriate hydrology.
2. Hydric soils may be a *seedbank* of wetland plants still established in the soil.
3. Hydric soils may have the appropriate soil chemistry for enhancing certain wetland processes. For example, mineral hydric soils generally have higher soil carbon than do mineral nonhydric soils. This soil carbon, in turn, stimulates wetland processes such as denitrification and methane production.

Otherwise, it is possible to create wetlands on upland soils, and in the long run those soils will develop characteristics typical of hydric soils, such as higher carbon content and seedbanks (see the following box).

Hydric Soil Development and Soil Carbon Sequestration in Created Wetlands

It was not well known how long it would take upland soils to develop wetland conditions; it was thought to be over a decade or even a century, depending on the soil types and the hydrology. In the study of the two wetlands at the Olentangy River Wetland Research Park in Ohio that were created in 1994, hydric soil conditions were shown to develop after only two years of continued flooding (Mitsch et al., 2005c). Before the basins were first flooded, the most prevalent hue was 10YR, and the value/chroma soil color varied between 3/3 and 3/4. The chroma of 3 to 4 indicates nonhydric soils. In 1995, about 18 months after flooding, chromas of 3 or less were common (median = 3/2). The mean value in the surface samples was 3/2; subsurface sample median values were 4/2. Chromas started to consistently be of 2 or below in 1996, two years after flooding began. As of 2006, almost all samples in these experimental wetlands will show chromas of 2 or less in the surface sediments.

Organic carbon in the soil is often illustrated as a shortcoming of wetland creation. The organic content of the upper soils in these experimental wetlands increased steadily over the first decade, from 1994 when water was first added (Fig. 12.18). The organic content of the surface (0–8 cm depth) soils increased from 5.3 ± 0.1 percent in 1993 (before water was added), to 6.1 ± 0.1 percent in 1995 (18 months after creation), to 8.9 ± 0.2 percent in 2004 (10 years after the wetland was created) (Anderson et al., 2005). Total carbon increased from 1.57 ± 0.04 percent in 1993, to 2.06 ± 0.12 percent in 1995, and to 3.76 ± 0.12 percent in 2004. In other words, the organic content and total carbon of these wetland surface sediments increased by 67 percent and 139 percent, respectively, over a decade.

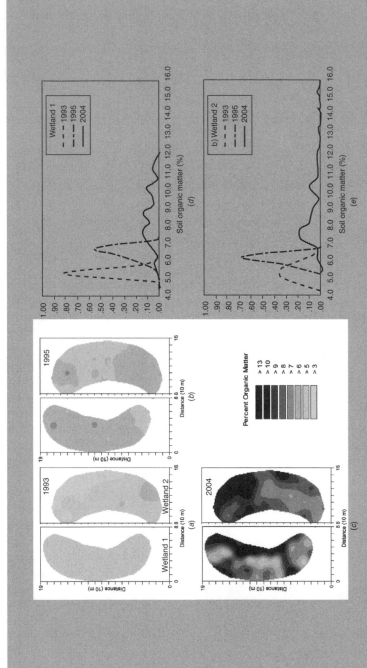

Figure 12.18 Soil organic matter development in two 1-ha created experimental wetlands over a 12-year period at the Olentangy River Wetland Research Park at The Ohio State University. Spatial distribution maps of soil organic matter for Wetlands 1 and 2 shown for (a) 1993, before water was added but after basins were excavated; (b) 1995, 16 months after flooding began; and (c) 2004, 10 years after flooding began; (d. and e.) frequency distribution curves for two experimental wetlands of soil organic matter in 1993, 1995, and 2004. Wetlands were excavated in nonhydric alluvial soils in 1993, and water was first added in March 1994 and continuously since that date. *(From Anderson et al., 2005.)*

414

Table 12.2 Mean carbon and nutrient accumulation rates in two experimental wetlands at the Olentangy River Wetland Research Park, Ohio State University, 1994–2004

Parameter	Mean Annual Accumulation Rates g m^{-2} yr^{-1}
Total C	181–193
Organic C	152–166
Inorganic C	23–26
Total N	16.2–16.6
Total P	3.3–3.5
Total Ca	80.8–86.3

Source: Anderson and Mitsch (2006).

While most of the carbon increase is believed to be due to algal and macrophyte productivity, it is clear from these numbers and from other studies (Wu and Mitsch, 1998; Tuttle, et al., in review) that a substantial amount of the carbon accumulation is due to inorganic calcium carbonate and calcite that precipitate at high rates in the growing season due to water column productivity in these wetlands.

Overall, these experimental wetlands sequestered about 180 to 190 g-C m^{-2} yr^{-1}, with about 87 percent of that carbon in organic form (Table 12.2). The wetlands also retained in the surface soils about 16 g-N m^{-2} yr^{-1} of total nitrogen and about 3.4 g-P m^{-2} yr^{-1} of total phosphorus. Carbon sequestration rates are higher than a weighted average of 22 g-C m^{-2} yr^{-1} of methane emissions estimated from these wetlands (Altor and Mitsch, 2006) by a factor of 8.5. Methane emissions in the shallow emergent plant zones were 32 percent of the emissions from the deepwater zones, suggesting that shallow wetlands could minimize methane emissions while maximizing carbon sequestration.

Rules of thumb from this study on newly created wetlands are important first estimates that can be used in carbon management and wetland restoration of created wetlands:

- Organic content in surface soils in newly created wetlands increases by about 1 percentage point every three years.
- Carbon sequestration in these systems should be expected to be about 200 g-C m^{-2} yr^{-1}.
- With a pulsing hydrology, carbon sequestration is an order of magnitude higher than the carbon lost through methane emissions.

It has been argued that upland soils often do not allow the development of a major diversity of plant communities but often become *Typha* marshes instead because of the absence of seed banks. This is as much because upland lands converted to wetlands have often been used for agriculture for many years and are thus quite eutrophic. The high-nutrient conditions invariably leads to high-productivity, low-diversity systems. Again, the main advantage of using hydric soils in wetland restoration and creation is that they are indicators of the appropriate hydrologic conditions.

Introducing Vegetation

The species of vegetation types to be introduced to created and restored wetlands depend on the type of wetland desired, the region, and the climate, as well as the design characteristics described previously. Table 12.3 summarizes some of the plant species used for wetland creation and restoration projects.

Table 12.3 Selected plant species planted in created and restored wetlands

Scientific Name	Common Name	Scientific Name	Common Name
FRESHWATER MARSH—EMERGENT			
Acorus calamus	sweet flag	Pontederia cordata	pickerelweed
Cladium jamaicense	Sawgrass	Sagittaria rigida	duck potato
Carex spp.	Sedges	Sagittaria latifolia	duck potato; arrowhead
Eleocharis spp.	spike rush	Saururus cernuus	lizard's tail
Glyceria spp.	manna grass	Schoenoplectus tabernaemontani*	soft-stem bulrush
Hibiscus spp.	rose mallow		
Iris pseudacorus	yellow iris	Scirpus acutus	hard-stem bulrush
Iris versicolor	blue iris	Scirpus americanus	three-square bulrush
Juncus effusus	soft rush	Scirpus cyperinus*	woolgrass
Leersia oryzoides	rice cutgrass	Scirpus fluviatilis	river bulrush
Panicum virgatum	Switchgrass	Sparganium eurycarpum	giant bur reed
Peltandra virginica	arrow arum	Spartina pectinata	prairie cordgrass
Phalaris arundinacea	reed canary grass	Typha angustifolia*	narrow-leaved cattail
Phragmites australis*	giant reed	Typha latifolia*	wide-leaved cattail
Polygonum spp.	Smartweed	Zizania aquatica	wild rice
FRESHWATER MARSH—SUBMERGED			
Ceratophyllum demersum	Coontail	Potamogeton pectinatus	Sago pondweed
Elodea nuttallii	Waterweed	Vallisneria spp.	wild celery; tape grass
Myriophyllum aquaticum	Milfoil		
FRESHWATER MARSH—FLOATING			
Azolla caroliniana	water fern	Nuphar luteum	spatterdock
Eichhornia crassipes	water hyacinth	Pistia stratiotes	water lettuce
Hydrocotyle umbellata	water pennywort	Salvinia rotundifolia	floating moss
Lemna spp.	Duckweed	Wolffia sp.	water meal
Nymphaea odorata	fragrant white water lily		

Table 12.3   (Continued).

Scientific Name	Common Name	Scientific Name	Common Name
BOTTOMLANDS/FORESTED WETLAND			
Acer rubrum	red maple	*Gordonia lasianthus*	loblolly bay
Acer floridanum	Florida maple	*Liquidambar styracifula*	sweetgum
Acer saccharinum	silver maple	*Platanus occidentalis*	sycamore
Alnus spp.	Alder	*Populus deltoides*	cottonwood
Carya illinoensis	Pecan	*Quercus falcata* var.	cherrybark oak
Celtis occidentalis	Hackberry	*pagodifolia*	
Cephalanthus occidentalis	Buttonbush	*Quercus nigra*	water oak
Cornus stolonifera	red-osier dogwood	*Quercus nuttallii*	Nuttall oak
Fraxinus caroliniana	water ash	*Quercus phellos*	willow oak
Fraxinus pennsylvanica	green ash	*Salix spp.*	willow
		Ulmus americana	American elm
DEEPWATER SWAMP			
Nyssa aquatica	swamp tupelo	*Taxodium distichum*	bald cypress
Nyssa sylvatica var.biflora	black gum	*Taxodium distichum* var. nutans	pond cypress
SALT MARSH			
Distichlis spicata	spike grass	*Spartina foliosa*	cordgrass (Western U.S.)
Salicornia sp.	Saltwort	*Spartina patens*	salt meadow grass
Spartina alterniflora	cordgrass (Eastern U.S.)	*Spartina townsendii*	cordgrass (Europe)
Spartina anglica	cordgrass (Europe; China)		
MANGROVE			
Rhizophera mangle	red mangrove		
Avicennia germinans	black mangrove		
Laguncularia racemosa	white mangrove		

*Commonly planted in wastewater wetlands

Freshwater Marshes

Common plants used for freshwater marshes include bulrush (*Scirpus* spp. and *Schoenoplectus* spp.), cattails (*Typha* spp.), sedges (*Carex* spp.), and floating-leaved aquatic plants such as white water lilies (*Nymphaea* spp.) and spatterdock (*Nuphar* spp.). Submerged plants are not common in wetland design, and their propagation is often hampered by turbidity and algal growth in the early years of wetland development.

Coastal Marshes

For coastal salt marshes, *Spartina alterniflora* is the primary choice for coastal marsh restoration in the Eastern United States. Both *Spartina townsendii* and *S. anglica* have been used to restore salt marshes in Europe and in China. The details of successful coastal wetland creation are site specific, but several generalizations seem to be valid in most situations:

1. *Sediment elevation is the most critical factor determining the successful establishment of vegetation and the plant species that will survive.* The site must be intertidal and, in general, the upper half of the intertidal zone is more rapidly vegetated than lower elevations. Sediment composition does not seem to be a critical factor in colonization by plants unless the deposits are almost pure sand that is subject to rapid desiccation at the upper elevations. Another important requirement is protection of the site from high wave energy. It is difficult or impossible to establish vegetation at high-energy sites.

2. *Most coastal restoration sites seem to revegetate naturally if the elevation is appropriate and the wave energy is moderate.* Sprigging live plants has been accomplished successfully in several cases, and seeding also has been successful in the upper half of the intertidal zone.

3. *Good stands of vegetation can be established during the first season of growth, although sediment stabilization does not occur until after several seasons.* Within four years, successfully planted sites are indistinguishable superficially from natural marshes.

Forested Wetlands

Forested wetland restoration and creation usually involve the establishment of seedlings. In the Southeastern United States, deciduous hardwood species typical of bottomland forests are planted. They include Nuttall oak (*Quercus nuttallii*), cherrybark oak (*Q. falcata* var. *pagodifolia*), willow oak (*Q. phellos*), water oak (*Q. nigra*), cottonwood (*Populus deltoides*), sycamore (*Platanus occidentalis*), green ash (*Fraxinus pennsylvanica*), sweetgum (*Liquidambar styracifula*), and pecan (*Carya illinoensis*). There is less use of deepwater plants such as bald cypress (*Taxodium distichum*) and water tupelo (*Nyssa aquatica*), although *Taxodium* spp. is the dominant genus of introduced species in many wetland restorations in Florida (Clewell, 1999). In Florida, a wide variety of wetland oaks, bays, gums, ashes, and pines are also used in forested wetland restoration.

Wetland Planting Techniques

Plants can be introduced to a wetland by transplanting roots, rhizomes, tubers, seedlings, or mature plants; by broadcasting seeds obtained commercially or from other sites; by importing substrate and its seed bank from nearby wetlands; or by relying completely on the seed bank of the original and surrounding site. If planting stocks rather than site seed banks are used, it is most desirable to choose plants from wild stock rather than nurseries because the former are generally better adapted to the environmental conditions they will face in constructed wetlands. The plants should come from nearby if possible and should be planted within 36 hours of collection. If nursery plants are used, they should be from the same general climatic conditions and should be shipped by express service to minimize losses. Marshes should be planted at densities to ensure rapid colonization, adequate seed source, and effective competition with undesirable plants such as *Typha* spp. Specifically, this could mean introducing from 2,000 to 5,000 plants/ha.

For emergent plants, the use of planting materials with at least 20 to 30 cm stems is recommended, and whole plants, rhizomes, or tubers rather than seeds have been most successful. In temperate climates, both fall and spring planting times are possible for certain species, but spring plantings are generally more successful, because it is a better time to minimize destructive grazing of plants in the winter by migratory animals and the uprooting of the new plants by ice.

Transplanting plugs or cores (8–10 cm in diameter) from existing wetlands is another technique that has been used with success, for it brings seeds, shoots, and roots of a variety of wetland plants to the newly restored or created wetland.

If seeds and seed banks are used for wetland vegetation, several precautions must be taken. The seed bank should be evaluated for seed viability and species present. The use of seed banks from other nearby sites can be an effective way to develop wetland plants in a constructed wetland if the hydrologic conditions in the new wetland are similar. Seed bank transplants have been successful for many different species, including sedges *(Carex* spp.), *Sagittaria* sp., *Scirpus acutus, S. validus,* and *Typha* spp. The disruption of the wetland site where the seed bank is obtained must also be considered.

When seeds are used directly to vegetate a wetland, they must be collected when they are ripe and stratified if necessary. If commercial stocks are used, the purity of the seed stock should be determined. The seeds can be added with commercial drills or by broadcasting from the ground, watercraft, or aircraft. Seed broadcasting is most effective when there is little to no standing water in the wetland.

Natural Succession vs. Horticulture

To develop a wetland that will ultimately be a low-maintenance one, natural successional processes need to be allowed to proceed. The best strategy is usually to introduce, by seeding and planting, as many native choices as possible to allow natural processes to sort out the species and communities in a timely fashion. Wetlands created or restored by this approach are called *self-design wetlands*. Providing some help to this selection process (e.g., selective weeding) may be necessary in the beginning, but ultimately the system needs to survive with its own successional patterns unless significant labor-intensive management is possible. A somewhat different approach, called *designer wetlands*, occurs when specified plant species are introduced, and the success or failure of those plants is used as indicators of success or failure of that wetland. This is akin to horticulture.

An important general consideration of wetland design is whether plant material is going to be allowed to develop naturally from some initial seeding and planting or whether continuous horticultural selection for desired plants will be imposed. W. E. Odum (1987) suggested "in many freshwater wetland sites it may be an expensive waste of time to plant species which are of high value to wildlife.... It may be wiser to simply accept the establishment of disturbance species as a cheaper although somewhat less attractive solution." Reinartz and Warne (1993) found that the way vegetation is established can affect the diversity and value of the mitigation wetland system. Their study showed that early introduction of a diversity of wetland plants may enhance the long-term diversity of vegetation in created wetlands. The study examined the natural colonization of plants in 11 created wetlands in southeastern Wisconsin. The wetlands under study were small, isolated, depressional wetlands. A two-year sampling program was conducted for the created wetlands, aged one to three years. Colonization was compared to five seeded wetlands where 22 species were introduced. The diversity and richness of plants in the colonized wetlands increased with age, size, and proximity to the nearest wetland source. In the colonized sites, *Typha* spp. comprised 15 percent of the vegetation for one-year wetlands and 55 percent for three-year wetlands, with the possibility of monocultures of *Typha* spp. developing over time in colonized wetlands. The seeded wetlands had a high species diversity and richness after two years. *Typha* cover in these sites was lower than in the colonized sites after two years.

Another study where the effects of planting versus not planting have been observed for several years is at two experimental wetlands at the Olentangy River Wetland Research Park at The Ohio State University (experimental wetlands are described in boxes in Chapters 4 and 5). In essence, both wetlands were different degrees of self-design because there were no expectations as to what the ultimate cover would be and there was no gardening to get to any endpoint. After three years, both wetlands were principally dominated by soft-stem bulrush *Schoenoplectus tabernaemontani* (= *Scirpus validus*) and were thought to be similar. Researchers found that both planted and unplanted wetlands converged in most of the 16 ecological indicators (8 biological measures; 8 biochemical measures) in those three years (Mitsch et al., 1998). After six years, however, several communities of vegetation continued to exist in the planted basin, but a highly productive monoculture of *Typha* dominated the unplanted basin (Mitsch et al., 2005a). This study has now completed over a decade of observations where one full-scale wetland was planted and an identical wetland remained unplanted. The wetlands showed that a few differences in wetland function persist a decade after planting in planted and unplanted (naturally colonizing) wetland basins that could be traced to effects of the initial planting (Mitsch et al., 2005c). Some values that we appreciate in wetlands (e.g., carbon sequestration and amphibian production) were higher in the naturally colonizing wetland, whereas other values such as macrophyte community diversity were higher in the planted wetland. Planting did have an effect, but whether to plant depends on the original objective of the wetland. If plant diversity is desired, then planting makes sense. If productivity and carbon sequestration are desired, it may be a waste of effort to plant unless there are no sources of plant propagules (e.g., seed banks or inflowing rivers). In any regard, there appears to be a lingering long-term effect on ecosystem function caused by planting.

Exotic or Undesirable Plant Species

In some cases, certain plants are viewed as desirable or undesirable because of their value to wildlife or their aesthetics. Reed grass *(Phragmites australis)* is often favored in constructed wetlands in Europe, and there is real concern for "reed die-back" around lakes and ponds in Europe. But reed grass is considered an invasive, undesirable plant in much of Eastern North America, particularly in coastal freshwater and brackish marshes (Philipp and Field, 2005). Some plants are considered undesirable in wetlands because they are aggressive competitors. In many parts of the tropics and subtropics, the floating aquatic plants water hyacinth *(Eichhornia crassipes)* and alligator weed *(Alternanthera philoxeroides)* are considered undesirable and, in Eastern North America, particularly around the Great Lakes, the emergent purple loosestrife *(Lythrum salicaria)* is considered an undesirable alien plant in wetlands. Throughout the United States, cattail *(Typha* spp.) is championed by some and disdained by others, because it is a rapid colonizer but is of limited wildlife value. In other parts of the world, *Typha* is considered a perfectly acceptable plant in restored wetlands. In New Zealand, several species of willow *(Salix)* are invading marshes and other wetlands, and programs to eradicate them are common.

Estimating Success

Few satisfactory methods are available to determine the success of a created or restored wetland or even a mitigation wetland created to replace the functions lost with the original wetland. Figure 12.1 illustrated conceptually how it should be done for replacement wetlands. It is clear from the many studies of created and restored wetlands that some cases are successes, but there are still far too many examples of failure of created and restored wetlands to meet expectations (Fig. 12.19; Table 12.4). In some cases, expectations were unreasonable, as when endangered species habitat was to be established in a heavily urbanized environment (Malakoff, 1998). In such cases, the original wetland should not have been lost to begin with. Where expectations are ecologically reasonable, there is optimism that wetlands can be created and restored and that wetland function can be replaced. The spotty record to date is due, in our opinion, to three factors:

1. Little understanding of wetland function by those constructing the wetlands
2. Insufficient time for the wetlands to develop
3. A lack of recognition or underestimation of the self-design capacity of nature (Mitsch and Wilson, 1996)

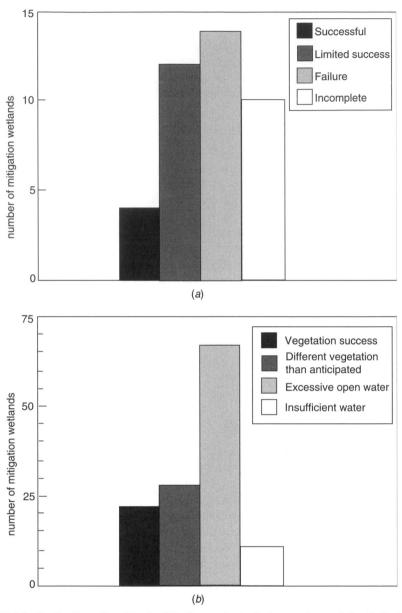

Figure 12.19 Evaluation of wetland mitigation projects in two regions of the United States: (a) 40 mitigation projects in south Florida involving wetland creation, mitigation, and preservation. The average age of the projects was less than three years. "Successful" meant that the project met all of its stated goals, whereas "failure" meant that few goals were met, or the created/restored wetland did not have functional equivalency to a reference wetland *(From Erwin, 1991)*; (b) 128 wetland mitigation sites required by 61 permits in the six-county region around Chicago, Illinois. The permits were issued between 1990 and 1994, and this study began in 1996. *(From Gallihugh and Rogner, 1998.)*

Table 12.4 Permit requirements and compliance for five replacement wetlands investigated in Ohio

location, country	Wetland Area (ha)			Percentage of Required Area Replaced
	Lost	Required	Happened	
Portage	0.4	0.6	0.6	100
Delaware	3.7	5.4	~4.0	74
Franklin	15.0	28.0	3.2	11
Jackson	4.8	7.2	7.5	105
Gallia	0.5	0.8	0.7	88
Total	24.4	42.0	~16.0	38

Source: Wilson and Mitsch (1996).

Understanding wetlands enough to be able to create and restore them requires substantial training in plants, soils, wildlife, hydrology, water quality, and engineering. Replacement projects and other restorations involving freshwater marshes need enough time, closer to 15 or 20 years than to 5 years, before success is apparent. Restoration and creation of forested wetlands, coastal wetlands, or peatlands may require even more time. Peatland restoration could take decades or more. Forested wetland restoration generally takes a lifetime. Finally, we should recognize that Nature remains the chief agent of self-design, ecosystem development, and ecosystem maintenance; humans are not the only participants in these processes. Sometimes we refer to these self-design and time requirements for successful ecosystem restoration and creation as invoking "Mother Nature and Father Time" (Mitsch and Wilson, 1996; Mitsch et al., 1998).

Wetland science will continue to make significant contributions to the process of reducing our uncertainty about predicting wetland success. Wetland creation and restoration need to become part of an applied ecological science, not a technique without theoretical underpinnings. Scientists need to make the connections between structure such as vegetation density, diversity, and productivity and functions such as wildlife use, organic sediment accretion, or nutrient retention, in quantitative and carefully designed experiments. Engineers and managers need to recognize that designing systems that emphasize the role of self-design and sustainable structures are more ecologically viable in the long run than are heavily managed systems.

Summary Principles for Successful Wetland Restoration and Creation

Robin Lewis and Kevin Erwin, two wetland consultants from Florida, have spent about five decades restoring and creating wetlands around the world. Their experience has led to the following 15 recommendations (Lewis et al., 1995). They serve as a fitting summary to this chapter on effective restoration and creation of wetlands.

1. Wetland restoration and creation proposals must be viewed with great care, particularly when promises are made to restore or recreate a natural system in exchange for a permit.

2. Multidisciplinary expertise in planning and careful project supervision at all project levels is needed.

3. Clear, site-specific measurable goals should be established.

4. A relatively detailed plan concerning all phases of the project should be prepared in advance to help evaluate the probability of success.

5. Site-specific studies should be carried out in the original system prior to wetland alteration if wetlands are being lost in the project.

6. Careful attention to wetland hydrology is needed in design.

7. Wetlands should, in general, be designed to be self-sustaining systems and persistent features of the landscape.

8. Wetland design should consider relationships of the wetland to the watershed, water sources, other wetlands in the watershed, and adjacent upland and deepwater habitat.

9. Buffers, barriers, and other protective measures are often needed.

10. Restoration should be favored over creation.

11. The capability for monitoring and midcourse corrections is needed.

12. The capability for long-term management is needed for some type of systems.

13. Risks inherent in restoration and creation, and the probability of success for restoring or creating particular wetland types and functions should be reflected in the standards and criteria for projects and project design.

14. Restoration for artificial or already altered systems requires special treatment.

15. Emphasis on ecological restoration of watersheds and landscape ecosystem management requires advanced planning.

Recommended Readings

Lewis, R. R., III. 2005. "Ecological engineering for successful management and restoration of mangrove forests." *Ecological Engineering* 24:403–418.

Mitsch, W. J., and S. E. Jørgensen. 2004. *Ecological Engineering and Ecosystem Restoration*. John Wiley & Sons, Hoboken, NJ.

Mitsch, W.J. (ed.). 2006. *Wetland Creation, Restoration, and Conservation: The State of the Science*. Elsevier, Amsterdam, 175 pp.

National Research Council. 2001. *Compensating for Wetland Losses under the Clean Water Act*. National Academy Press, Washington, DC, 158 pp.

Streever, W., ed. 1999. *An International Perspective on Wetland Rehabilitation*. Kluwer Academic Publishers, Dordrecht, The Netherlands. 338 pp.

Treatment Wetlands

Treatment wetlands have the singular goal of improving water quality. There are three types of treatment wetlands: natural wetlands and two types of constructed wetlands—surface-flow wetlands and subsurface-flow wetlands. Each type has advantages and disadvantages. Studies on the use of natural wetlands to treat wastewater in Florida and Michigan pioneered the use of surface-flow systems; subsurface-flow systems were started and developed in Europe and remain the dominant treatment wetland system there. Because wetlands can be sinks for almost any chemical, applications of treatment wetlands are quite varied, with thousands of applications worldwide to treat domestic wastewater, mine drainage, non–point source pollution, stormwater runoff, landfill leachate, and confined livestock operations. The design of treatment wetlands requires particular attention to hydrology, chemical loading, soil physics and chemistry, and wetland vegetation. After the wetlands are completed and wastewater is applied, management can include wildlife control and attraction, mosquito and pathogen control, and water-level management. Treatment wetlands are not inexpensive to build and operate, but they usually cost less than chemical and physical treatment processes and are often an effective wastewater treatment strategy for developing countries.

Wastewater and polluted water treatment by wetlands is an intriguing concept involving the forging of a partnership between humanity (our wastes) and an ecosystem (wetlands). Therefore, it is a good example of *ecological engineering* (see Mitsch and Jørgensen, 2004, for details of this field). In this chapter, we discuss the use of wetlands for removing unwanted chemicals from waters, be the waters municipal wastewater, non–point source runoff, or other forms of polluted waters.

As described in Chapter 5, wetlands can be sources, sinks, or transformers for a great number of chemicals. A wetland is a *sink* if it has a net retention of an element or a specific form of that element (e.g., organic or inorganic); that is, if the inputs are greater than the outputs (see Fig. 5.22a). If a wetland exports more of an element or material to a downstream or adjacent ecosystem than would occur without that wetland, it is considered a *source* (see Fig. 5.22b). If a wetland transforms a chemical from, say, dissolved to particulate form but does not change the amount going into or out of the wetland, it is considered to be a *transformer* (see Fig. 5.22c). The desired situation of treatment wetlands is to optimize the wetlands' ability to serve as chemical (and sometimes biological) sinks.

The fact that wetlands were observed to serve as a sinks for inorganic and even organic substances intrigued researchers in the 1960s and 1970s. Researchers in the United States in the 1970s investigated the role of natural wetlands, in regions where they are found in abundance, to treat wastewater and thus recycle clean water back to groundwater and surface water. Earlier, German scientists investigated the use of constructed basins with macrophytes (*höhere Pflanzen*) for purification of wastewater. The two different approaches, one utilizing natural wetlands and the other using artificial systems, have converged into the general field of *treatment wetlands*. The field now encompasses the construction and/or use of wetlands for a myriad of water quality applications, including municipal wastewater, small-scale rural wastewater, acid mine drainage, landfill leachate, and non–point source pollution from both urban and agricultural runoff. While water quality improvement by treatment wetlands is the primary goal of treatment wetlands, they also provide habitat for a wide diversity of plants and animals, and can support many of the other wetland functions described in this book.

Much has been written about treatment wetlands, with tomes by Kadlec and Knight (1996) and Kadlec and Wallace (2008) and the international review report coauthored by Kadlec and colleagues (IWA Specialist Group, 2000) providing the most comprehensive coverage of treatment wetlands. There are also many books and journal special issues covering the general subject of treatment wetlands in general (Godfrey et al., 1985; Reddy and Smith, 1987; Hammer, 1989; Cooper and Findlater, 1990; Knight, 1990; Johnston, 1991; U.S. EPA, 1993; Moshiri, 1993; Reed et al., 1995; Tanner et al., 1999; Vymazal, 2005), non–point source wetlands (Olson, 1992; Mitsch et al., 2000a; Reddy et al., 2006), and landfill leachate (Mulamoottil et al., 1999). Papers featuring such systems are frequently published in the journal *Ecological Engineering*. Much of the work given in this chapter is updated from a chapter on treatment wetlands in *Ecological Engineering and Ecosystem Restoration* (Mitsch and Jørgensen, 2004).

Classifications of Wastewater Treatment Wetlands

Three General Approaches

Three types of wetlands are used to treat wastewater. In the first approach, wastewater is purposefully introduced to existing natural wetlands rather than constructed wetlands (Fig. 13.1a). In the 1970s, studies involving application of wastewater to natural wetlands were carried out in locations of the United States such as Michigan and Florida where there were abundant wetlands. At that time, legal protection of wetlands had not been institutionalized. These pioneering studies elevated the importance of wetlands as "nature's kidneys" to the general public and governmental agencies. This importance was then translated, appropriately, into laws that protected wetlands. But these same laws now generally prohibit the addition of wastewater or polluted water to natural wetlands.

There are two types of *constructed wetlands* that are alternatives to using natural wetlands. *Surface-flow constructed wetlands* (Fig. 13.1b) mimic natural wetlands and can be a better habitat for certain wetland species because of standing water through most if not all of the year. *Subsurface-flow constructed wetlands* (Fig. 13.1c) more closely resemble wastewater treatment plants than wetlands. In these systems, the water flows horizontally through a porous medium, usually sand or gravel, supporting

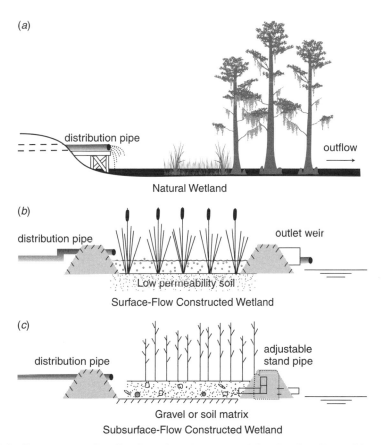

Figure 13.1   Three types of wetland treatment systems: (a) natural wetland, (b) surface-flow wetland, and (c) subsurface-flow wetland. (*After Kadlec and Knight, 1996.*)

one or two of a relatively narrow list of macrophytes such as *Phragmites australis*. There is little to no standing water in these systems as the wastewater passes laterally through the medium.

Subsurface treatment wetlands had their start in the Max-Planck Institute in Germany in the 1950s. Dr. Käthe Seidel performed many experiments with emergent macrophytes, *Schoenoplectus lacustris* in particular, and found that the plants contributed to the reduction of bacteria, and organic and inorganic chemicals (Seidel, 1964, 1966). This process was translated into a gravel-bed macrophyte system that became known as the *Max-Planck-Institute-Process* or the *Krefeld System* (Seidel and Happl, 1981, as cited in Brix, 1994). The development of subsurface wetlands continued in Europe using a system of subsurface-flow basins planted with *Phragmites australis*. These systems were called the *root-zone method* (Wurzelraumentsorgung). Subsurface wetland systems continued to be studied and refined through the work of DeJong (1976) in Holland, Brix (1987) in Denmark, and many other scientists in Europe. The appeal of these more "artificial" types of wetlands in Europe (as opposed to free-water surface wetlands in North America) is due to two factors: (1) there are fewer natural wetlands remaining in Europe, and those that are left are protected for nature; and (2) space is much more at a premium in Europe, and subsurface wetlands require less land area.

Classification According to Vegetation

Treatment wetlands can also be classified based on the life form of their vegetation. In this case, there are five systems based on herbaceous macrophytes:

1. Free-floating macrophyte systems, e.g., water hyacinth (*Eichhornia crassipes*), duckweed (*Lemna* spp.)
2. Emergent macrophyte systems, e.g., reed grass (*Phragmites australis*), cattails (*Typha* spp.)
3. Submerged macrophyte systems
4. Forested wetland systems
5. Multispecies algal systems, particularly algal-scrubber systems

Subsurface-flow constructed wetlands are limited to emergent macrophytes, whereas surface-flow constructed wetlands often utilize a combination of free-floating, emergent, and submerged macrophytes. Forested wetland treatment systems are generally not constructed wetlands at all, but are natural wetlands to which wastewater is applied. They will often develop extensive communities of all of the other vegetation types described in this classification.

Treatment Wetland Types

The type of wastewater being treated can classify treatment wetlands. While many of these systems are used for municipal wastewater and that is often thought as the conventional system, there has been much interest in the use of wetlands to treat storm water from urban areas, acid mine drainage from coal mines, non–point source pollution in rural landscapes, livestock and aquaculture wastewaters, and an array of industrial wastewaters.

Municipal Wastewater Wetlands

In Europe, most of the development of subsurface constructed wetlands was to replace both primary and secondary treatment to remove biochemical oxygen demand (BOD) and suspended solids as well as inorganic nutrients. Hundreds of subsurface wetland treatment systems for municipal wastewater have been constructed in Europe (Vymazal et al., 1998), particularly in the United Kingdom (Cooper and Findlater, 1990), Denmark (Brix and Schierup, 1989a,b; Brix, 1998; Brix and Arias, 2005), the Czech Republic (Vymazal, 1995, 1998, 2002; Vymazal and Kropfelova, 2005), Norway (Braskerud, 2002a,b), Spain (Solano et al., 2004), and Estonia (Teiter and Mander, 2005). There are also many applications of this technology in Australia (Mitchell et al., 1995; Greenway et al., 2003; Greenway, 2005; Headley et al., 2005; Davison et al., 2006), New Zealand (Cooke, 1992; Tanner, 1996; Nguyen et al., 1997; Nguyen, 2000), and Costa Rica (Nahlik and Mitsch, 2006).

In North America, most but certainly not all of the wetlands built for treatment of municipal wastewater treatment are surface-water wetlands. Locations of wastewater wetlands that have been studied in some detail include Florida (Knight et al., 1987; J. Jackson, 1989), California (Gerheart et al., 1989; Gerheart, 1992; Sartoris et al., 2000; Thullen et al., 2005), Louisiana (Boustany et al., 1997; Day et al., 2004), Arizona (Wilhelm et al., 1989), Kentucky (Steiner et al., 1987), Pennsylvania (Conway and Murtha, 1989), Ohio (Spieles and Mitsch, 2000a,b), North Dakota (Litchfield and Schatz, 1989), and Alberta, Canada (White et al., 2000). Created wetlands for treating wastewater have been most effective for controlling organic matter (BOD), suspended sediments, and nutrients. Their value for controlling trace metals and other toxic materials is more controversial, not because these chemicals are not retained in the wetlands, but because of concerns that they might concentrate in wetland substrate and fauna.

CASE STUDY 1: Houghton Lake, Michigan

Much of the interest in using surface-flow wetlands for water quality management was sparked by several studies begun in the early 1970s. In one of those studies, peatlands in Michigan were investigated by researchers from the University of Michigan for the wetlands' capacity to treat wastewater (Fig. 13.2). A pilot operation for disposing of up to 380 m^3/day (100,000 gallons per day) of secondarily treated wastewater in a rich fen at Houghton Lake led to significant reductions in ammonia nitrogen and total dissolved phosphorus as the wastewater passed from the point of discharge through the wetlands. Inert materials such as chloride did not change as the wastewater passed through the wetland. In 1978, the flow was increased to approximately 5,000 m^3 day^{-1} over a much larger area, essentially all the wastewater from the local treatment plant. Data after more than 22 years of operation at this high flow show that although the area of influence of the wastewater on the peatland has grown from 23 to 77 ha (Fig. 13.2a,b), the effectiveness of the wetland in removing both inorganic nitrogen and total phosphorus remained extremely high (Fig. 13.2c,d).

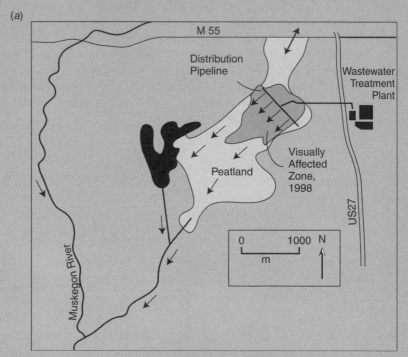

Figure 13.2 Houghton Lake treatment wetland in Michigan, where treated sewage effluent has been applied to a natural peatland since 1978: (a) map of site showing visually affected area in 1998; (b) area of visually affected zone in peatland, primarily where vegetation changes have occurred, 1981–1998; (c) dissolved inorganic nitrogen of influent and outlet stream, 1978–1999; (d) total phosphorus of influent and outlet stream. *(After Kadlec and Knight, 1996; R. H. Kadlec, personal communication, January 2000.)*

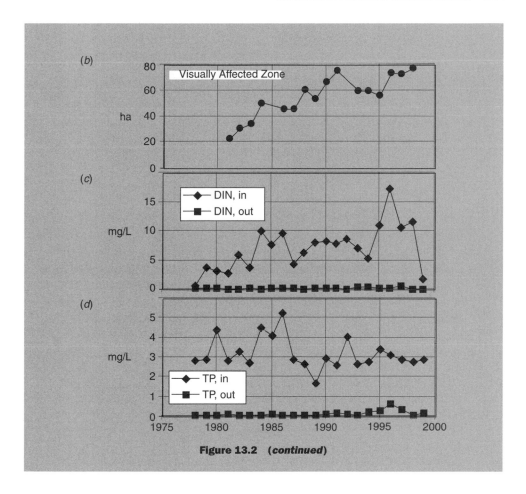

Figure 13.2 (*continued*)

Mine Drainage Wetlands

Wetlands have been frequently used as downstream treatment systems for mineral mines (Figure 13.3). Acid mine drainage water, with its low pH and high concentrations of iron, sulfate, aluminum, and trace metals, is a major water pollution problem in many coal mining regions of the world, and constructed wetlands are a viable treatment option. The use of wetlands for coal mine drainage control was probably first considered when volunteer *Typha* wetlands were observed near acid seeps in a harsh environment where no other vegetation could grow. By the 1980s, more than 140 wetlands had been constructed in the eastern United States alone to treat mine drainage water (Wieder, 1989). The most common goal of these systems is usually the removal of iron from the water column to avoid its discharge downstream, but sulfate reduction and the alleviation of extremely acidic conditions are also appropriate goals (Wieder and Lang, 1984; Brodie et al., 1988; Fennessy and Mitsch, 1989; Mitsch and Wise, 1998; Tarutis et al., 1999).

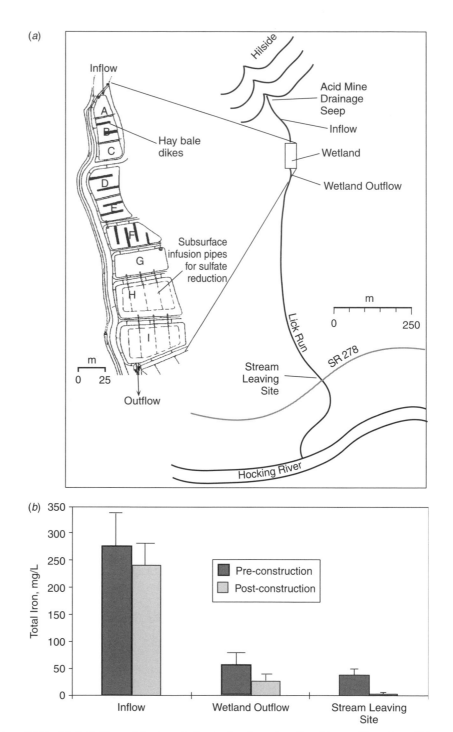

Figure 13.3   (a) A 0.4-ha acid mine drainage treatment wetland in southeastern Ohio; (b) total iron in stream before wetland was built and after it was constructed. *(After Mitsch and Wise, 1998.)*

Design criteria for these wetlands have been developed, but they are neither consistent from site to site nor generally accepted. Stark and Williams (1995) found design features that enhanced iron removal and decreased acidity included broad drainage basins, nonchannelized flow patterns, high plant diversity, southern exposure, low flow rates and loadings, and shallow depths. It is not always cost effective to construct wetlands when extremely high (>85–90 percent) iron removal efficiencies are necessary or when the pH of the mine drainage water is less than 4. Hydraulic loading rates as high as 29 cm day^{-1} have been suggested for wetlands designed for acid mine drainage, although Fennessy and Mitsch (1989) recommended 5 cm/day as a conservative loading rate for this type of wetland and a minimum detention time of one day, with much longer periods for more effective iron removal (Table 13.1). Loading rates of 2 to 10 g-Fe m^{-2} day^{-1} were suggested for circumneutral mine drainage (Fennessy and Mitsch, 1989) and 0.72 g-Fe m^{-2} day^{-1} by Brodie et al. (1988) for mine drainage with pH < 5.5. Manyin et al. (1997) ran a series of mesocosm experiments and found that a rate less than 2.5 g-Fe m^{-2} day^{-1} would be necessary to achieve a consistent iron concentration of less than the water quality standard of 3.5 mg-Fe/L.

The long-term suitability of wetland treatment systems is poorly understood, although it appears that *Typha*-dominated systems can survive decades in a mine-drainage system. For surface mining restoration, the restoration of the larger site, probably through terrestrial succession either imposed or natural, means that the post-mining hydrology needs to be considered in the project design as much as the hydrology when the wetland is built (Kalin, 2001). With mine drainage systems, the accumulation of iron hydroxides can eventually cause the system to begin to export

Table 13.1   Suggested design parameters for constructed wetlands used for controlling coal mine drainage

Parameter	Design	Reference
Hydrologic Loading Rate, cm/day	5	Fennessy and Mitsch, 1989
Retention Time, days	>1	Fennessy and Mitsch, 1989
Iron Loading, g-Fe m^{-2} day^{-1}		
pH < 5.5	0.72	Brodie et al., 1988
pH > 5.5	1.29	Brodie et al., 1988
For 90% removal, pH = 6	2–10	Fennessy and Mitsch, 1989
For 50% removal, pH = 6	20–40	Fennessy and Mitsch, 1989
pH = 3.5, ouflow < 3.5 mg-Fe/L	2.5	Manyin et al., 1997
Basin characteristics		
Depth, m	<0.3	
Number of cells	>3	
Plant material	*Typha* spp.	
Substrate material	organic peat over clay seal; spent mushroom material	

materials, unless the design and management includes adequate storage capacity and/or material removal. Some researchers suggest that these wetlands, after several decades, can become mineral mines in their own sense, effectively recycling minerals that otherwise would be lost to downstream watersheds back to the economy. Furthermore, effluent from constructed wetlands does not always meet strict regulatory requirements. Nevertheless, where no other alternative is feasible, the use of wetlands to reduce this type of water pollution is viewed as a low-cost alternative to costly chemical treatment or to downstream water pollution.

Urban Stormwater Treatment Wetlands

The control of stormwater pollution with wetlands is a valid application of ecological engineering of wetlands. Unlike municipal wastewater, stormwater and other non–point source pollution are seasonal, often quite sporadic, and variable in quality, depending on season and recent land use. Wetlands are one of several choices for systems to control urban runoff. More conventional approaches involve either dry detention ponds that fill only during storms or wet detention ponds that are usually deepwater systems, where the edge is usually stabilized with rocks and plant growth is actually discouraged. Wetlands have been designed for capturing stormwater in urban areas in Florida (Johengen and LaRock, 1993), Washington (Reinelt and Horner, 1995), and England (Shutes et al., 1997).

Stormwater from urban areas is particularly rapid as it comes from impervious sources such as roofs, parking lots, and highways. One of the features of stormwater wetland systems is that severe storms have a dramatic effect on treatment efficiency. High flows resulting from high-intensity rainstorms usually result in lower nutrient and other chemical retention as percent of inflow and, sometimes, the storms cause a net release of nutrients. The very nature of the sudden but short stormwater pulses makes management of these systems particularly difficult.

The pollutants that are transported to treatment wetlands also can create problems. In addition to nutrients and organic wastes, stormwater runoff can either be very high in sediments if coming from construction sites or relatively low in sediments if coming from roofs. If parking lots and highways are part of the wetland watershed, then pollutants result from vehicle exhaust, asphalt erosion, road salts, rubber, oil, grease, metals, and even large rubbish. It is not uncommon to see these wetlands severely stressed with some of these materials, and frequent maintenance can sometimes be necessary, if for no other reason than to remove urban litter from the basins.

Design approaches to these types of wetlands are few, despite several studies that have been done on these systems since the early 1980s. A layout of an ideal stormwater treatment wetland (Figure 13.4) illustrates that a combination of deep ponds and marshes may be most appropriate, with the former pond dampening the rapid stormwater pulse, allowing the wetland to "treat" the runoff in a more effective manner. Multiple cells of marshes and a small outflow deepwater pond can contribute to the system's effectiveness. A summary of almost 60 stormwater treatment wetland systems provides some indication of the long-term pollution

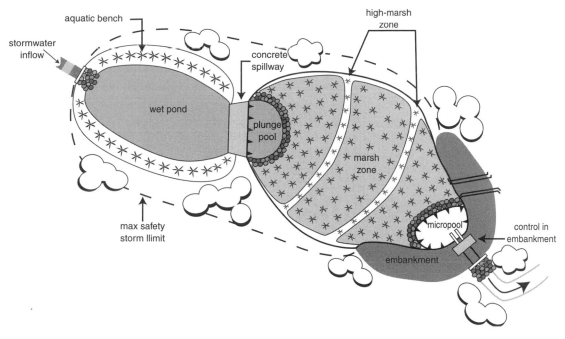

Figure 13.4   General design of a stormwater treatment wetland. *(From Schueler, 1992.)*

Table 13.2   Average retention of chemicals in stormwater wetlands

Pollutant	% reduction
Suspended solids	75
Total nitrogen	25
Total phosphorus	45
Organic carbon	15
Lead	75
Zinc	50
Bacterial count	10^{-2} decrease

Source: Schueler (1992).

capabilities of these systems (Table 13.2). Sediment retention capability is the strong point of these wetlands, but if any significant construction projects occur upstream, even this capacity can be temporarily or permanently overwhelmed. There is some nutrient and organic matter retention, but it is generally not greater than 50 percent.

Agricultural Runoff Wetlands

One of the most important applications of wetland treatment systems—yet an application that is far behind municipal treatment wetlands in understanding design issues—is the use of non–point source wetlands for treating subsurface and surface runoff from agricultural fields. Research projects illustrating the effects and functioning of these types of wetlands in agricultural watersheds have been carried out in southeastern Australia (Raisin and Mitchell, 1995; Raisin et al., 1997), northeastern Spain (Comin et al., 1997), Illinois (Kadlec and Hey, 1994; Phipps and Crumpton, 1994; Mitsch et al., 1995; Kovacic et al., 2000; Larson et al., 2000; Hoagland et al., 2001), Florida (Moustafa, 1999; Reddy et al., 2006), Ohio (Fink and Mitsch, 2004; see CASE STUDY 2), and Sweden (Leonardson et al., 1994; Jacks et al., 1994; Arheimer and Wittgren, 1994). Several wetland sites have received the equivalent of non–point source pollution but under somewhat controlled hydrologic conditions (e.g., river overflow to riparian basins) over several years of study. Bony Marsh, a constructed wetland located along the Kissimmee River in southern Florida, was investigated for nutrient retention of river water for nine years (1978–1986) by the South Florida Water Management District (Moustafa et al., 1996) and found to be a consistent sink of nitrogen and phosphorus but at relatively low levels.

As described in Chapter 9, the water quality of some of the Florida Everglades has been threatened by agricultural runoff. Treatment wetlands are being created to serve as sinks of phosphorus coming from upstream agricultural fields to prevent high-phosphorus water from entering the Everglades. These wetlands are described in the River Diversion Wetland section below.

CASE STUDY 2: Agricultural Runoff Wetland

An agricultural runoff wetland (Fig. 13.5) was constructed in the spring of 1998 in Logan County, Ohio, United States, several kilometers upstream of a popular recreational lake in northwestern Ohio called Indian Lake. The multicelled Indian Lake wetland was 1.2 ha and receives drainage from a 17-ha watershed, 14.2 ha of which was used for intensive row-crop agriculture and 2.8 ha of which was forested. Thus, the wetland had a watershed ratio of 14:1. It was estimated that surface inflow in 2000 was 646 cm yr^{-1}, and groundwater discharge at multiple locations within the site amounted to almost the same amount of inflow (Fink and Mitsch, 2004). Surface water levels of a two-year period of study varied over 40 cm in depth (Fig. 13.6); muskrat activity in one of the cells actually led to a 30-cm water-level decrease in the second year of study. Overall, the wetlands retained 59 percent of total phosphorus, 59 percent of soluble reactive phosphorus, and 40 percent of nitrate/nitrite-nitrogen (Table 13.3).

Major storm events led to dramatic but short increases in water level of over 20 cm; these storm events, primarily in the late winter and early spring, led to rapid flow. Investigation of selected storm events showed reductions of 28 percent, 74 percent, and 41 percent for total phosphorus, soluble reactive

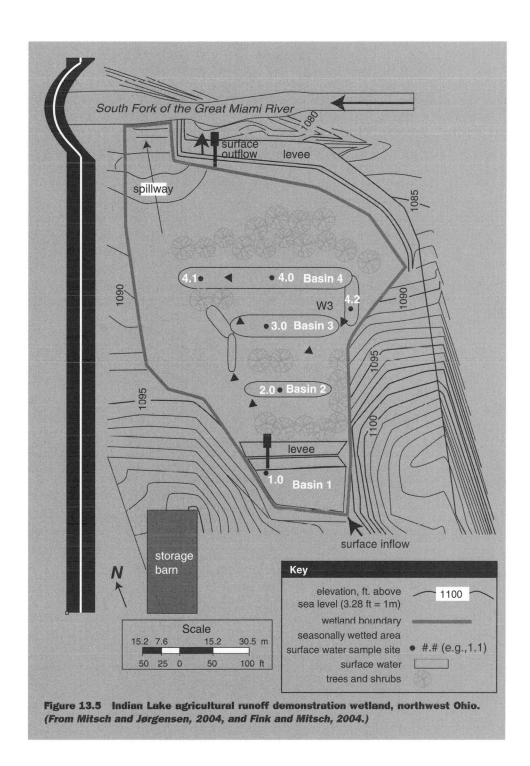

Figure 13.5 Indian Lake agricultural runoff demonstration wetland, northwest Ohio.
(From Mitsch and Jørgensen, 2004, and Fink and Mitsch, 2004.)

totals in cm	Oct	Nov	Dec	Jan	Feb	Mar	Apr	May	Jun	Jul	Aug	Sep	Total
1998-1999	1.6	2.0	2.5	—	4.7	4.5	4.7	2.8	3.1	7.0	2.6	1.6	40.7
1999-2000	6.5	4.9	7.8	5.0	5.4	5.7	9.3	8.4	9.9	8.0	7.3	6.5	84.6

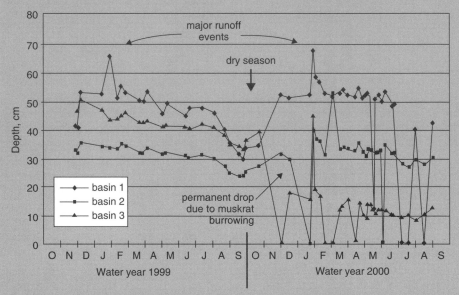

Figure 13.6   Water-level fluctuations in a created wetland downstream of a farm field in Indian Lake agricultural wetland, northwest Ohio. Note short peaks associated with runoff events and seasonal changes in water levels. Also note effect of muskrat activity on water level in one cell. (From Fink and Mitsch, 2004.)

Table 13.3   Percent reduction of nutrients at each sampling site within the Indian Lake agricultural runoff wetland in northwestern Ohio relative to the inflow concentration (A minus sign indicates a net export of nutrient.)

Nutrient Type	Site[a]					
	1.0	2.0	3.0	4.2	4.0	4.1
soluble reactive phosphorus	15	72	71	64	49	56
total phosphorus	−41	56	65	48	30	59
nitrate/nitrite-nitrogen	20	34	55	37	43	40

[a]Sampling sites are shown in Fig. 13.5.
Source: Fink and Mitsch (2004) and Misch and Jørgensen (2004).

phosphorus, and nitrite/nitrate-nitrogen, respectively (Fink and Mitsch, 2004). The design of this wetland, with multiple cells and a watershed:wetland ratio of 14:1, appeared to be appropriate to receive storm pulses of surface runoff coupled with more consistent yet also variable amounts of groundwater inflow. It was also able to accommodate some self-design imposed on the constructed basins by muskrats.

Agricultural Wastewater Wetlands

In addition to the nonpoint sources from agriculture that were discussed previously, serious water pollution problems occur in many parts of the world resulting from runoff from confined animals, particularly dairy, cattle, and swine operations (Tanner et al., 1995; Cronk, 1996; CH2 M-Hill and Payne Engineering, 1997). As more animals are concentrated per unit area to increase food production, the concentrations and volumes of effluents are becoming more noticeable, both by the public and by water pollution control authorities. Concentrations of organic matter, organic nitrogen, ammonia-nitrogen, phosphorus, and fecal coliforms from animal feedlots far exceed concentrations in most municipal sewer systems. Two examples from the Eastern United States of the effectiveness of wetlands for treating wastewater from dairy milkhouses (Table 13.4) showed significant reductions in most pollutants in the treated water, although ammonia-nitrogen increased substantially in the Connecticut wetland, and nitrate-nitrogen increased by 80 percent in the Maryland case.

Table 13.4 Hydrology and water quality of two wetlands constructed to deal with heavily polluted dairy milkhouse effluent

	Connecticut[a]		Maryland[b]	
Wetland area, m^2	400		1,160	
Flow, m^3/wk	18.8		—	
retention time, days	41			
	Inflow	Outflow	Inflow	Outflow
BOD, mg/L	2,680	611	1,914	59
Total N	103	74	170	13
Ammonia-N	8	52	72	32
Nitrate-N	0.3	0.1	5.5	10.0
Total P, mg/L	26	14	53	2.2
TSS, mg/L	1,284	130	1,645	65
coliform, #/100 mL	557,000	13,700	—	—

[a]Newman et al., 2000.
[b]Schaafsma et al., 2000.

In addition to livestock waste from land-based agriculture, constructed wetlands have been used to treat effluent from several aquaculture operations, including shrimp ponds in Thailand and tilapia fish ponds in the United Kingdom.

River Diversion Wetlands

A somewhat different approach to cleaning up water is to pass river water through wetlands built on adjacent floodplains or backwaters. These are analogs of riverine oxbows or billabongs found throughout the world, and they have been shown to consistently improve water quality. These wetlands also are simulations of agriculture runoff wetlands, but with usually lower concentrations of nutrients. However, river sediment concentrations can be high, sometimes in excess of that found in agricultural runoff. These river diversions have been done on a large scale in the Mississippi River Delta in Louisiana (see Case Study 3) and on a much smaller scale for research and water quality improvement in Midwestern United States (see Case Study 4) in the Florida Everglades (see Case Study 5). In all cases, significant improvement in river water quality has been observed as the water is pumped or otherwise distributed to wetlands on the adjacent floodplains.

CAST STUDY 3: Diverting the Mississippi River in the Louisiana Delta

In Louisiana, the diversion of the Mississippi River at Caernarvon (Fig. 13.7; See also CASE STUDY 6 in Chapter 12) is one of the largest diversions in operation on the river aimed at restoring deteriorating wetlands in the Mississippi delta. The diversion structure on the east bank of the river south of New Orleans has a maximum flow of 280 m^3 sec^{-1}. River diversion began in August 1991, and average minimum and maximum flows are 14 and 114 m^3 sec^{-1}, respectively, with summer flow rates generally near the minimum and winter flow rates of 50 to 80 percent of the maximum (Lane et al., 1999, 2004). The diversion delivers river water to the 260-km^2 Caernarvon freshwater wetland, which eventually discharges into the brackish Breton Sound estuary, which is part of coastal Gulf of Mexico. Diversions such as this one in the Louisiana Delta are estimated to be key facilities for the restoration of the Louisiana Delta wetlands as described in Chapter 12 by diverting much needed sediments to the backwater regions of the delta.

An intriguing issue is that these wetlands also retain nutrients, which is particularly important given the hypoxia in the Gulf of Mexico discussed elsewhere in this book (Chapters 5 and 12). The Caernarvon wetland retained 39 to 92 percent of nitrate by mass and concentration, depending on the sampling location in downstream Breton Sound (Fig. 13.7). At the Caernarvon Louisiana sampling station that was most comparable to the Ohio diversion wetlands (described later) for loading rates, the nitrate-nitrogen retention was 55 percent by mass and concentration (Mitsch et al., 2005b).

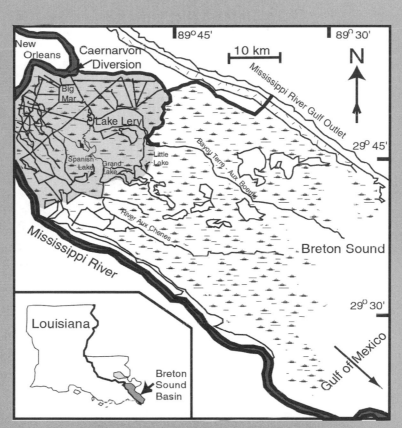

Figure 13.7 Caernarvon diversion from the Mississippi River and downstream Breton Sound in the Louisiana Delta immediately downstream of New Orleans. The shaded area indicates the area over which nitrate-nitrogen retention was 55 percent. *(From Mitsch et al., 2005b; Copyright Elsevier, reprinted with permission.)*

CAST STUDY 4: Ohio State's Kidney Wetlands Act Like Kidneys

In the Midwestern United States, created riparian wetlands at the Des Plaines River Wetlands in northeastern Illinois (Kadlec and Hey, 1994; Phipps and Crumpton, 1994; Mitsch et al., 1995) and the Olentangy River Wetland Research Park at The Ohio State University in central Ohio (Mitsch et al., 1998, 2005a, b, c; Fig. 13.8) have shown patterns of nutrient and sediment retention over multiple years of study of these systems. Both received pumped or overflow river water, thus simulating oxbow wetlands receiving dilute

Figure 13.8 Wilma H. Schiermeier Olentangy River Wetland Research Park experimental wetlands at The Ohio State University, Columbus. These kidney-snaped wetland basins, constructed on a floodplain of the Olentangy River in 1993–94, have received pumped water from the river since March 1994; water flows into kidney basins at the bottom of the kidneys shown here and discharges at the top of the kidneys and return to the river in a swale. Boardwalks enable researchers to reach remote locations in these wetland basins.

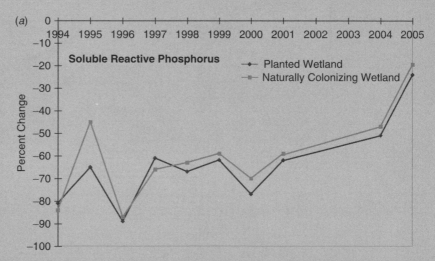

Figure 13.9 Percent reduction in (a) soluble reactive phosphorus, (b) total phosphorus, and (c) nitrite + nitrate nitrogen in the two experimental wetlands at the Olentangy River Wetland Research Park shown in Figure 13.8 for 1994-2005. Each data point represents the average annual decrease in concentrations from inflow to outflow based on weekly sampling.

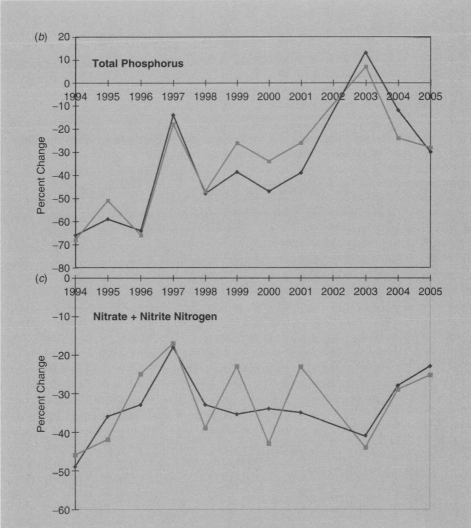

Figure 13.9 (*continued*)

non–point source pollution. For twelve years (1994–2005) at Ohio State's kidney-shaped experimental wetlands (Fig. 13.8), soluble reactive phosphorus and nitrate + nitrite-nitrogen were consistently reduced by about 50 percent and 30 percent, respectively (Fig. 13.9). Both soluble reactive phosphorus and total phosphorus showed trends of decreased retention over the 12 years, with two experimental wetlands actually exporting total phosphorus in one year, 2003. Nitrate-nitrogen retention has shown a steadier pattern. There has been little difference in nutrient retention between the two experimental wetlands since they were created in 1994, even though one of the wetland basins was planted in 1994 (see Chapter 7) and the other was allowed to colonize naturally.

CAST STUDY 5: Creating Treatment Wetlands to Protect Downstream Wetlands in Florida's Everglades

An ambitious plan of stream diversion is occurring in the Florida Everglades where 16,000 ha of created wetlands, called Stormwater Treatment Areas (STAs), are being planned for phosphorus control from upstream agricultural areas. As described in Chapter 9, the main cause of the spread of cattail (*Typha domingensis*) in the otherwise nutrient-poor Florida Everglades dominated by sawgrass (*Cladium jamaicense*) is nutrient enrichment—especially by phosphorus emanating from agricultural areas in the basin. A prototype of the STAs, a 1,544-ha treatment wetland complex called the Everglades Nutrient Removal (ENR) project, has operated since mid-1994 (Reddy et al., 2006). Water is pumped to the ENR wetland complex from adjacent drainage canals, so it is classified as a river diversion wetland here. It was originally designed to reduce phosphorus to 50 µg-P/L, or a 63 to 74 percent reduction in phosphorus (Chimney and Goforth, 2006). That effluent goal has been decreased to 10 µg-P/L, essentially background concentrations of phosphorus in the oligotrophic Everglades. Over its first six-year operating schedule (1994–99), the wetland decreased total phosphorus and total nitrogen by 79 and 26 percent, respectively (Gu et al., 2006), with an average outflow concentration of 21 µg-P/L over that period (Kadlec, 2006). This ENR wetland has been sufficiently successful so that plans are now underway in Florida to expand the amount of treatment wetlands similar to the ENR to 16,000 ha or even more to serve as buffer systems between the Everglades Agricultural Area and the downstream Everglades.

Landfill Leachate Wetlands

Impermeable liners are used to collect groundwater that has passed through the landfill. This leachate is often quite variable in water quality but generally has very high concentrations of ammonium-nitrogen and chemical oxygen demand (Kadlec, 1999). This wastewater has always presented a problem to landfill operators, and stricter water quality standards are making it necessary for advanced treatment. Wetlands are one of several options for management of leachate; other options include spray irrigation, physical/chemical treatment, biological treatment, and piping to a wastewater treatment plant. Mulamoottil et al. (1999) present a summary of results from several dozen constructed wetlands that are treating landfill leachate in Canada, the United States, and Europe.

Wetland Design

The need for rigor in designing a wetland varies widely depending on the site and application. In general, a design that uses natural processes to achieve the objectives yields a less expensive and more satisfactory solution in the long run. However, "naturally" designed wetlands may not develop as predictably as more tightly designed systems should. The choice of design is strongly affected by the site and the objectives. In Europe and many parts of North America, subsurface wetlands are designed in rectangular basins to very specific design criteria. In coastal Louisiana, by contrast, there are now several projects where wetlands are being used as tertiary treatment systems for the removal of nutrients from wastewater. In the following sections, we focus on rigidly designed wetlands, in part because this kind of wetland creation requires much greater ecotechnological sophistication.

Hydrology

Hydrology is an important variable in any wetland design. If the proper hydrologic conditions are developed, chemical and biological conditions will respond accordingly. Improper hydrology leads to the failure of many created wetlands because it will not always correct itself, as will the more forgiving biological components of the system. Ultimately, hydrologic conditions determine wetland function. Several parameters used to describe the hydrologic conditions of treatment wetlands include hydroperiod, depth, seasonal pulses, hydraulic loading rates, and retention time.

Hydroperiod and Depth

In wetlands, hydroperiod is the depth of water in a wetland over time (see Chapter 4). Wetlands that have a seasonal fluctuation of water depth have the most potential for developing a diversity of plants, animals, and biogeochemical processes. In a constructed wastewater wetland with a similar inflow of wastewater every day, water levels often vary little seasonally unless storm water is part of the treatment inflow. During the startup period of constructed wetlands, low water levels are needed to avoid flooding newly emerged plants. Startup periods for the establishment of vegetation may take two to three years of careful attention to water levels.

While storms and seasonal patterns of floods rarely affect constructed wastewater wetlands built for municipal treatment (except when storm sewers are part of the inflow), they can significantly affect the performance of wetlands designed for the control of non–point source runoff. A variable hydroperiod, exhibiting dry periods interspersed with flooding, is a natural cycle in non–point source wetlands, and fluctuating water levels should be considered a natural feature. A fluctuating water level could provide needed oxidation of organic sediments and can, in some cases, rejuvenate a system to higher levels of chemical retention. A typical water-level variation in a newly constructed multicell wetland built in an agricultural region to control nutrients is illustrated in Figure 13.6. There is a definitive seasonal cycle, plus sudden bursts of water levels during winter and spring storms if they occur during nonfrozen conditions. Furthermore, noting the importance of biology in wetland types, the water level in one basin drops almost 30 cm in late fall because of burrowing activity by muskrats (*Ondatra zibethicus*).

Hydraulic Loading Rate

The hydraulic loading rate, one of the most important variables in treatment wetlands, is defined as:

$$q = 100(Q/A) \qquad (13.1)$$

where,

> q = hydraulic loading rate (HLR), the inflow volume per unit time per unit area, cm day^{-1}
> Q = inflow rate, m^3 day^{-1}
> A = wetland surface area, m^2

Table 13.5 summarizes hydrologic loading rates (HLR) of some of the many surface-flow and subsurface-flow wastewater wetlands in North America and Europe. Loading rates to surface-flow constructed wetlands for wastewater treatment from small municipalities ranged from 1.4 to 22 cm day^{-1} (average = 5.4 cm day^{-1}), while rates to subsurface-flow constructed wetlands varied between 1.3 and 26 cm day^{-1} (average = 7.5 cm day^{-1}). Knight (1990) reviewed several dozen wetlands constructed for wastewater treatment and found loading rates to vary between 0.7 and 50 cm day^{-1}. He recommended a rate of 2.5 to 5 cm/day for surface-flow constructed wetlands and 6 to 8 cm day^{-1} for subsurface-flow wetlands.

Detention Time

Detention time of treatment wetlands is calculated as:

$$t = V\,p/Q \qquad (13.2)$$

where,

> t = theoretical detention time, day
> V = volume of wetland basin, m^3
>   = volume of water for surface-flow wetlands
>   = volume of medium through which the wastewater flows for subsurface-flow
> p = porosity of medium (e.g., sand or gravel for subsurface-flow wetlands)
>
> ~1.0 for surface-flow wetlands
>
> Q = flow rate through wetland, m^3 day^{-1}
>   = $(Q_i + Q_o)/2$, where Q_i = inflow, Q_o = outflow

Table 13.5   Hydrologic loading rates for treatment wetlands in North America and Europe

Type of wetland	Loading rate, cm/day
Surface-flow treatment wetlands (n = 15)	5.4 ± 1.7
Subsurface-flow treatment wetlands (n = 23)	7.5 ± 1.0

The optimum detention time (or nominal residence time) has been suggested to be from 5 to 14 days for treatment of municipal wastewater. Florida regulations on wetlands require that the volume in the permanent pools of the wetland must provide for a residence time of at least 14 days. M. T. Brown (1987) suggested a detention time of a riparian treatment wetland system in Florida of 21 days in the dry season and more than 7 days in the wet season.

Calculation of detention time or nominal residence time with Equation 13.2 is not always realistic because of short-circuiting and the ineffective spreading of the waters as they pass through the wetland. Tracer studies of flow through wetlands have illustrated the importance of not over-relying on the theoretical detention time to design treatment wetlands. Not all parcels of water that enter at a certain time leave the wetland at the same time. In some instances, water will short-circuit through the wetland, whereas other water will remain in backwater locations for considerably more time than the theoretical detention time.

Basin Morphology

Several aspects related to the morphology of constructed wetland basins need to be considered when designing wetlands. For example, Florida regulations for the Orlando area require, for littoral zones, a shelf with a gentle slope of 6:1 or flatter to a point of from 60 to 77 cm below the water surface. Slopes of 10:1 or flatter are even better. A flat littoral zone maximizes the area of appropriate water depth for emergent plants, thus allowing more wetland plants to develop more quickly and allowing wider bands of different plant communities. Plants will also have room to move "uphill" if water levels are raised in the basins because of flows being higher than predicted or to enhance treatment. Bottom slopes of less than 1 percent are recommended for wetlands built to control runoff, whereas a substrate slope, from inlet to outlet, of 0.5 percent or less has been recommended for surface-flow wetlands used to treat wastewater.

Flow conditions should be designed so that the entire wetland is effective in nutrient and sediment retention if these are desired objectives. This may necessitate several inflow locations and a wetland configuration to avoid channelization of flows. Steiner and Freeman (1989) suggested a length-to-width ratio (L/W) (called the *aspect ratio*) of at least 10:1 if water is purposely introduced to the system. A minimum aspect ratio of 2:1 to 3:1 has been recommended for surface-flow wastewater wetlands.

Providing a variety of deep and shallow areas is optimum. Deep areas (>50 cm), while too deep for continuous emergent vegetation, offer habitat for fish, increase the capacity of the wetland to retain sediments, can enhance nitrification as a prelude to later denitrification if nitrogen removal is desired, and can provide low-velocity areas where water flow can be redistributed. Shallow depths (<50 cm) provide maximum soil–water contact for certain chemical reactions such as denitrification and can accommodate a greater variety of emergent vascular plants.

Individual wetland cells, placed in series or parallel, often offer an effective design to create different habitats or establish different functions. Cells can be parallel so that alternate drawdowns can be accomplished for mosquito control or redox enhancement, or they can be in a series to enhance biological processes.

Chemical Loadings

When water flows into a wetland, it brings chemicals that may be beneficial or possibly detrimental to the functioning of that wetland. In an agricultural watershed, this inflow will include nutrients such as nitrogen and phosphorus, as well as sediments and possibly pesticides. Wetlands in urban areas can have all of these chemicals plus other contaminants such as oils and salts. Wastewater, when added to wetlands, has high concentrations of nutrients and, with incomplete primary treatment, high concentrations of organic matter (BOD) and suspended solids. At one time or another, wetlands have been subjected to all of these chemicals, and they often serve as effective sinks. Wetlands can be sized using design graphs, standard retention rates, or empirical models.

Design Graphs

The simplest model available to estimate the retention of nutrients or other chemicals by wetlands is to use design graphs that give some measure of chemical retention versus chemical loading, either areal (e.g., $g\ m^{-2}\ yr^{-1}$) or volumetric (e.g., $g\ m^{-3}\ yr^{-1}$). If a wetland were designed to retain nutrients, for example, it would be desirable to know how well that retention would occur for various nutrient inflows. Data compiled from a large number of wetland sites in North America and Europe provide an indication of the nutrient retention of wetlands. For example, Figure 13.10, compiled from wetlands in the Mississippi River Basin, illustrates the percent removal of nitrate-nitrogen versus loading for the Midwestern United States in two ways: (1) mass retention per unit area, and (2) percent retention by concentration. Each of the data points is based on one year's data at one wetland basin in either the Midwestern United States or the river delta in Louisiana.

Retention Rates

Another approach to estimating the retention of nutrients is to simply compare several studies and estimate the chemical retention that consistently happens in wetlands. Averages from data from many constructed wastewater wetlands are shown in Table 13.6. In general, as suggested by the HLR data in Table 13.5, subsurface wetlands

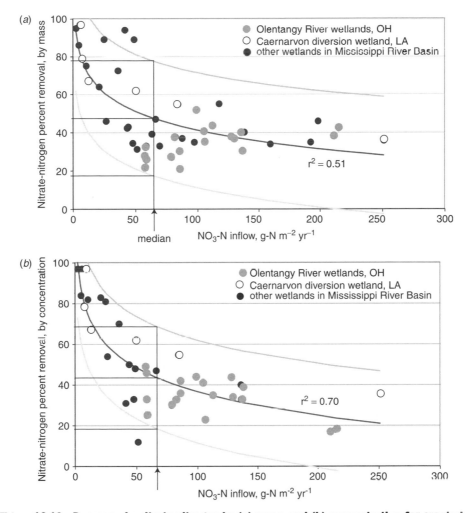

Figure 13.10 Decrease in nitrate-nitrogen by (a) mass, and (b) concentration for created and managed wetlands in the Mississippi River Basin. Each data point represents data for a complete year for a wetland. Outside lines are 95 percent confidence intervals. Vertical lines in graphs indicate median loading rate of 60 g-N m^{-2} yr^{-1}. (Mitsch et al., 2005b, Copyright Elsevier reprinted with permission.)

receive more wastewater and thus receive greater loadings of chemicals and sediments. The high average mass retention of nitrate-nitrogen in subsurface wetlands is due more to these high loading rates in subsurface-flow wetlands than it is to any ability of these systems to sequester more nitrate-nitrogen. Note that the percent nitrate-nitrogen retention is much higher in the surface-flow than in the subsurface-flow wetlands. The retention of phosphorus tends to be more variable in subsurface wetlands than in surface wetlands.

Table 13.6 Nutrient and sediment removal rates and efficiency in constructed wastewater wetlands

	Loading (g m^{-2} yr^{-1})	Retention (g m^{-2} yr^{-1})	Percentage Retention
SURFACE-FLOW CONSTRUCTED WETLANDS			
Nitrate + nitrate nitrogen	29	13	44.4
Total nitrogen	277	126	45.6
Total phosphorus	4.7–56	2.1–45	46–80
Suspended solids	107–6,520	65–5,570	61–98
SUBSURFACE-FLOW CONSTRUCTED WETLANDS			
Nitrate + nitrate nitrogen	5,767	547	9.4
Total nitrogen	1,058	569	53.8
Total phosphorus	131–631	11–540	8–89
Suspended solids	1,500–5,880	1,100–4,930	49–89

Source: Knight (1990) and Kadlec and Knight (1996).

Summaries of retention rates for several wetlands intercepting non–point source pollution are given in Table 13.7. Rules of thumb for this type of wetland is that wetlands can consistently retain phosphorus in amounts of 1 to 2 g-P m^{-2} yr^{-1} and nitrogen in amounts of about 10 to 20 g-N m^{-2} yr^{-1} (Mitsch et al., 2000b). Nitrate-nitrogen retention capabilities of freshwater marshes receiving non–point source pollution in seasonal to cold climates shows a range of nitrogen retention from 3 to 93 g-N m^{-2} yr^{-1} and a phosphorus retention rate of 0.1 to 6 g-P m^{-2} yr^{-1}. Low retention numbers are generally from wetlands that are "under-fed" nutrients. High numbers are usually only periodic, and therefore would be inappropriate for design purposes.

Empirical Models

A third method for estimating the ability of wetlands to retain chemicals is to use equations that either have a theoretical base or are empirically determined from large databases of existing wastewater wetlands. One such general model, developed by Kadlec and Knight (1996) and others, is based on a mass-balance approach called the "k-C* model" and is given as:

$$q(dC/dy) = k_A(C - C^*) \tag{13.3}$$

where

y = fractional distance from inlet to outlet, unitless
C = chemical concentration, g m^{-3}
k_A = areal removal rate constant, m yr^{-1}
C^* = residual or background chemical concentration, g m^{-3}

Table 13.7 Nutrient retention in constructed and natural wetlands receiving low-concentration (I.e., non-wastewater, nutrient loading from rivers, overflows, or non-point source pollution)

Wetland location and type	Wetland size, ha	Nitrogen $\text{g-N m}^{-2}\text{ yr}^{-1}$	Phosphorus $\text{g-P m}^{-2}\text{ yr}^{-1}$	Reference
WARM CLIMATE				
Everglades marsh, S. Florida	8000	—	*0.4–0.6	Richardson and Craft, 1993; Richardson et al., 1997
Boney Marsh, S. Florida	49	4.9	0.36	Moustafa et al., 1996
Everglades Nutrient Removal Project, S. Florida	1545	10.8	0.94	Moustafa, 1999
Restored marshes, Mediterranean delta, Spain	3.5	69	—	Comin et al., 1997
Constructed rural wetland, Victoria, Australia	0.045	23	2.8	Raisin et al., 1997
COLD CLIMATE				
Constructed wetlands, NE Illinois				Mitsch, 1992; Phipps and Crumpton, 1994; Mitsch et al., 1995
river-fed and high-flow	2	**11–38	1.4–2.9	
river-fed and low-flow	2–3	**3–13	0.4–1.7	
Artificially flooded meadows, southern Sweden	180	43–46	—	Leonardson et al., 1994
Constructed wetland basins, Norway	0.035–0.09	50–285	26–71	Braskerud, 2002a,b
Palustrine freshwater wetlands, NW Washington				Reinhelt and Horner, 1995
urban area	2	—	0.44	
rural area	15	—	3.0	
Created instream wetland, OH	6	—	2.9	Niswander and Mitsch, 1995
Created river wetlands, OH (2)	1	**58–66	5.2–5.6	Mitsch et al., 1998; Spieles and Mitsch, 2000a; Nairn and Mitsch, 2000
Created river diversion wetland, OH	3	32	4.5	Fink and Mitsch, 2007
Agricultural wetlands, OH	1.2	**39	6.2	Fink and Mitsch, 2004
Agricultural wetlands, IL (3)	0.3–0.8	**33	0.1	Kovacic et al., 2000
Natural marsh, Alberta, Canada	360	—	*0.43	White et al., 2000

*estimated by phosphorus accumulation in sediment
**nitrate-nitrogen only

This equation is based on an assumption that processes can be described on an areal basis. Thus the coefficient k_A has units of velocity and can be recognized as being similar to a "settling velocity" coefficient used in sedimentation models. C* represents a background concentration of a chemical or constituent, below which it is generally agreed that treatment wetlands cannot go. Integrating this equation over the entire length of the wetland, the solution can be expressed as a first-order areal model:

$$(C_o - C^*)/(C_i - C^*) = \exp[-k_A/q] \tag{13.4}$$

where,

$$C_o = \text{outflow concentration, g m}^{-3}$$
$$C_i = \text{inflow concentration, g m}^{-3}$$
$$q = \text{hydraulic loading rate, m yr}^{-1}$$

Estimates of the two parameters needed for this model, C* and k_A, are listed in Table 13.8. This equation does not work equally well for all parameters, but it does provide a way of estimating the area of a wetland necessary for achieving a certain removal. Rearranging Equations 13.4 and 13.1 gives the following calculation of wetland area for given results:

$$A = Q\ln[(C_o - C^*)/(C_i - C^*)]/k_A \tag{13.5}$$

where

$$Q = \text{flow rate through wetland, m}^3 \text{ yr}^{-1}$$

Where this model is insufficiently backed with good data or does not work properly, strictly empirical relationships of the outflow concentration C_o as a function of the inflow concentration C_i and the hydraulic loading rate (q) have been developed (Table 13.9). Several regression equations of this nature can be used, with other approaches, to estimate outflow concentrations and, in one case, wetland area. For

Table 13.8 Parameters for first-order areal model given in equations 13.3 to 13.5 for several constituents of wastewater wetlands (Subsurface-flow constructed wetlands and surface-flow constructed wetlands are given as wetland type where appropriate.)

Constituent and wetland type	k_A (m yr^{-1})	C* (g m^{-3})
BOD, surface-flow	34	$3.5 + 0.053C_i$
BOD, subsurface-flow	180	$3.5 + 0.053C_i$
Suspended solids, surface-flow	1,000	$5.1 + 0.16C_i$
Total phosphorus, surface and subsurface-flow	12	0.02
Total nitrogen, surface-flow	22	1.5
Total nitrogen, subsurface-flow	27	1.5
Ammonia nitrogen, surface-flow	18	0
Ammonia nitrogen, subsurface-flow	34	0
Nitrate nitrogen, surface-flow	35	0
Nitrate nitrogen, subsurface-flow	50	0

Source: Kadlec and Knight (1996).

Table 13.9 Empirical equations for the estimation of outflow concentrations or wetland area based on inflow concentrations and hydraulic retention time (Correlation coefficient (R^2) and number of wetlands used in analysis (n) are also given.)

Constituent	Equation[a]	R^2 (n)
BOD		
Surface-flow wetlands	$C_o = 4.7 + 0.173\ C_i$	0.62 (440)
Subsurface-flow, soil	$C_o = 1.87 + 0.11\ C_i$	0.74 (73)
Subsurface-flow, gravel	$C_o = 1.4 + 0.33\ C_i$	0.48 (100)
Suspended solids		
Surface-flow wetlands	$C_o = 5.1 + 0.158\ C_i$	0.23 (1,582)
Subsurface-flow wetlands	$C_o = 4.7 + 0.09\ C_i$	0.67 (77)
Ammonia nitrogen		
Surface-flow wetlands	$A = 0.01Q/\exp[1.527\ \ln C_o - 1.05\ \ln C_i + 1.69]$	
Surface-flow marshes	$C_o = 0.336\ C_i^{0.728} q^{0.456}$	0.44 (542)
Subsurface-flow wetlands	$C_o = 3.3 + 0.46\ C_i$	0.63 (92)
Nitrate nitrogen		
Surface-flow marshes	$C_o = 0.093\ C_i^{0.474} q^{0.745}$	0.35 (553)
Subsurface-flow wetlands	$C_o = 0.62\ C_i$	0.80 (95)
Total nitrogen		
Surface-flow marshes	$C_o = 0.409\ C_i + 0.122q$	0.48 (408)
Subsurface-flow wetlands	$C_o = 2.6 + 0.46\ C_i + 0.124q$	0.45 (135)
Total phosphorus		
Surface-flow marshes	$C_o = 0.195\ C_i^{0.91} q^{0.53}$	0.77 (373)
Surface-flow swamps	$C_o = 0.37\ C_i^{0.70} q^{0.53}$	0.33 (166)
Surface-flow wetlands	$C_o = 0.51\ C_i^{1.10}$	0.64 (90)

[a] C_i, inflow concentration (g m^{-3}); C_o, outflow concentration (g m^{-3}); A, area of wetland (ha); Q, wetland inflow, (m^3/day); q, hydraulic retention time, (cm/day).
Source: Kadlec and Knight (1996).

example, the equation estimating total phosphorus efflux in a surface-flow wastewater marsh is:

$$C_o = 0.195\ C_i^{0.91}\ q^{0.53} \tag{13.6}$$

where

$$C_o \text{ and } C_i = \text{outflow and inflow concentrations, respectively, g m}^{-3}$$
$$q = \text{hydraulic loading rate, cm day}^{-1}$$

Other Chemicals

Although most evaluations of the efficiency of wetlands have been concerned with this capacity to remove nutrients, sediments, and organic carbon (BOD), there is some literature on other chemicals, such as iron, cadmium, manganese, chromium, copper, lead, mercury, nickel, and zinc. Wetland soils or biota or both often easily sequester metals. That is the basic problem in using wetlands as sinks for such chemicals: they can accumulate in the food chain.

Soils

The topsoil is important to the overall function of a constructed wetland (Figure 13.11). It is the primary medium supporting rooted vegetation and, particularly for subsurface wetlands, it is part of the treatment system. The sediments retain certain chemicals and provide the habitat for micro-and macro-flora and fauna that are involved in chemical transformations. Constructed wetland soil texture depends on whether surface flow over the substrate or subsurface flow through the substrate is being considered. Surface-flow wetland soils are generally less effective in removing pollutants per unit area but are closer in design to natural wetlands. Their ability to provide structure and nutrition to the wetland plants is important. Clay material, although favored as a subsurface liner, limits root and rhizome penetration and may prevent water from reaching plant roots. Silt clay or loam soils are preferable for the overlying soils in constructed wetlands. Sandy soil is less preferred for surface-flow

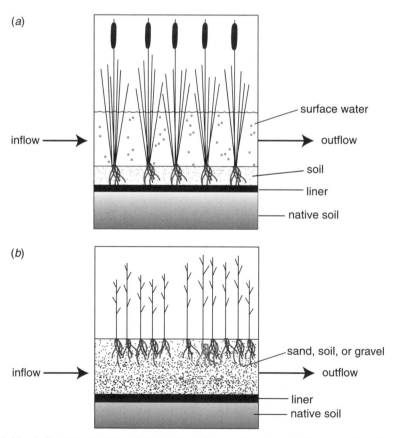

Figure 13.11 Soil cross-section of (a) surface-flow wetland, and (b) subsurface-flow wetland. (After Knight, 1990.)

wetlands. For subsurface-flow wetlands, high permeability is preferred. The material needs to be sand, gravel, or some other highly permeable media.

The subsoil of constructed wetlands (usually below the root zone and referred to as a *liner*) must have permeability low enough to cause standing water or saturated soils. If clay is not available on site, it may be advisable to add a layer of clay to minimize percolation. Studies have also been undertaken to investigate other materials as liners for constructed wetlands. The most frequently used liners for constructed wetlands are clays, clay bentonite mixtures, or synthetic materials such as polyvinyl chloride (PVC) and high-density polyethylene (HDPE) (Kadlec and Knight, 1996). Experiments have been conducted in recycling materials such as coal combustion waste products. As it turns out, using calcium-rich sulfur-scrubber waste material may actually increase the phosphorus-retention capability of the wetlands (Ahn et al., 2001; Ahn and Mitsch, 2001), but care must be taken that the material completely seals the wetland because leachate from this liner material is highly alkaline.

Subsurface flow through subsurface wetlands can be through soil media *(root-zone method)* or through rocks, gravel, or sand *(rock-reed filters)*. Flow in both cases is 15 to 30 cm below the surface. Gravel is sometimes added to the substrate of subsurface-flow wetlands *(gravel-bed)* to provide a relatively high permeability that allows water to percolate into the root zone of the plants where microbial activity is high. In a survey of several hundreds of wetlands built in Europe for sewage treatment in rural settings, Cooper and Hobson (1989) reported that gravel has been used in combination with soil but that the substrate remains the greatest uncertainty in artificial reed *(Phragmites)* wetlands. Gravel can be silica-based or limestone-based; the former has less capacity for phosphorus retention. Another evaluation of the European-design subsurface wetlands indicated that they often decrease in hydrologic conductivity after several years and become clogged, essentially becoming partial-surface-flow wetlands.

Organic Content

The organic content of soils has some significance for the retention of chemicals in a wetland. Mineral soils generally have lower cation exchange capacity than organic soils do; the former is dominated by various metal cations, and the latter is dominated by the hydrogen ion. Organic soils can therefore remove some contaminants (e.g., certain metals) through ion exchange and can enhance nitrogen removal by providing an energy source and anaerobic conditions appropriate for denitrification. Organic matter in wetland soils varies between 5 and 75 percent, with higher concentrations in peat-building systems such as bogs and fens and lower concentrations in mineral-soil wetlands such as riparian bottomland wetlands subject to mineral sedimentation or erosion. When wetlands are constructed, especially subsurface-flow wetlands, organic matter such as composted mushrooms, peat, or detritus is often added in one of the layers. For construction of many wetlands, however, organic soils are avoided because they are low in nutrients, can cause low pH, and often provide inadequate support for rooted aquatic plants.

Depth and Layering of Soil

The depth of substrate is an important design consideration for wastewater wetlands, particularly those that use subsurface flow. The depth of suitable topsoil or substrate should be adequate to support and hold vegetation roots. A common substrate depth for constructed wetlands is 60 to 100 cm. In some cases, layering more elaborate than that shown in Figure 13.11 is suggested. Meyer (1985) described a layered substrate in wetlands to control stormwater runoff as having the following materials from the base upward from the liner: 60 cm of 1.9 cm limestone; 30 cm of 2 mm crushed limestone to raise pH and aid in the precipitation of dissolved heavy metals and phosphate; 60 cm of coarse to medium sand as filter; and 50 cm organic soil (see the Vegetation discussion later).

Soil Chemistry

Although exact specifications of nutrient conditions in a wetland soil necessary to support aquatic plants are not well known, low nutrient levels characteristic of organic, clay, or sandy soils can cause problems for initial plant growth. Although fertilization may be necessary in some cases to establish plants and enhance growth, it should be avoided if possible in wetlands that eventually will be used as sinks for the same macronutrients. When fertilization is required to get plants started in constructed wetlands, slow-release granular and tablet fertilizers are often useful.

When soils are submerged and anoxic conditions result, iron is reduced from ferric (Fe^{+++}) to the ferrous (Fe^{++}) ions, releasing phosphorus that was previously held as insoluble ferric phosphate compounds. The Fe-P compound can be a significant source of phosphorus to overlying and interstitial waters after flooding and anaerobic conditions occur, particularly if the wetland was constructed on previously agricultural land. After an initial pulse of released phosphorus in such constructed wetlands, the iron and aluminum contents of a wetland soil exert significant influences on the ability of that wetland to retain phosphorus. All things being equal, soils with higher aluminum and iron concentrations are more desirable because their affinity for phosphorus is higher.

Vegetation

Just as the question "What plants should be used?" arises for creating and restoring wetlands as discussed in Chapter 12, vegetation choice is also a consideration for treatment wetlands. But there is at least one significant difference for treatment wetlands: While creation and restoration of wetlands are principally done to develop a diverse vegetation cover and provide habitat, treatment wetlands are constructed with the main goal of improving water quality. The plants in created and restored wetlands are part of the solution; in treatment wetlands, they are the partial cause of the solution. Furthermore, treatment wetlands invariably have higher concentrations of chemicals in the water, which by its very nature limits the number of plant species that will survive in those wetlands. Experience has shown that relatively few plants thrive in the high-nutrient, high-BOD wastewaters that are applied to treatment wetlands. Among those plants are cattails (*Typha* spp.), the bulrushes (*Schoenoplectus* spp., *Scirpus* spp.), and reed grass (*Phragmites australis*). The last is the preferred plant in subsurface-flow wetlands around the world but is not favored in many parts of North America because of its aggressive behavior in freshwater and brackish marshes (see Delaware Bay case study, Chapter 12).

When water is deeper than 30 cm, emergent plants often have difficulty growing. In these cases, surface-flow wetlands can become covered with duckweed (*Lemna* spp.) in temperate zones and water hyacinths (*Eichhornia crassipes*) and water lettuce (*Pistia* spp.) in the subtropics and tropics. While rooted floating aquatics such as *Nymphaea*, *Nuphar*, and *Nelumbo* are favored for their aesthetics, they thrive only in rare instances in treatment wetlands where, due to high-nutrient conditions, they are easily overwhelmed by duckweed and filamentous algae.

Tanner (1996) compared relative nutrient uptake and pollutant removal of eight macrophytes in gravel-bed wetland mesocosms fed by dairy wastes in New Zealand (Fig. 13.12). Greatest aboveground biomass was seen in this highly polluted wastewater by *Glyceria maxima* and *Zizania latifolia*, while greatest belowground biomass was seen with *Bolboschoenus fluviatilis* (*Scirpus fluviatilis* in the United States), which had belowground biomass 3.3 times its aboveground biomass (Fig. 13.12a). Total nitrogen removed from these mesocosms was linearly correlated with total plant biomass (Fig. 13.12b). Based on key growth characteristics of the plants in this wastewater, three productive gramminoids (*Zizania latifolia*, *Glyceria maxima*, and *Phragmites australis*) had the highest overall scores. *Baumea articulata*, *Cyperus involucratus*, and *Schoenoplectus validus* had medium scores, while *Scirpus fluviatilis* and *Juncus effusus* had the lowest scores and are least likely to be effective plants in wastewater wetlands.

Establishing Vegetation

Vegetation can be established through the same general procedures outlined in Chapter 12—that is, by planting roots and rhizomes directly or by seeding. Because these wetlands are usually constructed on former upland with no connection to rivers or streams, reliance on nature bringing in plant propagules generally does not work. Field harvested plants or nursery-grown stock can be used for plantings. The former have the advantage of establishing vegetation cover more quickly than would smaller nursery stock. Also, these plants, if harvested nearby, are adapted to the local climate and may be from the proper genotype for the region. Conversely, harvesting plants in large numbers from natural wetlands may threaten those wetlands. Nursery plants are easier to plant because of their generally small size, and a greater diversity and number of plants can be obtained from good nurseries. However, it is often unclear what genetic stock was used to start these plants, and they may not be from stock adapted to the local climate.

Water, either too much or too little, is the major reason why macrophytes do not become well established in wetlands constructed for wastewater treatment. When plants are first establishing themselves, the optimum conditions are moist soils or very shallow (<5 cm) water depth. If water is too deep, the new plants will be flooded out.

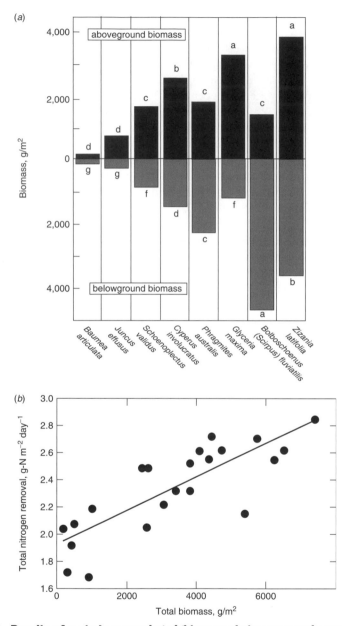

Figure 13.12 Results of a study comparing eight macrophytes commonly used in waste-water wetlands in New Zealand after 124 days of culture in dairy farm wastewater. (a) Mean aboveground and belowground biomass accumulation of the eight macrophytes. Different letters indicate significant differences. (b) Relationship between total nitrogen removal from ammonium-rich dairy farm wastewater and total biomass. Regression coefficient $r^2 = 0.66$. (After Tanner, 1996.)

If there is inadequate water and topsoil dries out, the plants will not survive. If the wastewater can be used in measured amounts to irrigate the plants, this is optimum. If not, artificial irrigation might be required to make sure that plants are successful.

Wetland Management after Construction

Wildlife Control

Although the development of wildlife is a welcomed and often desired aspect of treatment wetlands, managing plant and animal populations often becomes necessary maintenance of constructed wetlands. In North America, beaver (*Castor canadensis*) and muskrat (*Ondatra zibethicus*) create obstructions to inflows and outflows, destroy vegetation, or burrow into dikes. (This is one reason why dikes should not be built up around constructed wetlands if they can be avoided.) Major vegetation removal, particularly by herbivorous muskrats that use the plant material both for food and shelter, can take a fully vegetated marsh and turn it into a plant-devoid pond in the matter of weeks or months. These events are referred to as eat-outs. There is very little that can be done to prevent these eat-outs except to trap the animals, which is a laborious task.

In other cases, animals such as beaver and muskrat and large birds such as Canada geese (*Branta canadensis*) and snow geese (*Chen* spp.) grazing on newly planted perennial herbs and seedlings are particularly destructive. The timing of planting is important, especially if migratory animals are involved in destructive grazing in the winter. Using gunshot devices and the extract of grape juice as a "hot foot" material on the adjacent landscape have all been suggested but without any permanent success. Probably the easiest approach we have noted in many years of observing geese is to have a wide band of emergent vegetation between where they land (on the water) and where they like to graze (upland lawns). But of course, you will have to get the local muskrats to cooperate and not remove the vegetation.

Trapping of muskrats and beavers can alleviate their impact on constructed wetland basins, but it can also be time consuming and often ineffective. Grazing by geese is more difficult to control, although their impact is most significant soon after vegetation is planted. If vegetation can get established, the grazing effects of geese can be controlled.

Similarly, deeper wetlands often become havens for undesirable fish such as carp (*Cyprinus carpio*), which can cause excessive turbidity and uproot vegetation. Carnivores such as northern pike have been discussed as a potential control of carp. Total removal of fish by drawdown is probably necessary if carp begin to degrade outflow water quality excessively. The problem is that this fish removal might affect mosquito control (see the following discussion).

Attracting Wildlife

Just as many animals can cause maintenance headaches, the attraction of wildlife to constructed wetlands is one of the reasons why public support for such projects can be high in the first place. So every attempt should be made to have a diverse ecosystem and not just a pond with water flowing through it. Weller (1994) recommends a 50:50 ratio of open water to vegetation cover in marshes to attract water birds, and this kind of ratio, with proper development of the initial bathymetry of the ponds, is quite easy. Also, creating diverse habitats with live and dead vegetation, islands, and floating structures is desirable.

In many cases of wetland construction, wildlife enhancement begins soon after construction. At a constructed wetland at Pintail Lake in Arizona, the area's waterfowl population increased dramatically by the second year of use; duck nest density increased 97 percent over the first year (Wilhelm et al., 1989). A considerable increase in avian activity was also noted at the Des Plaines River Wetlands Demonstration Project in northeastern Illinois. Migrating waterfowl increased from 3 to 15 species and from 13 to 617 individuals between 1985 (preconstruction) and 1990 (one year after water was introduced to the wetlands). The number of wetland-dependent breeding birds increased from 8 to 17 species, and two state-designated endangered birds, the least bittern and the yellow-headed blackbird, nested at the site after wetland construction (Hickman and Mosca, 1991).

One interesting question related to attracting birds to treatment wetlands is whether the birds might have an effect on the treatment capacity of the wetlands and, specifically, if birds are in high numbers, their excreta could undermine the effectiveness of the wetlands for nutrient and organic removals. Anderson et al. (2003) presented a several-year study done on a 10-ha treatment wetland in northern California on that possible effect. Bird use peaked at 12,000 individuals during the second year during a four-month period. They found average daily inputs by birds of 2.5 g N m^{-2} yr^{-1} and 0.9 g P m^{-2} yr^{-1}, which represented less than 10 percent of the mean daily loading rates to the wetland. The authors conclude that bird use "does not lead to a significant reduction in treatment performance" (Anderson et al., 2003).

Mosquito Control

The subject of mosquito control will always be brought up when wetlands are being constructed, particularly when the wetlands receive runoff or wastewater. In general, it has been concluded that properly managed wastewater treatment wetlands pose no more mosquito threat than do natural wetlands (Knight et al., 2003). Mosquitoes can be controlled in constructed wetlands by changing the hydrologic conditions of the wetland to inhibit mosquito larvae development (flow-through conditions discourage mosquitoes), or by using chemical or biological control. Many researchers have proposed mosquito control by fish, especially the air-gulping mosquito fish *(Gambusia affinis)* or similar small fish. One reason to maintain some deeper areas in wetlands in temperate zones is to allow fish such as *Gambusia* and other top minnows and sunfish to survive the winter and feed on mosquito larvae. Little is known about the role that water quality has on encouraging or discouraging mosquitoes directly, but the effect of poor water quality removing fish can have a dramatic effect in causing mosquito increases. Bacterial insecticides (e.g., *Bacillus sphaericus* and *Bacillus thuringiensis* var. *israelensis)* and the fungus *Lagenidium giganteum* are known pathogens of mosquito larvae, but they have not been extensively tested, and there is always the possibility of resistance induction in the mosquitoes (Knight et al. 2003). Constructing boxes to encourage nesting by swallows (Hirundinidae), swifts (Apodidae), and bats have also been used to control adult mosquito populations at constructed wetlands.

Some studies have evaluated the relative importance of different macrophyte species on the propensity of mosquito survival (Table 13.10). In general, the more dense the plant stands, the more difficult it is for both predators and mosquito control efforts to reach the mosquitoes. Thus, the highly productive plants (*Typha*, several *Schoenoplectus* = *Scirpus*, *Phragmites*, and *Eichhornia crassipes*) have the highest mosquito scores in Table 13.10. Knight et al. (2003) suggest the following strategy for treatment wetland design to minimize mosquitoes:

- Select plant species that optimize both wastewater treatment performance and mosquito control.
- Include deepwater zones that are free of emergent and aquatic plants to provide fish habitat and access to vegetated areas.
- Limit the width of emergent plant zones to facilitate access by predaceous fish and for application of chemical control agents.
- Design wetlands with steep embankments (although this is effective for mosquito control, it is not a good strategy to develop a diverse littoral zone around the wetland).

Pathogens

Because many treatment wetlands are specifically built to deal with human and animal wastewater, proper sanitary engineering techniques should be used to minimize human exposure to pathogens. Treatment wetlands are meant to be biologically rich systems, and microbial activity is a major part of the treatment process. Measurements of indicator organisms such as fecal and total coliforms should be part of the monitoring of municipal wastewater treatment wetlands. Nearby wells should also be sampled, because water seeping from a wastewater wetland near potable water supplies should be carefully monitored. If a wetland is being used as tertiary treatment to a conventional treatment plant, consideration in the design of the conventional treatment plant has to be given to the disinfection system. Chlorine disinfection and the resulting chlorine residual would cause significant problems in treatment wetlands, so other means of disinfection (ozonation or UV radiation) should be used if there is disinfection before the wastewater enters the wetland.

Water-Level Management

The water level of surface-flow treatment wetlands is the key to both water quality enhancement and vegetation success. Most constructed municipal wastewater

Table 13.10 Estimated mosquito production propensity of various wetland plant species

Plant group	Plant species	Common name	Mosquito production score*
Rooted emergent plants			
	Alisma geyeri	Water-plantain	7
	Alisma trivale	Water-plantain	7
	Alopercurus howellii	Foxtail	9
	Carex obnupta	Sedge	11
	Carex rostrata	Sedge	14
	Carex stipata	Sedge	13
	Cyperus aristatus	Flat sedge	9
	Cyperus difformis	Flat sedge	11
	Cyperus esculentus	Flat sedge	13
	Cyperus niger	Flat sedge	12
	Deschampsia danthonides	Grass	11
	Echinochloa crusgalli	Barnyard grass	11
	Echinodorus berteroi	Burhead	10
	Eleocharis palustris	Spikerush	10
	Equisetum arvense	Horsetail	14
	Frankenia grandifolia	Alkali heath	14
	Glyceria leptostachya	Mannagrass	12
	Juncus acutus	Softrush	13
	Juncus effusus	Softrush	10
	Jussiaea repens	Primrose	16
	Leersia oryzoides	Rice cutgrass	11
	Leptochloa fasicularis	Salt-meadow grass	10
	Ludwigia spp.	Primrose willow	9
	Lythrum californicum	Loosestrife	13
	Oryza sativa	Rice	9
	Phalaris arundinacea	Reed canary grass	14
	Phragmites australis	Common reed	17
	Plantago major	Common plantain	9
	Polygonum amphibium	Water smartweed	14
	Polygonum hydropiperoides	Smartweed	12
	Polygonum pennsylvanicum	Pinkweed	12
	Polygonum punctatum	Smartweed	12
	Polypogon elongatus	Rabbitfoot grass	11
	Potentilla palustris	Cinquefoil	11
	Pterididum aquilinum	Fern	13
	Sagittaria latifolia	Duck-potato	7
	Sagittaria longiloba	Arrowhead	7
	Sagittaria montevidensis	Giant arrowhead	8
	Scirpus acutus	Bulrush	15
	Scirpus americanus	Three-square bulrush	10
	Scirpus californicus	Giant bulrush	15
	Scirpus olneyi	Alkali bulrush	12
	Sparganium eurycarpum	Burreed	13
	Typha angustifolia	Narrowleaf cattail	16
	Typha glauca	Cattail	16
	Typha latifolia	Common cattail	17
	Zizania aquatica	Wildrice	13

Table 13.10 (*continued*)

Plant group	Plant species	Common name	Mosquito production score*
Floating aquatic plants			
	Azolla filiculoides	Water fern	10
	Bacopa nobsiana	Water hyssop	13
	Brasenia schreberi	Water shield	12
	Eichhornia crassipes	Water hyacinth	18
	Hydrocotyle ranunculoides	Pennywort	15
	Hydrocotyle umbellata	Pennywort	15
	Lemna gibba	Duckweed	9
	Lemna minima	Duckweed	9
	Nasturtium officinale	Water cress	15
	Nuphar polysepalum	Spatterdock	11
	Pistia stratiotes	Water lettuce	18
	Potamogeton crispus	Curled pondweed	8
	Potamogeton diversifolius	Pondweed	8
	Ranunculus aquatilis	Buttercup	16
	Ranunculus flammula	Buttercup	15
	Spirodela polyhyiza	Duckmeat	9
	Wolffiella lingulata	Bog mat	9
Submerged aquatic plants			
	Callitriche longipedunculata	Water starwort	11
	Ceratophyllum demersum	Coontail	15
	Eleocharis acicularis	Spikerush	8
	Elodea canadensis	Waterweed	8
	Elodea densa	Waterweed	11
	Isoetes howellii	Quillwort	7
	Isoetes orcuttii	Quillwort	7
	Lilaeopsis occidentalis	Lilaeosis	7
	Myriophyllum spicatum	Water milfoil	14
	Najas flexilis	Naiad	11
	Najas graminea	Naiad	11
	Potamogeton filiformis	Pondweed	13
	Potamageton pectinatus	Sago pondweed	13
	Ruppia spiralis	Ditchgrass	11
	Utricularia gibba	Bladderwort	12
	Utricularia vulgaris	Bladderwort	13
	Zannichellia palustris	Horned pondweed	10

*Scores less than 9 indicate minimal mosquito breeding problems; scores between 9 and 13 indicate a need to maintain a low coverage for this plant species; and scores of 14 and above indicate a need to minimize the occurrence of the plant species in the wetland to avoid mosquito issues (Collins and Resh, 1989).
Source: Knight et al. (2003) and Collins and Resh (1989).

wetlands have little control on the overall inflow of wastewater. Flow and hence depth are first controlled by designing the basin large enough to create the proper hydraulic loading rates (HLR). Most constructed wetlands have a control structure such as a flume or weir to control outflow; these structures should be flexibly designed so they can be manipulated to control water depth. Too much water stresses macrophytes as much as too little water. Water depths of 30 cm or less are optimum for most herbaceous macrophytes used in treatment wetlands. Water depths greater than 30 cm can lead to vegetation reduction.

Compounding the effect that water level has on vegetation is the effect that it has on wastewater treatment. High water levels favor a high HLR and sediment and phosphorus retention associated with sedimentation and similar processes; it also leads to less resuspension and longer retention time. Shallow water leads to closer proximity of sediments and overlying water, often causing anaerobic or near-anaerobic conditions during the growing season. Low water thus favors the reduction of nitrate-nitrogen through denitrification. Optimizing wastewater treatment and vegetation success is a continual balancing act.

Economics and Values of Treatment Wetlands

It is generally believed that treatment wetlands are less expensive to build and maintain than conventional wastewater treatment, and that is the appeal of these systems to many people. However, cost comparisons should be carefully made before investing in these systems. Any estimate of the cost of a new wetland's development should include the following items:

1. Engineering plan
2. Preconstruction site preparation
3. Construction costs (e.g., labor, equipment, materials, supervision, indirect and overhead charges)
4. Cost of land

Capital Costs

An equation estimating the cost of constructing wastewater wetlands (not including the cost of land) is:

$$C_A = \$196,336 \; A^{-0.511} \tag{13.7}$$

where

C_A = capital cost of wetland construction per unit area, $\$ha^{-1}$
A = wetland area, ha

This relationship suggests that a 1-ha wetland would cost almost $200,000, a 10-ha wetland would cost $60,000 per ha, and a 100-ha wetland would cost $19,000 per ha. The data clearly suggest that there is an economy of scale involved in wetland construction.

Operating and Maintenance Costs

Operating and maintenance costs vary according to the wetland's use and to the amount and complexity of mechanical parts and plumbing that the wetland contains.

Fewer data on operational costs are available. Kadlec and Knight (1996) estimate the operation and maintenance costs for one wastewater wetland to be about $85,500 per year. That estimate included $50,000 per year for personnel to be in charge of the 175-ha wetland. A wide range of $5,000 to $50,000 per year of operating and maintenance costs was estimated by Kadlec and Knight (1996) from smaller wetlands. Gravity-fed wetlands are far less expensive to maintain than highly mechanized wetlands that need significant plumbing and pumps.

Other Benefits of Treatment Wetlands

Subsurface treatment wetlands provide little additional benefit beyond the water quality improvement they were designed to provide, but surface-flow treatment wetlands have a variety of additional benefits. The watery habitat that is created can be a major ancillary benefit of these systems. In addition to providing habitat for mammals such as nutria, beavers, muskrats, amphibians, fish, and voles, surface-flow treatment wetlands are often a haven for waterfowl and wading birds. Human uses such as trapping and hunting are not incompatible with some wastewater wetlands. If designed properly in an urban area, wetlands are locations where the public can visit and learn about their important water quality role. This message is a powerful one to the uninitiated, and they will often become ardent wetland conservationists as a result of seeing "wetlands at work."

Another benefit of using both natural and constructed wetlands for water quality improvement relates to areas where "land-building" is needed. In the subsiding environment of Louisiana's Gulf Coast, nutrients are permanently retained in the peat of wetlands receiving high-nutrient wastewater as the wetland aggrades to match subsidence. In this case, wastewater discharge into a wetland can occur without saturating the system and simultaneously helps counteract the deleterious effects of land subsidence.

Comparing Wetlands and Conventional Technology

A comparison of the construction and operating costs of a proposed large (>2,000 ha) wetland that was to be constructed in the Florida Everglades with conventional chemical treatment is illustrated in Table 13.11. In this example, if land costs are not considered, the wetland alternative has an 11 percent lower capital cost and 56 percent lower operating costs than the chemical treatment alternative. Although land costs can be significant for treatment wetlands, particularly in urban areas (in essence, solar energy is being substituted for fossil fuel energy), it is generally not appropriate to use the cost of land in comparison with technological solutions that require little land. This is because the land being used by the wetland can be sold after the life of the wetland is completed, while the salvage value of the worn-out equipment used for conventional treatment alternatives is generally zero (Kadlec and Knight, 1996).

One of the more clever calculations of the difference between using wetlands versus conventional mechanical systems for wastewater treatment is an illustration of

Table 13.11   Estimated cost comparison for phosphorus control in 760,000 m^3 day^{-1} agricultural runoff in Florida

	Treatment Wetland	Chemical Treatment
Land cost	$34,434,000	$2,140,000
Capital costs (land free)	$95,836,000	$108,260,000
Total annual operation/maintenance	$1,094,000	$2,490,000
O&M, present worth	$33,443,000	$76,153,000
Total present worth, without land cost	$129,279,000	$185,637,000
Total present worth, with land cost	$163,713,000	$187,777,000

Source: Kadlec and Knight (1996).

Table 13.12   Net atmospheric generation of carbon for a 3,800 m^3 day^{-1} wastewater treatment facility using treatment wetlands or conventional mechanical treatment

Carbon Flow	Treatment Wetland (metric tons C day^{-1})	Conventional Treatment (metric tons C day^{-1})
Atmospheric carbon from power generation	53	1,350
Carbon sequestration	−3	0
Net atmospheric carbon	50	1,350

Source: Ogden (1999).

the relative impact on the emission of the greenhouse gas CO_2. Normalizing estimates for a 3,800 m^3 day^{-1} (1 million gal/day) flow of wastewater, mechanical treatment leads to 27 times more emission of carbon dioxide to the atmosphere than does a treatment wetland (Table 13.12). The wetland system, in fact, has the additional benefit of sequestering a small amount of carbon. Conventional wastewater treatment uses 3.9 kg of fossil fuel carbon to remove 1 kg of carbon; a wetland treatment system uses 0.16 kg of fossil fuel carbon to remove 1 kg of carbon (Ogden, 1999).

Summary Considerations

Wastewater treatment wetlands are not the solution to all water quality problems and should not be viewed as such. Many pollution problems, such as excessive BOD or metal contamination, may require more conventional approaches. Yet the fact that thousands of wetlands have been constructed around the world for pollution control attests to their importance and value. Several considerations, both technical and institutional, must be considered as treatment wetlands are designed and built.

Technical Considerations

1. Values of the wetlands such as wildlife habitat should be considered in any treatment wetland development.

2. Acceptable pollutant and hydrologic loadings must be determined for the use of wetlands in wastewater management. Appropriate loadings, in turn, determine the size of the wetland to be constructed. Overloading a constructed wetland can be worse than not building it at all.

3. All existing characteristics of local natural wetlands, including vegetation, geomorphology, hydrology, and water quality, should be well understood so that natural wetlands can be "copied" in the construction of treatment wetlands.

4. Particular care should be taken in the wetland design to address public health, including mosquito control and protection of groundwater resources.

Institutional Considerations

1. Wastewater treatment by wetlands can often serve the dual purposes of both wetland habitat development and wastewater treatment and recycling. The creation of treatment wetlands as mitigation for lost wetlands is still generally unacceptable because of the lack of sustainability and the high level of pollution of treatment wetlands compared to restored wetlands, but opportunities for dual use of wetlands should continue to be explored.

2. Many permit processes in governments do not recognize treatment wetland systems as alternatives for wastewater treatment. In these cases, experimental systems should first be established for a given region. Modification of requirements for granting permits for pilot wetlands is needed to make effective progress in developing approaches.

It is useful to remember that wetland design is an inexact science and that perturbation and biological change are the only things we can be sure of in these created ecosystems. Traditional engineering approaches to wastewater wetlands, without an appreciation of self-design in ecosystems, are sure to cause disappointment. If a treatment wetland continues to function according to its main goal—that is, improving water quality—changes in plant species should not be viewed as that significant unless invasive non-native plants become dominant.

Recommended Readings

IWA Specialists Group on Use of Macrophytes in Water Pollution Control. 2000. Constructed Wetlands for Pollution Control. *Scientific and Technical Report No. 8*. International Water Association (IWA), London, England, 156 pp.

Kadlec, R., and R. Knight. 1996. *Treatment Wetlands*. CRC Press, Boca Raton, FL, 893 pp.

Kadlec, R., and S. Wallace. 2008. *Treatment Wetlands II*. CRC Press, Boca Raton, FL.

Wetland Laws and Protection

In earlier times, wetland drainage was considered the only policy for managing wetlands in many parts of the world. With the recognition of wetland values, wetland protection has been emphasized by many laws and international agreements. The U.S. federal government has relied on executive orders, a "no net loss" policy, and the Section 404 dredge-and-fill permit program of the Clean Water Act for wetland protection, augmented by wetland protection in agricultural programs and by individual states and aided by the development of wetland delineation procedures. Individual states are developing stronger wetland protection laws, as two Supreme Court decisions in the first six years of the 21st century have potentially limited the jurisdiction of federal protection of some wetlands. International cooperation in wetland protection, particularly through the Ramsar Convention and the North American Waterfowl Management Plan, has been emphasized in recent years as we enter a century of a global economy and policy makers realize that the functions of local wetlands cross international boundaries.

Wetlands are now the focus of institutional and legal protection efforts throughout the world, but, because of this focus, they are beginning to be defined by legal fiat as much as by the application of ecological principles. Chapter 2 reviewed the major definitions of wetlands that have developed in the United States and internationally. Some definitions are scientific, whereas others are principally to allow legal definition and thus protection of wetlands. Protection has been implemented through a variety of policies, laws, and regulations, ranging from animal and plant protection to land use and zoning restrictions to enforcement of dredge-and-fill laws. In the United States, wetland protection has historically been a national initiative, often with assistance and implementation provided by individual states. In the international arena, agreements to protect ecologically important wetlands throughout the world have been negotiated and ratified and are becoming more important every year with the globalization of the world's economies.

Legal Protection of Wetlands in the United States

The policy of the United States for more than 120 years was to drain wetlands. The Swamp Land Acts of 1849, 1850, and 1860 described in Chapter 9 were precursors to one of the most rapid and dramatic changes in the landscape that has ever occurred in history, even though the acts were deemed to be largely ineffective in their intended purpose (National Research Council, 1995). By the mid-1970s, about half of the wetlands in the lower 48 states were drained (see Chapter 3). In the early 1970s, interest in wetland protection began as scientists identified and quantified the many values of these ecosystems. This interest in wetland protection began to be translated at the federal level in the United States into interpretation of existing laws, regulations, and public policies. Prior to this time, federal policy on wetlands was vague and often contradictory. Policies in agencies such as the U.S. Army Corps of Engineers, the Soil Conservation Service (now the Natural Resources Conservation Service), and the Bureau of Reclamation encouraged the destruction of wetlands, whereas policies in the Department of the Interior, particularly in the U.S. Fish and Wildlife Service, encouraged their protection. Some states also developed inland and coastal wetland laws and policies during the 1970s as well.

The primary wetland protection mechanisms used by the U.S. federal government are summarized in Table 14.1. Some of the more significant activities of the federal government that led to a more consistent wetland protection policy have included presidential orders on wetland protection and floodplain management, implementation of a dredge-and-fill permit system to protect wetlands, coastal zone management policies, and initiatives and regulations issued by various agencies. Despite all of this activity related to federal wetland management, two major points should be emphasized:

1. *There is no specific national wetland law in the United States.* Wetland management and protection result from the application of many laws intended for other purposes. Jurisdiction over wetlands has also been spread over several agencies, and, overall, federal policy continually changes and requires considerable interagency coordination.

2. *Wetlands have been managed under regulations related to both land use and water quality.* Neither of these approaches, taken separately, can lead to a comprehensive wetland policy. The regulatory split mirrors the scientific split noted by many wetland ecologists, a split that is personified by people who have developed expertise in either aquatic or terrestrial systems. Rarely do individuals possess expertise in both areas.

Early Presidential Orders

President Jimmy Carter issued two executive orders in May 1977 that established the protection of wetlands and riparian systems as the official policy of the federal

Table 14.1 Major federal laws, directives, and regulations in the United States used for the management and protection of wetlands

	Date	Responsible Federal Agency
DIRECTIVE OR STATUTE		
Rivers and Harbors Act, Section 10	1899	U.S. Army Corps of Engineers
Fish and Wildlife Coordination Act	1967	U.S. Fish and Wildlife Service
Land and Water Conservation Fund Act	1968	U.S. Fish and Wildlife Service, Bureau of Land Management, Forest Service, National Park Service
National Environmental Policy Act	1969	Council on Environmental Quality
Federal Water Pollution Control Act (PL 92-500) as amended (Clean Water Act)	1972, 1977, 1982	
Section 404—Dredge-and-Fill Permit Program		U.S. Army Corps of Engineers with assistance from Environmental Protection Agency and U.S. Fish and Wildlife Service
Section 208—Areawide Water Quality Planning		U.S. Environmental Protection Agency
Section 303—Water Quality Standards		U.S. Environmental Protection Agency
Section 401—Water Quality Certification		U.S. Environmental Protection Agency (with state agencies)
Section 402—National Pollutant Discharge Elimination System		U.S. Environmental Protection Agency (or state agencies)
Coastal Zone Management Act	1972	Office of Coastal Zone Management, Department of Commerce
Flood Disaster Protection Act	1973, 1977	Federal Emergency Management Agency
Federal Aid to Wildlife Restoration Act	1974	U.S. Fish and Wildlife Service
Water Resources Development Act	1976, 1990	U.S. Army Corps of Engineers
Executive Order 11990—Protection of Wetlands	May 1977	All agencies
Executive Order 11988—Floodplain Management	May 1977	All agencies
Food Security Act, swampbuster provisions	1985	Department of Agriculture, Natural Resources Conservation Service
Emergency Wetland Resources Act	1986	U.S. Fish and Wildlife Service
Executive Order 12630—Constitutionally Protected Property Rights	1988	All agencies
Wetlands Delineation Manual (various revisions)	1987, 1989, 1991	All agencies

(*continued overleaf*)

Table 14.1 (*continued*)

	Date	Responsible Federal Agency
DIRECTIVE OR STATUTE		
"No Net Loss" Policy	1988	All agencies
North American Wetlands Conservation Act	1989	U.S. Fish and Wildlife Service
Coastal Wetlands Planning, Protection and Restoration Act	1990	U.S. Army Corps of Engineers
Wetlands Reserve Program	1991	Department of Agriculture, Natural Resources Conservation Service
Executive Order 12962—Conservation of Aquatic Systems for Recreational Fisheries	1995	All agencies
Federal Agriculture improvement and Reform Act	1996	Department of Agriculture, Natural Resources Conservation Service
POLICY AND TECHNICAL GUIDANCE		
Water Quality Standards Guidance	1990	U.S. Environmental Protection Agency
Non-Point Source Guidance	1990	U.S. Environmental Protection Agency
Mitigation/Mitigation Banking	1990, 1995	U.S. Army Corps of Engineers
Wetlands on Agricultural Lands, memo of agreement	1990, 1994	U.S. Army Corps of Engineers, Department of Agriculture
Wetlands and Forestry Guidance	1995	U.S. Army Corps of Engineers, Department of Agriculture
Regulatory Guidance Letter on Wetland Mitigation	2001, 2002	U.S. Army Corps of Engineers
Proposed Rules for Compensatory Mitigation	2006	U.S. Army Corps of Engineers, U.S. Environmental Protection Agency

government. Executive Order 11990, Protection of Wetlands, required all federal agencies to consider wetland protection as an important part of their policies:

> Each agency shall provide leadership and shall take action to minimize the destruction, loss or degradation of wetlands, and to preserve and enhance the natural and beneficial values of wetlands in carrying out the agency's responsibilities for (1) acquiring, managing, and disposing of Federal lands and facilities; and (2) providing federally undertaken, financed, or assisted construction and improvement; and (3) conducting Federal activities and programs affecting land use, including but not limited to water and related land resources planning, regulating, and licensing activities.

Executive Order 11988, Floodplain Management, established a similar federal policy for the protection of floodplains, requiring agencies to avoid activity in the floodplain wherever practicable. Furthermore, agencies were directed to revise their procedures to consider the impact that their activities might have on flooding and to avoid direct or indirect support of floodplain development when other alternatives are available.

Both of these executive orders were significant because they set in motion a review of wetland and floodplain policies by almost every federal agency. Several agencies, such as the U.S. Environmental Protection Agency (U.S. EPA) and the Soil Conservation Service, established policies of wetland protection prior to the issuance of these executive orders, but many other agencies, such as the Bureau of Land Management, were compelled to review or establish wetland and floodplain policies.

No Net Loss

A more significant initiative in developing a national wetlands policy was undertaken in 1987, when a National Wetlands Policy Forum was convened by the Conservation Foundation at the request of the U.S. EPA to investigate the issue of wetland management in the United States (National Wetlands Policy Forum, 1988; Davis, 1989). The 20 distinguished members of this forum (which included three governors, a state legislator, state and local agency heads, the chief executive officers of environmental groups and businesses, farmers, ranchers, and one of the co-authors of this book [JGG]) published a report that set significant goals for the nation's remaining wetlands. The forum formulated one overall objective:

> To achieve no overall net loss of the nation's remaining wetlands base and to create and restore wetlands, where feasible, to increase the quantity and quality of the nation's wetland resource base (National Wetlands Policy Forum, 1988).

The group recommended as an interim goal that the holdings of wetlands in the United States should decrease no further, and as a long-term goal that the number and quality of the wetlands should increase—*net gain*. In his 1988 presidential campaign and in his 1990 budget address to Congress, President George Bush echoed the "no net loss" concept as a national goal, shifting the activities of many agencies such as the Department of the Interior, the U.S. EPA, the U.S. Army Corps of Engineers, and the Department of Agriculture toward achieving a unified and seemingly simple goal. Nevertheless, it was not anticipated that there would be a complete halt of wetland loss in the United States when economic or political reasons dictated otherwise. Consequently, implied in the no net loss concept is wetland creation and restoration to replace destroyed wetlands. The no net loss concept became a cornerstone of wetland conservation in the United States and remains so to this day.

The Clean Water Act

The primary vehicle for wetland protection and regulation in the United States now is Section 404 of the Federal Water Pollution Control Act (FWPCA) amendments of

1972 (PL 92-500) and subsequent amendments (also known as the Clean Water Act). Section 404 required that anyone dredging or filling in "waters of the United States" must request a permit from the U.S. Army Corps of Engineers. This requirement was an extension of the 1899 Rivers and Harbors Act, in which the Corps had responsibility for regulating the dredging and filling of navigable waters.

The use of Section 404 for wetland protection has been controversial and the subject of continued lower and Supreme Court actions and revisions of regulations. The surprising point about the importance of the Clean Water Act in wetland protection is that wetlands are not directly mentioned in Section 404, and at first this directive was interpreted narrowly by the Corps to apply only to navigable waters. The definition of waters of the United States was expanded to include wetlands in two 1974–1975 court decisions, *United States v. Holland* and *Natural Resources Defense Council v. Calloway*. These decisions, along with Executive Order 11990, Protection of Wetlands, put the Army Corps of Engineers squarely in the center of wetland protection in the United States. On July 25, 1975, the Corps issued revised regulations for the Section 404 program that enunciated the policy of the United States on wetlands:

> As environmentally vital areas, [wetlands] constitute a productive and valuable public resource, the unnecessary alteration or destruction of which should be discouraged as contrary to the public interest (Federal Register, July 25, 1975).

Wetlands were defined in these regulations to encompass coastal wetlands ("marshes and shallows and . . . those areas periodically inundated by saline or brackish waters and that are normally characterized by the prevalence of salt or brackish water vegetation capable of growth and reproduction") and freshwater wetlands ("areas that are periodically inundated and that are normally characterized by the prevalence of vegetation that requires saturated soil conditions for growth and reproduction") (*Federal Register*, July 25, 1975). By these actions, the jurisdiction of the Corps has been extended to include 60 million ha of wetlands, 45 percent of which are in Alaska. Several times since 1975, the Corps has issued revised regulations for the dredge-and-fill permit program, and in 1985 the U.S. Supreme Court, in *United States v. Riverside Bayview Homes*, rejected the contention that Congress did not intend to include wetland protection as part of the Clean Water Act.

The procedure for obtaining a "404 permit" for dredge-and-fill activity in wetlands is complex (Fig. 14.1). As a starting point, no discharge of dredged or fill material can be permitted in wetlands if a practicable alternative exists. So in the initial screening of a project that involves potential effects on wetlands, the following three approaches are evaluated in sequence:

1. *Avoidance*—taking steps to avoid wetland impacts where practicable
2. *Minimization*—minimizing potential impacts to wetlands
3. *Mitigation*—providing compensation for any remaining, unavoidable impacts through the restoration or creation of wetlands (see Chapter 12)

An individual Section 404 permit is usually required for potentially significant impacts, but for many activities that have minimal adverse effects, the Army Corps of Engineers used to issue general permits. The decision to issue a permit rests with the Corps' district engineer, and it must be based on several considerations, including conservation, economics, aesthetics, and other factors listed in Figure 14.1. Assistance to the Corps on the dredge-and-fill permit process in wetland cases is provided by the U.S. EPA, the U.S. Fish and Wildlife Service, the National Marine Fisheries Service, and state agencies. The U.S. EPA has statutory authority to designate wetlands subject to permits, and also has veto power on the Corps' decisions. Some states require state permits as well as Corps permits for wetland development. The district engineer, according to Corps regulations, should not grant a permit if a wetland is identified as performing important functions for the public, such as biological support, wildlife sanctuary, storm protection, flood storage, groundwater recharge, or water purification. An exception is allowed when the district engineer determines "that the benefits of the proposed alteration outweigh the damage to the wetlands resource and the proposed alteration is necessary to realize those benefits" (*Federal Register*, July 19, 1977). The effectiveness of the Section 404 program has varied since the program began and has also varied from district to district.

Swampbuster

Normal agricultural and silvicultural activities were exempted from the Section 404 permit requirements for the first decade of the Section 404 permit program, thereby still allowing wetland drainage on farms and in commercial forests. Allowing such exemptions created conflict within the federal government: The U.S. Army Corps of Engineers and the U.S. EPA were encouraging wetland conservation through the Clean Water Act, and the Department of Agriculture was encouraging wetland drainage by providing federal subsidies for drainage projects. The conflict ended when Congress passed, as part of the 1985 Food Security Act, "swampbuster" provisions that denied federal subsidies to any farm owner who knowingly converted wetlands to farmland after the act became effective. The "swampbuster" provisions of the act drew the U.S. Soil Conservation Service (now the Natural Resources Conservation Service, or NRCS) into federal wetland management, primarily as an advisory agency helping farmers identify wetlands on their farms. The NRCS also administers a Wetlands Reserve Program (WRP) that was set up in 1990 to acquire federal easements (program was described in Chapter 12).

In August 1993, the Clinton administration released a document entitled "Protecting America's Wetlands: A Fair, Flexible, and Effective Approach." The document reaffirmed no net loss, established that 21.5 million ha (53 million acres) of previously converted wetlands would not be subject to regulations, and established the NRCS as the lead agency for identifying wetlands on agricultural land under both the Clean Water Act and the Food Security Act "swampbuster" provisions. The policy was agreed to in a January 6, 1994, memorandum of agreement among the four principal

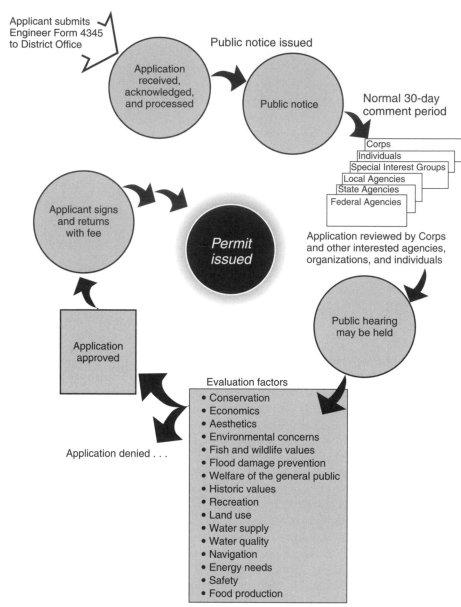

Applicant submits Engineer Form 4345 to District Office

Application received, acknowledged, and processed

Public notice issued

Public notice

Normal 30-day comment period

Corps
Individuals
Special Interest Groups
Local Agencies
State Agencies
Federal Agencies

Application reviewed by Corps and other interested agencies, organizations, and individuals

Applicant signs and returns with fee

Permit issued

Public hearing may be held

Application approved

Application denied . . .

Evaluation factors
• Conservation
• Economics
• Aesthetics
• Environmental concerns
• Fish and wildlife values
• Flood damage prevention
• Welfare of the general public
• Historic values
• Recreation
• Land use
• Water supply
• Water quality
• Navigation
• Energy needs
• Safety
• Food production

Figure 14.1 Typical U.S. Army Corps of Engineers review process for Section 404 dredge-and-fill permit request. (From Kusler, 1983.)

federal agencies involved in wetland policy in the United States (i.e., U.S. Fish and Wildlife Service, Natural Resources Conservation Service, U.S. Army Corps of Engineers, and U.S. EPA). Since that time, some of that collaboration has diminished, and the agencies' programs are diverging again.

Wetland Delineation

To be able to determine whether a particular piece of land was a wetland and, therefore, if it was necessary to obtain a Section 404 permit to dredge or fill that wetland, federal agencies, beginning with the Army Corps of Engineers, began to develop guidelines for the demarcation of wetland boundaries in a process that came to be called *wetland delineation*. In 1987, the Army Corps of Engineers published a technical manual for wetland delineation (*1987 Wetlands Delineation Manual*). This manual specified three mandatory technical criteria—hydrology, soils, and vegetation—for a parcel of land to be declared a wetland [see the following box for details]. Subsequent to that, the U.S. EPA, the Soil Conservation Service, and the U.S. Fish and Wildlife Service developed separate documents for their respective roles in wetland protection.

Finally, in January 1989, after several months of negotiation, a single *Federal Manual for Identifying and Delineating Jurisdictional Wetlands* was published by the four federal agencies to unify the government's approach to wetlands. This *1989 Wetlands Delineation Manual*, while also requiring the three mandatory technical criteria for a parcel of land to be declared a wetland, allowed one criterion to infer the presence of another (e.g., the presence of hydric soils to infer hydrology). The manual also provided some guidance about how to use field indicators such as water marks on trees or stains on leaves to determine recent flooding, wetland vegetation (from published lists), and hydric soil indicators such as mottling. The manual was used by developers and agencies alike to prove or disprove the presence of wetlands in the Section 404 permit process. Consulting firms specializing in wetland delineation sprung up overnight, and short courses on the methodology became very popular.

Beginning in early 1991, modifications of the manual were proposed in response to heavy lobbying by developers, agriculturalists, and industrialists for a relaxing of the wetland definitions, in order to lessen the regulatory burden on the private sector. A new manual was published for public comment in August 1991 (the *1991 Wetlands Delineation Manual*) but was quickly and heavily criticized for its lack of scientific credibility and unworkability (Environmental Defense Fund and World Wildlife Fund, 1992), it was eventually abandoned in 1992. At present, the 1987 Corps technical manual, which was generally agreed to be a version between the "liberal" (politically speaking) 1989 manual and the "conservative" 1991 manual, has been used since 1992 as the official way in which wetlands are determined, and there is no reason to believe that this practice will change in the near future.

Delineating Wetlands in the United States

Guidelines follow from the Section 404 definition of wetlands (*Federal Register*, 1980; Federal Register, 1982): "those areas that are inundated or saturated by surface or ground water [*hydrology*] at a frequency and duration sufficient to support, and that under normal circumstances do support, a prevalence of vegetation [*vegetation*] typically adapted for life in saturated soil conditions [*soil*]'' (bracketed words added by us for emphasis). The definition refers to (1) wetland-adapted vegetation, (2) soil, and (3) flooding or saturating hydrology. Wetland delineation according to the 1987 manual (U.S. Army Corps of Engineers, 1987) depends on determining the boundaries of the area for which these three parameters are met.

Vegetation

Wetland vegetation is defined as macrophytes typically adapted to inundated or saturated conditions. Plants are grouped into five categories (Table 14.2):

Table 14.2 Plant indicator status categories used in wetland delineation

Indicator Category	Indicator Symbol	Definition
Obligate wetland plants	OBL	Plants that occur almost always (estimated probability >99%) in wetlands under natural conditions, but that may also occur rarely (estimated probability <1%) in nonwetlands. Examples: *Spartina alterniflora, Taxodium distichum.*
Facultative wetland plants	FACW	Plants that occur usually (estimated probability >67–99%) in wetlands, but also occur (estimated probability 1–33%) in nonwetlands. Examples: *Fraxinus pennsylvanica, Cornus stolonifera.*
Facultative plants	FAC	Plants with a similar likelihood (estimated probability 33–67%) of occurring in both wetlands and nonwetlands. Examples: *Gleditsia triaconthos, Smilax rotundifolia.*
Facultative upland plants	FACU	Plants that occur sometimes (estimated probability 1–<33%) in wetlands, but occur more often (estimated probability >67–99%) in nonwetlands. Examples: *Quercus rubra, Potentilla arguta.*
Obligate upland plants	UPL	Plants that occur rarely (estimated probability <1%) in wetlands, but occur almost always (estimated probability >99%) in nonwetlands under natural conditions. Examples: *Pinus echinata, Bromus mollis.*

Source: U.S. Army Corps of Engineers (1987).

(1) obligate wetland plants, OBL, (2) facultative wetland plants, FACW, (3) facultative plants, FAC, (4) facultative upland plants, FACU, and (5) obligate upland plants, UPL. To meet the wetland vegetation requirement, more than 50 percent of the dominant species must be OBL, FACW, or FAC. Species lists of plants in these categories are available from several sources (U.S. Army Corps of Engineers, 1987). Other indicators of wetland plants may also be used, including morphological, physiological, and reproductive adaptations, such as buttressed tree trunks, pneumatophores, adventitious roots, and enlarged lenticels. Furthermore, the technical literature may provide additional information about the ability of plants to endure saturated soils.

Hydric Soils

A hydric soil is a soil that is saturated, flooded, or ponded long enough during the growing season to develop anaerobic conditions that favor the growth and regeneration of hydrophytic vegetation (see Chapter 5). All histosols (organic soils) except folists are hydric, as well as soils in a few other groups, particularly aquic soils that are poorly drained, are saturated, or have shallow (typically less than 15 cm) water tables for a significant period (usually more than one week) during the growing season. In general, the hydric condition of mineral soils is determined by their color as determined by a Munsell Color Chart as described in Chapter 5. When a hydric soil is drained, it may not be referred to as hydric, unless the vegetation is hydrophytic and indicators of hydrology support the designation as a hydric soil. Hydric soil designation can be supported by additional indicators (defined in detail in the manual), such as low permeability, appropriate soil chroma, development of mottles, and iron or manganese concretions.

Wetland Hydrology

Areas with evident characteristics of wetland hydrology are those in which the presence of water has an overriding influence on characteristics of vegetation and soils caused by anaerobic and reducing conditions, respectively. Generally, determination of wetland hydrology depends on the frequency, timing, and duration of inundation, or soil saturation, as presented in Table 14.3 for nontidal areas. Zone I is aquatic and Zone VI is upland. Zones II through IV are wetlands. Zone V may or may not be considered wetland, depending on other indicators. Additional indicators of wetland hydrology use recorded data from stream, lake, or tidal gauges, flood predictions, and historical data on flooding. Visual observations are also indicators, such as soil saturation, watermarks on trees or other structures, drift lines, sediment deposits, and drainage patterns.

Table 14.3 Hydrologic zones for nontidal areas used in hydrology determinations for wetland delineation

Zone	Name	Duration[a]	Comments
I	Permanently inundated	100%	Inundation >2 m mean water depth. Aquatic, not wetlands
II	Semipermanently to nearly permanently inundated or saturated	>75%–<100%	Inundation defined as <2 m mean water depth
III	Regularly inundated or saturated	>25%–75%	
IV	Seasonally inundated or saturated	>12.5%–25%	
V	Irregularly inundated or saturated	≥5%–12.5%	Many areas having these hydrologic characteristics are not wetlands
VI	Intermittently or never inundated or saturated	<5%	Areas with these hydrologic characteristics are not wetlands

[a]Refers to duration of inundation and/or soil saturation during the growing season.
Source: U.S. Army Corps of Engineers, 1987.

Delineation Procedure

Routine delineation methods require a combination of office gathering and synthesis of available data on the site, combined with on-site inspection and additional data generation. Comprehensive delineation methods are reserved for particularly sensitive cases and usually require significant time and effort to obtain the needed quantitative data.

All methods begin with accumulation of available data on the site to be delineated, data such as U.S. Geological Survey (USGS) quadrangle maps, National Wetlands Inventory (NWI) wetland maps, plant surveys, soil surveys, gauge data, environmental assessments or impact statements, remotely sensed data local expertise, and the applicant's survey plans and engineering designs (often with topographic surveys). These data are synthesized into a preliminary determination of whether the information is adequate to make a wetland delineation of the entire tract in question. A flowchart (Fig. 14.2) shows the steps to determine first if on-site inspection is necessary, and second, if unnecessary, to determine whether the area is a jurisdictional wetland.

The 1987 manual also details methods, depending on the size of the project area, for on-site evaluation when available data are inadequate. These methods may include, for example, the use of transects when the area is

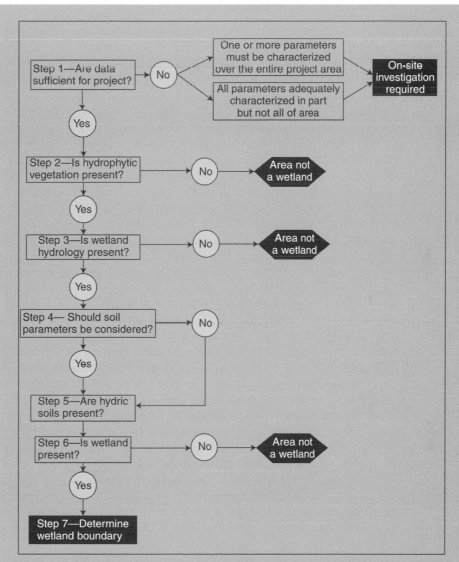

Figure 14.2 Flowchart of steps involved in making a wetland determination when an on-site inspection is unnecessary. (U.S. Army Corps of Engineers, 1987.)

too large to survey in its entirety. The intensity of the on-site investigation is determined by the available information on the site, the type of project anticipated, the ecological sensitivity of the area, and other factors. The objective of on-site investigations is to obtain adequate data to determine whether all or part of the project area fits the criteria for wetlands and, if so, where the wetland boundaries lie.

National Academy of Science Studies

Two notable studies related to wetlands were carried out, at the request of the federal government, by the National Academy of Science (NAS) and its operating arm, the National Research Council (NRC) during the 1990s. The NAS is a nongovernmental agency set up in the 19th century by Abraham Lincoln to provide scientific reviews of subjects chosen and paid for by the federal government. It has its strengths in its independence from the government and in its ability to recruit scientists and engineers from anywhere in the country for its committees.

The first NAS study dealt with the proper procedures for delineating wetlands. The development of a wetland delineation manual that everyone could agree on led to a contentious period in wetland policy between 1989 and 1991 (see Fig. 14.3), when the 1989 and proposed 1991 manuals were introduced. About that time, many scientists began to call for the NAS to answer the question: "what is a wetland?" In April 1993, the U.S. EPA, at the request of the U.S. Congress, asked the NRC to appoint a committee to undertake a scientific review of scientific aspects of wetland characterization. The 17-member committee was selected in the summer of 1993 and met over a 2-year period. The committee was charged with considering (1) definition of wetlands; (2) adequacy of science for evaluating hydrologic, biological, and other ways that wetlands function; and (3) regional variation. The report from that committee (NRC, 1995) presented a new definition of wetlands (see Chapter 2) and gave 80 recommendations on topics such as fine-tuning the delineation procedure, dealing with especially controversial wetlands, regionalization, mapping, modeling, administrative issues, and functional assessment of wetlands. The report, in essence, suggested that use of the 1987 manual was appropriate with a few minor modifications. The report was released in early 1995, just as the U.S. Congress was considering two bills on wetlands (House Bill 961 and Senate Bill 851) that would have drastically changed the definitions and management of wetlands in the United States (Fig. 14.4). It may have been a result of the release of the NRC report or just by coincidence, but neither bill became law.

The second NRC study in the late 1990s was in response to questions about whether ecological function was being replaced in wetlands created and restored to mitigate wetland loss in compliance with the "no net loss" policy (see Chapter 12). That study report (NRC, 2001) concluded the following:

- The goal of no net loss of wetlands in not being met for wetland functions by the mitigation program, despite progress in the last 20 years;
- A watershed approach would improve permit decision making; and
- Performance expectations in Section 404 permits have often been unclear, and compliance has often not been assured or attained.

The U.S. Army Corps of Engineers, as the lead agency with interest in both wetland delineation and mitigation of wetland loss, responded to both NRC reports by tightening up both delineation procedures and replacement wetland standards.

"*AHA!* BY AUTHORITY OF THE EPA, I HEREBY DECLARE THIS PROPERTY A FEDERAL WETLAND!"

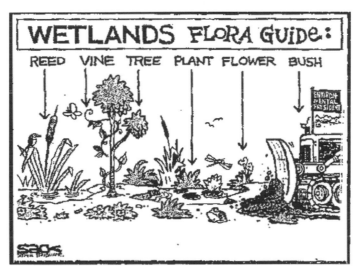

Figure 14.3   Political cartoons that appeared in the early 1990s in the United States when wetland protection and regulation were front-page stories. Top, by Henry Payne, copyright 1991 by United Media. Bottom, by Steve Sack, copyright by Minneapolis Star Tribune.

But soon the Corps' hands would be tied again, at least on defining wetlands, by decisions coming from the U.S. Supreme Court (pp. 485–486).

Other Federal Activity

Several other federal laws and activities have led to wetland protection since the 1970s. The Coastal Zone Management Program, established by the Coastal Zone Management Act of 1972, has provided up to 80 percent of matching-funds grants to states to develop plans for coastal management based on establishing a high priority to

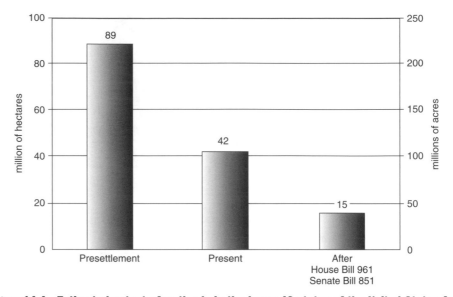

Figure 14.4 Estimated extent of wetlands in the lower 48 states of the United States for presettlement times (1780s) and present day. The numbers in the first two bars, already presented in Chapter 3, are compared with an estimate of the extent of wetlands that would have remained legally protected if House Bill 961 or Senate Bill 851 in the U.S. Congress had been passed in 1995. Each proposed law contained formal definitions of wetlands. These proposed laws would have protected only 11 million to 15 million ha of "legal" wetlands in the United States. Neither law passed, but this illustrates that wetlands can be lost either by drainage or by legal fiat that redefines wetlands.

protecting wetlands. The National Flood Insurance Program offers some protection to riparian and coastal wetlands by offering federally subsidized flood insurance to state and local governments that enact local regulations against development in flood-prone areas. The Clean Water Act, in addition to supporting the Section 404 program, supported the U.S. Fish and Wildlife Service to complete its inventory of wetlands of the United States (see Chapter 8). The Emergency Wetlands Resource Act passed by Congress in 1986 required the U.S. Fish and Wildlife Service to update its report on the status of and trends in wetlands every 10 years (see Chapter 3 for the conclusions of these reports to date).

The North American Wetlands Conservation Act's purpose was to encourage voluntary, public-private partnerships to conserve North American wetland ecosystems. This law, passed in 1989, provides grants, primarily to state agencies and private and public organizations, to manage, restore, or enhance wetland ecosystems to benefit wildlife. From 1991 through mid-1999, almost 650 projects in Canada, Mexico, and the United States have been approved for funding. Approximately 3.5 million ha (8.6 million acres) of wetlands and associated uplands have been acquired, restored, or enhanced in the United States and Canada. The act also paid for a significant amount of wetland conservation education and management plan projects in Mexico.

Wetland mitigation and wetland mitigation banking became an interest of the U.S. Congress and, although regulations were promulgated for wetland banks in the mid-1990s, no specific laws were enacted at the federal level.

The "Takings" Issue

One of the dilemmas of valuing and protecting wetlands is that the values accrue to the public at large but rarely to individual landowners who happen to have a wetland on their property. If government laws that protect wetlands or other natural resources lead to a loss of the use of that land by the private landowner, the restriction on that use has been referred to as a "taking" (denial of an individual's right to use his or her property). Many legal scholars believed that wetland and other land use laws could result in "takings" and thus be against the Fifth Amendment of the U.S. Constitution. In a major ruling in June 1992 (*Lucas v. South Carolina Coastal Council*), the U.S. Supreme Court ruled that regulations denying "economically viable use of land" require compensation to the landowner, no matter how great the public interest served by the regulations (Runyon, 1993). This case was referred back to the state of South Carolina to determine if the developer, David Lucas, was denied all economically viable use of his land (beachfront property that was rezoned by South Carolina in response to the 1980 Coastal Zone Management Act). The ultimate result of this Supreme Court decision on wetland legal protection was originally thought to be important but mostly turned out to be inconclusive. The major days for wetlands in the U.S. Supreme Court were yet to come.

U.S. Supreme Court Decisions in the 21st Century

In January 2001, the U.S. Supreme Court, in a 5–4 decision in the case *Solid Waste Agency of Northern Cook County (SWANCC) v. U.S. Army Corps of Engineers*, limited the scope of the Corps' Section 404 authority applied to "isolated wetlands." The case was brought forward by SWANCC, a consortium of Chicago suburban municipalities, when they were prohibited from using a 216-ha landfill site that had become a wooded wetland complex with more than 200 permanent and seasonal ponds and wetlands and substantial wildlife, including 121 species of birds. The basic issue brought up by SWANCC was that the wetlands were not specifically connected to interstate streams and should not fall under the authority of the federal government, but rather should be the responsibility of the state of Illinois. The Corps of Engineers denied a permit request from SWANCC for a landfill, partially because the wetland had become the second-largest heron rookery in northeastern Illinois and because the landfill could have an impact on a drinking water aquifer below the site (Downing et al., 2003).

In that case, the Supreme Court held that the Corps' "migratory bird rule" from 1986 regulation exceeded its authority. In 1996, the U.S. Army Corps of Engineers had adopted a migratory bird rule, which stated that areas fell under Section 404 jurisdiction as interstate waters (a) that are or would be used as habitat by birds protected by migratory bird treaties; or (b) that are or would be used as habitat by other migratory birds which cross state lines. Before the Supreme Court disallowed this rule, the U.S. Army Corps of Engineers was using both water and birds to show that wetlands were related to interstate commerce. The real issue of this court decision was that it reintroduced the connection of wetlands to "navigable waters of the United States" that was the original basis of Section 404 of the Clean Water Act (Downing et al., 2003).

A new term called "significant nexus" to navigable bodies of water entered the general wetland vocabulary as a result of this case. It came into more prominent use with the Supreme Court decision described next.

In a second U.S. Supreme Court decision on wetlands in the 21st century, the Supreme Court agreed to hear two "waters of the United States" cases from Michigan—*Rapanos v. United States* and *Carabell v. U.S. Army Corps of Engineers*—and ruled on these cases in June 2006. By a 5–4 vote, the Supreme Court continued to question the Corps' regulation of isolated wetlands under the Clean Water Act. The 5–4 vote remanded the case back to the lower courts in Michigan. The ruling has caused more confusion than clarity because it had three dominate opinions. Four justices took a narrow view of interstate wetlands in the Clean Water Act and believed that the act should consider "only those wetlands with a continuous surface connection to [other regulated waters]" (Justice Scalia opinion, *Rapanos v. United States*, 126 S. Ct. 2208, 2006). Four other justices "took a broad view of the Act's jurisdiction, deferring to the Corps' current categorical regulation of all tributaries and their adjacent wetlands" (Murphy, 2006).

The ninth judge, Justice Kennedy, took the middle road, rejecting both of these positions and finding that waters need to have a significant nexus to navigable waters and that this nexus needs to be determined on a case-by-case basis. The definition of nexus was given by Justice Kennedy:

> Wetlands possess the requisite nexus, and thus come within the statutory phrase 'navigable waters' if the wetlands, either alone or in combination with similarly situated lands in the region, significantly affect the chemical, physical, and biological integrity of other covered waters more readily understood as "navigable." When, in contrast, wetlands' effects on water quality are speculative or insubstantial, they fall outside the one fairly encompassed by the statutory term "navigable waters. (Justice Kennedy opinion, *Rapanos v. United States*, 126 S. Ct. 2208, 2006)

Because his was a middle opinion, Justice Kennedy's opinion will get the most attention. The overall effect of this decision remains unclear, although "significant nexus" will be the test for many decisions in the future on specific wetland cases. As pointed out in a review of this decision by Murphy (2006), "the Court's decision was, to use a phrase only water attorneys could love, quite turbid."

State and Local Management of Wetlands

Many individual states have issued wetland protection statutes or regulations, particularly in light of the Supreme Court decisions described previously. State wetland programs may become more important as the federal government attempts to delegate or abdicate much of its authority to local and state governments. Although local communities may also have wetland protection programs, states are much more probable governmental units for wetland protection for the following reasons:

1. Wetlands cross local government boundaries, making local control difficult.
2. Wetlands in one part of a watershed affect other parts that may be in different jurisdictions.
3. There is usually a lack of expertise and resources at the local level to study wetland values and hazards.
4. Many of the traditional functions of states, such as fish and wildlife protection, are related to wetland protection.

Yet there are many cases where local laws were passed to regulate and protect wetlands, such as that which occurred in Village of New Albany, Ohio, where a Sub-Division Development Ordinance allowed off-site mitigation of wetland loss only as a last resort and required all developers to verify whether they had jurisdictional wetlands on their property (Bill Resch, personnel communication). That village has some of the toughest—or at least closely monitored—wetland protection in the country.

Many states that contain coastlines initially paid more attention to managing their coastal wetlands than to managing their inland wetlands as a result of an earlier interest in coastal wetland protection at the federal level and of the development of coastal zone management programs. In general, state programs can be divided into those that are based on specific coastal wetland laws and those that are designed as a part of broader regulatory programs such as coastal zone management. Several coastal states have coastal dredge-and-fill permit programs, whereas other states have specific wetland regulations administered by a state agency.

State programs for inland wetlands, although in an earlier stage of development, are more diverse, ranging from comprehensive laws to a lack of concern for inland wetlands. Comprehensive laws have been enacted in several states. Other states have few regulations governing inland wetlands. Between these two extremes, many states rely on federal–state cooperation programs or on state laws that indirectly protect wetlands. Michigan has assumed responsibility from the federal government to issue Section 404 permits, although the Army Corps of Engineers retains control of the permit program for navigable waters, and the U.S. EPA retains federal oversight of the program. Floodplain protection laws or scenic and wild river programs are being implemented in more than half of the states and are often effective in slowing the destruction of riparian wetlands. States are also involved in wetland protection through wetland acquisition programs, conservation easement programs, preferential tax treatment for landowners who protect wetlands, and enforcement of state water quality standards as required by the Clean Water Act (Meeks and Runyon, 1990).

In the aftermath of the two Supreme Court decisions on wetlands in the 21st century, most states have developed wetland acts that allow the state government to fill the gap in the protection of isolated wetlands that are now unregulated by the

U.S. Army Corps of Engineers. In most states in the United States, wetlands are now protected by a joint program of federal and state regulations, with the former regulating wetlands adjacent to streams and rivers, and the latter taking the lead in regulating isolated wetlands.

International Wetland Conservation

The Ramsar Convention

Intergovernmental cooperation on wetland conservation has been spearheaded by the Convention on Wetlands of International Importance, more commonly referred to as the *Ramsar Convention* because it was initially adopted at an international conference held in Ramsar, Iran, in 1971. The global treaty provides the framework for the international protection of wetlands as habitats for migratory fauna that do not observe international borders and for the benefit of human populations dependent on wetlands. The convention's mission is the conservation and wise use of all wetlands through local, regional, and national actions and international cooperation, as a contribution towards achieving sustainable development throughout the world" (*www.ramsar.org*, 2006). A permanent secretariat headquartered at the International Union of Conservation of Nature and Natural Resources (IUCN) in Switzerland was established in 1987 to administer the convention, and a budget based on the United Nations scale of contributions was adopted.

The specific obligations of countries that have ratified the Ramsar Convention are the following:

1. Member countries shall formulate and implement their planning so as to promote the "wise use" of all wetlands in their territory, and develop national wetland policies.
2. Member countries shall designate at least one wetland in their territory for the "List of Wetlands of International Importance." The so-called Ramsar sites should be developed based on their international significance in terms of ecology, botany, zoology, limnology, or hydrology.
3. Member countries shall establish nature reserves at wetlands.
4. Member countries shall cooperate over shared species and development assistance affecting wetlands.

Early in the Ramsar process, the emphasis was on the protection of migratory fauna, particularly waterfowl. The importance of wetlands for many other biological functions has more recently been recognized, and currently eight criteria are used to evaluate potential wetland sites for formal designation as "wetlands of international importance" (Table 14.4). Group A sites must meet criterion 1 that they contain representative, rare, or unique wetland types. Group B sites, internationally important for conserving biological diversity, are judged on seven criteria involving questions of rare and endangered communities, biodiversity, habitat for waterfowl, or habitat or food source for indigenous fish species.

Table 14.4 Ramsar Convention criteria for identifying wetlands of international importance

GROUP A. SITES CONTAINING REPRESENTATIVE, RARE, OR UNIQUE WETLAND TYPES

Criterion 1: A wetland should be considered internationally important if it contains a representative, rare, or unique example of a natural or near-natural wetland type found within the appropriate biogeographic region.

GROUP B. SITES OF INTERNATIONAL IMPORTANCE FOR CONSERVING BIOLOGICAL DIVERSITY

Criteria based on species and ecological communities

Criterion 2: A wetland should be considered internationally important if it supports vulnerable, endangered, or critically endangered species or threatened ecological communities.

Criterion 3: A wetland should be considered internationally important if it supports populations of plant and/or animal species important for maintaining the biological diversity of a particular biogeographic region.

Criterion 4: A wetland should be considered internationally important if it supports plant and/or animal species at a critical stage in their life cycles, or provides refuge during adverse conditions.

Specific criteria based on waterbirds

Criterion 5: A wetland should be considered internationally important if it regularly supports 20,000 or more waterbirds.

Criterion 6: A wetland should be considered internationally important if it regularly supports 1 percent of the individuals in a population of one species or subspecies of waterbird.

Specific criteria based on fish

Criterion 7: A wetland should be considered internationally important if it supports a significant proportion of indigenous fish subspecies, species or families, life-history stages, species interactions, and/or populations that are representative of wetland benefits and/or values and thereby contributes to global biological diversity.

Criterion 8: A wetland should be considered internationally important if it is an important source of food for fishes, spawning ground, nursery, and/or migration path on which fish stock, either within the wetland or elsewhere, depend.

As of mid-2007, 154 countries have joined the Ramsar Convention, and they have registered 1,651 wetland sites totally 150 million ha (Fig. 14.5; see Appendix B for a complete list of countries and their area of wetlands of international importance). The program is advancing rapidly in international interest (Fig. 14.6). For example, in 2000, there were 117 member countries with 1,021 Ramsar wetland sites, totaling 74.8 million ha; in 1993, there were 582 Ramsar wetland sites, comprising almost 37 million ha in the world. The distribution of Ramsar wetlands is uneven from continent to continent, with 38 percent of the wetland area in Africa and 20 percent in the Neotropics (Fig. 14.7). Europe and North America have about 15 and 13 percent respectively of the world's Ramsar wetlands. Overall, the Ramsar program has done a credible job of bringing needed attention to wetland conservation and protection around the world.

North American Waterfowl Management Plan

The United States and Canada, partially as a result of collaboration begun by the Ramsar Convention, established the *North American Waterfowl Management Plan*

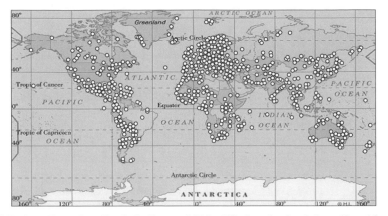

Figure 14.5 Location of wetlands in the world identified as having international importance by the Ramsar Convention. (From Wetlands International, Wageningen, The Netherlands.)

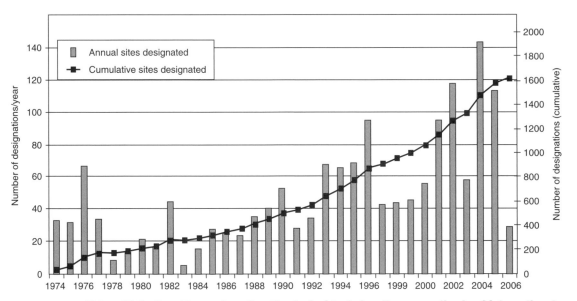

Figure 14.6 Trend in number of wetlands designated as Ramsar wetlands of international importance annually and cumulative number of sites so designated. (From Ramsar Convention on Wetlands Web site: www.ramsar.org.)

in 1986 to conserve and restore about 2.4 million ha of waterfowl wetland habitat in Canada and the United States. This treaty was formulated as a partial response to the steep decline in waterfowl in Canada and the United States that had become apparent in the early 1980s (see Chapter 11). This bilateral treaty is jointly administered by the U.S. Fish and Wildlife Service and the Canadian Wildlife Service, but also involves

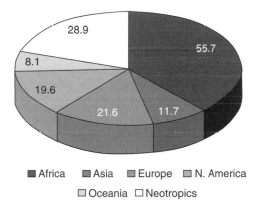

Figure 14.7 Distribution of Ramsar wetlands of international signifiance by continent. Numbers indicate Ramsar wetland area x million ha for each continent. Total Ramsar wetland area is 145.6 million ha in early 2007. *(From Ramsar Convention on Wetlands Web site: www.ramsar.org.)*

public and private participation by groups such as Ducks Unlimited. Mexico became a participant in the plan in 1994. To date, the plan has developed joint private-public ventures that have invested $4.5 billion to protect, restore, or enhance about 6.3 million ha of waterfowl habitat, mostly wetlands. Major emphasis has been placed on sites that cross international borders, including the prairie pothole region, the lower Great Lakes–St. Lawrence River basin, and the Middle–Upper Atlantic Coastline.

Recommended Readings

Millenium Ecosystem Assessment. 2005. *Ecosystems and Human Well-Being: Wetlands and Water Synthesis*. World Resources Institute, Washington, DC.

National Research Council (NRC). 1995. *Wetlands: Characteristics and Boundaries.* National Academy Press, Washington, DC. 306 pp.

National Research Council. 2001. *Compensating for Wetland Losses under the Clean Water Act*. National Academy Press, Washington, DC. 158 pp.

The Ramsar Convention on Wetlands Web site: *www.ramsar.org*, The Ramsar Convention Secretariat, Gland, Switzerland.

Wetland Losses by State in the United States

State	Original Wetlands Circa 1780 (x1,000 HA)	National Wetlands Inventory, mid-1980s (x1,000 HA)	Change (percent)
Alabama	3,063	1,531	−50
Alaska	68,799	68,799	−0.1
Arizona	377	243	−36
Arkansas	3,986	1,119	−72
California	2,024	184	−91
Colorado	809	405	−50
Connecticut	271	70	−74
Delaware	194	90	−54
Florida	8,225	4,467	−46
Georgia	2,769	2,144	−23
Hawaii	24	21	−12
Idaho	355	156	−56
Illinois	3,323	508	−85
Indiana	2,266	304	−87
Iowa	1,620	171	−89
Kansas	340	176	−48
Kentucky	634	121	−81
Louisiana	6,554	3,555	46
Maine	2,614	2,104	−19
Maryland	668	178	−73
Massachusetts	331	238	−28
Michigan	4,533	2,259	50
Minnesota	6,100	3,521	−42
Mississippi	3,995	1,646	59
Missouri	1,960	260	−87
Montana	464	340	−27
Nebraska	1,178	771	−35
Nevada	197	96	−52
New Hampshire	89	81	−9
New Jersey	607	370	−39
New Mexico	291	195	−33
New York	1,037	415	−60
North Carolina	4,488	2,300	−44
North Dakota	1,994	1,008	−49
Ohio	2,024	195	−90
Oklahoma	1,150	384	−67
Oregon	915	564	−38
Pennsylvania	456	202	−56
Rhode Island	42	26	−37
South Carolina	2,596	1,885	−27
South Dakota	1,107	720	−35
Tennessee	784	318	−59
Texas	6,475	3,080	−52
Utah	325	226	−30
Vermont	138	89	−35
Virginia	748	435	−42

State	Original Wetlands Circa 1780 (x1,000 HA)	National Wetlands Inventory, mid-1980s (x1,000 HA)	Change[a] (percent)
Washington	546	380	−31
West Virginia	54	41	−24
Wisconsin	3,966	2,157	−46
Wyoming	809	506	−38
Total wetlands	158,395	111,060	−30
Total ''lower 48''	89,491	42,240	−53

[a]From original wetlands (1780s) to mid-1980s.
Source: Dahl (1990).

Contracting Parties to the Ramsar

Convention on Wetlands

Country	Entry into convention, day.month.year	Number of Ramsar sites	Surface area of wetlands of international importance, ha
Albania	29.02.96	3	83,062
Algeria	04.03.84	42	2,959,615
Antigua and Barbuda	02.10.05	1	3,600
Argentina	04.09.92	15	3,609,831
Armenia	06.11.93	2	492,239
Australia	21.12.75	64	7,371,873
Austria	16.04.83	19	122,277
Azerbaijan	21.05.01	2	99,560
Bahamas	07.06.97	1	32,600
Bahrain	27.02.98	2	6,810
Bangladesh	21.09.92	2	611,200
Barbados	12.04.06	1	33
Belarus	25.08.91	8	283,107
Belgium	04.07.86	9	42,938
Belize	22.08.98	2	23,592
Benin	24.05.00	2	139,100
Bolivia	27.10.90	8	6,518,073
Bosnia and Herzegovina	01.03.92	1	7,411
Botswana	09.04.97	1	6,864,000
Brazil	24.09.93	8	6,434,086
Bulgaria	24.01.76	10	20,306
Burkina Faso	27.10.90	3	299,200
Burundi	05.10.02	1	1,000
Cambodia	23.10.99	3	54,600
Cameroon	20.07.06	2	600,415
Canada	15.05.81	37	13,066,571
Cape Verde	18.11.05	3	...
Central African Rep	05.04.06	1	101,300
Chad	13.10.90	5	9,879,068
Chile	27.11.81	9	159,154
China	31.07.92	30	2,937,481
Colombia	18.10.98	3	447,888
Comoros	09.06.95	3	16,030
Congo	18.10.98	1	438,960
Costa Rica	27.04.92	11	510,050
Côte d'Ivoire	27.06.96	6	127,344
Croatia	25.06.91	4	80,455
Cuba	12.08.01	6	1,188,411
Cyprus	11.11.01	1	1,585
Czech Republic	01.01.93	1	43,432
Dem. Rep. of Congo	18.05.96	2	866,000
Denmark	02.01.78	38	2,078,823
Djibouti	22.03.03	1	3,000
Dominican Republic	15.09.02	1	20,000
Ecuador	07.01.91	12	170,771
Egypt	09.09.88	2	105,700

Country	Entry into convention, day.month.year	Number of Ramsar sites	Surface area of wetlands of international importance, ha
El Salvador	22.05.99	3	125,769
Equatorial Guinea	02.10.03	3	136,000
Estonia	29.07.94	11	218,344
Fiji	11.08.06	1	615
Finland	21.12.75	49	799,518
France	01.12.86	23	828,585
Gabon	30.04.87	3	1,080,000
Gambia	16.01.97	1	20,000
Georgia	07.06.97	2	34,223
Germany	26.06.76	32	839,327
Ghana	22.06.88	6	178,410
Greece	21.12.75	10	163,501
Guatemala	26.10.90	6	593,390
Guinea	18.03.93	14	5,587,861
Guinea-Bissau	14.05.90	1	39,098
Honduras	23.10.93	6	223,320
Hungary	11.08.79	26	203,645
Iceland	02.04.78	3	58,970
India	01.02.82	25	677,131
Indonesia	08.08.92	3	656,510
Iran	21.12.75	22	1,481,147
Ireland	15.03.85	45	66,994
Israel	12.03.97	2	366
Italy	14.04.77	46	57,137
Jamaica	07.02.98	3	37,765
Japan	17.10.80	33	130,293
Jordan	10.05.77	1	7,372
Kenya	05.10.90	5	101,849
Kyrgyz Republic	12.03.03	2	639,700
Latvia	25.11.95	6	148,363
Lebanon	16.08.99	4	1,075
Lesotho	01.11.04	1	434
Liberia	02.11.03	5	95,879
Libyan Arab Jamahiriya	05.08.00	2	83
Liechtenstein	06.12.91	1	101
Lithuania	20.12.93	5	50,451
Luxembourg	15.08.98	2	17,213
Madagascar	25.01.99	5	785,593
Malawi	14.03.97	1	224,800
Malaysia	10.03.95	5	55,355
Mali	25.09.87	1	4,119,500
Malta	30.01.89	2	16
Marshall Islands	13.11.04	1	69,000
Mauritania	22.02.83	3	1,231,100
Mauritius	30.09.01	1	26
Mexico	04.11.86	65	5,263,891

Country	Entry into convention, day.month.year	Number of Ramsar sites	Surface area of wetlands of international importance, ha
Monaco	20.12.97	1	10
Mongolia	08.04.98	11	1,439,530
Morocco	20.10.80	24	272,010
Mozambique	03.12.04	1	688,000
Myanmar	17.03.05	1	256
Namibia	23.12.95	4	629,600
Nepal	17.04.88	4	23,488
Netherlands	23.09.80	49	818,908
New Zealand	13.12.76	6	39,068
Nicaragua	30.11.97	8	405,691
Niger	30.08.87	12	4,317,869
Nigeria	02.02.01	1	58,100
Norway	21.12.75	37	116,369
Pakistan	23.11.76	19	1,343,627
Palau	18.02.03	1	493
Panama	26.11.90	4	159,903
Papua New Guinea	16.07.93	2	594,924
Paraguay	07.10.95	6	785,970
Peru	30.03.92	11	6,779,393
Philippines	08.11.94	4	68,404
Poland	22.03.78	13	125,760
Portugal	24.03.81	17	73,784
Republic of Korea	28.07.97	5	4,550
Republic of Moldova	20.10.00	3	94,705
Republic of Montenegro	27.04.92	1	20,000
Republic of Serbia	27.04.92	4	20,837
Romania	21.09.91	4	682,166
Russian Federation	11.02.77	35	10,323,767
Rwanda	01.04.06	1	...
Saint Lucia	19.06.02	2	85
Samoa	06.02.05	1	...
Sao Tome and Principe	21.12.06	1	23
Senegal	11.11.77	4	99,720
Seychelles	22.03.05	1	121
Sierra Leone	13.04.00	1	295,000
Slovak Republic	01.01.93	14	40,414
Slovenia	25.06.91	3	8,205
South Africa	21.12.75	17	498,721
Spain	04.09.82	49	173,126
Sri Lanka	15.10.90	3	8,522
Sudan	07.05.05	2	6,784,600
Suriname	22.11.85	1	12,000
Sweden	21.12.75	51	514,500
Switzerland	16.05.76	11	8,676
Syrian Arab Republic	05.07.98	1	10,000
Tajikistan	18.11.01	5	94,600
Thailand	13.09.98	10	370,600

Country	Entry into convention, day.month.year	Number of Ramsar sites	Surface area of wetlands of international importance, ha
The FYR of Macedonia	08.09.91	1	18,920
Togo	04.11.95	2	194,400
Trinidad & Tobago	21.04.93	3	15,919
Tunisia	24.03.81	1	12,600
Turkey	13.11.94	12	179,482
Uganda	04.07.88	11	354,803
Ukraine	01.12.91	33	744,651
United Kingdom	05.05.76	165	893,486
United Rep. of Tanzania	13.08.00	4	4,868,424
United States of America	18.04.87	22	1,306,265
Uruguay	22.09.84	2	424,904
Uzbekistan	08.02.02	1	31,300
Venezuela	23.11.88	5	263,636
Viet Nam	20.01.89	2	25,759
Zambia	28.12.91	3	593,000
former USSR		4	929,700

Source: www.ramsar.org

Useful Conversion Factors

Multiply	By	To Obtain
LENGTH		
centimeters (cm)	0.3937	inches
feet	0.3048	meters (m)
inches	2.54	centimeters (cm)
kilometers (km)	0.6214	miles
meters (m)	3.2808	feet
meters (m)	39.37	inches
meters (m)	1.0936	yards
miles	1.6093	kilometers (km)
yards	0.9144	meters (m)
AREA		
acres	0.4047	hectares (ha)
hectares (ha)	2.47	acres
hectares (ha)	10,000	square meters (m^2)
acres	4047	square meters (m^2)
hectares (ha)	0.01	square kilometers (km^2)
square kilometers (km^2)	100	hectares (ha)
square meters (m^2)	0.0001	hectares (ha)
VOLUME		
cubic feet	0.02834	cubic meters (m^3)
cubic meters (m^3)	35.31	cubic feet
cubic centimeters (cm^3)	10^{-3}	liters (L)
acre-feet	1223.5	cubic meters (m^3)
Gallons	3.785	liters (L)
Gallons	0.003785	cubic meters (m^3)
cubic meters (m^3)	264.2	gallons
liters (L)	0.2642	gallons
FLOW		
cubic feet per second (cfs)	0.002832	cubic meters per second ($m^3\ s^{-1}$)
cubic feet per second (cfs)	10.1952	cubic meters per hour ($m^3\ hr^{-1}$)
cubic feet per second (cfs)	448.86	gallons per minute (gpm)
cubic meters per second ($m^3\ s^{-1}$)	35.31	cubic feet per second (cfs)
cubic meters per second ($m^3\ s^{-1}$)	3600	cubic meters per hour ($m^3\ hr^{-1}$)
gallons per minute (gpm)	0.002228	cubic feet per second (cfs)
gallons per minute (gpm)	0.06308	liters per second ($L\ s^{-1}$)
gallons per minute (gpm)	0.00379	cubic meters per minute ($m^3\ min^{-1}$)

Multiply	By	To Obtain
MASS		
grams (g)	0.002205	pounds
grams (g)	0.001	kilograms (kg)
kilograms (kg)	2.2046	pounds
kilograms (kg)	1000	grams (g)
pounds	453.6	grams (g)
pounds	0.4536	kilograms (kg)
metric tons (t)	2205	pounds
metric tons (t)	1000	kilograms (kg)
PRESSURE		
Atmosphere (atm)	1.01325×10^5	pascal (Pa)
Atmosphere (atm)	760	millimeters of mercury (mm Hg)
FLUX OF MASS		
grams per square meter per year (g m^{-2} yr^{-1})	10	kilograms per hectare per year (kg ha^{-1} yr^{-1})
grams per square meter per year (g m^{-2} yr^{-1})	8.924	pounds per acre per year
kilograms per hectare per year (kg ha^{-1} yr^{-1})	0.1	grams per square meter per year (g m^{-2} yr^{-1})
pound per acre per year	1.12	kilograms per hectare per year (kg ha^{-1} yr^{-1})
pounds per acre per year	0.112	grams per square meter per year (g m^{-2} yr^{-1})
ENERGY		
British Thermal Units (BTU)	0.2530	kilocalories (kcal)
British Thermal Units (BTU)	1054	joules (J)
calories (cal)	4.1869	joules (J)
calories (cal)	0.001	kilocalories (kcal)
joules (j)	0.239	calories (cal)
joules (J)	2.390×10^{-4}	kilocalories (kcal)
kilocalories (kcal)	1000	calories (cal)
kilocalories (kcal)	3.968	British Thermal Units (BTU)
kilocalories (kcal)	4183	joules (J)
kilocalories (kcal)	4.183	kilojoules (kJ)
kilocalories (kcal)	0.001162	kilowatt-hours (kWhr)
kilojoule (kJ)	0.239	kilocalories (kcal)
kilowatt-hours (kWhr)	860.5	kilocalories (kcal)

Multiply	By	To Obtain
kilowatt-hours (kWhr)	3.6×10^6	joules (J)
langley (ly)*	1	calories per square centimeter (cal cm^{-2})
langley (ly)	10	kilocalories per square meter (kcal m^{-2})

POWER

horsepower	0.7457	kilowatts (kW)
horsepower	10.70	kilocalories per minute (kcal/min)
kilocalories/day (kcal/day)	6.4937×10^{-5}	horsepower
kilocalories/day (kcal/day)	4.8417×10^{-5}	kilowatts (kW)
kilowatts (kW)	1.341	horsepower
kilowatts (kW)	14.34	kilocalories per minute (kcal/min)
kilowatts (kW)	1000	watts (W)
watt (W)	1	joule per second (J/sec)

PRIMARY PRODUCTIVITY/ENERGY FLOW

grams dry weight (g-dw)	4.5	kilocalories (kcal)
grams dry weight (g-dw)	0.45	grams C (g-C)
grams O_2 (g-O_2)**	3.7	kilocalories (kcal)
grams O_2 (g-O_2)**	0.375	grams C (g-C)
grams C (g-C)**	10	kilocalories (kcal)
grams C (g-C)**	2.67	grams O_2 (g-O_2)
kilocalories (kcal)	4.18	kilojoules (kJ)
kilocalories (kcal)**	0.1	grams C (g-C)

STOICHIOMETRY OF ORGANIC MATTER***

molar ratio	106 C:16 N:1P	
weight ratio	41 C:7.2 N:1P	

CONCENTRATIONS IN WATER

part per thousand (ppt)	1	grams per liter (g L^{-1})
part per million (ppm)	1	milligrams per liter (mg L^{-1})
parts per million (ppm)	1	grams per cubic meter (g m^{-3})
parts per million (ppm)	1000	parts per billion (ppb)
parts per billion (ppb)	1	micrograms per liter (μg L^{-1})
milligrams per liter (mg L^{-1})	1000	micrograms per liter (μg L^{-1})
millimolarity (m mole/L)	molecular weight	milligrams per liter (mg L^{-1})

Multiply	By	To Obtain
micromolarity (μ mole/L)	molecular weight	micrograms per liter ($\mu g\ L^{-1}$)
microgram-atoms per liter (μg-atom L^{-1})	molecular weight	micrograms per liter ($\mu g\ L^{-1}$)
milligrams per liter ($mg\ L^{-1}$)	$\left[\dfrac{ioniccharge}{molecularweight}\right]$	milliequivatents per liter ($meq\ L^{-1}$)
milliequivatents per liter ($meq\ L^{-1}$)	$\left[\dfrac{molecularweight}{ioniccharge}\right]$	milligrams per liter ($mg\ L^{-1}$)
micromhos per cm (μmho cm^{-1})	1	microSiemens per centimeter ($\mu S\ cm^{-1}$)

*solar constant = radiant energy at outer limit of earth's atmosphere $\sim$2.00 langleys per minute (ly min^{-1})
**based generally on the production of glucose: $6CO_2 + 12\ H_2O + (118 \times 6)$ kcal $\rightarrow C_6H_{12}O_6 + 6\ O_2 + 6\ H_2O$
***based on Redfield (1958) molecule of plankton organic matter: $[CH_2O)_{106}(NH_3)_{16}(H_2PO_4)]$

Useful Atomic Weights

H	1	Al	27	Ca	40
C	12	Si	28	Mn	55
N	14	P	31	Fe	56
O	16	S	32	Ni	59
Na	23	Cl	35.5	Cu	63.5
Mg	24	K	39	Se	79

Glossary

Aapa peatlands— Also called string bogs and patterned fens; peatlands identified by watertracks of long, narrow alignment of the high peat hummocks (strings) that form ridges perpendicular to the slope of the peatland and are separated by deep pools (flarks).

ADH— Alcohol dehydrogenase, the enzyme that catalyzes the reduction of acetaldehyde to ethanol in fermentation.

Adventitious roots— Roots that develop from some part of a vascular plant other than the seed. Usually they originate from the stem, and while common in most plants, they also develop as adaptations to anoxia in both flood-tolerant trees (e.g., *Salix* and *Alnus*) and herbaceous species, and flood-intolerant (e.g., tomato) plants just above the anaerobic zone when these plants are flooded.

Aerenchyma— Large air spaces in roots and stems of some wetland plants that allow the diffusion of oxygen from the aerial portions of the plant into the roots.

Alcohol dehydrogenase— *See* ADH.

Allochthonous— Pertains to material that is imported into an ecological system of interest from outside that system; usually refers to organic material and/or nutrients and minerals.

Allogenic succession— Ecosystem development whereby the distribution of species is governed by their individual responses to their environment with little or no feedback from organisms to their environment. Also called individualistic hypothesis, continuum concept, and Gleasonian succession.

Alluvial plain— The floodplain of a river, where the soils are alluvial deposits carried in by the overflowing river.

Ammonia volatilization— NH_3 released to the atmosphere as a gas.

Anadromous— Refers to marine species that spawn in freshwater streams.

Anammox— Abbreviation for anaerobic ammonium oxidation that leads to conversion of nitrite-nitrogen to dinitrogen gas.

Anaerobic— Refers to oxygenless conditions.

Anoxia— Waters or soils with no dissolved oxygen.

Artificial wetland— *See* Constructed wetland.

Assimilatory nitrate reduction— Nitrate (NO_3) assimilated by plants or microbes and converted into biomass.

Assimilatory sulfate reduction— Process in the sulfur cycle whereby sulfur-reducing obligate anaerobes such as *Desulfovibrio* bacteria utilize sulfates as terminal electron acceptors in anaerobic respiration.

Autochthonous— Pertains to material that is produced within the ecological system of interest (e.g., organic material produced by photosynthesis). *See also* Allochthonous.

Autogenic succession— Clementian theory of succession of ecosystems whereby vegetation occurs in recognizable and characteristic communities; community change through time is brought about by the biota; changes are linear and directed toward a mature stable climax ecosystem.

Bankfull discharge— Streamflow at which a river begins to overflow onto its floodplain.

Basin wetland— A wetland that is hydrologically isolated with little or no flooding from streams, rivers, or tides.

Billabong— Australian term for a riparian wetland that is periodically flooded by the adjacent stream or river.

Biogeochemical cycling— The transport and transformation of chemicals in ecosystems.

Blanket bogs— In humid climates, peat that blankets large areas far from the site of the original peat accumulation, through the process of paludification.

BOD— Biochemical oxygen demand, a biological test for degradable organic matter in water.

Bog— A peat-accumulating wetland that has no significant inflows or outflows and supports acidophilic mosses, particularly *Sphagnum*.

Bottomland— Lowland along streams and rivers, usually on alluvial floodplains, that is periodically flooded.

Bottomland hardwood forest— Term used principally in southeastern and eastern United States to mean a mesic riparian forested ecosystem along a higher order stream or river that is subject to intermittent to frequent flooding from that stream or river; dominated by oaks and other deciduous hardwood tree species.

Bulk density— Dry weight of a known volume of soil, divided by that volume.

Buttress— Swollen bases of tree trunks growing in water.

Cajun— Term used for culture of former French-speaking immigrants who have lived for several centuries in the swamps of the Louisiana delta.

Carr— Term used in Europe for forested wetlands characterized by alders (*Alnus*) and willows (*Salix*).

Cat clays— When coastal wetlands are drained, soil sulfides often oxidize to sulfuric acid, and the soils become too acidic to support plant growth.

Cation exchange capacity— The sum of exchangeable cations (positive ions) that a soil can hold.

Cheia— Annual period of flooding from March through May in the Pantanal region of South America that supports luxurious aquatic plant and animal life. *See also* Enchente, Seca, and Vazante.

Clay depletions— Clay is selectively removed along root channels after iron and manganese oxides have been depleted in wetland soils, only to redeposit as clay-coatings on soil particles below the clay depletions.

Coliforms— A quantitative biological test for the presence of colon bacteria or related forms; because of their ubiquitous presence, they are used as a presumptive index of general bacterial contamination of water.

Concentric domed bog— A concentric pattern of pools and peat communities formed around the most elevated part of a bog.

Constructed wetland— A wetland developed on a former uplands to create poorly drained soils and wetland flora and fauna for the primary purpose of contaminant or pollution removal from wastewater or runoff.

Consumer surplus— In economics, the net benefit of a good to the consumer.

Continuum concept— *See* Allogenic succession.

Co-precipitation of phosphorus— Some calcium phosphate is precipitated along with the major precipitation of calcium carbonate in alkaline waters.

Created wetland— A wetland constructed where one did not exist before.

Cumbungi swamp— Cattail (*Typha*) marsh in Australia.

Cumulative loss— When ecosystems such as wetlands are lost, usually as a result of human development, one small piece at a time, with the cumulative loss being substantial.

Cypress domes— Also called cypress ponds or cypress heads; poorly drained to permanently wet depressions dominated by pond cypress (*Taxodium distichum* var. *nutans*). Called domes because the cypress grows more vigorously in the center than around the perimeter of the dome, giving it a domed appearance from a distance.

Cypress strand— A diffuse freshwater stream flowing through a shallow forested depression (dominated by *Taxodium*) on a gently sloping plain.

Dalton's law— Flux is proportional to a pressure gradient. An example of a process that follows Dalton's law is evaporation, which is proportional to the difference between the vapor pressure at the water surface and the vapor pressure in the overlying air.

Dambo— A seasonally waterlogged and grass-covered linear depressions in headwater zone of rivers with no marked stream channel or woodland vegetation. Term is ChiChewa (Central Africa) dialect meaning "meadow grazing".

Darcy's law— Groundwater equation that states that flow of groundwater is proportional to a hydraulic gradient and the hydraulic conductivity, or permeability, of the soil or substrate.

Delineation— Technique of determining an exact boundary of a wetland. Used for identifying jurisdictional wetlands in United States.

Delta— Location where rivers meet the sea and deposit sediments, often in a broad alluvial fan; there are also examples of inland deltas such as the Peace-Athabasca Delta in Canada and the Okavango Delta in Botswana where the water never reaches the sea.

Demand curve— Economist's estimate of consumer benefits

Denitrification— Process in the nitrogen cycle carried out by microorganisms in anaerobic conditions, where nitrate acts as a terminal electron acceptor, resulting in the loss of nitrogen as it is converted to nitrous oxide (N_2O) and nitrogen gas (N_2).

Designer wetland— Created or restored wetland in which certain plant species or other organisms are introduced and the success or failure of those plants or organisms is used as the indicator of success or failure of that wetland.

Detention time— A measure of the length of time a parcel of water stays in a wetland; equivalent to the turnover time or retention time and the inverse of the turnover rate. *See also* Retention time. Detention time is the term used most frequently for designing treatment wetlands.

Discharge wetland— Wetland that has surface water (or groundwater) level lower hydrologically than the surrounding water table, leading to an inflow of groundwater.

Dissimilatory nitrate reduction to ammonia (DNRA)— Conversion of nitrate-nitrogen to ammonium-nitrogen.

Dissimilatory nitrogen reduction— Several pathways of nitrate reduction, particularly nitrate reduction to ammonia and denitrification. It is called dissimilatory as the nitrogen is not assimilated into a biological cell.

Diversion wetland— Wetland created or enhanced by diversion of an adjacent body of water, usually a river. Created diversion wetlands along rivers in upper watersheds are similar to oxbows or billabongs. River diversions in deltas are means to re-establish deltaic distributaries.

DMS— Dimethyl sulfide, one of the gases given off by wetlands.

Drop roots— *See* Prop roots.

Duck stamps— Stamps sold in several countries to hunters to help pay for the protection of waterfowl habitat. The Duck Stamp program in the United States started in 1934.

Eat-out— A major wetland vegetation removal by herbivory, often by geese or muskrats.

Ebullitive flux— Flux of gases from wetland soils as bubbles or diffusion to the surface of the water and then to the atmosphere.

Ecological engineering— The design, creation, and restoration of ecosystems for the benefit of humans and nature.

Ecosystem engineers— Plants, animals, and microbes that carry out essential biological feedbacks in ecosystems. Examples in wetlands are beavers and muskrats.

Embodied energy— The total energy required to produce a commodity.

Emergy— Calculation of total energy requirement for any product in nature or humanity based on using transformities. Short for "energy memory." See Odum, H.T. (1996).

Enchente— Period of rising waters from December through February in Pantanal region of South America. *See also* Cheia, Seca, and Vanzante.

Ericaceous plants— Flowering plants of the family Ericaceae, which, as a group, are acid-loving or acid-tolerant plants that often dominate bogs and other sites with acidic substrates.

Estuary— General location where rivers meet the sea and freshwater mixes with saltwater.

Eutrophic— Nutrient rich; generally used in lake classification, but is also applicable to peatlands.

Eutrophication— Process of aquatic ecosystem development whereby an ecosystem such as a lake, estuary, or wetland goes from an oligotrophic (nutrient poor) to eutrophic (nutrient rich) condition. If caused by humans, it is called cultural eutrophication.

Excentric raised bogs— Bogs that form from previously separate basins on sloping land and form elongated hummocks and pools aligned perpendicular to the slope.

Facultative— Adapted equally to either wet or dry condition. Usually used in the context of vegetation adapted to growing in saturated soils or upland soils.

Fen— A peat-accumulating wetland that receives some drainage from surrounding mineral soil and usually supports marshlike vegetation.

Fermentation— Partial oxidation of organic matter, when organic matter itself is the terminal electron acceptor in anaerobic respiration by microorganisms; forms various low-molecular-weight acids and alcohols and CO_2. Also called glycolysis.

Fibrists— *See* Peat.

Flarks— *See* Aapa peatlands.

Flood duration— The amount of time that a wetland is in standing water.

Flood frequency— The average number of times that a wetland is flooded during a given period.

Flood peak— Peak runoff into a wetland caused by a specific rainfall event.

Flood pulse concept (FPC)— Pulsing river discharge as the major force controlling biota in river floodplains, including the lateral exchange between floodplains and river channels.

Fluted trunk— Flared tree trunks at the ground surface that occurs on some trees growing in wet conditions.

Folists— Organic soils caused by excessive moisture (precipitation > evapotranspiration) that accumulate in tropical and boreal mountains; these soils are not classified as hydric soils as saturated conditions are the exception rather than the rule.

Functional guild— Categorization of plant communities into functional groups that can be defined by measurable traits.

Gardians— "Cowboys" who ride horses through the wetlands of southern France's Camargue.

Gator holes— Deep sloughs and solution holes that hold water during the dry season and that serve as wildlife refuges; term mostly used in the Florida Everglades.

Geogenous— Peatland subject to external flows.

Gleying— Development of black, gray, or sometimes greenish or blue-gray color in soils when flooded.

Glycolysis— *See* Fermentation.

Guild— A group of functionally similar species in a community.

Halophiles— "Salt-loving" organisms.

Halophytes— Salt-tolerant plants.

Hammock— Slightly raised tree islands, such as tree island freshwater hammocks or mangrove islands in the Florida Everglades.

Hatch–Slack–Kortschak pathway— Biochemical pathway of photosynthesis for C_4 plants.

Hemists— Mucky peat or peaty muck; conditions between saprist and fibrist soil.

HGM— *See* Hydrogeomorphic classification.

High marsh— Upper zone of a salt marsh that is flooded irregularly and generally is located between mean high water and extreme high water. Called inland salt marsh in Gulf of Mexico coastline.

Histosols— Organic soils that have organic soil material in more than half of the upper 80 cm, or that are of any thickness if they overlie rock or fragmental materials that have interstices filled with organic soil material.

HLR— *See* Hydraulic loading rate.

Hydrarch succession— Development of a terrestrial forested climax community from a shallow lake with wetland as an intermediate sere. In this view, lakes gradually fill in as organic material from dying plants accumulates and minerals are carried in from upslope.

Hydraulic conductivity— *See* Permeability.

Hydraulic loading rate (HLR)— Amount of water added to a wetland, generally described as the depth of water (volume of flooding per wetland area) per unit time; generally used for treatment wetlands.

Hydric soils— Soils that formed under conditions of saturation, flooding, or ponding long enough during the growing season to develop anaerobic conditions in the upper part.

Hydrochory— Seed dispersal by water.

Hydrodynamics— An expression of the fluvial energy that drives a system.

Hydrogeomorphic classification (HGM)— Wetland classification system based on type and direction of hydrologic conditions, local geomorphology and climate.

Hydrogeomorphology— Combination of climate, basin geomorphology, and hydrology that collectively influences a wetland's function.

Hydroperiod— The seasonal pattern of the water level of a wetland. This approximates the hydrologic signature of each wetland type.

Hydrophyte— Plant adapted to the wet conditions.

Hydrophytic vegetation— Plant community dominated by hydrophytes.

Hypoxia— Waters with dissolved oxygen less than 2 mg/L.

Interception— Precipitation that is retained in the overlying vegetation canopy.

Intermittently exposed— Refers to nontidal wetlands that are flooded throughout the year, except during periods of extreme drought.

Intermittently flooded— Refers to nontidal wetlands that are usually exposed, with surface water present for variable periods without detectable seasonal patterns.

Intertidal— Part of coastal wetland flooded periodically with tidal water.

Intrariparian continuum— The structure and function of riparian communities along a river system.

Irregularly exposed— Refers to coastal wetlands with surface exposed by tides less often than daily.

Irregularly flooded— Refers to coastal wetlands with surface flooded by tides less often than daily.

Isolated wetland— Legal term used in the United States to define wetlands that do not have an obvious surface-water connection to a navigable stream or river (*see also* Significant nexus).

Jurisdictional wetland— Term used in the United States to refer to wetlands that fall under the jurisdiction of federal laws for the purpose of permit issuance or other legal matters.

Kahikatea— Refers to both the tree (*Dacrycarpus dacrydiodes*) and the forested wetlands found throughout New Zealand. Referred to as "white pine" forests by locals.

Karst— A topography formed over limestone, dolomite, or gypsum.

Krefeld system or Max-Planck-Institute process— Gravel bed macrophyte subsurface flow treatment wetlands.

Lacustrine— Pertaining to lakes or lake shores.

Lagoon— Term frequently used in Europe to denote deepwater enclosed or partially opened aquatic system, especially in coastal delta regions.

Lentic— Related to slow-moving or standing water systems; usually refers to lake (lacustrine) and stagnant swamp systems.

Lenticels— Small pores found on mangrove tree prop roots and pneumatophores above low tide and presumed to be sites of oxygen influx for anaerobic roots survival.

Limnogenous peatland— Geogenous peatland that develops along a slow-flowing stream or a lake.

Littoral— Zone between high and low tide in coastal waters or the shoreline of a freshwater lake.

Loading rate— The amount of a material (e.g., a chemical) applied to a wetland, measured either per unit area (e.g., $g\ m^{-2}\ yr^{-1}$) or volumetrically (e.g., $g\ m^{-3}\ yr^{-1}$).

Lotic— Pertaining to running water (i.e., rivers and streams).

Low marsh— Intertidal or lower marsh in salt marsh that is located in the intertidal zone and is flooded daily. Called streamside salt marsh in coastal Gulf of Mexico.

Mangal— Same as mangrove.

Mangrove— Subtropical and tropical coastal ecosystem dominated by halophytic trees, shrubs, and other plants growing in brackish to saline tidal waters. The word "mangrove" also refers to the dozens of tree and shrub species that dominate mangrove wetlands.

Marginal value— The value of an additional increment of a commodity in a free market.

Marsh— A frequently or continually inundated wetland characterized by emergent herbaceous vegetation adapted to saturated soil conditions. In European terminology, a marsh has a mineral soil substrate and does not accumulate peat. *See* also Tidal freshwater marsh and Salt marsh.

Mesotrophic peatlands— Also called transition or poor fens. Peatlands intermediate between minerotrophic and ombrotrophic.

Methane emissions— Amount of methane released from a landscape as net result of methanogenesis minus methane oxidation.

Methane oxidation— Conversion by methane to methanol, formaldehyde, and carbon dioxide by obligate methanotrophic bacteria.

Methanogenesis— Carbon process under extremely reduced conditions when certain bacteria (methanogens) use CO_2 or low-molecular-weight organic compounds as electron acceptors for the production of gaseous methane (CH_4).

Methanogens— Bacteria that carry out methanogenesis.

Methanotrophs— Aerobic bacteria that oxidize methane.

Mineral soil— Soil that has less than 20 to 35 percent organic matter.

Minerotrophic peatlands— Also called rheotrophic peatlands or rich fens; peatlands that receive water that has passed through mineral soil.

Mire— Synonymous with any peat-accumulating wetland (European definition); from the Norse word "myrr." The Danish and Swedish word for peatland is now "mose."

Mitigate— To lessen or compensate for an impact. Used here in the context of mitigating wetland loss by restoring or creating wetlands.

Mitigation ratio— The ratio of restored or created wetland to wetland lost to development.

Mitigation wetland— *See* Replacement wetland.

Moor— Synonymous with peatland (European definition). A highmoor is a raised bog; a lowmoor is a peatland in a basin or depression that is not elevated above its perimeter. The primitive sense of the Old Norse root is "dead" or barren land.

Mottles (or redox concentrations)— Orange/reddish-brown (because of iron oxides) or dark reddish-brown/black (because of manganese oxides) accumulations in hydric soils throughout an otherwise gray (gleyed) soil matrix. Mottles suggest intermittently exposed soils and are relatively insoluble, enabling them to remain in soil long after it has been drained.

Muck— Sapric organic soil material with virtually all of the organic matter decomposed, not allowing for the identification of plant forms. Bulk density generally greater than $0.2 \, g/cm^3$ (more than peat).

Munsell soil color chart— Book of standard color chips for determining soil color value and chroma. Used to identify hydric soils.

Muskeg— Large expanse of peatlands or bogs; particularly used in Canada and Alaska.

NAD— Nicotinamide adenine dinucleotide, an enzyme that accumulates in anaerobic conditions.

NADP— NAD phosphate.

Nitrification— Ammonium nitrogen oxidized by microbes to nitrite nitrogen and nitrate nitrogen.

Nernst equation— Equation based on a hydrogen scale showing how redox potential is related to the concentrations of oxidants and reductants in a redox reaction.

Nitrogen fixation— Process in the nitrogen cycle whereby N_2 gas is converted to organic nitrogen through the activity of certain organisms in the presence of the enzyme nitrogenase.

No net loss— Wetland policy in the United States that began in the late 1980s and means that if wetlands are lost they must be replaced so that there is no "net loss" of wetlands overall.

Nutrient budget— Mass balance of a nutrient in an ecosystem.

Nutrient spiraling— The process whereby resources (organic carbon, nutrients, etc.) are temporarily stored, then released as they "spiral" downstream from organic to inorganic form and back again.

Obligate— Requiring a specific environment to grow, as in adapted only a wet environment. In the context of wetlands, obligate generally refers to plants requiring saturated soils.

Oligotrophic— Nutrient poor; generally used in lake classification, but is also applicable to peatlands.

Oligotrophication— Often the process of peatland development whereby a peatland eventually elevates itself above the surrounding landscape and goes from eutrophic (nutrient rich) to oligotrophic (nutrient poor).

Ombrogenous— Peatland with inflow from precipitation only; also called ombrotrophic.

Ombrotrophic— Literally rain fed, referring to wetlands that depend on precipitation as the sole source of water.

Opportunity cost— The net worth of a non–free market resource in its best alternative use; that is, the net benefit of the area in its best alternative use that has to be foregone in order to keep it in its natural state.

Organic soil— Soil that has more than 12 to 18 percent organic carbon, depending on clay content (*see* Fig. 6.1).

Osmoconformers— Marine animals in which the internal cell environment follows closely the osmotic concentration of the external medium.

Osmoregulators— Marine animals that control their internal cell environment despite a different osmotic concentration of the external medium.

Outwelling— Function of coastal wetlands as "primary production pumps" that feed large areas of adjacent waters with organic material and nutrients; analogous to upwelling of deep ocean water, which supplies nutrients to some coastal waters from deep water.

Overland flow— Nonchannelized sheet flow that usually occurs during and immediately following rainfall or a spring thaw, or as tides rise in coastal wetlands.

Oxbow— Abandoned river channel, often developing into a swamp or marsh, on a river floodplain.

Oxidation— Chemical process of giving up an electron (e.g., $Fe^{2+} \rightarrow Fe^{3+} + e^-$). Special cases involve uptake of oxygen or removal of hydrogen (e.g., $H_2S \rightarrow S^{2-} + 2H^+$).

Oxidized pore linings— *See* Oxidized rhizosphere.

Oxidized rhizosphere (also called oxidized pore linings)— Thin traces of oxidized soils through an otherwise dark matrix indicating where roots of hydrophytes were once found.

Paalsa peatlands— Peatlands found in the southern limit of the tundra biome; large plateaus of peat (20 to 100 m in breadth and length and 3 m high) generally underlain by frozen peat and silt.

Pakihi— Peatland in southwestern New Zealand dominated by sedges, rushes, ferns, and scattered shrubs. Most pakihi form on terraces or plains of glacial or fluvial outwash origin and are acid and exceedingly infertile.

Palmer Drought Severity Index (PDSI)— A relative measure of climatic "wetness." Used primarily to estimate the severity of droughts.

Paludification— The blanketing of terrestrial ecosystems by overgrowth of bog vegetation. *See also* Blanket bog.

Palustrine— Nontidal wetlands.

Panne— Bare, exposed, or water-filled depression in a salt marsh.

Patterned fens— *See* Aapa peatlands.

Peat— Fibric organic soil material with virtually all of the organic matter allowing for the identification of plant forms. Bulk density generally less than $0.1 \, g/cm^3$ (less than muck).

Peatland— A generic term of any wetland that accumulates partially decayed plant matter (peat).

Penman equation— Empirical equation for estimating evapotranspiration using an energy budget approach.

Perched wetland— Wetland that holds water well above the groundwater table.

Permanently flooded— Refers to nontidal wetlands that are flooded throughout the year in all years.

Permeability— The capacity of soil to conduct water flow Also known as hydraulic conductivity. *See also* Darcy's law.

Phreatophytes— Plants that obtain their water from phreatic sources (i.e., groundwater or the capillary fringe of the groundwater table).

Physiognomy— The appearance or life form of vegetation.

Piezometers— Groundwater wells that are only partially screened and thus measure the piezometric head of an isolated part of the groundwaer.

Playa— An arid- to semiarid-region wetland that has distinct wet and dry seasons. Term used in the southwest United States for shallow depressional recharge wetlands occurring in the Great Plains region of North America that are formed through a combination of wind, wave, and dissolution processes.

Pneumatophores— "Air roots" that protrude out of the mud from the main roots of wetland plants such as black mangroves (*Avicennia*) and cypress (*Taxodium distichum*) and are thought to be organs for transport of oxygen and other gases to and from the roots of the plant. Called "knees" for cypress.

Pocosin— Peat-accumulating, nonriparian freshwater wetland, generally dominated by evergreen shrubs and trees and found on the southeastern Coastal Plain of the United States. The term comes from the Algonquin for "swamp on a hill."

Porosity— Total pore space in soil, generally expressed as a percentage.

Pothole— Shallow marshlike pond, particularly as found in the Dakotas and central Canadian provinces; the so-called prairie pothole region.

Producer surplus or economic rent— The area over a good's supply curve bounded by price.

Prop roots— Above-ground arched roots that aid in support of some wetland trees such as the mangrove *Rhizophora*.

Pulse stability concept— Concept that pulses can be both a subsidy and a stress to an ecosystem, depending on their strength, with subsidies occurring with moderate pulses, while both weak and excessive pulses can result in stress responses.

Quaking bog— *Schwingmoor* in German. Bog in which the peat layer and plant cover is only partially attached in the basin bottom or is floating like a raft.

Quickflow— Direct runoff component of streamflow during a storm that causes an immediate increase in streamflow.

Raised bogs— Peat deposits that fill entire basins, are raised above groundwater levels, and receive their major inputs of nutrients from precipitation. *See* Ombrogenous and Ombrotrophic.

Ramsar Convention— International treaty originally started in Ramsar, Iran, in the early 1970s to protect wetland habitat around the world, especially for migratory waterfowl.

Raupo swamp – Cattail (*Typha*) marsh in New Zealand.

Recharge wetland— Wetland that has surfacewater (or groundwater) level higher hydrologically than the surrounding water table, leading to an outflow of groundwater.

Recurrence interval— The average interval between the recurrence of floods at a given or greater magnitude.

Redox concentrations— Bodies of accumulated iron and manganese oxides in wetland soils such as nodules and concretions, masses (formerly called "reddish mottles"), and pore linings (formerly called "oxidized rhizospheres").

Redox depletions— Bodies of low chroma (2 or less) where the natural (gray or black) color of the parent sand, silt, or clay results when soluble forms of iron, manganese, or clay are leached out of the soil. Generally have Munsell color values of 4 or greater. *See also* Clay depletions.

Redoximorphic features— Features formed by the reduction, translocation, and/or oxidation of iron and manganese oxides; used to identify hydric soils. Formerly called mottles and low-chroma colors.

Redox potential— Reduction-oxidation potential, a measure of the electron pressure (or availability) in a solution or measure of the tendency of soil solution to oxidize or reduce substances. Low redox potential indicates reduced conditions; high redox potential indicates oxidized conditions.

Reduced matrix— Soil that has low chroma and high value but whose color changes in hue or chroma when exposed to air.

Reduction— Chemical process of gaining an electron (e.g., $Fe^{3+} + e^- \rightarrow Fe^{2+}$). Special cases involve releasing oxygen or gaining hydrogen (hydrogenation) (e.g., $S^{2-} + 2H^+ \rightarrow H_2S$).

Reedmace swamp— Cattail (*Typha*) marsh in the United Kingdom.

Reedswamp— Marsh dominated by *Phragmites* (common reed); term used particularly in Europe.

Reference wetland— Natural wetland used as a reference or control site to judge the condition of another created, restored, or impacted wetland.

Regularly flooded— Refers to coastal wetlands with surface flooded and exposed by tides at least once daily.

Regulators (or avoiders)— In reference to biological adaptations to stress, organisms that actively avoid stress or modify it to minimize its effects.

Rehabilitation— Less than full restoration of an ecosystem to its predisturbance condition.

Renewal rate— *See* Turnover rate.

Replacement value— The sum of the cheapest way of replacing all the various services performed by a natural ecosystem area.

Replacement wetland— A wetland constructed to replace the functions lost by human development, usually in the same or an adjacent watershed.

Residence time— *See* Retention time.

Resource spiraling— *See* Nutrient spiraling.

Restoration— To return a site to an approximation of it condition before alteration. *See also* Wetland restoration.

Retention rate— The amount of a material retained in a wetland per unit time and area; usually refers to material more or less removed from water flowing over or through a wetland, as contrasted to *detention*, which is transitory.

Retention time— A measure of the average time that water remains in the wetland. Nominal residence time or retention time refers to the theoretical time that water stays in a wetland as calculated from the flowthrough and the water volume in the wetland.

Rheotrophic peatlands— *See* Minerotrophic peatlands.

Riparian— Pertaining to the bank of a body of flowing water; the land adjacent to a river or stream that is, at least periodically, influenced by flooding.

Riparian ecosystem— Ecosystem with a high water table because of proximity to an aquatic ecosystem, usually a stream or river. Also called bottomland hardwood forest, floodplain forest, bosque, riparian buffer, and streamside vegetation strip.

River continuum concept (RCC)— Theory to describe the longitudinal patterns of biota found in streams and rivers.

Root zone method (Wurzelraumentsorgung)— Subsurface flow wetland basins, almost always found in Europe, and generally planted with *Phragmites australis*.

Runoff— Nonchannelized surfacewater flow.

Salt exclusion— A salinity adaptation by some wetland plants by which plants prevent salt from entering the plant at the roots.

Salt marsh— A halophytic grassland on alluvial sediments bordering saline water bodies where water level fluctuates either tidally or nontidally.

Salt secretion— A salinity adaptation by which some wetland plants excrete salt from specialized organs in the leaves.

Saprists— *See* Muck.

Saturated soils— Refers to nontidal wetlands where the soil or substrate is saturated for extended periods in the growing season, but standing water is rarely present.

Sclerophylly— Refers to the thickening of the plant epidermis.

Seasonally flooded— Refers to nontidal wetlands that are flooded for extended periods in the growing season, but with no surface water by the end of the growing season.

Seca— Dry period in Pantanal region of South America from September through November when the wetland reverts to vegetation typical of dry savannas. *See also* Cheia, Enchente, and Vazante.

Secondary treatment— Treatment of wastewater to remove organic material.

Sedge meadow— Very shallow wetland dominated by several species of sedges (e.g., *Carex, Scirpus, Cyperus*).

Seed bank— Seeds stored in soils, often for many years. In wetlands changing hydroperiod, as in wetland restoration or wetland drainage, can often lead to germination.

Self-design— The application of self-organization in the design of ecosystems. The process of ecosystem development whereby the continual or periodic introduction of species propagules (plants, animals, microbes) by humans or nature and their subsequent survival (or non-survival) provide the essence of the successional and functional development of an ecosystem.

Semipermanently flooded— Refers to nontidal wetlands that are flooded in the growing season in most years.

Serial discontinuity concept— Describes the effects that floodplains, dams, and the transverse dimension in general has on the functioning of a river system.

Shrub–scrub— Wetlands dominated by woody, low-stature vegetation such as freshwater buttonwood (*Cephalanthus*) or saltwater dwarf mangrove (*Rhizophora*) swamps.

Significant nexus— Legal term used in the United States to describe the connection of a wetland to an adjacent navigable water. The wetland should, by itself or in combination with other lands, significantly affect the chemical, physical, and biological integrity of the adjacent navigable water (*see also* Isolated wetland).

Sink— Term used in the context of wetland nutrient budgets to define a wetland that imports more of a certain nutrient than it exports.

Slough— An elongated swamp or shallow lake system, often adjacent to a river or stream. A slowly flowing shallow swamp or marsh in the southeastern United States (e.g., cypress slough). From the Old English word "sloh" meaning a watercourse running in a hollow. *See* also Cypress strand.

Soligenous peatland— Geogenous peatland that develops with regional interflow and surface runoff.

Source— Term used in the context of wetland nutrient budgets to define a wetland that exports more of a certain nutrient than it imports.

Spit— A neck of land along a coastline behind which coastal wetlands sometimes develop.

SRP— Soluble reactive phosphorus, similar to orthophosphate; a measure of biologically available phosphorus.

Stem hypertrophy— Noticeable swelling of lower stem of vascular plant, usually caused by water or saturated soils. Includes tree buttresses and fluted trunks.

Stemflow— Precipitation that passes down the stems of vegetation. Used generally in connection with forests and forested wetlands.

Streamflow— Channelized surfacewater flow.

Stream order— Numerical system that classifies stream and river segments by size according to the order of its tributaries.

String bogs— *See* Aapa peatlands.

Strings— *See* Aapa peatlands.

Subsidence— Sinking of ground level, caused by natural and artificial settling of sediments over time.

Subsurface-flow constructed wetlands— Constructed wetlands through which water flows beneath the surface rather than over the surface. *See* also Root zone method.

Subtidal— Coastal wetland permanently flooded with tidal water.

Supply curve— Economist's estimate of producer benefits.

Surface-flow constructed wetlands— Constructed wetlands that mimic many natural wetlands with flow on surface rather than below the surface.

Swamp— Wetland dominated by trees or shrubs (U.S. definition). In Europe, forested fens and wetlands dominated by reed grass (*Phragmites*) are also called swamps (*see* Reedswamp).

Swampbuster— Provision of the U.S. Food Security Act that encourages farmers not to drain wetlands and thereby lose their farm subsidies.

Swamp gas (or marsh gas)— Methane.

Taking— The legal denial of an individual's right to use all or part of the area or structure (trees, wildlife, etc.) of his or her property.

Telmatology— A term originally coined to mean "bog science." From the Greek word "telma" for bog.

Temporarily flooded— Refers to nontidal wetlands that are flooded for brief periods in the growing season, but otherwise the water table is well below the surface.

Terrestrialization— Generally in reference to succession of peatlands, the infilling of shallow lakes until they become, in appearance, a peat basin supporting terrestrial vegetation.

Tertiary treatment— Advanced treatment of wastewater after secondary treatment to remove inorganic nutrients and other trace materials. Wetlands are often used for this purpose.

Thornthwaite equation— Empirical equation for estimating potential evapotranspiration as a function of air temperature.

Throughfall— Precipitation that passes through vegetation cover to the water or substrate below. Used particularly in forests and forested wetlands.

Tidal creeks— Small streams that serve as important conduits for material and energy transfer between salt marshes and adjacent bodies of water.

Tidal freshwater marsh— Marsh along rivers and estuaries close enough to the coastline to experience significant tides by nonsaline water. Vegetation is often similar to nontidal freshwater marshes.

Tolerators (also called resisters)— In reference to biological adaptations to stress, organisms that have functional modifications that enable them to survive and often to function efficiently in the presence of stress.

Topogenous— Refers to peatland development when the peatland modifies the pattern of surface water flow.

Total suspended solids— *See* TSS.

Transformer— Term used in the context of wetland nutrient budgets to define a wetland that imports and exports the same amount of a certain nutrient but changes it from one form to another.

Transitional ecosystems— Subirrigated riparian sites such as inactive floodplains, terraces, toeslopes, and meadows that have seasonally high water tables that recede to below the rooting zone in late summer.

Translocation— Movement of nutrients between below-ground and above-ground portions of plants.

Treatment wetland— Wetland constructed to treat wastewater or polluted runoff. *See also* Constructed wetland.

TSS— Total suspended solids, a measure of the sediments in a unit volume of water.

Turlough— Term is specific for these types of wetlands found mostly in western Ireland. Areas seasonally flooded by karst groundwater with sufficient frequency and duration to produce wetland characteristics. They generally flood in winter and are dry in summer and fill and empty through underground passages.

Turnover rate— Ratio of throughput of water to average volume of water within a wetland. This is the inverse of turnover time, residence time, or retention time of a wetland.

Turnover time— *See* Retention time.

Value— Something worthy, desirable, or useful to humanity; although the term is used often in ecology to refer to processes (e.g., primary production) or ecological structures (e.g., trees) as they are "valuable" to the way an ecosystem functions, the term generally should be limited to an anthropocentric connotation. Humans decide what is of "value" in an ecosystem.

Vazante— Period of declining water in Pantanal region of South America from June through August. *See also* Cheia, Enchente, and Seca.

Vernal pool— Shallow, intermittently flooded wet meadow, generally typical of Mediterranean-type climate with dry season for most of the summer and fall. Term is now used to indicate wetlands temporarily flooded during the spring throughout the United States.

Viviparity— The production of young in a living state.

Viviparous seedlings— Seedlings of trees germinate while still attached to the tree canopy, as with the mangrove genera *Rhizophora*. A specific case of viviparity.

Vleis— Seasonal wetland similar to a Dambo; term used in southern Africa.

Wad (pl. Wadden)— Unvegetated tidal flat originally referring to the northern Netherlands and northwestern German coastline. Now used throughout the world for coastal areas.

Watertracks— *See* Aapa peatlands.

Wetland— *See* various wetland definitions in Chapter 2. Generally, wetlands have the presence of shallow water or flooded soils for part of the growing season, have organisms adapted to this wet environment, and have soil indicators of this flooding such as hydric soils.

Wetland creation— The conversion of a persistent upland or shallow water area into a wetland by human activity.

Wetland delineation— The demarcation of wetland boundaries for legal purposes. *See* Jurisdictional wetlands.

Wetlanders— People who live in proximity to wetlands and whose culture is linked to the wetlands.

Wetland restoration— The return of a wetland from a condition disturbed or altered by human activity to a previously existing condition.

Wetlands of international importance— Wetlands designated by the Ramsar Convention as important international wetlands because they contain rare wetland types, support biological diversity, waterfowl, and fish.

Wet meadow— Grassland with waterlogged soil near the surface but without standing water for most of the year.

Wet prairie— Similar to a marsh, but with water levels usually intermediate between a marsh and a wet meadow.

Willingness-to-pay, or net willingness-to-pay— A hypothetical market that establishes the amount society would be willing to pay to produce and/or use a good beyond that which it actually does pay.

References

Adamus, P. R. 1983. A Method for Wetland Functional Assessment, Vol. I: Critical Review and Evaluation Concepts, and Vol. II: FHWA Assessment Method. Federal Highway Reports FHWA-IP-82–83 and FHWA-IP-82–84, U.S. Department of Transportation, Washington, DC. 16 pp. and 134 pp.

Abraham, K.F. and C.J. Keddy. 2005. *The Hudson Bay Lowland*. In L.A. Fraser and P.A. Keddy, eds. The World's Largest Wetlands: Ecology and Conservation, Cambridge University Press, Cambridge, UK, pp. 118–148.

Ahn, C., and W. J. Mitsch. 2001. Chemical analysis of soil and leachate from experimental wetland mesocosms lined with coal combustion products. *Journal of Environmental Quality* 30: 1457–1463.

Ahn, C., W. J. Mitsch, and W. E. Wolfe. 2001. Effects of recycled FGD liner material on water quality and macrophytes of constructed wetlands: A mesocosm experiment. *Water Research* 35: 633–642.

Alper, J. 1998. Ecosystem "engineers" shape habitats for other species. *Science* 280: 1195–1196.

Alphin, T. D., and M. H. Posey. 2000. Long-term trends in vegetation dominance and infaunal community composition in created marshes. *Wetlands Ecology and Management* 8: 317–325.

Altinbilek, D. 2004. Development and management of the Euphrates–Tigris basin. *Water Resources Development* 20: 15–33.

Altor, A. E., and W. J. Mitsch. 2006. Methane flux from created riparian marshes: Relationship to intermittent versus continuous inundation and emergent macrophytes. *Ecological Engineering* 28: 224–234.

Anderson, C. J., W. J. Mitsch, and R. W. Nairn. 2005. Temporal and spatial development of surface soil conditions in two created riverine marshes. *Journal of Environmental Quality* 34: 2072–2081.

Anderson, C. J., and W. J. Mitsch. 2006. Sediment, carbon, and nutrient accumulation at two 10-year-old created riverine marshes. *Wetlands* 26: 779–792.

Anderson, D. C., J. J. Sartoris, J. S. Thullen, and P. G. Reusch. 2003. The effects of bird use on nutrient removal in a constructed wastewater-treatment wetland. *Wetlands* 23: 423–435.

Arheimer, B., and H. B. Wittgren. 1994. Modelling the effects of wetlands on regional nitrogen transport. *Ambio* 23: 378–386.

Aselmann, I., and P. J. Crutzen. 1989. Global distribution of natural freshwater wetlands and rice paddies, their net primary productivity, seasonality and possible methane emissions. *Journal of Atmospheric Chemistry* 8: 307–358.

Aselmann, I., and P. J. Crutzen. 1990. *A global inventory of wetland distribution and seasonality, net primary productivity, and estimated methane emissions.* In A. F. Bouwman, ed. Soils and the Greenhouse Effect. John Wiley & Sons, New York, pp. 441–449.

Atchue, J. A., III, H. G. Marshall, and F. P. Day, Jr. 1982. Observations of phytoplankton composition from standing water in the Great Dismal Swamp. *Castanea* 47: 308–312.

Atkinson, R. B., J. E. Perry, and J. Cairns, Jr. 2005. Vegetation communities of 20-year-old created depressional wetlands. *Wetlands Ecology and Management* 13: 469–478.

Australian Nature Conservation Agency. 1996. Wetlands Are Important. Two-page flyer, National Wetlands Program, ANCA, Canberra, Australia.

Baker, J. M. 1973. Recovery of salt marsh vegetation from successive oil spillages. *Environmental Pollution* 4: 223–230.

Baldassarre, G.A. and E.G. Bolen. 2006. Waterfowl Ecology and Management, 2nd ed., Krieger Publishing Company, Malabar, Florida, 567 pp.

Balmford, A., A. Bruner, P. Cooper, R. Costanza, S. Farber, R. E. Green, M. Jenkins, P. Jefferiss, V. Jessamy, J. Madden, K. Munro, N. Myers, S. Naeem, J. Paavola, M. Rayment, S. Rosendo, J. Roughgarten, K. Trumper, and R. K. Turner. 2002. Economic reasons for conserving wild nature. *Science* 297: 950–953.

Bardecki, M. J. 1987. Wetland Evaluation: Methodology Development and Pilot Area Selection. Report 1, Canadian Wildlife Service and Wildlife Habitat Canada, Toronto.

Bardi, E., and M. T. Brown, 2001. *Emergy evaluation of ecosystems: A basis for environmental decision making.* In M. T. Brown, ed. Emergy Synthesis: Theory and Applications of the Emergy Methodology. Proceedings of a conference held at Gainesville, FL, September 1999. The Center for Environmental Policy, University of Florida, Gainesville, pp. 81–98.

Barry, J. M. 1997. Rising Tide: The Great Mississippi Flood of 1927 and How It Changed America. Simon & Schuster, New York.

Bartlett, C. H. 1904. Tales of Kankakee Land. Charles Scribner's Sons, New York. 232 pp.

Bay, R. R. 1967. Groundwater and vegetation in two peat bogs in northern Minnesota. *Ecology* 48: 308–310.

Bazilevich, N. I., L. Ye. Rodin, and N. N. Rozov. 1971. Geophysical aspects of biological productivity. *Soviet Geography* 12: 293–317.

Beaumont, P. 1975. *Hydrology*. In B. Whitton, ed. River Ecology. Basil Blackwell, Oxford, UK, pp. 1–38.

Bellamy, D. J. 1968. *An Ecological Approach to the Classification of the Lowland Mires in Europe*. In Proceedings of the Third International Peat Congress, Quebec, Canada, pp. 74–79.

Benthem, W., L. P. van Lavieren, and W. J. M. Verheugt. 1999. *Mangrove rehabilitation in the coastal Mekong Delta, Vietnam*. In W. Streever, ed., An International Perspective on Wetland Rehabilitation. Kluwer Academic Publishers, Dordrecht, The Netherlands, pp. 29–36.

Bernatowicz, S., S. Leszczynski, and S. Tyczynska. 1976. The influence of transpiration by emergent plants on the water balance of lakes. *Aquatic Botany* 2: 275–288.

Bertani, A., I. Bramblila, and F. Menegus. 1980. Effect of anaerobiosis on rice seedlings: Growth, metabolic rate, and rate of fermentation products. *Journal of Experimental Botany* 3: 325–331.

Beschel, R. E., and P. J. Webber. 1962. Gradient analysis in swamp forests. *Nature* 194: 207–209.

Bhowmik, N. G., A. P. Bonini, W. C. Bogner, and R. P. Byrne. 1980. Hydraulics of flow and sediment transport to the Kankakee River in Illinois. *Illinois State Water Survey Report of Investigation* 98, Champaign, IL. 170 pp.

Bishop, J. M., J. G. Gosselink, and J. M. Stone. 1980. Oxygen consumption and hemolymph osmolality of brown shrimp, *Penaeus aztecus*. *Fisheries Bulletin* 78: 741–757.

Black, C. C., Jr. 1973. Photosynthetic carbon fixation in relation to net CO_2 uptake. *Annual Review of Plant Physiology* 24: 253–286.

Blom, C. W. P. M., G. M. Bögemann, P. Laan, A. J. M. van der Sman, H. M. van de Steeg, and L. A. C. J. Voesenek. 1990. Adaptations to flooding in plants from river areas. *Aquatic Botany* 38: 29–47.

Boon, P. I. 1999. Carbon cycling in Australian wetlands: The importance of methane. *Verhandlungen Internationale Vereinigung für Limnologie* 27: 1–14.

Boon, P. I., and B. K. Sorrell. 1991. Biogeochemistry of billabong sediments. I. The effect of macrophytes. *Freshwater Biology* 26: 209–226.

Boon, P. I., and A. Mitchell. 1995. Methanogenesis in the sediments of an Australian freshwater wetland: Comparison with aerobic decay and factors controlling methanogenesis. *FEMS Microbiology Ecology* 18: 174–190.

Boon, P. I., and Sorrell, B. K. 1995. Methane fluxes from an Australian floodplain wetland: The importance of emergent macrophytes. *Journal of North American Benthological Society* 14: 582–598.

Boulé, M. E. 1988. *Wetland creation and enhancement in the Pacific Northwest*. In J. Zelazny and J. S. Feierabend, eds. Proceedings of the Conference on Wetlands: Wetlands: Increasing Our Wetland Resources. Corporate Conservation Council, National Wildlife Federation, Washington, DC, pp. 130–136.

Boustany, R. G., C. R. Crozier, J. M. Rybczyk, and R. R. Twilley. 1997. Denitrification in a south Louisiana wetland forest receiving treated sewage effluent. *Wetlands Ecology and Management* 4: 273–283.

Boutin, C., and P. A. Keddy. 1993. A functional classification of wetland plants. *Journal of Vegetation Science* 4: 591–600.

Bradshaw, A. D. 1996. Underlying principles of restoration. *Canadian Journal of Fisheries and Aquatic Sciences* 53 (Suppl. 1): 3–9.

Braskerud, B. C. 2002a. Factors affecting nitrogen retention in small constructed wetlands treating agricultural non-point source pollution. *Ecological Engineering* 18: 351–370.

Braskerud, B. C. 2002b. Factors affecting phosphorus retention in small constructed wetlands treating agricultural non-point source pollution. *Ecological Engineering* 19: 41–61.

Bridgham, S. D., S. P. Faulkner, and C. J. Richardson. 1991. Steel rod oxidation as a hydrologic indicator in wetland soils. *Soil Science Society of America Journal* 55: 856–862.

Brinson, M. M. 1987. Cumulative Increases in Water Table as a Dimension for Quantifying Hydroperiod in Wetlands. In Estuarine Research Federation Meeting, October 26, 1987, New Orleans.

Brinson, M. M. 1989. Fringe Wetlands in Albemarle and Pamlico Sounds: Landscape Position, Fringe Swamp Structure, and Response to Rising Sea Level. In Albemarle-Pamlico Estuarine Study, Raleigh, NC.

Brinson, M. M. 1993a. A Hydrogeomorphic Classification for Wetlands. Wetlands Research Program Technical Report WRP-DE-4, U.S. Army Corps of Engineers Waterways Experiment Station, Vicksburg, MS.

Brinson, M. M. 1993b. Changes in the functioning of wetlands along environmental gradients. *Wetlands* 13: 65–74.

Brinson, M. M., W. Kruczynski, L. C. Lee, W. L. Nutter, R. D. Smith, and D. F. Whigham. 1994. *Developing an approach for assessing the functions of wetlands*. In W. J. Mitsch, ed. Global Wetlands: Old World and New. Elsevier, Amsterdam, The Netherlands, pp. 615–624.

Brix, H. 1987. Treatment of wastewater in the rhizosphere of wetland plants: The root-zone method. *Water Science and Technology* 19: 107–118.

Brix, H. 1989. Gas exchange through dead culms of reed, *Phragmites australis*(Cav.) Trin. ex Steudel. *Aquatic Botany* 35: 81–98.

Brix, H. 1994. Use of constructed wetlands in water pollution control: Historical development, present status, and future perspectives. *Water Science and Technology* 30: 209–223.

Brix, H. 1998. *Denmark*. In J. Vymazal, H. Brix, P. F. Cooper, M. D. Green, and R. Haberl, eds. Constructed Wetlands for Wastewater Treatment in Europe. Backhuys Publishers, Leiden, The Netherlands, pp. 123–152.

Brix, H., and C. A. Arias. 2005. The use of vertical flow constructed wetlands for on-site treatment of domestic wastewater: New Danish guidelines. *Ecological Engineering* 25: 491–500.

Brix, H., and H.-H. Schierup. 1989a. The use of aquatic macrophytes in water-pollution control. *Ambio* 18: 100–107.

Brix, H., and H.-H. Schierup. 1989b. Sewage treatment in constructed reed beds—Danish experiences. *Water Science and Technology* 21: 1655–1668.

Brix, H., B. K. Sorrell, and P. T. Orr. 1992. Internal pressurization and convective gas flow in some emergent freshwater macrophytes. *Limnology and Oceanography* 37: 1420–1433.

Brodie, G. A., D. A. Hammer, and D. A. Tomljanovich. 1988. *An evaluation of substrate types in constructed wetland drainage treatment systems.* In Mine Drainage and Surface Mine Reclamation, Vol. I: Mine Water and Mine Waste. U.S. Department of the Interior, Pittsburgh, PA, pp. 389–398.

Brooks, R. P., D. E. Samuel, and J. B. Hill, eds. 1985. Wetlands and Water Management on Mined Lands. Proceedings of a Conference, October 23–24, 1985. Pennsylvania State University Press, University Park. 393 pp.

Broome, S. W., E. D. Seneca, and W. W. Woodhouse, Jr. 1988. Tidal salt marsh restoration. *Aquatic Botany* 32: 1–22.

Brown, M. T. 1987. Conceptual Design for a Constructed Wetlands System for the Renovation of Treated Effluent. Report from the Center for Wetlands, University of Florida, Gainesville. 18 pp.

Brown, M. T. 2005. Landscape restoration following phosphate mining: 30 years of co-evolution of science, industry, and regulation. *Ecological Engineering* 24: 309–329.

Brown, S. L. 1981. A comparison of the structure, primary productivity, and transpiration of cypress ecosystems in Florida. *Ecological Monographs* 51: 403–427.

Bruins, R. J. F., S. Cai, S. Chen, and W. J. Mitsch. 1998. Ecological engineering strategies to reduce flooding damage to wetland crops in central China. *Ecological Engineering* 11: 231–259.

Burton, J. D., and P. S. Liss. 1976. Estuarine Chemistry. Academic Press, London. 229 pp.

Cahoon, D. R., D. J. Reed, and J. W. Day, Jr. 1995. Estimating shallow subsidence in microtidal salt marshes of the southeastern United States—Kaye and Barghoorn revisted. *Marine Geology* 128: 1–9.

Calloway, J. C., and J. B. Zedler. 2004. Restoration of urban salt marshes: Lessons from southern California. *Urban Ecosystems* 7: 107–124.

Cameron, C. C. 1970. *Peat deposits of northeastern Pennsylvania.* Bulletin 1317-A. U.S. Geological Survey, Washington, DC. 90 pp.

Campbell, D. 2005. *The Congo River basin.* In L.A. Fraser and P.A. Keddy, eds. The World's Largest Wetlands: Ecology and Conservation, Cambridge University Press, Cambridge, UK, pp. 149–165.

Carroll, P., and P. Crill. 1997. Carbon balance of a temperate poor fen. *Global Biogeochemical Cycles* 11: 349–356.

CH2M-Hill and Payne Engineering. 1997. Constructed Wetlands for Livestock Wastewater Management. Prepared for Gulf of Mexico Program, Nutrient Enrichment Committee, CH2M-Hill, Gainesville, FL.

Chambers, J. M., and A. J. McComb. 1994. *Establishment of wetland ecosystems in lakes created by mining in Western Australia*. In W. J. Mitsch, ed. Global Wetlands: Old World and New. Elsevier, Amsterdam, The Netherlands, pp. 431–441.

Chapman, V. J. 1938. Studies in salt marsh ecology. I-III. *Journal of Ecology* 26: 144–221.

Chapman, V. J. 1940. Studies in salt marsh ecology. VI-VII. *Journal of Ecology* 28: 118–179.

Check, E. 2005. Roots of recovery. *Nature* 438: 910–911.

Chen, Y. 1995. *Study of Wetlands in China*. Jilin Sciences Technology Press, Changchun, China. [in Chinese with English summaries.]

Chimner, R. A., and K. C. Ewel. 2005. A tropical freshwater wetland: II. Production, decomposition, and peat formation. *Wetlands Ecology and Management* 13: 671–684.

Chimney, M. J., and G. Goforth. 2006. History and description of the Everglades Nutrient Removal Project, a subtropical constructed wetland in south Florida (USA). *Ecological Engineering* 27: 268–278.

Chow, V. T., ed. 1964. Handbook of Applied Hydrology. McGraw-Hill, New York. 1453 pp.

Christensen, T. 1991. Arctic and sub-Arctic soil emissions: Possible implications for global climate change. *The Polar Record* 27: 205–210.

Chung, C. H. 1982. *Low marshes, China*. In R. R. Lewis III, ed. Creation and Restoration of Coastal Plant Communities. CRC Press, Boca Raton, FL, pp. 131–145.

Chung, C. H. 1989. *Ecological engineering of coastlines with salt marsh plantations*. In W. J. Mitsch and S. E. Jörgensen, eds. Ecological Engineering: An Introduction to Ecotechnology. John Wiley & Sons, New York, pp. 255–289.

Chung, C. H. 1994. *Creation of Spartina plantations as an effective measure for reducing coastal erosion in China*. In W. J. Mitsch, ed. Global Wetlands: Old World and New. Elsevier, Amsterdam, The Netherlands, pp. 443–452.

Cintrón, G., A. E. Lugo, and R. Martinez. 1985. *Structural and functional properties of mangrove forests*. In W. G. D'Arcy and M. D. Corma, eds. The Botany and Natural History of Panama, IV Series: Monographs in Systematic Botany, Vol. 10. Missouri Botanical Garden, St. Louis, pp. 53–66.

Clarkson, B. R. 1997. Vegetation recovery following fire in two Waikato peatlands at Whangamarino and Moanatuatua. *New Zealand Journal of Botany* 35: 167–179.

Clements, F. E. 1916. Plant Succession. Publication 242. Carnegie Institution of Washington. 512 pp.

Clewell, A. F. 1999. Restoration of riverine forest at Hall Branch on phosphate-mined land, Florida. *Restoration Ecology* 7: 1–14.

Clewell, A. F., L. F. Ganey, Jr., D. P. Harlos, and E. R. Tobi. 1976. Biological Effects of Fill Roads across Salt Marshes. Report FL-E.R-1–76. Florida Department of Transportation, Tallahassee.

Clymo, R. S. 1963. Ion exchange in *Sphagnum* and its relation to bog ecology. *Annals of Botany (London) New Series* 27: 309–324.

Clymo, R. S. 1965. Experiments on breakdown of *Sphagnum* in two bogs. *Journal of Ecology* 53: 747–758.

Clymo, R. S. 1983. *Peat*. In A. J. P. Gore, ed. Ecosystems of the World, Vol. 4A: Mires: Swamp, Bog, Fen, and Moor. Elsevier, Amsterdam, The Netherlands, pp. 159–224.

Coastal Protection and Restoration Authority of Louisiana. 2007. Integrated Ecosystem Restoration and Hurricane Protection: Louisiana's Comprehensive Master Plan for a Sustainable Coast. CPRA, Baton Rouge, LA, 140 pp.

Cohen, A. D., D. J. Casagrande, M. J. Andrejko, and G. R. Best, eds. 1984. The Okefenokee Swamp: Its Natural History, Geology, and Geochemistry. Wetland Surveys, Los Alamos, NM. 709 pp.

Coles, B., and J. Coles. 1989. People of the Wetlands, Bogs, Bodies and Lake-Dwellers. Thames & Hudson, New York. 215 pp.

Collins, J. N., and V. H. Resh. 1989. Guidelines for the ecological control of mosquitoes in non-tidal wetlands of the San Francisco Bay area. California Mosquito Vector Control Association and University of California Mosquito Research Program, Sacramento, CA.

Comin, F. A., J. A. Romero, V. Astorga, and C. Garcia. 1997. Nitrogen removal and cycling in restored wetlands used as filters of nutrients for agricultural runoff. *Water Science and Technology* 35: 255–261.

Conner, W. H., and J. W. Day, Jr. 1982. *The ecology of forested wetlands in the southeastern United States*. In B. Gopal, R. E. Turner, R. G. Wetzel, and D. F. Whigham, eds. Wetlands: Ecology and Management. National Institute of Ecology and International Scientific Publications, Jaipur, India, pp. 69–87.

Conrad, R. 1993. *Controls of methane production in terrestrial ecosystems*. In M. O. Andreae and D. S. Schimel, eds. Exchange of Trace Gases Between Terrestrial Ecosystems and the Atmosphere. John Wiley & Sons, New York, pp. 39–58.

Contreras-Espinosa, F., and B. G. Warner. 2004. Ecosystem characteristics and management considerations from coastal wetlands in Mexico. *Hydrobiologia* 511: 233–245.

Conway, T. E., and J. M. Murtha. 1989. *The Iselin Marsh Pond Meadow*. In D. A. Hammer, ed. Constructed Wetlands for Wastewater Treatment. Lewis Publishers, Chelsea, MI, pp. 139–144.

Cooke, J. G. 1992. Phosphorus removal processes in a wetland after a decade of receiving a sewage effluent. *Journal of Environmental Quality* 21: 733–739.

Cooper, P. F., and J. A. Hobson. 1989. *Sewage treatment by reed bed systems: The present situation in the United Kingdom*. In D. A. Hammer, ed. Constructed Wetlands for Wastewater Treatment. Lewis Publishers, Chelsea, MI, pp. 153–172.

Cooper, P. F., and B. C. Findlater, eds. 1990. Constructed Wetlands in Water Pollution Control. Pergamon Press, Oxford, UK. 605 pp.

Costanza, R. 1980. Embodied energy and economic evaluation. *Science* 210: 1219–1224.

Costanza, R. 1984. *Natural resource valuation and management: Toward ecological economics*. In A. M. Jansson, ed. Integration of Economy and Ecology—An Outlook for the Eighties. University of Stockholm Press, Stockholm, pp. 7–18.

Costanza, R., S. C. Farber, and J. Maxwell. 1989. Valuation and management of wetland ecosystems. *Ecological Economics* 1: 335–361.

Costanza, R., R. d'Arge, R. de Groot, S. Farber, M. Grasso, B. Hannon, K. Limburg, S. Naeem, R. V. O'Neill, J. Paruelo, R. G. Raskin, P. Sutton, and M. van den Belt. 1997. The value of the world's ecosystem services and natural capital. *Nature* 387: 253–260.

Costanza, R., W. J. Mitsch, and J. W. Day. 2006. A new vision for New Orleans and the Mississippi delta: Applying ecological economics and ecological engineering. *Frontiers in Ecology and the Environment* 4: 465–472.

Cowardin, L. M., V. Carter, F. C. Golet, and E. T. LaRoe. 1979. Classification of Wetlands and Deepwater Habitats of the United States. FWS/OBS-79/31. U.S. Fish and Wildlife Service, Washington, DC. 103 pp.

Cowles, H. C. 1899. The ecological relations of the vegetation on the sand dunes of Lake Michigan. *Botanical Gazette* 27: 95–117, 167–202, 281–308, 361–369.

Cowles, H. C. 1901. The physiographic ecology of Chicago and vicinity. *Botanical Gazette* 31: 73–108, 145–182.

Cowles, H. C. 1911. The causes of vegetative cycles. *Botanical Gazette* 51: 161–183.

Craft, C., S. Broome, and C. Campbell. 2002. Fifteen years of vegetation and soil development after brackish-water marsh creation. *Restoration Ecology* 10: 248–258.

Crill, P. M., K. B. Bartlett, R. C. Harriss, E. Gorham, E. S. Verry, D. I. Sebacher, L. Madzar, and W. Sanner. 1988. Methane flux from Minnesota peatlands. *Global Biogeochemical Cycles* 2: 371–384.

Cronk, J. K. 1996. Constructed wetlands to treat wastewater from dairy and swine operations: A review. *Agriculture, Ecosystems and Environment* 58: 97–114.

Cronk, J. K., and W. J. Mitsch. 1994. Aquatic metabolism in four newly constructed freshwater wetlands with different hydrologic inputs. *Ecological Engineering* 3: 449–468.

Cui, J., C. Li, and C. Trettin. 2005. Analyzing the ecosystem carbon and hydrologic characteristics of forested wetland using a biogeochemical process model. *Global Change Biology* 11: 278–289.

Curtis, J. T. 1959. The Vegetation of Wisconsin. University of Wisconsin Press, Madison. 657 pp.

Cushing, E. J. 1963. Late-wisconsin pollen stratigraphy in East-Central Minnesota. Ph.D. dissertation, University of Minnesota, Minneapolis.

Dacey, J. W. H. 1980. Internal winds in water lilies: An adaption for life in anaerobic sediments. *Science* 210: 1017–1019.

Dacey, J. W. H. 1981. Pressurized ventilation in the yellow waterlily. *Ecology* 62: 1137–1147.

Dachnowski-Stokes, A. P. 1935. Peat land as a conserver of rainfall and water supplies. *Ecology* 16: 173–177.

Dahl, T. E. 1990. Wetlands losses in the United States, 1780s to 1980s. U.S. Department of Interior, Fish and Wildlife Service, Washington, DC. 21 pp.

Dahl, T. E. 2000. Status and trends of wetlands in the conterminous United States 1986 to 1997. U.S. Department of the Interior, Fish and Wildlife Service, Washington, DC, 82 pp.

Dahl, T. E. 2006. Status and trends of wetlands in the conterminous United States 1998 to 2004. U.S. Department of the Interior, Fish and Wildlife Service, Washington, DC, 112 pp.

Dahl, T. E., and C. E. Johnson. 1991. Wetlands status and trends in the conterminous United States mid-1970s to mid-1980s. U.S. Department of Interior, Fish and Wildlife Service, Washington, DC. 28 pp.

Danielsen, F., M. K. Sørensen, M. F. Olwig, V. Selvam, F. Parish, N. D. Burgess, T. Hiraishi, V. M. Karunagaran, M. S. Rasmussen, L. B. Hansen, A. Quarto, and N. Suryadiputra. 2005. The Asian tsunami: A protective role for coastal vegetation. *Science* 310: 643.

da Silva, C. J., and P. Girard. 2004. New challenges in the management of the Brazilian Pantanal and catchment area. *Wetlands Ecology and Management* 12: 553–561.

Davidson, E. A., and I. A. Janssens. 2006. Temperature sensitivity of soil carbon decomposition and feedbacks to climate change. *Nature* 440: 165–173.

Davis, C. A. 1907. Peat: Essays on its origin, uses, and distribution in Michigan. In Report State Board Geological Survey Michigan for 1906, pp. 95–395.

Davis, D. G. 1989. No net loss of the nation's wetlands: A goal and a challenge. *Water Environment and Technology* 4: 513–514.

Davis, J. H. 1940. The ecology and geologic role of mangroves in Florida. Publication 517. Carnegie Institution of Washington, pp. 303–412.

Davis, J. H. 1943. The natural features of southern Florida, especially the vegetation and the Everglades. *Florida Geological Survey Bulletin* 25. 311 pp.

Davis, O. K., T. Minckley, T. Moutox, T. Jull, and B. Kalin. 2002. The transformation of Sonoran Desert wetlands following the historical decrease of burning. *Journal of Arid Environments* 50: 393–412.

Davis, S. N., and R. J. M. DeWiest. 1966. Hydrogeology. John Wiley & Sons, New York. 463 pp.

Davison, L., D. Pont, K. Bolton, and T. Headley. 2006. Dealing with nitrogen in subtropical Australia: Seven case studies in the diffusion of ecotechnological innovation. *Ecological Engineering* 28: 213–223.

Day, J. W., Jr., C. A. S. Hall, W. M. Kemp, and A. Yánez-Arancibia. 1989. Estuarine Ecology. John Wiley & Sons, New York. 558 pp.

Day, J. W., Jr., and P. H. Templet. 1989. Consequences of sea level rise: Implications from the Mississippi Delta. *Coastal Management* 17: 241–257.

Day, J. W., J. Ko, J., Rybczyk, D. Sabins, R. Bean, G. Berthelot, C. Brantley, L. Cardoch, W. Conner, J. N. Day, A. J. Englande, S. Feagley, E. Hyfield, R. Lane, J. Lindsey, J. Mistich, E. Reyes, and R. Twilley. 2004. The use of wetlands in the Mississippi delta for wastewater assimilation: A review. *Ocean and Coastal Management* 47: 671–691.

Day, J. W., Jr., J. Barras, E. Clairain, J. Johnston, D. Justic, G. P. Kemp, J.-Y. Ko, R. Lane, W. J. Mitsch, G. Steyer, P. Templet, and A. Yañez-Arancibia. 2005. Implications of global climatic change and energy cost and availability for the restoration of the Mississippi Delta. *Ecological Engineering* 24: 253–265.

Day, J.W., Jr., D. F. Boesch, E. J. Clairain, G. P. Kemp, S. B. Laska, W. J. Mitsch, K. Orth, H. Mashriqui, D. R. Reed, L. Shabman, C. A. Simenstad, B. J. Streever, R. R. Twilley, C. C. Watson, J. T. Wells, and D. F. Whigham. 2007. Restoration of the Mississippi Delta: Lessons From Hurricanes Katrina and Rita. Science 315: 1679–1684.

De Jong, J. 1976. *The purification of wastewater with the aid of rush or reed ponds*. In J. Tourbier and R. W. Pierson, Jr., eds. Biological Control of Water Pollution. University of Pennsylvania, Philadelphia, pp. 133–139.

Delaune, R. D., R. H. Baumann, and J. G. Gosselink. 1983. Relationships among vertical accretion, coastal submergence, and erosion in a Louisiana Gulf Coast marsh. *Journal of Sedimentary Petrology* 53: 147–157.

Delaune, R. D., and S. Pezeshki. 2003. The role of soil organic carbon in maintaining surface elevation in rapidly subsiding U.S. Gulf of Mexico coastal marshes. *Water, Air and Soil Pollution* 3: 167–179.

de Leon, R. O. D., and A. T. White. 1999. *Mangrove rehabilitation in the Philippines*. In W. Streever, ed. An International Perspective on Wetland Rehabilitation. Kluwer Academic Publishers, Dordrecht, The Netherlands, pp. 37–42.

Dennis, J. V. 1988. The Great Cypress Swamps. Louisiana State University Press, Baton Rouge. 142 pp.

Denny, P. 1993. *Wetlands of Africa: Introduction*. In D. F. Whigham, D. Dykyjová, and S. Hejny, eds. Wetlands of the World, I: Inventory, Ecology, and Management. Kluwer Academic Publishers, Dordrecht, The Netherlands, pp. 1–31.

Dinger, E. C., A. E. Cohen, A. Dean, A. Hendrickson, and J.C. Marks. 2005. Aquatic invertebrates of Cuatro Ciénegas, Coahila, Mexico: Native and exotics. *Southwestern Naturalist* 50: 237–281.

Douglas, M. S. 1947. The Everglades: River of Grass. Ballantine, New York. 308 pp.

Downing, D. M., C. Winer, and L. D. Wood, 2003. Navigating Through Clean Water Act Jurisdiction: A Legal Review. *Wetlands* 23: 475–493.

Duever, M. J. 1988. *Hydrologic processes for models of freshwater wetlands*. In W. J. Mitsch, M. Straskraba, and S. E. Jørgensen, eds. Wetland Modelling. Elsevier, Amsterdam, The Netherlands, pp. 9–39.

Dugan, P. 1993. Wetlands in Danger. Michael Beasley, Reed International Books, London. 192 pp.

Dunne, T., and L. B. Leopold. 1978. Water in Environmental Planning. W. H. Freeman and Company, New York. 818 pp.

Du Rietz, G. E. 1949. Huvudenheter och huvudgränser i Svensk myrvegetation. *Svensk Botanisk Tidkrift* 43: 274–309.

Du Rietz, G. E. 1954. Die Mineralbodenwasserzeigergrenze als Grundlage Einer Natürlichen Zweigleiderung der Nord-und Mitteleuropäischen Moore. *Vegetatio* 5– 6: 571–585.

Edwards, K. R., and C. D. Proffitt. 2003. Comparison of wetland structural characteristics between created and natural salt marshes in southwest Louisiana, USA. *Wetlands* 23: 344–356.

Eggelsmann, R. 1963. Die Potentielle und Aktuelle Evaporation eines Seeklimathochmoores. Publication 62, International Association of Hydrological Sciences, pp. 88–97.

Eisenlohr, W. S. 1976. Water loss from a natural pond through transpiration by hydrophytes. *Water Resources Research* 2: 443–453.

Ellison, A. M., E. J. Farnsworth, and R. R. Twilley. 1996. Facultative mutualism between red mangroves and root-fouling sponges in Belizean mangal. *Ecology* 77: 2431–2444.

Ellison, A. M. 2004. Wetlands of Central America. *Wetlands Ecology and Management* 12: 3–55.

Emery, K. O., and E. Uchupi. 1972. Western North Atlantic Ocean: Topography, rocks, structure, water, life, and sediments. *Memoirs of the American Association of Petroleum Geologists* 17. 1532 pp.

Environmental Defense Fund and World Wildlife Fund. 1992. How Wet Is a Wetland? The Impact of the Proposed Revisions to the Federal Wetlands Delineation Manual. Environmental Defense Fund and World Wildlife Fund, Washington, DC. 175 pp.

Ernst, W. H. O. 1990. Ecophysiology of plants in waterlogged and flooded environments. *Aquatic Botany* 38: 73–90.

Errington, P. L. 1957. Of Men and Marshes. The Iowa State University Press, Ames, IA.

Erwin, K. L. 1991. An Evaluation of Wetland Mitigation in the South Florida Water Management District, Vol. I. Final Report to South Florida Water Management District, West Palm Beach, FL. 124 pp.

Euliss, N. H., R. A. Gleason, A. Olness, R. L. McDougal, H. R. Murkin, R. D. Robarts, R. A. Bourbonniere, and B. G. Warner. 2006. North American prairie wetlands are important nonforested land-based carbon storage sites. *Science of the Total Environment* 361: 179–188.

Evink, G. L. 1980. Studies of Causeways in the Indian River, Florida. Report FL-ER-7–80. Florida Department of Transportation, Tallahassee. 140 pp.

Farber, S., and R. Costanza. 1987. The economic value of wetlands systems. *Journal of Environmental Management* 24: 41–51.

Faulkner, S. P., W. H. Patrick, Jr., and R. P. Gambrell. 1989. Field techniques for measuring wetland soil parameters. *Soil Science Society of America Journal* 53: 883–890.

Feierabend, S. J., and J. M. Zelazny. 1987. Status Report on Our Nation's Wetlands. National Wildlife Federation, Washington, DC. 50 pp.

Fennessy, M. S., and W. J. Mitsch. 1989. Treating coal mine drainage with an artificial wetland. *Research Journal of the Water Pollution Control Federation* 61: 1691–1701.

Fink, D. F., and W. J. Mitsch. 2004. Seasonal and storm event nutrient removal by a created wetland in an agricultural watershed. *Ecological Engineering* 23: 313–325.

Fink, D. F., and W. J. Mitsch. 2007. Hydrology and nutrient biogeochemistry in a created river diversion oxbow wetland. *Ecological Engineering*. 30: 93–162.

Finlayson, M., and M. Moser, eds. 1991. Wetlands. Facts on File, Oxford, UK. 224 pp.

Finlayson, M. and N.C. Davidson. 1999. Global Review of Wetland Resources and Priorities for Wetland Inventory. Ramsar Bureau Contract 56, Ramsar Convention Bureau, Gland, Switzerland.

Fitter, A. H., and R. K. M. Hay. 1987. Environmental Physiology of Plants. Academic Press, New York. 423 pp.

Fleischer, S., A. Gustafson, A. Joelsson, C. Johansson, and L. Stibe. 1994. *Restoration of wetlands to counteract coastal eutrophication in Sweden*. In W. J. Mitsch, ed. Global Wetlands: Old World and New. Elsevier, Amsterdam, The Netherlands, pp. 901–907.

Folke, C. 1991. *The societal value of wetland life-support*. In C. Folke and T. Kaberger, eds. Linking the Natural Environment and the Economy. Kluwer Academic Publishers, Dordrecht, The Netherlands, pp. 141–171.

Food and Agriculture Organization of the United Nations (FAO), The State of World Fisheries and Aquaculture, 1996 (FAO, Rome, 1997), p. 12.

Forsyth, J. L. 1960. The Black Swamp. Ohio Department of Natural Resources, Division of Geological Survey, Columbus, OH. 1 p.

Frayer, W. E., T. J. Monahan, D. C. Bowden, and F. A. Graybill. 1983. Status and trends of wetlands and deepwater habitat in the conterminous United States, 1950s to 1970s. Department of Forest and Wood Sciences, Colorado State University, Fort Collins. 32 pp.

Fredrickson, L. H., and F. A. Reid. 1990. *Impacts of hydrologic alteration on management of freshwater wetlands*. In J. M. Sweeney, ed. Management of Dynamic Ecosystems, North Central Section. Wildlife Society, West Lafayette, IN, pp. 71–90.

Freiberger, H. J. 1972. Streamflow Variation and Distribution in the Big Cypress Watershed During Wet and Dry Periods. Map Series 45, Bureau of Geology, Florida Department of Natural Resources, Tallahassee.

Gallihugh, J. L., and J. D. Rogner. 1998. Wetland Mitigation and 404 Permit Compliance Study, Vol. I. U.S. Fish and Wildlife Service, Region III, Burlington, IL, and U.S. Environmental Protection Agency, Region V, Chicago. 161 pp.

Galloway, J. N., J. D. Aber, J. W. Erisman, S. P. Seitzinger, R. W. Howarth, E. B. Cowling, and B. J. Cosby. 2003. The nitrogen cascade. *BioScience* 53: 341–356.

Gambrell, R. P., and W. H. Patrick, Jr. 1978. *Chemical and microbiological properties of anaerobic soils and sediments*. In D. D. Hook and R. M. M. Crawford, eds.

Plant Life in Anaerobic Environments. Ann Arbor Science, Ann Arbor, MI, pp. 375–423.

Garbisch, E. W. 1977. Recent and Planned Marsh Establishment Work Throughout the Contiguous United States: A Survey and Basic Guidelines. CR D-77–3, U.S. Army Corps of Engineers Waterways Experiment Station, Vicksburg, MS.

Garbisch, E. 2005. Hambleton Island restoration: Environmental Concern's first wetland creation project. *Ecological Engineering* 24: 289–307.

Garbisch, E. W., P. B. Woller, and R. J. McCallum. 1975. Salt Marsh Establishment and Development. Technical Memo 52. U.S. Army Coastal Engineering Research Center, Fort Belvoir, VA.

Gerheart, R. A. 1992. Use of constructed wetlands to treat domestic wastewater, city of Arcata, California. *Water Science and Technology* 26: 1625–1637.

Gerheart, R. A., F. Klopp, and G. Allen. 1989. *Constructed free surface wetlands to treat and receive wastewater: Pilot project to full scale.* In D. A. Hammer, ed. Constructed Wetlands for Wastewater Treatment. Lewis Publishers, Chelsea, MI, pp. 121–137.

Gibbons, A. 1997. Did birds fly through the K–T extinction with flying colors? *Science* 275: 1068.

Gilman, K. 1982. *Nature conservation in wetlands: Two small fen basins in western Britain.* In D. O. Logofet and N. K. Luckyanov, eds. Ecosystem Dynamics in Freshwater Wetlands and Shallow Water Bodies, Vol. I. SCOPE and UNEP Workshop, Center of International Projects, Moscow, pp. 290–310.

Gilvear, D. J., J. H. Tellam, J. W. Lloyd, and D. N. Lerner. 1989. The Hydrodynamics of East Anglian Fen Systems. University of Birmingham Press, Edgbaston, UK.

Glaser, P. H., P. C. Bennett, D. I. Siegel, and E. A. Romanowicz. 1997. Palaeoreversals in groundwater flow and peatland development at Lost River, Minnesota, USA. *Holocene* 6: 413–421.

Gleason, H. A. 1917. The structure and development of the plant association. *Torrey Botanical Club Bulletin* 44: 463–481.

Godfrey, P. J., E. R. Kaynor, S. Pelczarski, and J. Benforado, eds. 1985. Ecological Considerations in Wetlands Treatment of Municipal Wastewaters. Van Nostrand Reinhold, New York. 474 pp.

Golet, F. C., A. J. K. Calhoun, W. R. DeRagon, D. J. Lowry, and A. J. Gold. 1993. Ecology of Red Maple Swamps in the Glaciated Northeast: A Community Profile. Biological Report 12, U.S. Fish and Wildlife Service, Washington, DC. 151 pp.

Goodman, G. T., and D. F. Perkins. 1968. The role of mineral nutrients in *Eriophorum* communities. IV. Potassium supply as a limiting factor in an E. *vaginatum* community. *Journal of Ecology* 56: 685–696.

Goodman, P. J., and W. T. Williams. 1961. Investigations into "die-back." *Journal of Ecology* 49: 391–398.

Gorham, E. 1956. The ionic composition of some bogs and fen waters in the English lake district. *Journal of Ecology* 44: 142–152.

Gorham, E. 1961. Factors influencing supply of major ions to inland waters, with special references to the atmosphere. *Geological Society of America Bulletin* 72: 795–840.

Gorham, E. 1967. Some Chemical Aspects of Wetland Ecology. Technical Memorandum 90, Committee on Geotechnical Research, National Research Council of Canada, pp. 2–38.

Gorham, E. 1991. Northern peatlands: Role in the carbon cycle and probable responses to climatic warming. *Ecological Applications* 1: 182–195.

Gorham, E. and L. Rochefort. 2003. Peatland restoration: A brief assessment with special reference to *Sphagnum* bogs. *Wetlands Ecology and Management* 11: 109–119.

Gosselink, J. G., E. P. Odum, and R. M. Pope. 1974. The Value of the Tidal Marsh. Publication LSU-SG-74–03. Center for Wetland Resources, Louisiana State University, Baton Rouge. 30 pp.

Gosselink, J. G., and R. E. Turner. 1978. *The role of hydrology in freshwater wetland ecosystems*. In R. E. Good, D. F. Whigham, and R. L. Simpson, eds. Freshwater Wetlands: Ecological Processes and Management Potential. Academic Press, New York, pp. 63–78.

Gosselink, J. G., L. C. Lee, and T. A. Muir, eds. 1990a. Ecological Processes and Cumulative Impacts. Illustrated by Bottomland Hardwood Wetland Ecosystems. Lewis Publishers, Chelsea, MI. 708 pp.

Gosselink, J. G., L. C. Lee, and T. A. Muir, eds. 1990b. Ecological Processes and Cumulative Impacts: Illustrated by Bottomland Hardwood Wetland Ecosystems. Lewis Publishers, Chelsea, MI. 708 pp.

Gottgens, J. F., B. P. Swartz, R. W. Kroll, and M. Eboch. 1998. Long-term GIS-based records of habitat changes in a Lake Erie coastal marsh. *Wetlands Ecology and Management* 6: 5–17.

Gottgens, J. F., J. E. Perry, R. H. Fortney, J. Meyer, M. Benedict, and B. E. Rood. 2001. The Paraguay-Paraná Hidrovia: Protecting the Pantanal with lessons from the past. *BioScience* 51: 301–308.

Grant, M. 1962. Myths of the Greeks and Romans. Mentor/New American Library, New York.

Gray, L. C., O. E. Baker, F. J. Marschner, B. O. Weitz, W. R. Chapline, W. Shepard, and R. Zon. 1924. *The utilization of our lands for crops, pasture, and forests*. In U.S. Department of Agriculture Yearbook, 1923. Government Printing Office, Washington. DC, pp. 415–506.

Greenway, M. 2005. The role of constructed wetlands in secondary effluent treatment and water reuse in subtropical and arid Australia. *Ecological Engineering* 25: 501–509.

Greenway, M., P. Dale, and H. Chapman. 2003. An assessment of mosquito breeding and control in four surface flow wetlands in tropical-subtropical Australia. *Water Science & Technology* 48: 249–256.

Grime, H. H. 1979. Plant Strategies and Vegetation Processes. John Wiley & Sons, New York.

Gross, W. J. 1964. Trends in water and salt regulation among aquatic and amphibious crabs. *Biological Bulletin* 127: 447–466.

Grosse, W., and P. Schröder. 1984. Oxygen supply of roots by gas transport in alder trees. *Zeitschrift für Naturforsch.* 39C: 1186–1188.

Grosse, W., S. Sika, and S. Lattermann. 1990. *Oxygen supply of roots by thermo-osmotic gas transport in Alnus-glutinosa and other wetland trees.* In D. Werner and P. Müller, eds. Fast Growing Trees and Nitrogen Fixing Trees. Gustav Fischer Verlag, New York, pp. 246–249.

Grosse, W., J. Frye, and S. Lattermann. 1992. Root aeration in wetland trees by pressurized gas transport. *Tree Physiology* 10: 285–295.

Grosse, W., and D. Meyer. 1992. The effect of pressurized gas transport on nutrient uptake during hypoxia of alder roots. *Botany Acta* 105: 223–226.

Grosse, W., H. B. Büchel, and S. Lattermann. 1998. *Root aeration in wetland trees and its ecophysiological significance.* In A. D. Laderman, ed. Coastally Restricted Forests. Oxford University Press, New York, pp. 293–305.

Grunwald, M. 2006. The Swamp: The Everglades, Florida, and the Politics of Paradise. Simon & Schuster, New York, 450 pp.

Gu, B., M. J. Chimney, J. Newman, and M. K. Nungesser. 2006. Limnological characteristics of a subtropical constructed wetland in south Florida (USA). *Ecological Engineering* 27: 345–360.

Hadi, A., K. Inubushi, Y. Furukawa, E. Purnomo, M. Rasmadi, and H. Tsuruta. 2005. Greenhouse gas emissions from tropical peatlands of Kalimantan, Indonesia. *Nutrient Cycling in Agroecosystems* 71: 73–80.

Hall, F. R., R. J. Rutherford, and G. L. Byers. 1972. The Influence of a New England Wetland on Water Quantity and Quality. New Hampshire Water Resource Center Research Report 4. University of New Hampshire, Durham. 51 pp.

Hammer, D. A., ed. 1989. Constructed Wetlands for Wastewater Treatment. Lewis Publishers, Chelsea, MI. 831 pp.

Hammer, D. E., and R. H. Kadlec. 1983. Design Principles for Wetland Treatment Systems. EPA-600/2–83–26. U.S. Environmental Protection Agency, Ada, OK. 244 pp.

Hanby, J., and D. Bygott. 1998. Ngorongoro Conservation Area. Kibuyu Partners, Karatu, Tanzania.

Hare, F. K. 1980. Long-term annual surface heat and water balances over Canada and the United States south of 60°N: Reconciliation of precipitation, runoff, and temperature fields. *Atmosphere-Ocean* 18: 127–153.

Harper, D. M., C. Adams, and K. Mavuti. 1995. The aquatic plant communities of the Lake Naivasha wetland, Kenya: Pattern, dynamics, and conservation. *Wetlands Ecology and Management* 3: 111–123.

Harris, L. D., and J. G. Gosselink. 1990. *Cumulative impacts of bottomland hardwood forest conversion on hydrology, water quality, and terrestrial wildlife.* In J. G. Gosselink, L. C. Lee, and T. A. Muir, eds. Ecological Processes and Cumulative Impacts. Illustrated by Bottomland Hardwood Wetland Ecosystems. Lewis Publishers, Chelsea, MI, pp. 259–322.

Harris, M. B., W. Tomas, G. Mourao, C. J. DaSilva, E. Guimaraes, F. Sonoda, and E. Fachim. 2005. Safeguarding the Pantanal wetlands: Threats and conservation initiatives. *Conservation Biology* 19: 714–720.

Harris, T. 1991. Death in the Marsh. Island Press, Washington, DC. 245 pp.

Harriss, R. C., D. I. Sebacher, and F. P. Day, Jr. 1982. Methane flux in the Great Dismal Swamp. *Nature* 297: 673–674.

Harter, S. K., and W. J. Mitsch. 2003. Patterns of short-term sedimentation in a freshwater created marsh. *Journal of Environmental Quality* 32: 325–334.

He, W., R. Feagin, J. Lu, W. Liu, Q. Yan, and Z. Xie. 2007. Impacts of introduced *Spartina alterniflora* along an elevation gradient at the Jiuduansha Shoals in the Yangtze Estuary, suburban Shanghai, China. *Ecological Engineering* 29: 245–248.

Headley, T. R., E. Herity, and L. Davison. 2005. Treatment at different depths and vertical mixing within a 1-m deep horizontal subsurface-flow wetland. *Ecological Engineering* 25: 567–582.

Healey, M. C. 1994. *Effects of dams and dikes on fish habitat in two Canadian river deltas*. In W. J. Mitsch, ed. Global Wetlands: Old World and New. Elsevier, Amsterdam, The Netherlands, pp. 385–398.

Hefner, J. M., and J. D. Brown. 1985. Wetland trends in the southeastern United States. *Wetlands* 4: 1–11.

Heimburg, K. 1984. *Hydrology of north-central Florida cypress domes*. In K. C. Ewel and H. T. Odum, eds. Cypress Swamps. University Presses of Florida, Gainesville, pp. 72–82.

Heinselman, M. L. 1970. Landscape evolution and peatland types, and the Lake Agassiz Peatlands Natural Area, Minnesota. *Ecological Monographs* 40: 235–261.

Hemond, H. F. 1980. Biogeochemistry of Thoreau's Bog, Concord, Mass. *Ecological Monographs* 50: 507–526.

Hemond, H. F., and J. L. Fifield. 1982. Subsurface flow in salt marsh peat: A model and field study. *Limnology and Oceanography* 27: 126–136.

Hemond, H. F., and J. C. Goldman. 1985. On non-Darcian water flow in peat. *Journal of Ecology* 73: 579–584.

Hernandez, M. E., and W. J. Mitsch. 2006. Influence of hydrologic pulses, flooding frequency, and vegetation on nitrous oxide emissions from created riparian marshes. *Wetlands* 26: 862–877.

Hernandez, M. E., and W. J. Mitsch. 2007. Denitrification in created riverine wetlands: Influence of hydrology and season. *Ecological Engineering* 30: 78–88.

Hey, D. L., and N. S. Philippi. 1995. Flood reduction through wetland restoration: The Upper Mississippi River Basin as a case study. *Restoration Ecology* 3: 4–17.

Hickman, S. C., and V. J. Mosca. 1991. Improving Habitat Quality for Migratory Waterfowl and Nesting Birds: Assessing the Effectiveness of the Des Plaines River Wetlands Demonstration Project. Technical Paper 1, Wetlands Research, Chicago. 13 pp.

Hinkle, R., and W. J. Mitsch. 2005. Salt marsh vegetation recovery at salt hay farm wetland restoration sites on Delaware Bay. *Ecological Engineering* 25: 240–251.

Hiraishi, T., and K. Harada. 2003. Greenbelt tsunami prevention in south-Pacific region. Report of the Port and Airport Research Institute, Vol. 24, No. 2. EqTAP Project, Japan. *http://eqtap.edm.bosai.go.jp.*

Hoagland, C. R., L. E. Gentry, M. B. David, and D. A. Kovacic. 2001. Plant nutrient uptake and biomass accumulation in a constructed wetland. *Journal of Freshwater Ecology* 16: 527–540.

Hollands, G. G. 1987. *Hydrogeologic classification of wetlands in glaciated regions.* In J. Kusler, ed. Wetland Hydrology. Proceedings of a National Wetland Symposium. Association of State Wetland Managers, Berne, NY.

Houghton, J. T., Y. Ding, D. J. Griggs, M. Noguer, P. J. van der Linden, X. Dai, K. Maskell, and C. A. Johnson. 2001. Climate Change 2001: The Scientific Basis. Cambridge University Press, Cambridge, UK.

Hunt, R. J., D. P. Krabbenhoft, and M. P. Anderson. 1996. Groundwater inflow measurements in wetland systems. *Water Resources Research* 32: 495–507.

Hyatt, R. A., and G. A. Brook. 1984. *Groundwater flow in the Okefenokee Swamp and hydrologic and nutrient budgets for the period August, 1981 through July, 1982.* In A. D. Cohen, D. J. Casagrande, M. J. Andrejko, and G. R. Best, eds. The Okefenokee Swamp: Its Natural History, Geology, and Geochemistry. Wetland Surveys, Los Alamos, NM, pp. 229–245.

Ingram, H. A. P. 1967. Problems of hydrology and plant distribution in mires. *Journal of Ecology* 55: 711–724.

Ingram, H. A. P. 1983. *Hydrology.* In A. J. P. Gore, ed. Ecosystems of the World, Vol. 4A: Mires: Swamp, Bog, Fen, and Moor. Elsevier, Amsterdam, The Netherlands, pp. 67–158.

IPPC. 2001. Climate Change 2001: The Scientific Basis. Published for the Intergovernmental Panel on Climate Change. Cambridge University Press, UK.

IPPC. 2007a. Climate Change 2007, 4th Assessment Report. Published for the Intergovernmental Panel on Climate Change. Cambridge University Press, UK.

IPPC. 2007b. Climate Change 2007: The Physical Science Basis. Summary for Policymakers. IPPC Secretariat, Geneva, Switzerland., 18 pp.

IUCN. 1993. The Wetlands of Central and Eastern Europe. IUCN, Gland, Switzerland, and Cambridge, UK. 83 pp.

IWA Specialists Group on Use of Macrophytes in Water Pollution Control. 2000. Constructed Wetlands for Pollution Control. Scientific and Technical Report No. 8. International Water Association, London, England, 156 pp.

Jacks, G., A. Joelsson, and S. Fleischer. 1994. Nitrogen retention in forested wetlands. *Ambio* 23: 358–362.

Jackson, J. 1989. *Man-made wetlands for wastewater treatment: Two case studies.* In D. A. Hammer, ed. Constructed Wetlands for Wastewater Treatment. Lewis Publishers, Chelsea, MI, pp. 574–580.

Jackson, S. T., R. P. Rutyma, and D. A. Wilcox. 1988. A paleoecological test of a classical hydrosere in the Lake Michigan dunes. *Ecology* 69: 928–936.

James, P. S. B. R. 1999. Shrimp farming development in India—An overview of environmental, socio-economic, legal and other implications. *Aquaculture Magazine, www.aquaculturemag.com*.

Jasinski, S. M. 1999. *Peat*. In Minerals Yearbook 1999: Volume I—Metals and Minerals. Minerals and Information, U.S. Geological Survey, Reston, VA.

Johengen, T. H., and P. A. LaRock. 1993. Quantifying nutrient removal processes within a constructed wetland designed to treat urban stormwater runoff. *Ecological Engineering* 2: 347–366.

Johnson, D. C. 1942. The Origin of the Carolina Bays. Columbia University Press, New York. 341 pp.

Johnson, P. and P. Gerbeaux. 2004. Wetland Types in New Zealand. New Zealand Department of Conservation, Wellington, 184 pp.

Johnson, W. C., B. V. Millett, T. Gilmanov, R. A. Voldseth, G. R. Guntenspergen, and D. E. Naugle. 2005. Vulnerability of northern prairie wetlands to climate change. *BioScience* 55: 863–872.

Johnston, C. A. 1991. Sediment and nutrient retention by freshwater wetlands: Effects on surface water quality. *Critical Reviews in Environmental Control* 21: 491–565.

Johnston, C. A. 1994. *Ecological engineering of wetlands by beavers*. In W. J. Mitsch, ed. Global Wetlands: Old World and New. Elsevier, Amsterdam, The Netherlands, pp. 379–384.

Johnston, C. A., S. D. Bridgham, and J. P. Schubauer-Berigan. 2001. Nutrient dynamics in relation to geomorphology of riverine wetlands. *Soil Science of America Journal* 65: 557–577.

Jones, C. G., J. H. Lawton, and M. Shachak. 1994. Organisms as ecosystem engineers. *Oikos* 69: 373–386.

Jones, C. G., J. H. Lawton, and M. Shachak. 1997. Positive and negative effects of organisms as physical ecosystem engineers. *Ecology* 78: 1946–1957.

Josselyn, M., J. Zedler, and T. Griswold. 1990. *Wetland mitigation along the Pacific coast of the United States*. In J. A. Kusler and M. E. Kentula, eds. Wetland Creation and Restoration. Island Press, Washington, DC, pp. 3–36.

Junk, W. J. 1982. *Amazonian floodplains: Their ecology, present and potential use*. In D. O. Logofet and N. K. Luckyanov, eds. Ecosystem Dynamics in Freshwater Wetlands and Shallow Water Bodies, Vol. I. SCOPE and UNEP Workshop, Center of International Projects, Moscow, pp. 98–126.

Junk, W. J. 1993. *Wetlands of tropical South America*. In D. F. Whigham, D. Dykyjová, and S. Hejny, eds. Wetlands of the World, I: Inventory, Ecology, and Management. Kluwer Academic Publishers, Dordrecht, The Netherlands, pp. 679–739.

Junk, W. J., P. B. Bayley, and R. E. Sparks. 1989. *The flood pulse concept in river-floodplain systems*. In D. P. Dodge, ed. Proceedings of the International Large River Symposium. Special Issue of the Journal of Canadian Fisheries and Aquatic Sciences 106: 11–127.

Junk, W. J., and M. T. F. Piedade. 2004. Status of knowledge, ongoing research, and research needs in Amazonian wetlands. *Wetlands Ecology and Management* 12: 597–609.

Junk, W. J., and M. T. F. Piedade. 2005. *The Amazon River basin*. In L.A. Fraser and P.A. Keddy, eds. The World's Largest Wetlands: Ecology and Conservation, Cambridge University Press, Cambridge, UK, pp. 63–117.

Junk, W. J., and C. Nunes de Cunha. 2005. Pantanal: A large South American wetland at a crossroads. *Ecological Engineering* 24: 391–401.

Kaatz, M. R. 1955. The Black Swamp: A study in historical geography. *Annals of the Association of American Geographers* 35: 1–35.

Kadlec, R. H. 1989. *Hydrologic factors in wetland water treatment*. In D. A. Hammer, ed. Constructed Wetlands for Wastewater Treatment. Lewis Publishers, Chelsea, MI, pp. 21–40.

Kadlec, R. H. 1999. *Constructed wetlands for treating landfill leachate*. In G. Mulam-oottil, E. A. McBean, and F. Rovers, eds. Constructed Wetlands for the Treatment of Landfill Leachates. Lewis Publishers, Boca Raton, FL, pp. 17–31.

Kadlec, R. H. 2006. Free surface wetlands for phosphorus removal: The position of the Everglades Nutrient Removal Project. 2006. *Ecological Engineering* 27: 361–379.

Kadlec, R. H., R. B. Williams, and R. D. Scheffe. 1988. *Wetland evapotranspiration in temperate and arid climates*. In D. D. Hook, W. H. McKee, Jr., H. K. Smith, J. Gregory, V. G. Burrell, M. R. DeVoe, R. E. Sojka, S. Gilbert, R. Banks, L. G. Stolzy, C. Brooks, T. D. Matthews, and T. H. Shear, eds. The Ecology and Management of Wetlands, Vol. 1: Ecology of Wetlands. Timber Press, Portland, OR, pp. 146–160.

Kadlec, R. H., and D. L. Hey. 1994. Constructed wetlands for river water quality improvement. *Water Science and Technology* 29: 159–168.

Kadlec, R. H., and R. L. Knight. 1996. Treatment Wetlands. CRC Press/Lewis Publishers, Boca Raton, FL. 893 pp.

Kadlec, R. H. and S. Wallace, 2008. *Treatment Wetlands II*. CRC Press, Boca Raton, FL.

Kalin, M. 2001. Biogeochemical and ecological considerations in designing wetland treatment systems in post-mining landscapes. *Waste Management* 21: 191–196.

Kantrud, H. A., J. B. Millar, and A. G. van der Valk. 1989. *Vegetation of wetlands of the prairie pothole region*. In A. G. van der Valk, ed. Northern Prairie Wetlands. Iowa State University Press, Ames, pp. 132–187.

Keddy, P. A. 1983. Freshwater wetland human-induced changes: Indirect effects must also be considered. *Environmental Management* 7: 299–302.

Keddy, P. A. 1992a. Assembly and response rules: Two goals for predictive community ecology. *Journal of Vegetation Science* 3: 157–164.

Keddy, P. A. 1992b. *Water level fluctuations and wetland conservation*. In J. Kusler and R. Smandon, eds. Wetlands of the Great Lakes. Proceedings of an International Symposium. Association of State Wetland Managers, Berne, NY, pp. 79–91.

Keddy, P. A. 2000. Wetland Ecology: Principles and Conservation. Cambridge University Press, Cambridge, UK. 614 pp.

Kesel, R., and D. J. Reed. 1995. *Status and trends in Mississippi River sediment regime and its role in Louisiana wetland development*. In D. J. Reed, ed. Status and Historical Trends of Hydrologic Modification, Reduction in Sediment Availability, and Habitat Loss/Modification in the Barataria–Terrebonne Estuarine System. BT/NEP Publication 20, Barataria–Terrebonne National Estuary Program, Thibodaux, LA, pp. 80–98.

King, D. M., and L. W. Herbert. 1997. The fungibility of wetlands. *National Wetlands Newsletter* 19: 10–13.

King, S. L., and B. D. Keeland. 1999. Evaluation of reforestation in the Lower Mississippi River alluvial valley. *Restoration Ecology* 7: 348–359.

Kirk, P. W., Jr. 1979. The Great Dismal Swamp. University Press of Virginia, Charlottesville. 427 pp.

Knight, R. L. 1990. *Wetland Systems*. In Natural Systems for Wastewater Treatment, Manual of Practice FD-16. Water Pollution Control Federation, Alexandria, VA, pp. 211–260.

Knight, R. L., B. H. Winchester, and J. C. Higman. 1984. Carolina Bays—Feasibility for effluent advanced treatment and disposal. *Wetlands* 4: 177–204.

Knight, R. L., T. W. McKim, and H. R. Kohl. 1987. Performance of a natural wetland treatment system for wastewater management. *Journal of the Water Pollution Control Federation* 59: 746–754.

Knight, R. L., W. E. Walton, G. F. O'Meara, W. K. Reisen, and R. Wass. 2003. Strategies for effective mosquito control in constructed treatment wetlands. *Ecological Engineering* 21: 211–232.

Koch, M. S., and I. A. Mendelssohn. 1989. Sulphide as a soil phytotoxin: Differential responses in two marsh species. *Journal of Ecology* 77: 565–578.

Koch, M. S., I. A. Mendelssohn, and K. L. McKee. 1990. Mechanism for the hydrogen sulfide-induced growth limitation in wetland macrophytes. *Limnology and Oceanography* 35: 399–408.

Koch, M. S., and K. R. Reddy. 1992. Distribution of soil and plant nutrients along a trophic gradient in the Florida Everglades. *Soil Science Society of America Journal* 56: 1492–1499.

Koreny, J. S., W. J. Mitsch, E. S. Bair, and X. Wu. 1999. Regional and local hydrology of a constructed riparian wetland system. *Wetlands* 19: 182–193.

Korgen, B. J. 1995. Seiches. *American Scientist* July-August 1995: 330–341.

Kovacic, D. A., M. B. David, L. E. Gentry, K. M. Starks, and R. A. Cooke. 2000. Effectiveness of constructed wetlands in reducing nitrogen and phosphorus export from agricultural tile drainage. *Journal of Environmental Quality* 29: 1262–1274.

Kroll, R. W., J. F. Gottgens, and B. P. Swartz. 1997. Wild rice to rip-rap: 120 years of habitat changes and management of a Lake Erie coastal marsh. *Transactions of the 62nd North American Wildlife and Natural Resources Conference* 62: 490–500.

Kulczynski, S. 1949. Peat bogs of Polesie. *Acad. Pol. Sci. Mem.*, Ser. B, No.15. 356 pp.

Kurz, H. 1928. Influence of *Sphagnum* and other mosses on bog reactions. *Ecology* 9: 56–69.

Kushlan, J. A. 1989. *Avian use of fluctuating wetlands*. In R. R. Sharitz and J. W. Gibbons, eds. Freshwater Wetlands and Wildlife. Department of Energy Symposium Series 61. Office of Scientific and Technical Information, Department of Energy, Oak Ridge, TN, pp. 593–604.

Kusler, J. A. 1983. Our National Wetland Heritage: A Protection Guidebook. Environmental Law Institute, Washington, DC. 167 pp.

Kusler, J., W. J. Mitsch, and J. S. Larson. 1994. Wetlands. *Scientific American* 270(1): 64–70.

Laanbroek, H. J. 1990. Bacterial cycling of minerals that affect plant growth in waterlogged soils: A review. *Aquatic Botany* 38: 109–125.

Lambert, J. M. 1964. The *Spartina* story. *Nature* 204: 1136–1138.

Lambou, V. W. 1990. Importance of bottomland forest zones to fishes and fisheries: A case history. In J. G. Gosselink, L. C. Lee, and T. A. Muir, eds. Ecological Processes and Cumulative Impacts: Illustrated by Bottomland Hardwood Wetland Ecosystems. Lewis Publishers, Chelsea, MI, pp. 125–193.

Lane, R. R., J.W. Day, Jr., and B. Thibodeaux. 1999. Water quality analysis of a freshwater diversion at Caernarvon, Louisiana. *Estuaries* 22: 327 336

Larson, A. C., L. E. Gentry, M. B. David, R. A. Cooke, and D. A. Kovacic. 2000. The role of seepage in constructed wetlands receiving tile drainage. *Ecological Engineering* 15: 91–104.

Larson, J. S., and J. A. Kusler. 1979. *Preface*. In P. E. Greeson, J. R. Clark, and J. E. Clark, eds. Wetland Functions and Values: The State of Our Understanding. American Water Resources Association, Minneapolis, MN.

Lashof, D. A., and D. R. Ahuja. 1990. Relative contributions of greenhouse gas emissions to global warming. *Nature* 344: 529–531.

Lavoie, C., and L. Rochefort. 1996. The natural revegetation of a harvested peatland in southern Quebec: A spatial and dentroecological analysis. *Ecoscience* 3: 101–111.

Lee, J. K., R. A. Park, and P. W. Mausel. 1991. Application of geoprocessing and simulation modeling to estimate impacts of sea level rise on northeastern coast of Florida. *Photogrammetric Engineering and Remote Sensing* 58: 1579–1586.

Lee, R. 1980. Forest Hydrology. Columbia University Press, New York. 349 pp.

Lefeuvre, J. C. 1990. *Ecological impact of sea level rise on coastal ecosystems of Mont-Saint-Michel Bay*. In J. J. Beukema, W. J. Wolff, and J. W. M. Brouns, eds. Expected Effects of Climatic Change on Marine Coastal Ecosystems. Kluwer Academic Publishers, Dordrecht, The Netherlands, pp. 139–153.

Lehner, B. and P. Döll. 2004. Development and validation of a global database of lakes, reservoirs, and wetlands. *Journal of Hydrology* 296: 1–22.

LeMer, J. and P. Roger. 2001. Production, oxidation, emission, and consumption of methane by soils: a review. *European Journal of Soil Biology* 37:25–50.

Leonardson, L., L. Bengtsson, T. Davidsson, T. Persson, and U. Emanuelsson. 1994. Nitrogen retention in artificially flooded meadows. *Ambio* 23: 332–341.

Leopold, L. B., M. G. Wolman, and J. E. Miller. 1964. Fluvial Processes in Geomorphology. W. H. Freeman, San Francisco. 522 pp.

Levitt, J. 1980. Responses of Plants to Environmental Stresses, Vol. II: Water, Radiation, Salt, and Other Stresses. Academic Press, New York. 607 pp.

Lewis, R. R. 1990a. *Wetland restoration/creation/enhancement terminology: Suggestions for standardization.* In J. A. Kusler and M. E. Kentula, eds. Wetland Creation and Restoration. Island Press, Washington, DC, pp. 1–7.

Lewis, R. R. 1990b. *Creation and restoration of coastal plain wetlands in Florida.* In J. A. Kusler and M. E. Kentula, eds. Wetland Creation and Restoration. Island Press, Washington, DC, pp. 73–101.

Lewis, R. R. 1990c. *Creation and restoration of coastal plain wetlands in Puerto Rico and the U.S. Virgin Islands.* In J. A. Kusler and M. E. Kentula, eds. Wetland Creation and Restoration. Island Press, Washington, DC, pp. 103–123.

Lewis, R. R. 2000. Ecologically based goal setting in mangrove forest and tidal marsh restoration. *Ecological Engineering* 15: 191–198.

Lewis, R. R., III. 2005. Ecological engineering for successful management and restoration of mangrove forests. *Ecological Engineering* 24: 403–418.

Lewis, R. R., J. A. Kusler, and K. L. Erwin. 1995. *Lessons learned from five decades of wetland restoration and creation in North America.* In C. Montes, G. Oliver, F. Molina, and J. Cobos, eds. Bases Ecologicas para la Restauracion de Humedales en la Cuenca Mediterránea. Consejeria de Medio Ambiente, Junta de Andalucía, Andalucía, Spain.

Lide, R. F., V. G. Meentemeyer, J. E. Pinder, and L. M. Beatty 1995. Hydrology of a Carolina Bay located on the Upper Coastal Plain of western South Carolina. *Wetlands* 15: 47–57.

Light, S. S., and J. W. Dineen. 1994. *Water control in the Everglades: A historical perspective.* In S. M. Davis and J. C. Ogden, eds. Everglades: The Ecosystem and Its Restoration. St. Lucie Press, Delray Beach, FL, pp. 47–84.

Likens, G. E., F. H. Bormann, R. S. Pierce, and J. S. Eaton. 1985. *The Hubbard Brook Valley.* In G. E. Likens, ed. An Ecosystem Approach to Aquatic Ecology: Mirror Lake and Its Environment. Springer-Verlag, New York, pp. 9–39.

Linacre, E. 1976. *Swamps.* In J. L. Monteith, ed. Vegetation and the Atmosphere, Vol. 2: Case Studies. Academic Press, London, pp. 329–347.

Lindeman, R. L. 1941. The developmental history of Cedar Creek Lake, Minnesota. *American Midland Naturalist* 25: 101–112.

Lindeman, R. L. 1942. The trophic-dynamic aspect of ecology. *Ecology* 23: 399–418.

Linsley, R. K., and J. B. Franzini. 1979. Water Resources Engineering, 3rd ed. McGraw-Hill, New York. 716 pp.

Liptak, M. A. 2000. Water column productivity, calcite precipitation, and phosphorus dynamics in freshwater marshes. Ph.D. dissertation, The Ohio State University, Columbus.

Litchfield, D. K., and D. D. Schatz. 1989. *Constructed wetlands for wastewater treatment at Amoco Oil Company's Mandan, North Dakota, refinery.* In D. A. Hammer, ed. Constructed Wetlands for Wastewater Treatment. Lewis Publishers, Chelsea, MI, pp. 101–119.

Littlehales, B., and W. A. Niering. 1991. Wetlands of North America. Thomasson-Grant, Charlottesville, VA. 160 pp.

Livingston, D. A. 1963. *Chemical composition of rivers and lakes.* Professional Paper 440G. U.S Geological Survey, Washington, DC. 64 pp.

Lockwood, C. C., and R. Gary. 2005. Marsh Mission: Capturing the Vanishing Wetlands. Louisiana State University Press, Baton Rouge. 106 pp.

Lodge, T. E. 2005. The Everglades Handbook: Understanding the Ecosystem, 2nd ed. CRC Press, Boca Raton, FL, 302 pp.

Lott, R. B., and R. J. Hunt. 2001. Estimating evapotranspiration in natural and constructed wetlands. *Wetlands* 21: 614–628.

Louisiana Coastal Wetlands Conservation and Restoration Task Force (LCWCR) and the Wetlands Conservation and Restoration Authority. 1998. Coast 2050: Toward a Sustainable Coastal Louisiana, an Executive Summary. Louisiana Department of Natural Resources, Baton Rouge. 12 pp.

Lu, J., ed. 1990. Wetlands in China. East China Normal University Press, Shanghai. 177 pp. [in Chinese].

Lu, J. 1995. Ecological significance and classification of Chinese wetlands. *Vegetatio* 118: 49–56.

Lugo, A. E., and S. C. Snedaker. 1974. The ecology of mangroves. *Annual Review of Ecology and Systematics* 5: 39–64.

Lugo, A. E., M. M. Brinson, and S. L. Brown, eds. 1990. Forested Wetlands: Ecosystems of the World 15. Elsevier, Amsterdam, The Netherlands. 527 pp.

Ma, X., X. Liu, and R. Wang. 1993. China's wetlands and agro-ecological engineering. *Ecological Engineering* 2: 291–330.

Maguire, C., and P. O. S. Boaden. 1975. Energy and evolution in the thiobios: An extrapolation from the marine gastrotrich *Thiodasys sterreri. Cahiers de Biologia* March 16: 635–646.

Malakoff, D. 1998. Restored wetlands flunk real world test. *Science* 280: 371–372.

Malmer, N. 1975. *Development of bog mires.* In A. D. Hasler, ed. Coupling of Land and Water Systems. Ecology Studies 10. Springer-Verlag, New York, pp. 85–92.

Maltby, E., and R. E. Turner. 1983. Wetlands of the world. *Geographic Magazine* 55: 12–17.

Manyin, T., F. M. Williams, and L. R. Stark. 1997. Effects of iron concentration and flow rate on treatment of coal mine drainage in wetland mesocosms: An experimental approach to sizing of constructed wetlands. *Ecological Engineering* 9: 171–185.

Martin, A. C., N. Hutchkiss, F. M. Uhler, and W. S. Bourn. 1953. Classification of Wetlands of the United States. Special Science Report—Wildlife 20, U.S. Fish and Wildlife Service, Washington, DC. 14 pp.

Massey, B. 2000. Wetlands Engineering Manual. Ducks Unlimited, Southern Regional Office, Hackson, MS, 16 pp.

Matthews, E. 1990. *Global distribution of forested wetlands*. In A. E. Lugo, M. Brinson, and S. Brown, eds. Addendum to Forested Wetlands. Elsevier, Amsterdam, The Netherlands.

Matthews, E., and I. Fung. 1987. Methane emissions from natural wetlands: Global distribution, area, and environmental characteristics of sources. *Global Biogeochemical Cycles* 1: 61–86.

Mauchamp, A., P. Chauvelon, and P. Grillas. 2002. Restoration of floodplain wetlands: Opening polders along a coastal river in Mediterranean France, Vistre marshes. *Ecological Engineering* 18: 619–632.

McCarthy, J. M., T. Gumbricht, T. McCarthy, P. Frost, K. Wessels, and F. Seidel. 2004. Flooding patterns in the Okavango wetland in Botswana between 1972 and 2000. *Ambio* 32: 453–457.

McCaffrey, R. J. 1977. A record of the accumulation of sediment and trace metals in a Connecticut, U.S.A. salt marsh. Ph.D. dissertation, Yale University, New Haven, CT. 156 pp.

McComb, A. J., and P. S. Lake. 1990. Australian Wetlands. Angus and Robertson, London. 258 pp.

McCoy, M. B., and J. M. Rodriguez. 1994. *Cattail* (Typha domingensis) *eradication methods in the restoration of the tropical seasonal freshwater marsh*. In W. J. Mitsch, ed. Global Wetlands: Old World and New. Elsevier, Amsterdam, The Netherlands, pp. 469–482.

McIntosh, R. P. 1985. The Background of Ecology, Concept and Theory. Cambridge University Press, Cambridge, UK. 383 pp.

McKee, K. L., I. A. Mendelssohn, and M. W. Hester. 1988. Reexamination of pore water sulfide concentrations and redox potentials near the aerial roots of *Rhizophora mangle* and *Avicennia germinans. American Journal of Botany* 75: 1352–1359.

McKee, K. L., and I. A. Mendelssohn. 1989. Response of a freshwater marsh plant community to increased salinity and increased water level. *Aquatic Botany* 34: 301–316.

McLeod, K. W., L. S. Donovan, and N. J. Stumpff. 1988. *Responses of woody seedlings to elevated flood water temperatures*. In D. D. Hook et al., eds. The Ecology and Management of Wetlands, Vol. 1: Ecology of Wetlands. Timber Press, Portland, OR, pp. 441–451.

Meeks, G., and L. C. Runyon. 1990. Wetlands Protection and the States. National Conference of State Legislatures, Denver, CO. 26 pp.

Megonigal, J. P., W. H. Conner, S. Kroeger, and R. R. Sharitz. 1997. Aboveground production in southeastern floodplain forests: A test of the subsidy–stress hypothesis. *Ecology* 78: 370–384.

Megonigal, J. P., M. E. Hines, and P. T. Visscher. 2004. *Anaerobic metabolism: Linkages to trace gases and aerobic processes*. In W. H. Schlesinger, ed. Biogeochemistry. Elsevier-Pergamon, Oxford, UK, pp. 317–424.

Meijer, L. E., and Y. Avnimelech. 1999. On the use of micro-electrodes in fish pond sediments. *Aquaculture Engineering* 21: 71–83.

Mendelsohn, J., and S. el Obeid. 2004. Okavango River: The Flow of a Lifeline. Struik Publishers, Cape Town, South Africa. 176 pp.

Mendelssohn, I. A., K. L. McKee, and M. L. Postek. 1982. *Sublethal stresses controlling Spartina alterniflora productivity*. In B. Gopal, R. E. Turner, R. G. Wetzel, and D. F. Whigham, eds. Wetlands: Ecology and Management. National Institute of Ecology and International Science Publications, Jaipur, India, pp. 223–242.

Mendelssohn, I. A., and M. L. Postek. 1982. Elemental analysis of deposits on the roots of *Spartina alterniflora* Loisel. *American Journal of Botany* 69: 904–912.

Mendelssohn, I. A., and D. M. Burdick. 1988. *The relationship of soil parameters and root metabolism to primary production in periodically inundated soils*. In D. D. Hook, W. H. McKee, Jr., H. K. Smith, J. Gregory, V. G. Burrell, M. R. DeVoe, R. E. Sojka, S. Gilbert, R. Banks, L. G. Stolzy, C. Brooks, T. D. Matthews, and T. H. Shear, eds. The Ecology and Management of Wetlands, Vol. 1: Ecology of Wetlands. Timber Press, Portland, OR, pp. 398–428.

Mesléard, F., P. Grillas, and L. T. Ham. 1995. Restoration of seasonally flooded marshes in abandoned ricefields in the Camargue (southern France)—Preliminary results on vegetation and use by ducks. *Ecological Engineering* 5: 95–106.

Meyer, A. H. 1935. The Kankakee "Marsh" of northern Indiana and Illinois. *Michigan Academy of Science, Arts, and Letters Papers* 21: 359–396.

Meyer, J. L. 1985. A detention basin/artificial wetland treatment system to renovate stormwater runoff from urban, highway, and industrial areas. *Wetlands* 5: 135–145.

Millar, J. B. 1971. Shoreline-area as a factor in rate of water loss from small sloughs. *Journal of Hydrology* 14: 259–284.

Millenium Ecosystem Assessment. 2005. Ecosystems and Human Well-Being: Wetlands and Water Synthesis. World Resources Institute, Washington, DC.

Minnemeyer, S. 2002. An analysis of access into Central Africa's rainforests. World Resources Institute, Washington DC.

Mitchell, D. S., A. J. Chick, and G. W. Rasin. 1995. The use of wetlands for water pollution control in Australia: An ecological perspective. *Water Science and Technology* 32: 365–373.

Mitchell, J. G., R. Gehman, and J. Richardson. 1992. Our disappearing wetlands. *National Geographic* 182(4): 3–45.

Mitra, S., R. Wassmann, and P. L. G. Vlek. 2005. An appraisal of global wetland area and its organic carbon stock. *Current Science* 88: 25–35.

Mitsch, W. J. 1979. *Interactions between a riparian swamp and a river in southern Illinois*. In R. R. Johnson and J. F. McCormick, tech. coords. Strategies for the Protection and Management of Floodplain Wetlands and Other Riparian Ecosystems. Proceedings of the Symposium, Calaway Gardens, GA, December 11–13, 1978. General Technical Report WO-12, U.S. Forest Service, Washington, DC, pp. 63–72.

Mitsch, W. J., ed. 1989. Wetlands of Ohio's Coastal Lake Erie: A Hierarchy of Systems. NTIS, OHSU-BS-007, Ohio Sea Grant Program, Columbus. 186 pp.

Mitsch, W. J. 1992. Combining ecosystem and landscape approaches to Great Lakes wetlands. *Journal of Great Lakes Research* 18: 552–570.

Mitsch, W. J. 1998. *Self-design and wetland creation: Early results of a freshwater marsh experiment*. In A. J. McComb and J. A. Davis, eds. Wetlands for the Future. Contributions from INTECOL's Fifth International Wetlands Conference. Gleneagles Publishing, Adelaide, Australia, pp. 635–655.

Mitsch, W.J. (ed.). 2006. Wetland Creation, Restoration, and Conservation: The State of the Science. Elsevier, Amsterdam, 175 pp.

Mitsch, W. J., and K. C. Ewel. 1979. Comparative biomass and growth of cypress in Florida wetlands. *American Midland Naturalist* 101: 417–426.

Mitsch, W. J., C. L. Dorge, and J. R. Wiemhoff. 1979a. Ecosystem dynamics: A phosphorus budget of an alluvial cypress swamp in southern Illinois. *Ecology* 60: 1116–1124.

Mitsch, W. J., W. Rust, A. Behnke, and L. Lai. 1979b. Environmental Observations of a Riparian Ecosystem during Flood Season. Research Report 142, Illinois University Water Resources Center, Urbana. 64 pp.

Mitsch, W. J., M. D. Hutchison, and G. A. Paulson. 1979c. The Momence Wetlands of the Kankakee River in Illinois—An Assessment of Their Value. Document 79/17, Illinois Institute of Natural Resources, Chicago. 55 pp.

Mitsch, W. J., J. R. Taylor, and K. B. Benson. 1983a. *Classification, modelling and management of wetlands—A case study in western Kentucky*. In W. K. Lauenroth, G. V. Skogerboe, and M. Flug, eds. Analysis of Ecological Systems: State-of-the-Art in Ecological Modelling. Elsevier, Amsterdam, The Netherlands, pp. 761–769.

Mitsch, W. J., J. R. Taylor, K. B. Benson, and P. L. Hill, Jr. 1983b. Atlas of Wetlands in the Principal Coal Surface Mine Region of Western Kentucky. FWS/OBS-82/72, U.S. Fish and Wildlife Service, Washington, DC. 135 pp.

Mitsch, W. J., J. R. Taylor, K. B. Benson, and P. L. Hill, Jr. 1983c. Wetlands and coal surface mining in western Kentucky—A regional impact assessment. *Wetlands* 3: 161–179.

Mitsch, W. J., and W. G. Rust. 1984. Tree growth responses to flooding in a bottomland forest in northeastern Illinois. *Forest Science* 30: 499–510.

Mitsch, W. J., and J. G. Gosselink. 1986. Wetlands. Van Nostrand Reinhold, New York.

Mitsch, W. J., and B. C. Reeder. 1992. Nutrient and hydrologic budgets of a Great Lakes coastal freshwater wetland during a drought year. *Wetlands Ecology and Management* 1(4): 211–223.

Mitsch, W. J., and J. G. Gosselink. 1993. Wetlands, 2nd ed. John Wiley & Sons, New York. 722 pp.

Mitsch, W. J., R. H. Mitsch, and R. E. Turner. 1994a. *Wetlands of the Old and New Worlds—Ecology and Management*. In W. J. Mitsch, ed. Global Wetlands: Old and New Elsevier, Amsterdam, The Netherlands, pp. 3–56.

Mitsch, W. J., B. C. Reeder, and D. M. Robb. 1994. *Modelling ecosystem and landscape scales of Lake Erie coastal wetlands*. In W. J. Mitsch, ed. Global Wetlands: Old World and New. Elsevier, Amsterdam, The Netherlands, pp. 563–574.

Mitsch, W. J., J. K. Cronk, X. Wu, R. W. Nairn, and D. L. Hey. 1995. Phosphorus retention in constructed freshwater riparian marshes. *Ecological Applications* 5: 830–845.

Mitsch, W. J., and X. Wu. 1995. *Wetlands and global change*. In R. Lal, J. Kimble, E. Levine, and B. A. Stewart, eds. Advances in Soil Science, Soil Management, and Greenhouse Effect. CRC Press/Lewis Publishers, Boca Raton, FL, pp. 205–230.

Mitsch, W. J., and R. F. Wilson. 1996. Improving the success of wetland creation and restoration with know-how, time, and self-design. *Ecological Applications* 6: 77–83.

Mitsch, W. J., and K. M. Wise. 1998. Water quality, fate of metals, and predictive model validation of a constructed wetland treating acid mine drainage. *Water Research* 32: 1888–1900.

Mitsch, W. J., X. Wu, R. W. Nairn, P. E. Weihe, N. Wang, R. Deal, and C. E. Boucher. 1998. Creating and restoring wetlands: A whole-ecosystem experiment in self-design. *BioScience* 48: 1019–1030.

Mitsch, W. J., A. Horne, and R. W. Nairn, eds. 2000a. *Nitrogen and phosphorus retention in wetlands*. Special Issue of *Ecological Engineering* 14: 1–206.

Mitsch, W. J., A. J. Horne, and R. W. Nairn. 2000b. Nitrogen and phosphorus retention in wetlands—Ecological approaches to solving excess nutrient problems. *Ecological Engineering* 14: 1–7.

Mitsch, W. J., and J. G. Gosselink. 2000a. The value of wetlands: Importance of scale and landscape setting. *Ecological Economics* 35: 25–33.

Mitsch, W. J., and J. G. Gosselink. 2000b. Wetlands, 3rd ed. John Wiley & Sons, New York.

Mitsch, W. J., J. W. Day, Jr., J. W. Gilliam, P. M. Groffman, D. L. Hey, G. W. Randall, and N. Wang. 2001. Reducing nitrogen loading to the Gulf of Mexico from the Mississippi River Basin: Strategies to counter a persistent ecological problem. *BioScience* 51: 373–388.

Mitsch, W. J., and S. E. Jørgensen. 2004. Ecological Engineering and Ecosystem Restoration. John Wiley & Sons, Hoboken, NJ.

Mitsch, W. J., N. Wang, L. Zhang, R. Deal, X. Wu, and A. Zuwerink. 2005a. *Using ecological indicators in a whole-ecosystem wetland experiment*. In S. E. Jørgensen, F-L. Xu, and R. Costanza, eds. Handbook of Ecological Indicators for Assessment of Ecosystem Health. CRC Press, Boca Raton, FL, pp. 211–235.

Mitsch, W. J., J. W. Day, Jr., L. Zhang, and R. Lane. 2005b. Nitrate-nitrogen retention by wetlands in the Mississippi River Basin. *Ecological Engineering* 24: 267–278.

Mitsch, W. J., L. Zhang, C. J. Anderson, A. Altor, and M. Hernandez. 2005c. Creating riverine wetlands: Ecological succession, nutrient retention, and pulsing effects. *Ecological Engineering* 25: 510–527.

Mitsch, W. J., and J. W. Day, Jr. 2006. Restoration of wetlands in the Mississippi-Ohio-Missouri (MOM) River Basin: Experience and needed research. *Ecological Engineering* 26: 55–69.

Moon, G. J., B. F. Clough, C. A. Peterson, and W. G. Allaway. 1986. Apoplastic and symplastic pathways in Avicennia marina (Forsk.) Vierh. roots revealed by fluorescent tracer dyes. *Australian Journal of Plant Physiology* 13: 637–648.

Moore, P. D., and D. J. Bellamy. 1974. Peatlands. Springer-Verlag, New York. 221 pp.

Moore, T. R., and R. Knowles. 1989. The influence of water table levels on methane and carbon dioxide emissions from peatland soils. *Canadian Journal of Soil Science* 69: 33–38.

Moore, T. R., and N. T. Roulet. 1995. *Methane emissions from Canadian peatlands*. In R. Lal, J. Kimble, E. Levine, and B. A. Stewart, eds. Advances in Soil Science: Soils and Global Change. CRC Press, Boca Raton, FL, pp. 153–164.

Moshiri, G. A., ed. 1993. Constructed Wetlands for Water Quality Improvement. Lewis Publishers, Boca Raton, FL.

Moustafa, M. Z. 1999. Nutrient retention dynamics of the Everglades nutrient removal project. *Wetlands* 19: 689–704.

Moustafa, M. Z., M. J. Chimney, T. D. Fontaine, G. Shih, and S. Davis. 1996. The response of a freshwater wetland to long-term "low level" nutrient loads—Marsh efficiency. *Ecological Engineering* 7: 15–33.

Mulamoottil, G., E. A. McBean, and F. Rovers, eds. 1999. Constructed Wetlands for the Treatment of Landfill Leachates. Lewis Publishers, Boca Raton, FL. 281 pp.

Mulholland, P. J., and E. J. Kuenzler. 1979. Organic carbon export from upland and forested wetland watersheds. *Limnology and Oceanography* 24: 960–966.

Murphy, J. E., 2006. Rapanos v. United States: Wading Through Murky Waters. *National Wetlands Newsletter* 28(5): 1.

Nahlik, A. M., and W. J. Mitsch. 2006. Tropical treatment wetlands dominated by free-floating macrophytes for water quality improvement in the Caribbean coastal plain of Costa Rica. *Ecological Engineering* 28: 246–257.

Naiman, R. J., T. Manning, and C. A. Johnston. 1991. Beaver population fluctuations and tropospheric methane emissions in boreal wetlands. *Biogeochemistry* 12: 1–15.

Nairn, R. W., and W. J. Mitsch. 2000. Phosphorus removal in created wetland ponds receiving river overflow. *Ecological Engineering* 14: 107–126.

National Research Council (NRC). 1992. Restoration of Aquatic Ecosystems. National Academy Press, Washington, DC. 552 pp.

National Research Council (NRC). 1995. Wetlands: Characteristics and Boundaries. National Academy Press, Washington, DC. 306 pp.

National Research Council. 2000. Clean Coastal Waters: Understanding and Reducing the Effects of Nutrient Pollution. National Academy Press, Washington, DC.

National Research Council. 2001. Compensating for Wetland Losses under the Clean Water Act. National Academy Press, Washington, DC, 158 pp.

National Wetlands Policy Forum. 1988. Protecting America's Wetlands: An Action Agenda. Conservation Foundation, Washington, DC. 69 pp.

National Wetlands Working Group. 1988. *Wetlands of Canada*. Ecological and Classification Series 24, Environment Canada, Ottawa, Ontario, and Polyscience Publications, Montreal, Quebec. 452 pp.

Natural Resources Conservation Service (NRCS). 1998. Field Indicators of Hydric Soils in the United States, Version 4.0. G. W. Gurt, P. M. Whited, and R. F. Pringle, eds. USDA, NRCS, Ft. Worth, TX. 30 pp.

Newman, J. M., J. C. Clausen, and J. A. Neafsey. 2000. Seasonal performance of a wetland constructed to process dairy milkhouse wastewater in Connecticut. *Ecological Engineering* 14: 181–198.

Nguyen, L. M. 2000. Phosphate incorporation and transformation in surface sediments of a sewage-impacted wetland as influenced by sediment sites, sediment pH and added phosphate concentration. *Ecological Engineering* 14: 139–155.

Nguyen, L. M., J. G. Cooke, and G. B. McBride. 1997. Phosphorus retention and release characteristics of sewage-impacted wetland sediments. *Water, Air, and Soil Pollution* 100: 163–179.

Nicholls, R. J. 2004. Coastal flooding and wetland loss in the 21st century: Changes under the SRES climate and socio-economic scenarios. *Global Environmental Change* 14: 69–86.

Nichols, D. S. 1983. Capacity of natural wetlands to remove nutrients from wastewater. *Journal of the Water Pollution Control Federation* 55: 495–505.

Niering, W. A. 1985. Wetlands. Alfred A. Knopf, New York. 638 pp.

Niering, W. A. 1988. *Endangered, threatened and rare wetland plants and animals of the continental United States*. In D. D. Hook et al., eds. The Ecology and Management of Wetlands, Vol. 1: Ecology of Wetlands. Timber Press, Portland, OR, pp. 227–238.

Niering, W. A. 1989. *Wetland vegetation development*. In S. K. Majumdar, R. P. Brooks, F. J. Brenner, and J. R. W. Tiner, eds. Wetlands Ecology and Conservation: Emphasis in Pennsylvania. Pennsylvania Academy of Science, Easton, pp. 103–113.

Niswander, S. F., and W. J. Mitsch. 1995. Functional analysis of a two-year-old created in-stream wetland: Hydrology, phosphorus retention, and vegetation survival and growth. *Wetlands* 15: 212–225.

Nixon, S. W., and C. A. Oviatt. 1973. Ecology of a New England salt marsh. *Ecological Monographs*. 43: 463–498.

Nixon, S. W., and V. Lee. 1986. Wetlands and Water Quality. Technical Report Y-86-2, U.S. Army Corps of Engineers Waterways Experiment Station, Vicksburg, MS.

Norgress, R. E. 1947. The history of the cypress lumber industry in Louisiana. *Louisiana Historical Quarterly* 30: 979–1059.

Novacek, J. M. 1989. *The water and wetland resources of the Nebraska Sandhills*. In A. G. van der Valk, ed. Northern Prairie Wetlands. Iowa State University Press, Ames, pp. 340–384.

Novitzki, R. P. 1979. *Hydrologic characteristics of Wisconsin's wetlands and their influence on floods, stream flow, and sediment.* In P. E. Greeson, J. R. Clark, and J. E. Clark, eds. Wetland Functions and Values: The State of Our Understanding. American Water Resources Association, Minneapolis, MN, pp. 377–388.

Novitzki, R. P. 1982. Hydrology of Wisconsin Wetlands. University of Wisconsin Extension Geological Natural History Survey Circular 40, University of Wisconsin, Madison. 22 pp.

Novitzki, R. P. 1985. *The effects of lakes and wetlands on flood flows and base flows in selected northern and eastern states.* In H. A. Groman et al., eds. Proceedings of a Conference—Wetlands of the Chesapeake. Environmental Law Institute, Washington, DC, pp. 143–154.

O'Brien, A. J., and W. S. Motts. 1980. Hydrogeologic evaluation of wetland basins for land use planning. *Water Resources Bulletin* 16: 785–789.

Odum, E. P. 1961. The role of tidal marshes in estuarine production. *New York State Conservation* 15(6): 12–15.

Odum, E. P. 1969. The strategy of ecosystem development. *Science* 164: 262–270.

Odum, E. P. 1971. Fundamentals of Ecology, 3rd ed. W. B. Saunders, Philadelphia. 544 pp.

Odum, E. P. 1979a. *Ecological importance of the riparian zone.* In R. R. Johnson and J. F. McCormick, tech. coords. Strategies for Protection and Management of Floodplain Wetlands and Other Riparian Ecosystems. Proceedings of the Symposium, Callaway Gardens, GA, December 11–13, 1978. General Technical Report WO-12, U.S. Forest Service, Washington, DC, pp. 2–4.

Odum, E. P. 1979b. *The value of wetlands: A hierarchical approach.* In P. F. Greeson, J. R. Clark, and J. E. Clark, eds. Wetland Functions and Values: The State of Our Understanding. American Water Resources Association, Minneapolis, MN, pp. 1–25.

Odum, E. P., and G. W. Barrett. 2005. Fundamentals of Ecology, 5th ed. Thomson Brooks/Cole, Belmont, CA, 598 pp.

Odum, H. T. 1951. The Carolina Bays and a Pleistocene weather map. *American Journal of Science* 250: 262–270.

Odum, H. T. 1971. Environment, Power and Society. John Wiley & Sons, New York.

Odum, H. T. 1988. Self-organization, transformity, and information. *Science* 242: 1132–1139.

Odum, H. T. 1989. *Ecological engineering and self-organization.* In W. J. Mitsch and S. E. Jørgensen, eds. Ecological Engineering. John Wiley & Sons, New York, pp. 79–101.

Odum, H. T. 1996. Environmental Accounting: Energy and Environmental Decision Making. John Wiley & Sons, New York. 370 pp.

Odum, H. T., B. J. Copeland, and E. A. McMahan, eds. 1974. Coastal Ecological Systems of the United States. Conservation Foundation, Washington, DC. 4 vols.

Odum, H. T., P. Kangas, G. R. Best, B. T. Rushton, S. Leibowitz, J. R. Butner, and T. Oxford. 1981. Studies on Phosphate Mining, Reclamation, and Energy. Center for Wetlands, University of Florida, Gainesville. 142 pp.

Odum, W. E. 1987. *Predicting ecosystem development following creation and restoration of wetlands*. In J. Zelazny and J. S. Feierabend, eds. Wetlands: Increasing Our Wetland Resources. Proceedings of the Conference Wetlands: Increasing Our Wetland Resources. Corporate Conservation Council, National Wildlife Federation, Washington, DC, pp. 67–70.

Odum, W. E., E. P. Odum, and Odum, H. T. 1995. Nature's pulsing paradigm. *Estuaries* 18: 547–555.

Office of Technology Assessment. 1984. Wetlands: Their Use and Regulation. Report O-206, Office of Technology Assessment, U.S. Congress, Washington, DC. 208 pp.

Ogawa, H., and J. W. Male. 1983. The Flood Mitigation Potential of Inland Wetlands. Publication 138, Water Resources Research Center, University of Massachusetts, Amherst. 164 pp.

Ogawa, H., and J. W. Male. 1986. Simulating the flood mitigation role of wetlands. *Journal of Water Resource Planning and Management* 112: 114–128.

Ogden, M. H. 1999. Constructed wetlands for small community wastewater treatment. Paper presented at Wetlands for Wastewater Recycling Conference, November 3, 1999, Baltimore. Environmental Concern, St. Michaels, MD.

Ohlendorf, H. M., D. J. Hoffman, M. K. Saiki, and T. W. Aldrich. 1986. Embryonic mortality and abnormalities of aquatic birds: Apparent impacts of selenium from irrigation drainwater. *Science of the Total Environment* 52: 49–63.

Ohlendorf, H. M., R. L. Hothem, C. M. Bunck, and K. C. Marois. 1990. Bioaccumulation of selenium in birds at Kesterson Reservoir, California. *Archives of Environmental Contamination and Toxicology* 19: 495–507.

Olson, R. K., ed. 1992. *The role of created and natural wetlands in controlling nonpoint source pollution*. Special Issue of Ecological Engineering 1: 1–170.

Ovenden, L. 1990. Peat accumulation in northern wetlands. *Quaternary Research* 33: 377–386.

Ozesmi, S. L., and M. E. Bauer. 2002. Satellite remote sensing of wetlands. *Wetlands Ecology and Management* 10: 381–402.

Page, S. E., R. A. J. Wust, D. Weiss, J. O. Rieley, W. Shotyk, and S. H. Limin. 2004. A record of late Pleistocene and Holocene carbon accumulation and climate change from an equatorial peat bog (Kalimantan, Indonesia): Implications for past, present, and future carbon dynamics. *Journal of Quaternary Science* 19: 625–635.

Parish, D., and C. Elliott. 1990. *Foreword*. In J. Lu. Wetlands in China. East China Normal University, Shanghai, p. x.

Park, R. A., J. K. Lee, P. W. Mausel, and R. C. Howe. 1991. Using remote sensing for modeling the impacts of sea level rise. *World Resource Review* 3: 184–205.

Patrick, W. H., Jr., and R. D. Delaune. 1972. Characterization of the oxidized and reduced zones in flooded soil. *Proceedings of the Soil Science Society of America* 36: 573–576.

Patrick, W. H., Jr., and R. D. Delaune. 1990. Subsidence, accretion, and sea level rise in south San Francisco Bay marshes. *Limnology and Oceanography* 35: 1389–1395.

Pearsall, W. H. 1920. The aquatic vegetation of the English lakes. *Journal of Ecology* 8: 163–201.

Pederson, R. L., and L. M. Smith. 1988. *Implications of wetland seed bank research: A review of Great Britain and prairie marsh studies*. In D. A. Wilcox, ed. Interdisciplinary Approaches to Freshwater Wetlands Research. Michigan State University Press, East Lansing, pp. 81–95.

Penman, H. L. 1948. Natural evaporation from open water, bare soil and grass. *Proceedings of the Royal Society of London* 93: 120–145.

Pérez-Arteaga, A., K. J. Gaston, and M. Kershaw. 2002. Undesignated sites in Mexico qualifying as wetlands of international importance. *Biological Conservation* 107: 47–57.

Peterson, S. B., J. M. Teal, and W. J. Mitsch, eds. 2005. Delaware Bay Salt Marsh Restoration. Special Issue of Ecological Engineering 25: 199–314.

Philipp, K. R., and R. T. Field. 2005. *Phragmites australis* expansion in Delaware Bay salt marshes. *Ecological Engineering* 25: 275–291.

Phillips, J. D. 1989. Fluvial sediment storage in wetlands. *Water Resources Bulletin* 25: 867–873.

Phipps, R. G., and W. G. Crumpton. 1994. Factors affecting nitrogen loss in experimental wetlands with different hydrologic loads. *Ecological Engineering* 3: 399–408.

Por, F. D. 1995. The Pantanal of Mato Grosso (Brazil). Kluwer Academic Publishers, Dordrecht, The Netherlands. 122 pp.

Post, W. M. 1990. Report of a Workshop on Climate Feedbacks and the Role of Peatlands, Tundra, and Boreal Ecosystems in the Global Carbon Cycle. Publication 3289, Environmental Science Division, Oak Ridge National Laboratory, Oak Ridge, TN.

Potonie, R. 1908. Aufbau und Vegetation der Moore Norddeutschlands. *Englers. Bot. Jahrb*. 90. Leipzig.

Prance, G. T. 1979. Notes on the vegetation of Amazonia. III. The terminology of Amazonian forest types subject to inundation. *Brittonia* 31: 26–38.

Prasad, V. P., D. Mason, J. E. Marburger, and C. R. A. Kumar. 1996. Illustrated Flora of Keoladeo National Park, Bharatpur, Rajasthan. Bombay Natural History Society, Mumbai, India. 435 pp.

Presser, T. S., and H. M. Ohlendorf. 1987. Biogeochemical cycling of selenium in the San Joaquin Valley. *Environmental Management* 11: 805–821.

Price, J., L. Rochefort, and F. Quinty. 1998. Energy and moisture considerations on cutover peatlands: Surface microtopography, mulch cover and *Sphagnum* regeneration. *Ecological Engineering* 10: 293–312.

Pride, R. W., F. W. Meyer, and R. N. Cherry. 1966. Hydrology of Green Swamp Area in Central Florida. Report 42, Florida Division of Geology, Tallahassee. 137 pp.

Prouty, W. F. 1952. Carolina Bays and their origin. *Geological Society of America Bulletin* 63: 167–224.

Qin, P., M. Xie, and Y. Jiang. 1998. *Spartina* green food ecological engineering. *Ecological Engineering* 11: 147–156.

Quinty, F., and L. Rochefort. 1997. *Plant reintroduction on a harvested peat bog.* In C. C. Trettin, M. F. Jurgensen, D. F. Grigal, M. R. Gale, and J. K. Jeglum, eds. Northern Forested Wetlands: Ecology and Management. CRC Press/Lewis Publishers, Boca Raton, FL, pp. 133–145.

Rabalais, N. N., W. J. Wiseman, R. E. Turner, B. K. Sengupta, and Q. Dortch. 1996. Nutrient changes in the Mississippi River and system responses on the adjacent continental shelf. *Estuaries* 19: 386–407.

Rabalais, N. N., R. E. Turner, W. J. Wiseman, and Q. Dortch. 1998. Consequences of the 1993 Mississippi River flood in the Gulf of Mexico. *Regulated Rivers* 14: 161–177.

Rabalais, N. N., R. E. Turner, and W. J. Wiseman, Jr. 2001. Hypoxia in the Gulf of Mexico. *Journal of Environmental Quality* 30: 320–329.

Rabenhorst, M. C. 2005. Biological zero: A soil temperature concept. *Wetlands* 25: 616–621.

Raisin, G. W., and D. S. Mitchell. 1995. The use of wetlands for the control of non-point source pollution. *Water Science and Technology* 32: 177–186.

Raisin, G. W., D. S. Mitchell, and R. L. Croome. 1997. The effectiveness of a small constructed wetland in ameliorating diffuse nutrient loadings from an Australian rural catchment. *Ecological Engineering* 9: 19–35.

Ramberg, L., P. Wolski, and M. Krah. 2006a. Water balance and infiltration in a seasonal floodplain in the Okavango Delta, Botswana. *Wetlands* 26: 677–690.

Ramberg, L., P. Hancock, M. Lindholm, T. Meyer, S. Ringrose, J. Silva., J. Van As, and C. VanderPost. 2006b. Species diversity of the Okavango Delta, Botswana. *Aquatic Sciences* 68: 310–337.

Ramsar Convention Secretariat. 2004. Ramsar Handbook for the Wise Use of Wetlands, 2nd ed. Handbook 10, Wetland Inventory: A Ramsar framework for wetland inventory. Ramsar Secretariat, Gland, Switzerland.

Ranwell, D. S. 1967. World resources of *Spartina townsendii* and economic use of *Spartina* marshland. *Coastal Zone Management Journal* 1: 65–74.

Ranwell, D. S. 1972. Ecology of Salt Marshes and Sand Dunes. Chapman & Hall, London. 258 pp.

Reddy, K. R., and W. H. Patrick, Jr. 1984. Nitrogen transformations and loss in flooded soils and sediments. *CRC Critical Reviews in Environmental Control* 13: 273–309.

Reddy, K. R., and W. H. Smith, eds. 1987. Aquatic Plants for Water Treatment and Resource Recovery. Magnolia Publishing, Orlando, FL.

Reddy, K. R., and D. A. Graetz. 1988. *Carbon and nitrogen dynamics in wetland soils.* In D. D. Hook, W. H. McKee, Jr., H. K. Smith, J. Gregory, V. G. Burrell, M. R. DeVoe, R. E. Sojka, S. Gilbert, R. Banks, L. G. Stolzy, C. Brooks, T. D. Matthews, and T. H. Shear, eds. The Ecology and Management of Wetlands, Vol. 1: The Ecology of Wetlands. Timber Press, Portland, OR, pp. 307–318.

Reddy, K. R., and E. M. D'Angelo. 1994. *Soil processes regulating water quality in wetlands.* In W. J. Mitsch, ed. Global Wetlands: Old World and New. Elsevier, Amsterdam, The Netherlands, pp. 309–324.

Reddy, K. R., R. H. Kadlec, E. Flaig, and P. M. Gale. 1999. Phosphorus retention in streams and wetlands: A review. *Critical Reviews in Environmental Science and Technology* 29: 83–146.

Reddy, K. R., R. H. Kadlec, M. J. Chimney, and W. J. Mitsch, eds. 2006. *The Everglades Nutrient Removal Project.* Special Issue of *Ecological Engineering* 27: 265–379.

Redfield, A. C. 1958. The biological control of chemical factors in the environment. *American Scientist* 46: 206–226.

Redfield, A. C. 1965. Ontogeny of a salt marsh estuary. *Science* 147: 50–55.

Reed, S. C., R. W. Crites, and E. J. Middlebrooks. 1995. Natural Systems for Waste Management and Treatment, 2nd ed. McGraw-Hill, New York. 433 pp.

Reinartz, J. A., and E. L. Warne. 1993. Development of vegetation in small created wetlands in southeast Wisconsin. *Wetlands* 13: 153–164.

Reinelt, L. E., and R. R. Horner. 1995. Pollutant removal from stormwater runoff by palustrine wetlands based on comprehensive budgets. *Ecological Engineering* 4: 77–97.

Revenga, C., J. Brunner, N. Henninger, K. Kassem, and R. Payne. 2000. Pilot Analysis of Global Ecosystems: Freshwater Systems. World Resources Institute, Washington, DC. 65 pp.

Rezendes, P., and P. Roy. 1996. Wetlands: The Web of Life. Sierra Club Books, San Francisco, CA. 156 pp.

Rheinhardt, R. D., M. M. Brinson, and P. M. Farley. 1997. Applying wetland reference data to functional assessment, mitigation, and restoration. *Wetlands* 17: 195–215.

Richardson, C. J., ed. 1981. Pocosin Wetlands. Hutchinson Ross Publishing, Stroudsburg, PA. 364 pp.

Richardson, C. J. 1983. Pocosins: Vanishing wastelands or valuable wetlands. *BioScience* 33: 626–633.

Richardson, C. J., R. Evans, and D. Carr. 1981. *Pocosins: An ecosystem in transition.* In C. J. Richardson, ed. Pocosin Wetlands. Hutchinson Ross Publishing, Stroudsburg, PA, pp. 3–19.

Richardson, C. J., and C. B. Craft. 1993. *Effective phosphorus retention in wetlands— Fact or fiction?* In G. A. Moshiri, ed. Constructed Wetlands for Water Quality Improvement. CRC Press, Boca Raton, FL, pp. 271–282.

Richardson, C. J., S. Qian, C. B. Craft, and R. G. Qualls. 1997. Predictive models for phosphorus retention in wetlands. *Wetlands Ecology and Management* 4: 159–175.

Richardson C. J., P. Reiss, N. A. Hussain, A. J. Alwash, and D. J. Pool. 2005. The restoration potential of the Mesopotamian marshes of Iraq. *Science* 307: 1307–1311.

Richardson, C. J., and N. A. Hussain. 2006. Restoring the garden of Eden: An ecological assessment of the marshes of Iraq. *BioScience* 56: 447–489.

Richardson, J., P. A. Straub, K. C. Ewel, and H. T. Odum. 1983. Sulfate-enriched water effects on a floodplain forest in Florida. *Environmental Management* 7: 321–326.

Riley, J. P., and G. Skirrow. 1975. Chemical Oceanography, 2nd ed., Vol. 2. Academic Press, New York. 647 pp.

Ringrose, S., C. Vanderpost, V. Matheson, P. Wolski, P. Huntsman-Mapila, M. Murray-Hudson, and A. Jellema. 2007. Indicators of desiccation-driven change in the distal Okavango Delta, Botswana. *Journal of Arid Environments* 68: 88–112.

Rivers, J. S., D. I. Siegel, L. S. Chasar, J. P. Chanton, P. H. Glaser, N. T. Roulet, and J. M. McKenzie. 1998. A stochastic appraisal of the annual carbon budget of a large circumboreal peatland, Rapid River Watershed, northern Minnesota. *Global Biogeochemical Cycles* 12: 715–727.

Robb, J. T. 2002. Assessing wetland compensatory mitigation sites to aid in establishing mitigation ratios. *Wetlands* 22: 435–440.

Rochefort, L., and S. Campeau. 1997. *Rehabilitation work on post-harvested bogs in south eastern Canada*. In L. Parkyn, R. E. Stoneman, and H. A. P. Ingram, eds. Conserving Peatlands. CAB International, Walingford, UK, pp. 287–284.

Rodgers, H. L., F. P. Day, and R. Atkinson. 2004. Root dynamics in restored and naturally regenerated Atlantic white cedar wetlands. *Restoration Ecology* 16: 401–411.

Roe, H. B., and Q. C. Ayres. 1954. Engineering for Agricultural Drainage. McGraw-Hill, New York. 501 pp.

Romanov, V. V. 1968. Hydrophysics of Bogs. Translated from Russian by N. Kaner; edited by Prof. Heimann. Israel Program for Scientific Translation, Jerusalem. Available from Clearinghouse for Federal Scientific and Technical Information, Springfield, VA. 299 pp.

Rosenberry, D. O., D. I. Stannard, T. C. Winter, and M. L. Martinez. 2004. Comparison of 13 equations for determining evapotranspiration from a prairie wetland, Cottonwood Lake Area, North Dakota, USA. *Wetlands* 24: 483–497.

Rosgen, D. L. 1985. *A stream classification system*. In R. R. Johnson, C. D. Ziebell, D. R. Patton, P. F. Ffolliott, and R. H. Hamre, eds. Riparian Ecosystems and Their Management: Reconciling Conflicting Uses. General Technical Report RM-120, Rocky Mountain Forest and Range Experiment Station, Forest Service, U.S. Department of Agriculture, Fort Collins, CO.

Roulet, N.T. 2000. Peatlands, carbon storage, greenhouse gases, and the Kyoto Protocol: Prospects and significance for Canada. *Wetlands* 20: 605–615.

Roulet, N. T., R. Ash, and T. R. Moore. 1992a. Low boreal wetlands as a source of atmospheric methane. *Journal of Geophysical Research* 97: 3739–3749.

Roulet, N. T., T. R. Moore, J. Bubier, and P. Lafleur. 1992b. Northern fens: CH_4 flux and climate change. *Tellus* 44B: 100–105.

Runyon, L. C. 1993. The Lucas Court Case and Land-Use Planning. National Conference of State Legislators, Denver, CO, Supplement to State Legislatures, Vol. 1, No. 10 (March).

Russell, H. S. 1976. A Long, Deep Furrow: Three Centuries of Farming in New England. University Press of New England, Hanover, NH. 671 pp.

Rycroft, D. W., D. J. A. Williams, and H. A. E. Ingram. 1975. The transmission of water through peat. I. Review. *Journal of Ecology* 63: 535–556.

Rykiel, E. J., Jr. 1984. *General hydrology and mineral budgets for Okefenokee Swamp: Ecological significance.* In A. D. Cohen, D. J. Casagrande, M. J. Andrejko, and G. R. Best, eds. The Okefenokee Swamp: Its Natural History, Geology, and Geochemistry. Wetland Surveys, Los Alamos, NM, pp. 212–228.

Sabin, T.J. and V.T. Holliday. 1995. Playas and lunettes on the Southern High Plains: Morphometric and spatial relationships. Annals of the Association American Geographers 85: 286–305.

Saltonstall, K., P. M. Peterson, and R. J. Soreng. 2004. Recognition of *Phragmites australis* subsp. *americanus* (*Poaceae: Arundinoideae*) in North America: Evidence from morphological and genetic analyses. *Brit. Org/SIDA* 21: 683–692.

Sanville, W., and W. J. Mitsch, eds. 1994. Creating Freshwater Marshes in a Riparian Landscape: Research at the Des Plaines River Wetland Demonstration Project. Special Issue of *Ecological Engineering* 3(4): 315–521.

Sartoris, J. J., J. S. Thullen, L. B. Barber, and D. E. Salas. 2000. Investigation of nitrogen transformations in a southern California constructed wastewater treatment wetland. *Ecological Engineering* 14: 49–65.

Sass, R. L., F. M. Fisher, Y. B. Wang, F. T. Turner, and M. F. Jund. 1992. Methane emission from rice fields: The effect of floodwater management. *Global Biogeochemical Cycles* 6: 249–262.

Sasser, C. E., M. D. Dozler, J. G. Gosselink, and J. M. Hill. 1986. Spatial and temporal changes in Louisiana's Barataria Basin marshes. *Environmental Management* 10: 671–680.

Savage, H. 1983. The Mysterious Carolina Bays. University of South Carolina Press, Columbia. 121 pp.

Schaafsma, J. A., A. H. Baldwin, and C. A. Streb. 2000. An evaluation of a constructed wetland to treat wastewater from a dairy farm in Maryland, USA. *Ecological Engineering* 14: 199–206.

Schamberger, M. L., C. Short, and A. Farmer. 1979. *Evaluation wetlands as a wildlife habitat.* In P. E. Greeson, J. R. Clark, and J. E. Clark, eds. Wetland Functions and Values: The State of Our Understanding. American Water Resources Association, Minneapolis, MN, pp. 74–83.

Scheffe, R. D. 1978. Estimation and prediction of summer evapotranspiration from a northern wetland. Master's Thesis, University of Michigan, Ann Arbor. 69 pp.

Schimel, J. 2000. Rice, microbes and methane. *Nature* 403: 375–376.

Schmidt, K. F. 2001. A true-blue vision for the Danube. *Science* 294: 1444–1447.

Schröder, P. 1989. Characterization of a thermo-osmotic gas transport mechanism in Alnus glutinosa (L.) Gaertn. *Trees* 3: 38–44.

Schueler, T. R. 1992. Design of stormwater wetland systems: Guidelines for creating diverse and effective stormwater wetlands in the mid-Atlantic region. Metropolitan Washington Council of Governments, Washington, DC, 133 pp.

Schutz, H., A. Holzapfel-Pschorn, R. Conrad, H. Rennenberg, and W. Seiler. 1989. A three year continuous record on the influence of daytime, season and fertilizer treatment on methane emission rates from an Italian rice paddy field. *Journal of Geophysical Research* 94: 16405–16416.

Scodari, P. F. 1990. Wetlands Protection: The Role of Economics. Environmental Law Institute, Washington, DC. 89 pp.

Scott, D. A., and T. A. Jones. 1995. Classification and inventory of wetlands: A global overview. *Vegetatio* 118: 3–16.

Seidel, K. 1964. Abbau von Bacterium Coli durch höhere Wasserpflanzen. *Naturwissenschaften* 51: 395.

Seidel, K. 1966. Reinigung von Gewässern durch höhere Pflanzen. *Naturwissenschaften* 53: 289–297.

Seidel, K., and H. Happl. 1981. Pflanzenkläranlage "Krefelder system." *Sicherheit in Chemic und Umbelt* 1: 127–129.

Seliskar, D. 1995. *Exploiting plant genotypic diversity for coastal salt marsh creation and restoration.* In M. A. Khan and I. A. Ungar, eds. Biology of Salt-Tolerant Plants. Department of Botany, University of Karachi, Pakistan, pp. 407–416.

Shaw, S. P., and C. G. Fredine. 1956. Wetlands of the United States, Their Extent, and Their Value for Waterfowl and Other Wildlife. Circular 39, U.S. Fish and Wildlife Service, U.S. Department of Interior, Washington, DC. 67 pp.

Shearer, J. C., and B. R. Clarkson. 1998. Whangamarino wetland: Effects of lowered river levels on peat and vegetation. *International Peat Journal* 8: 52–65.

Sheffield, R. M., T. W. Birch, W. H. McWilliams, and J. B. Tansey. 1998. *Chamaecyparis thyoides(Atlantic white cedar) in the United States.* In A. D. Laderman, ed. Coastally Restricted Forests. Oxford University Press, New York, pp. 111–123.

Shelford, V. E. 1907. Preliminary note on the distribution of the tiger beetle (*Cicindela*) and its relation to plant succession. *Biological Bulletin* 14.

Shelford, V. E. 1911. Ecological succession. II. Pond fishes. *Biological Bulletin* 21: 127–151.

Shelford, V. E. 1913. Animal Communities in Temperate America as Illustrated in the Chicago Region. University of Chicago Press, Chicago.

Shiel, R. J. 1994. *Death and life of the billabong.* In X. Collier, ed. Restoration of Aquatic Habitats. Selected Papers from New Zealand Limnological Society 1993 Annual Conference, Department of Conservation, pp. 19–37.

Shjeflo, J. B. 1968. Evapotranspiration and the water budget of prairie potholes in North Dakota. Professional Paper 585-B. U.S. Geological Survey, Washington, DC. 49 pp.

Shuman, C. S., and R. F. Ambrose. 2003. A comparison of remote sensing and ground-based methods for monitoring wetland restoration success. *Restoration Ecology* 11: 325–333.

Shutes, R. B. E., D. M. Revitt, A. S. Mungar, and L. N. L. Scholes. 1997. The design of wetland systems for the treatment of urban runoff. *Water Science and Technology* 35: 19–25.

Sikora, W. B. 1977. The ecology of *Palaemonetes pugio* in a southeastern salt marsh ecosystem with particular emphasis on production and trophic relationship. Ph.D. dissertation, University of South Carolina. 122 pp.

Simberloff, D., and T. Dayan. 1991. The guild concept and the structure of ecological communities. *Annual Review of Ecology and Systematics* 22: 115–143.

Singer, D. K., S. T. Jackson, B. J. Madsen, and D. A. Wilcox. 1996. Differentiating climatic and successional influences on long-term development of a marsh. *Ecology* 77: 1765–1778.

Singer, P. C., and W. Stumm. 1970. Acidic mine drainage: The rate-determining step. *Science* 167: 1121–1123.

Sjörs, H. 1948. Myrvegetation i bergslagen. *Acta Phytogeographica Suecica* 21: 1–299.

Sjörs, H. 1950. On the relationship between vegetation and electrolytes in North Swedish mire waters. *Oikos* 2: 239–258.

Smith, L.M. 2003. Playas of the Great Plains. University of Texas Press, Austin, Texas, 257 pp.

Smith, L. M., and J. A. Kadlec. 1985. Predictions of vegetation change following fire in a Great Salt Lake marsh. *Aquatic Botany* 21: 43–51.

Smith, R. C. 1975. *Hydrogeology of the experimental cypress swamps.* In H. T. Odum and K. C. Ewel, eds. Cypress Wetlands for Water Management, Recycling and Conservation. Second Annual Report to NSF and Rockefeller Foundation, Center for Wetlands, University of Florida, Gainesville, pp. 114–138.

Söderqvist, T., W. J. Mitsch, and R. K. Turner, eds, 2000. *The Values of Wetlands: Landscape and Institutional Perspectives.* Special Issue of Ecological Economics 35: 1–132.

Solano, M. L., P. Soriano, and M. P. Ciria. 2004. Constructed wetlands as a sustainable solution for wastewater treatment in small villages. *Biosystems Engineering* 87: 109–118.

Solomeshch, A.I. 2005. *The West Siberian Lowland.* In L.A. Fraser and P.A. Keddy, eds. The World's Largest Wetlands: Ecology and Conservation, Cambridge University Press, Cambridge, UK, pp. 11–62.

Sorrell, B. K., and P. I. Boon. 1992. Biogeochemistry of billabong sediments. II. Seasonal variations in methane production. *Freshwater Biology* 27: 435–445.

Sorrell, B. K., and P. I. Boon. 1994. Convective gas flow in *Eleocharis sphacelata* R. *Br.*: Methane transport and release from wetlands. *Aquatic Botany* 47: 197–212.

Sorrell, B. K., H. Brix, and P. I. Boon. 1994. Modelling of in situ oxygen transport and aerobic metabolism in the hydrophyte *Eleocharis sphacelata*. *R. Br. Proceedings of the Royal Society of Edinburgh* 102B: 367–372.

Souza, V, A. Escalante, L. Espinoza, A. Valera, A. Cruz, L. E. Eguilarte, F. Garía, and J. Elser. 2004. Cuatro Ciénegas un laboratorio de astrobiología. *Ciencias* 76: 4–11.

Spieles, D. J. 2005. Vegetation development in created, restored, and enhanced mitigation wetland banks of the United States. *Wetlands* 25: 51–63.

Spieles, D. J., and W. J. Mitsch. 2000a. The effects of season and hydrologic and chemical loading on nitrate retention in constructed wetlands: A comparison of low and high nutrient riverine systems. *Ecological Engineering* 14: 77–91.

Spieles, D. J., and W. J. Mitsch. 2000b. Macroinvertebrate community structure in high- and low-nutrient constructed wetlands. *Wetlands* 20: 716–729.

Stark, L. R., and F. M. Williams. 1995. Assessing the perforrmance indices and design parameters of treatment wetlands for H, Fe, and Mn retention. *Ecological Engineering* 5: 433–444.

Steever, E. Z., R. S. Warren, and W. A. Niering. 1976. Tidal energy subsidy and standing crop production of *Spartina alterniflora*. *Estuarine, Coastal and Marine Science* 4: 473–478.

Steiner, G. R., J. T. Watson, D. Hammer, and D. F. Harker, Jr. 1987. *Municipal wastewater treatment with artificial wetlands—A TVA/Kentucky demonstration*. In K. R. Reddy and W. H. Smith, eds. Aquatic Plants for Wastewater Treatment and Resource Recovery. Magnolia Publishing, Orlando, FL, p. 923.

Steiner, G. R., and R. J. Freeman, Jr. 1989. *Configuration and substrate design considerations for constructed wetlands for wastewater treatment*. In D. A. Hammer, ed. Constructed Wetlands for Wastewater Treatment. Lewis Publishers, Chelsea, MI, pp. 363–378.

Streever, W., ed. 1999. An International Perspective on Wetland Rehabilitation. Kluwer Academic Publishers, Dordrecht, The Netherlands. 338 pp.

Stumm, W., and J. J. Morgan. 1996. Aquatic Chemistry: Chemical Equilibria and Rates in Natural Waters, 3rd ed. John Wiley & Sons, New York. 1022 pp.

Sundareshwar, P. V., J. T. Morris, E. K. Koepfler, and B. Fornwalt. 2005. Phosphorus limitation of coastal ecosystem processes. *Science* 299: 563–565.

Swerhone, G. D. W., J. R. Lawrence, J. G. Richards, and M. J. Hendry. 1999. Construction and testing of a durable platinum wire electrode for *in situ* redox measurements in the subsurface. *Ground Water Monitoring and Remediation* 19(2): 132–136.

Tanner, C. C. 1996. Plants for constructed wetland treatment systems—A comparison of the growth and nutrient uptake of eight emergent species. *Ecological Engineering* 7: 59–83.

Tanner, C. C., J. S. Clayton, and M. P. Upsdell. 1995. Effect of loading rate and planting on treatment of dairy farm wastewaters in constructed wetlands. II. Removal of nitrogen and phosphorus. *Water Research* 29: 27–34.

Tanner, C. C., G. Raisin, G. Ho, and W. J. Mitsch, eds. 1999. *Constructed and Natural Wetlands for Pollution Control*. Special Issue of Ecological Engineering 12: 1–170.

Tarnocai, C., G. D. Adams, V. Glooschenko, W. A. Glooschenko, P. Grondin, H. E. Hirvonen, P. Lynch-Stewart, G. F. Mills, E. T. Oswald, F. C. Pollett, C. D. A. Rubec, E. D. Wells, and S. C. Zoltai. 1988. The Canadian wetland classification system. In National Wetlands Working Group, ed. Wetlands of Canada. Ecological Land Classification Series 24, Environment Canada, Ottawa, Ontario, and Polyscience Publications, Montreal, Quebec, pp. 413–427.

Tarnocai, C. 2006. The effect of climate change on carbon in Canadian peatlands. *Global and Planetary Change* 53: 222–232.

Tarutis, W. J., L. R. Stark, and F. M. Williams. 1999. Sizing and performance estimation of coal mine drainage wetlands. *Ecological Engineering* 12: 353–372.

Teal, J. M. 1958. Distribution of fiddler crabs in Georgia salt marshes. *Ecology* 39: 18–19.

Teal, J. M. 1962. Energy flow in the salt marsh ecosystem of Georgia. *Ecology* 43: 614–624.

Teal, J. M., and M. Teal. 1969. Life and Death of the Salt Marsh. Little, Brown, Boston. 278 pp.

Teal, J. M., and M. P. Weinstein. 2002. Ecological engineering, design, and construction considerations for marsh restorations in Delaware Bay, USA. *Ecological Engineering* 18: 607–618.

Thullen, J. S., J. J. Sartoris, and S. M. Nelson. 2005. Managing vegetation in surface-flow wastewater-treatment wetlands for optimal treatment performance. *Ecological Engineering* 25: 583–593.

Tieter, S., and Ü. Mander. 2005. Emission of N_2O, N_2, CH_4, and CO_2 from constructed wetlands for wastewater treatment and from riparian buffer zones. *Ecological Engineering* 25: 528–541.

Tilman, D. 1982. Resource Competition and Community Structure. Princeton University Press, Princeton, NJ. 296 pp.

Tiner, R. W. 1984. Wetlands of the United States: Current Status and Recent Trends. National Wetlands Inventory, U.S. Fish and Wildlife Service, Washington, DC. 58 pp.

Tiner, R. W. 1999. Wetland Indicators: A Guide to Witland Identification, Delineation, Classification, and Mapping. Lewis Publishers, Boca Raton, FL. 392 pp.

Tiner, R. W., and B. O. Wilen. 1983. The U.S. Fish and Wildlife Services National Wetlands Inventory Project. Unpublished Report, U.S. Fish and Wildlife Service, Washington, DC. 19 pp.

Titus, J. G. 1991. Greenhouse effect and coastal wetland policy: How Americans could abandon an area the size of Massachusetts at minimum cost. *Environmental Management* 15: 39–58.

Todd, D. K. 1964. *Groundwater*. In V. T. Chow, ed. Handbook of Applied Hydrology. McGraw-Hill, New York, pp. 13-1-13–55.

Toth, L. A., D. A. Arrington, M. A. Brady, and D. A. Muszick. 1995. Conceptual evaluation of factors potentially affecting restoration of habitat structure within the channelized Kissimmee River ecosystem. *Restoration Ecology* 3: 160–180.

Turner, R. E. 1982. *Protein yields from wetlands*. In B. Gopal, R. E. Turner, R. G. Wetzel, and D. F. Whigham, eds. Wetlands: Ecology and Management. National Institute of Ecology and International Scientific Publications, Jaipur, India, pp. 405–415.

Turner, R. E. 1997. Wetland loss in the northern Gulf of Mexico: Multiple working hypotheses. *Estuaries* 20: 1–13.

Turunen, J., E. Tomppo, K. Tolonen, and E. Reinkainen. 2002. Estimating carbon accumulation rates of undrained mires in Finland: Application to boreal and subarctic regions. *The Holocene* 12: 79–90.

Tuttle, C. L., L. Zhang, and W. J. Mitsch. In review. Aquatic metabolism as an indicator of the ecological effects of hydrologic pulsing in flow-through wetlands. *Ecological Indicators*.

Twilley, R. R. 1982. Litter dynamics and organic carbon exchange in black mangrove (*Avicennia germinans*) basin forests in a Southwest Florida estuary. Ph.D. dissertation, University of Florida, Gainesville.

United Nations Environmental Programme. 2001. The Mesopotamian Marshlands: Demise of an Ecosystem (UNEP/DEWA/TR.01 3 Rev.1), UNEP, Nairobi, Kenya.

Updegraff, K., S. D. Bridgham, J. Pastor, P. Weishampel, and C. Harth. 2001. Response of CO_2 and CH_4 emissions from peatlands to warming and water table manipulation. *Ecological Applications* 11: 311–326.

U.S. Army Corps of Engineers. 1972. Charles River Watershed, Massachusetts. New England Division, Waltham, MA. 65 pp.

U.S. Army Corps of Engineers. 1987. Corps of Engineers Wetlands Delineation Manual. Technical Report Y-87-1. U.S. Army Corps of Engineers Waterways Experiment Station, Vicksburg, MS. 100 pp. and appendices.

U.S. Department of Interior, Bureau of Reclamation. 1984. Water Measurement Manual, 2nd ed., revised reprint. U.S. Government Printing Office, Washington, DC. 327 pp.

U.S. Environmental Protection Agency. 1993. Constructed Wetlands for Wastewater Treatment and Wildlife Habitat: 17 Case Studies. EPA-832-R-93–005, U.S. Environmental Protection Agency, Washington, DC. 174 pp.

van der Valk, A. G. 1981. Succession in wetlands: A Gleasonian approach. *Ecology* 62: 688–696.

van der Valk, A. G. 1982. *Succession in temperate North American wetlands*. In B. Gopal, R. E. Turner, R. G. Wetzel, and D. F. Whigham, eds. Wetlands: Ecology and Management. National Institute for Ecology and International Science Publications, Jaipur, India, pp. 169–179.

van der Valk, A. G., ed. 1989. Northern Prairie Wetlands. Iowa State University Press, Ames. 400 pp.

van der Valk, A. G. 1998. *Succession theory and wetland restoration*. In A. J. McComb and J. A. Davis, eds. Wetlands for the Future. Contributions from INTECOL's Fifth International Wetland Conference. Gleneagles Publishing, Adelaide, Australia, pp. 657–667.

Vasander, H., E.-S. Tuittila, E. Lode., L. Lundin, M. Ilomets, T. Sallantaus, R. Heikkila, M.-L. Pitkanen, and J. Laine. 2003. Status and restoration of peatlands in northern Europe. *Wetlands Ecology and Management* 11: 51–63.

Vepraskas, M. J. 1995. Redoximorphic Features for Identifying Aquic Conditions. Technical Bulletin 301, North Carolina Agricultural Research Service, North Carolina State University, Raleigh. 33 pp.

Vernberg, F. J. 1981. *Benthic macrofauna*. In F. J. Vernberg and W. B. Vernberg, eds. Functional Adaptations of Marine Organisms. Academic Press, New York, pp. 179–230.

Vernberg, W. B., and F. J. Vernberg. 1972. Environmental Physiology of Marine Animals. Springer-Verlag, New York. 346 pp.

Vernberg, W. B., and B. C. Coull. 1981. *Meiofauna*. In F. J. Vernberg and W. B. Vernberg, eds. Functional Adaptations of Marine Organisms. Academic Press, New York, pp. 147–177.

Verry, E. S., and D. H. Boelter. 1979. *Peatland hydrology*. In P. E. Greeson, J. R. Clark, and J. E. Clark, eds. Wetland Functions and Values: The State of Our Understanding. American Water Resources Association, Minneapolis, MN, pp. 389–402.

Vidy, G. 2000. Estuarine and mangrove systems and the nursery concept: which is which. The case of the Sine Saloum system (Senegal). *Wetlands Ecology and Management* 8.1: 37–15.

Vitousek, P. M., J. Abner, R. W. Howarth, G. E. Likens, P. A. Matson, D. W. Schindler, W. H. Schlesinger, and G. D. Tilman. 1997. *Human alteration of the global nitrogen cycle: Causes and consequences*. Issues in Ecology 1. Ecological Society of America, Washington, DC.

Voesenek, L. A. C. J. 1990. Adaptation of *Rumex* in Flooding Gradients. Thesis, Catholic University of Nijmegen, The Netherlands. 159 pp.

Vymazal, J. 1995. Constructed wetlands for wastewater treatment in the Czech Republic—State of the art. *Water Science and Technology* 32: 357–364.

Vymazal, J. 1998. *Czech Republic*. In J. Vymazal, H. Brix, P. F. Cooper, M. B. Green, and R. Haberl, eds. Constructed Wetlands for Wastewater Treatment in Europe. Backhuys Publishers, Leiden, The Netherlands, pp. 95–121.

Vymazal, J. 2002. The use of sub-surface constructed wetlands for wastewater treatment in the Czech Republic: 10 years experience. *Ecological Engineering* 18: 633–646.

Vymazal, J., ed. 2005. *Constructed wetlands for wastewater treatment*. Special Issue of Ecological Engineering 25: 475–621.

Vymazal, J., H. Brix, P. F. Cooper, M. B. Green, and R. Haberl, eds. 1998. Constructed Wetlands for Wastewater Treatment in Europe. Backhuys Publishers, Leiden, The Netherlands.

Vymazal, J., and L. Kropfelova. 2005. Growth of *Phragmites australis* and *Phalaris arundinacea* in constructed wetlands for wastewater treatment in the Czech Republic. *Ecological Engineering* 25: 606–621.

Waddington, J. M., and N. T. Roulet. 1997. Groundwater flow and dissolved carbon movement in a boreal peatland. *Journal of Hydrology* 191: 122–138.

Walker, D. 1970. *Direction and rate in some British post-glacial hydroseres*. In D. Walker and R. G. West, eds. Studies in the Vegetational History of the British Isles. Cambridge University Press, Cambridge, UK, pp. 117–139.

Warner, B. G., and C. D. A. Rubec, eds. 1997. *The Canadian Wetland Classification System*. National Wetlands Working Group, Wetlands Research Centre, University of Waterloo, Ontario.

Weber, C. A. 1907. Autbau und Vegetation der Moore Norddutschlands. *Beibl. Bot. Jahrb*. 90: 19–34.

Weihe, P. E., and W. J. Mitsch. 2000. Garden wetland experiment demonstrates genetic differences in soft rush obtained from different regions (Ohio). *Ecological Restoration* 18: 258–259.

Weinstein, M. P., J. H. Balletto, J. M. Teal, and D. F. Ludwig. 1997. Success criteria and adaptive management for a large-scale wetland restoration project. *Wetlands Ecology and Management* 4: 111–127.

Weinstein, M. P., J. M. Teal, J. H. Balletto, and K. A. Strait. 2001. Restoration principles emerging from one of the world's largest tidal marsh restoration projects. *Wetlands Ecology and Management* 9: 387–407.

Weller, M. W. 1978. *Management of freshwater marshes for wildlife*. In R. E. Good, D. F. Whigham, and R. L. Simpson, eds. Freshwater Wetlands: Ecological Processes and Management Potential. Academic Press, New York, pp. 267–284.

Weller, M. W. 1981. Freshwater Marshes. University of Minnesota Press, Minneapolis. 146 pp.

Weller, M. W. 1994. Freshwater Marshes, 3rd ed. University of Minnesota Press, Minneapolis. 192 pp.

Weller, M. W. 1999. Wetland Birds. Cambridge University Press, Cambridge, UK.

West, R. G. 1964. Inter-relations of ecology and quaternary paleobotany. *Journal of Ecology* (Supplement) 52: 47–57.

Whalen, M. N., and W. S. Reeburgh. 1990. Consumption of atmospheric methane by tundra soils. *Nature* 346: 160–162.

Whalen, S. C. 2005. Biogeochemistry of methane exchange between natural wetlands and the atmosphere. *Environmental Engineering Science* 22: 73–94.

Wharton, C. H. 1970. The Southern River Swamp—A Multiple-Use Environment. Bureau of Business and Economic Research, Georgia State University, Atlanta. 48 pp.

White, J. S., S. E. Bayley, and P. J. Curtis. 2000. Sediment storage of phosphorus in a northern prairie wetland receiving municipal and agro-industrial wastewater. *Ecological Engineering* 14: 127–138.

Whooten, H. H., and M. R. Purcell. 1949. Farm Land Development: Present and Future by Clearing, Drainage, and Irrigation. Circular 825, U.S. Department of Agriculture, Washington, DC.

Wicker, K. M., D. Davis, and D. Roberts. 1983. Rockefeller State Wildlife Refuge and Game Preserve: Evaluation of Wetland Management Techniques. Coastal Environments, Baton Rouge, LA.

Widdows, J., B. L. Bayne, D. R. Livingstone. R. I. E. Newell, and E. Donkin. 1979. Physiological and biochemical responses of bivalve mollusks to exposure to air. *Comparative Biochemistry and Physiology* 62A(2): 301–308.

Wiebe, W. J., R. R. Christian, J. A. Hansen, G. King, B. Sherr, and G. Skyring. 1981. *Anaerobic respiration and fermentation.* In L. R. Pomeroy and R. G. Wiegert, eds. The Ecology of a Salt Marsh. Springer-Verlag, New York, pp. 137–159.

Wieder, R. K. 1989. A survey of constructed wetlands for acid coal mine drainage treatment in the eastern United States. *Wetlands* 9: 299–315.

Wieder, R. K., and G. E. Lang. 1983. Net primary production of the dominant bryophytes in a *Sphagnum*-dominated wetland in West Virginia. *Bryologist* 86: 280–286.

Wieder, R. K., and G. E. Lang. 1984. Influence of wetlands and coal mining on stream water chemistry. *Water, Air, and Soil Pollution* 23: 381–396.

Wilcox, D. A., and H. A. Simonin. 1987. A chronosequence of aquatic macrophyte communities in dune ponds. *Aquatic Botany* 28: 227–242.

Wilen, B. O., and H. R. Pywell. 1981. The National Wetlands Inventory. Paper presented at In-Place Resource Inventories: Principles and Practices—A National Workshop, Orono, ME, August 9–14. 10 pp

Wilhelm, M., S. R. Lawry, and D. D. Hardy. 1989. *Creation and management of wetlands using municipal wastewater in northern Arizona: A status report.* In D. A. Hammer, ed. Constructed Wetlands for Wastewater Treatment. Lewis Publishers, Chelsea, MI, pp. 179–185.

Wilson, L. R. 1935. Lake development and plant succession in Vilas County, Wisconsin. 1. The medium hard water lakes. *Ecological Monographs* 5: 207–247.

Wilson, R. F., and W. J. Mitsch. 1996. Functional assessment of five wetlands constructed to mitigate wetland loss in Ohio, USA. *Wetlands* 16: 436–451.

Wind- Mulder, H. L., L. Rochefort, and D. H. Vitt. 1996. Water and peat chemistry comparisons of natural and post-harvested peatlands across Canada and their relevance to peatland restoration. *Ecological Engineering* 7: 161–181.

Winter, T. C. 1977. Classification of the hydrologic settings of lakes in the north-central United States. *Water Resources Research* 13: 753–767.

Winter, T. C., and M.-K. Woo. 1990. *Hydrology of lakes and wetlands.* In M. G. Wolman and H. C. Riggs, eds. Surface Water Hydrology: The Geology of North America, Vol. 0–1. Geological Society of America, Boulder, CO, pp. 159–187.

Winter, T. C., and M. R. Llamas, eds. 1993. *Hydrogeology of Wetlands.* Special Issue of Journal of Hydrology 141: 1–269.

Wisheu, I. C., and P. A. Keddy. 1992. Competition and centrifugal organization of plant communities: Theory and tests. *Journal of Vegetation Science* 3: 147–156.

Wolff, W. J. 1993. *Netherlands—Wetlands*. In E. P. H. Best and J. P. Bakker, eds. Netherlands—Wetlands. Kluwer Academic Publishers, Dordrecht, The Netherlands, pp. 1–14.

Woo, M.-K., and T. C. Winter. 1993. The role of permafrost and seasonal frost in the hydrology of nothern wetlands in North America. *Journal of Hydrology* 141: 5–31.

Woodhouse, W. W., Jr. 1979. Building Salt Marshes Along the Coasts of the Continental United States. Special Report 4, U.S. Army Coastal Engineering Research Center, Fort Belvoir, VA.

World Wildlife Fund. 2001. *www.worldwildlife.org/wildworld/profiles/terrestrial*.

Wright, J. O. 1907. Swamp and Overflow Lands in the United States. Circular 76, U.S. Department of Agriculture, Washington, DC.

Wu, Y., F. H. Sklar, K. Gopu, and K. Rutchey. 1996. Fire simulations in the Everglades landscape using parallel programming. *Ecological Modelling* 93: 113–124.

Wu, X. and W. J. Mitsch. 1998. Spatial and temporal patterns of algae in newly constructed freshwater wetlands. *Wetlands* 18: 9–20.

Yagi, K., and K. Minami. 1990. Effect of organic matter application on methane emission from some Japanese paddy fields. *Soil Science and Plant Nutrition* 36: 599–610.

Zedler, J. B. 1988. *Salt marsh restoration: Lessons from California*. In J. Cairns, ed. Rehabilitating Damaged Ecosystems, Vol. I. CRC Press, Boca Raton, FL, pp. 123–138.

Zedler, J. B. 1996. Tidal Wetland Restoration: A Scientific Perspective and Southern California Focus. Report T-038, California Sea Grant College System, University of California, La Jolla. 129 pp.

Zedler, J. B. 2000a. Progress in wetland restoration ecology. *TREE* 15: 402–407.

Zedler, J. B. 2000b. Handbook for Restoring Tidal Wetlands. CRC Press, Boca Raton, FL, 464 pp.

Zedler, P. H. 1987. The Ecology of Southern California Vernal Pools: A Community Profile. Biological Report 85(7.11), U.S. Fish and Wildlife Service, Washington, DC.

Zobel, M., and V. Masing. 1987. Bog changing in time and space. Archiv für Hydrobiologie, Beiheft: *Ergebnisse der Limnologie* 27: 41–55.

Zoltai, S. C. 1988. *Wetland environments and classification*. In National Wetlands Working Group, ed. Wetlands of Canada. Ecological Land Classification Series 24, Environment Canada, Ottawa, Ontario, and Polyscience Publications, Montreal, Quebec, pp. 1–26.

Zoltai, S. C., S. Taylor, J. K. Jeglum, G. F. Mills, and J. D. Johnson. 1988. *Wetlands of boreal Canada*. In National Wetlands Working Group, ed. Wetlands of Canada. Ecological Land Classification Series 24, Environment Canada, Ottawa, Ontario, and Polyscience Publications, Montreal, Quebec, pp. 97–154.

Zoltai, S. C., and D. H. Vitt. 1995. Canadian wetlands: Environmental gradients and classification. *Vegetatio* 118: 131–137.

Zucker, L. A., and L. C. Brown, eds. 1998. *Agricultural Drainge: Water Quality Impacts and Subsurface Drainage Studies in the Midwest*. Ohio State University Extension Bulletin 871, Ohio State University, Columbus. 40 pp.

Index